BIOMASS NOW – SUSTAINABLE GROWTH AND USE

Edited by **Miodrag Darko Matovic**

Biomass Now – Sustainable Growth and Use

http://dx.doi.org/10.5772/2583
Edited by Miodrag Darko Matovic

Contributors

Petar Gvero, Semin Petrovic, Sasa Papuga, Milovan Kotur, Jan Křen, Tamara Dryšlová, Lubomír Neudert, Vojtěch Lukas, N. Abdullah, F. Sulaiman, Naoki Kano, Ademar Pereira Serra, Marlene Estevão Marchetti, Davi José Bungenstab, Maria Anita Gonçalves da Silva, Rosilene Pereira Serra, Franklyn Clawdy Nunes Guimarães, Vanessa Do Amaral Conrad, Henrique Soares de Morais, Mihai Caramihai, Irina Severin, Ebrahim Azarpour, Maral Moraditochaee, Dulce María Díaz- Montaño, Khalida Muda, Azmi Aris, Mohd Razman Salim, Zaharah Ibrahim, Enrique Javier Carvajal Barriga, Cristina Guamán-Burneo, Patricia Portero Barahona, Edgar Salas, Carolina Tufiño, Bernardo Bastidas, Le Duc Dung, Kayoko Morishita, Takayuki Takarada, Jin M. Triolo, Alastair J. Ward, Lene Pedersen, Sven G. Sommer, K.L. Chin, P.S. H'ng, Hongbin Cheng, Lei Wang, Adina-Elena Segneanu, Florentina Cziple, Paulina Vlazan, Paula Sfirloaga, Ioan Grozescu, Vasile Daniel Gherman, Shurong Wang, Werther Guidi, Frédéric E. Pitre, Michel Labrecque

Published by InTech

Janeza Trdine 9, 51000 Rijeka, Croatia

Publishing Process Manager Marina Jozipovic

Typesetting InTech Prepress, Novi Sad

Cover InTech Design Team

Additional hard copies can be obtained from orders@intechopen.com
Biomass Now – Sustainable Growth and Use, Edited by Miodrag Darko Matovic
p. cm.
ISBN 978-953-51-1105-4

For EU product safety concerns:
IN TECH d.o.o., Prolaz Marije Krucifikse Kozulić 3, 51000 Rijeka, Croatia,
info@intechopen.com or visit our website at intechopen.com.

Meet the editor

Dr. Miodrag Darko Matovic is professor at Queen's University in Kingston, Canada, Department of Mechanical and Materials Engineering. He traded his long experience with fossil fuel combustion with work on renewable energy resources, especially biomass. He looks at biomass as the sustainable source of energy, soil amendments, raw material for plastics and other organic materials and as means to capture atmospheric carbon and fix it away from the atmosphere. His current research includes various aspects of biomass for energy use, including storage safety, novel uses of ash and potentials of biochar use in soil remediation.

Contents

Preface

The increase in biomass related research and applications is driven by overall higher interest in sustainable energy and food sources, by increased awareness of potentials and pitfalls of using biomass for energy, by the concerns for food supply and by multitude of potential biomass uses as a source material in organic chemistry, bringing in the concept of bio-refinery. The present, two volume, Biomass book reflects that trend in broadening of biomass related research. Its total of 40 chapters spans over diverse areas of biomass research, grouped into 9 themes.

The first volume starts with the Biomass Sustainability and Biomass Systems sections, dealing with broader issues of biomass availability, methods for biomass assessment and potentials for its sustainable use. The increased tendency to take a second look at how much biomass is really and sustainably available is reflected in these sections, mainly applied to biomass for energy use. Similarly, Biomass for Energy section specifically groups chapters that deal with the application of biomass in the energy field. Notably, the chapters in this section are focused to those applications that deal with waste and second generation biofuels, minimizing the conflict between biomass as feedstock and biomass for energy. Next is the Biomass Processing section which covers various aspects of the second-generation bio-fuel generation, focusing on more sustainable processing practices. The section on Biomass Production covers short-rotation (terrestrial) energy crops and aquatic feedstock crops.

The second volume continues the theme of production with the Biomass Cultivation section, further expanding on cultivation methods for energy, the feedstock crops and microbial biomass production. It is followed by the Bio-reactors section dealing with various aspects of bio-digestion and overall bio-reactor processes. Two more chapters dealing with aquatic microbial and phytoplankton growth technologies are grouped into the Aquatic Biomass section, followed by the Novel Biomass Utilization section which concludes the second volume.

I sincerely hope that the wide variety of topics covered in this two-volume edition will readily find the audience among researchers, students, policy makers and all others with interest in biomass as a renewable and (if we are careful) sustainable source of organic material for ever wider spectrum of its potential uses. I also hope that further exploration of second-generation energy sources from biomass will help in resolving the conflict of biomass for food and biomass for energy.

Miodrag Darko Matovic,
Department of Mechanical and Materials Engineering,
Queen's University, Kingston,
Canada

Section 1

Biomass Sustainability

Biomass as Potential Sustainable Development Driver – Case of Bosnia and Herzegovina

Petar Gvero, Semin Petrovic, Sasa Papuga and Milovan Kotur

Additional information is available at the end of the chapter

http://dx.doi.org/10.5772/51551

1. Introduction

Bosnia and Herzegovina (B&H) is a country in southeastern Europe, on the western part of the Balkan Peninsula. B&H covers a total area of *51.129 km²* and it is almost landlocked, except for *26 km* of Adriatic Sea coastline. Bosnia and Herzegovina is a transition country in the process of European integrations. The result of the privatization and the war situation in nineties is devastated economy which has to find a new ways for the further development. One of the results of transition process is also that major part of the industry are small and medium enterprises (SME) and that also has to be taken into consideration, because development strategies and plans has to be adapted according to this fact. This paper gives the analysis of the potential connections between renewable energy sources (RES), particularly biomass and sustainable development of the B&H's economy, taking into consideration specific political structure of the state. Problem if the sustainable development and integration of RES in that is universal, and some of the analyzed issues and findings from this material can be interested not only for the people tries to establish some activities in Bosnia and Herzegovina, but also for the people dealing with bioenergy generally. Bosnia and Herzegovina is consists from two entities: Republic of Srpska (RS), Federation of Bosnia and Herzegovina (FBiH), and third administrative unit, Brcko District (BD). Energy sector, forestry, environmental and climate changes related issues are unider their jurisdiction.

2. Renewable energy sources in Bosnia and Herzegovina:

Bosnia and Herzegovina have significant physical potential regarding to renewable energy sources and belongs to the list of the countries which can develop their energy sector mainly based on that. Due to that hydro, biomass, geothermal, wind and solar potential can play important role in the whole state economy in the forthcoming period.

Regarding to small hydro, some analysis says that theoretically water power of B&H amounts 99,256 GWh/year, technical water power potential of 356 small and big HPP (which may be built) amounts to 23,395 GWh/year, out of which 2,599 GWh/year is in small HPP. From that amount around 77% is in Republic of Srpska (RS) and 23% is in Federation of Bosnia and Herzegovina (FB&H).

Real potential for wind energy in B&H is still not fully estimated. Some estimations related to 16 macro-locations under investigations goes says that total estimated installed capacity can be 720 to 950 MW, which can produce 1440 to 1950 GWh, annually [4]. It is important to emphasize that the existing infrastructure offers adequate conditions for connecting possible locations to the grid, as the high- and medium-voltage network is well developed.

Theoretical potential of the solar energy in B&H amounts 74.65 PWh. Technical potential amounts 190.277,80 GWh, that is 6.2 times more than quantity of energy out of totally balance needs for the primary energy in FB&H during 2000 [4]. Despite this, the use of solar energy is insignificant and the exploitation of solar energy with flat-plate collectors is also limited.

It is difficult to estimate total B&H's physical and technical geothermal potential. All estimates are mainly based on some experimental drills and theoretical investigations, and according to that temperatures at the known locations in north and central part of the country are between 54 and 85°C. This temperature level is relatively low for electricity generation, but it is interesting for district heating systems.

3. Potential of biomass

Looking from the time prospective, bioenergy interest has been greatly stimulated by the fuel price rises in the late 2000s. Bioenergy is seen as a way to protect against the rising fossil fuel prices, furthermore, biomass can act as a carbon sink and as a substitute for fossil fuels, due to that biomass is seen as one of the mechanisms mitigating climate changes.

Regarding definitions of biomass potentials, there are international practice and standards for that. Estimations can vary according to the calculation methodology and the assumptions made (e.g. land use patterns for food production, agricultural management systems, wood demand evolution, production technologies used, natural forest growth etc). In terms of biomass potentials, the following potential types are often discussed: theoretical, technical, economic, implementation potential and environmentally sustainable potential.

According to data from 1990, forests and forest land in BiH encompass an area of approximately 2,709.800 ha, which is around 53% of the territory of the country. Arable land accounts for 1,4 million ha and permanent meadows and pastures for 0,6 million ha [1,2]. Despite the fact that some 41% of the country comprises agriculture land, Bosnia and Herzegovina is relatively poor in agriculture resources, since some two thirds of the country is mountainous / hilly. Land is cultivated with various field crops, such as cereals, industrial crops, vegetables and fodder crops, represented just one quarter of the total agricultural land in 2008. On the contrary, meadows and pastures covered 49% of the agricultural land, while a significant part of the arable fields remained fallow or uncultivated during the same year. Finally, permanent crops, such as orchards and vineyards, covered 4% of the agricultural land or 86.000 ha [5]. The structure of agricultural sector is characterized by small family farms which to a large extent produce for home consumption. Over 50% of agriculture holdings are estimated to be less than 2 ha. State firms are much larger but are either operating under severe constraints or inoperable due to the incomplete process of privatization. As far as forest land is concerned, public forest land amounts to 73% in RS and 83% in FBiH of the total forest land, while the rest is private [5].

Regarding to the country distribution of biomass potentials, field crop residues are mostly found (70%) in the Republic of Srpska, while livestock manure, mostly cow and chicken manure, in the Federation of Bosnia and Herzegovina. Forest based biomass distribution between the two entities is quite balanced.

Different types of biomass have been analyzed, taking into consideration their theoretical and technical potential:

- Forest based biomass includes fuel wood, forest residues and wood industry residues.
- Agricultural biomass includes field crops, arboricultural residues, livestock and agro-industrial residues.
- Energy crops in this work are defined as crops specifically bred and cultivated for energy production either by direct conversion to heat and electricity or by production of bio-fuels (solid, gaseous or liquid).
- Municipal solid waste (MSW) refers to waste collected by or on behalf of municipalities.

3.1. Forest biomass

3.1.1. Forest sector and its characteristics

Forests represent one of the major natural resources in Bosnia and Herzegovina, due to their natural and diverse structure as well as their extensive natural regeneration. The main species found in BiH forests are mostly fir, spruce, Scotch and European pine, beech, different varieties of oak, and a less significant number of noble broadleaves along with fruit trees.

The professional development and management of the forestry sector has been dedicated to traditional systems and has recently (especially after a turbulent post-war period where forests have been neglected and misused) faced higher demands in terms of contributing more to the protection and enhancement of all forest functions, ranging from economical viability to social responsibility and environmental and ecological sustainability. Total forest area in Bosnia and Herzegovina amounts to 2,61 million ha, 1,59 ha in FBiH and 1,03 ha in RS, In BD, where there are approximately 11,000 ha of forests, of that 8,500 ha being privately owned and merely 2,500 ha within the public management system [4].

2,186.300 ha or 81% of forests and forest land is under state ownership, while private ownership consists of 523.500 ha or 19%. Most of these properties are very small in size (up to 2ha) and vastly scattered throughout the country, with outstanding issues in ownership due to population migration.

According to Constitutional provisions, the ownership of forests lies in authority of entities (FBiH, RS) and BD, where ministries of forestry are responsible for administrative management of these areas through the public forest management enterprises. Public forest land amounts to 73% in RS and 83% in FBiH of the total forest land, while the rest is private. Standing volume of forest biomass amounts to 350m m^3 in Bosnia and Herzegovina, however the real figure is higher since no data were available for private forests in FBiH. Furthermore, forests net annual increment is estimated to approximately 10m m^3 or 3% of the total woodstock. Although annual growth seems high, annual wood increment is constrained by inadequate local forest management practices [3].

In conformity with data shown above, almost 400,000 ha (186,141 ha for FBiH and 207,719 ha for RS) have been assumed as being bare lands with a productive function and in those terms could be potentially included in reforestation programs.

The customary management system of natural regeneration that has been practiced in BiH throughout the centuries has contributed to realizing significant forest diversity in this sense.

Nevertheless, some preceding studies (mostly based on the satellite surveys within the EU CORINE program) have shown that actual forest cover size might be lower by 10-15% than previously projected.

Due to activities such as illegal logging, ore mining, construction, forest fires and others, forested areas have been shrinking rapidly; furthermore, a significant part of the forest cover has been declared as area with land-mines (numbers indicate some 10%) and has evident damages due to war activities. In addition there are extensive unresolved property disputes and illegal land acquisition which await resolution due to complex legal mechanisms and administration.

In the recent years, significant progress has been made in the area of forest certification, where three of the forest management public enterprises have undergone scrutiny of international auditing against the Forest Stewardship Council (FSC) certification, while several others are presently preparing to undergo the same procedure and promote sustainable forest management within their practices. Currently around 50% of state managed forests in BiH have been certified according to FSC Standards.

As mentioned before, forestry legal and institutional framework has been structured through two entities. In FBiH there are cantonal forest management companies, whereas in RS, the forestry management operations are led by a single public enterprise. This decentralization of forest management authority, legal framework (two separate laws on forests) and administration has led to further difficulties in establishing appropriate mechanisms for controlling forest operations, especially illegal logging and land acquisition in bordering areas [4].

3.1.2. Biomass potential from forestry and wood processing industry

Bosnia and Herzegovina has abundant forests, with 46 % of the country area covered by forests. The production, harvesting and processing of timber is one of the country's oldest economic activities, and currently has major strategic importance for the country's economic development.

High forest predominates and deciduous species are the most dominant with beech (Fagus spp.) accounting for almost 40% of all species cover in the country. Oaks (Quercus spp.) contribute another 20%. Spruce and fir, located in the higher elevations and generally on the steepest terrain comprise an additional twenty percent of the forest cover in BiH. Annual allowable cut is calculated to 7,44 million m^3 according to an ongoing UNDP Project, while actual harvest was 5,60 million m^3 in 2008 [3]. From the 4,33 million m^3 of roundwood that were produced in 2008, 1,69 million m^3 were used as fuel wood (~40%), while 2,64 million m^3 were directed towards the wood industry (~60%). Furthermore, around 1,18 million m^3 of forest residues were produced at the logging sites.

The tradition of use biomass as energy source in Bosnia and Herzegovina has existed for a long time, but that use is characterized with a very low rate of utilization, mainly in rural and sub-urban areas as primary source for heating and cooking purposes in households and buildings. According to the recent findings from the total 77.19 PJ of final energy consumption in households, biomass makes 45.84 PJ. However, since energy demand and prices of fossil fuels rise rapidly other forest based biomass resources apart from fuelwood are also being considered for energy exploitation. These include forest residues and bark as well as residues/by-products arising from the processing of industrial wood.

Forest residues in BiH that can be utilized for energy production include tops, branches and stumps that are left at the logging sites. According to forest expert's estimation, forest residues that are available for energy purposes amount to 20% and 10% of the harvested volume of industrial roundwood and fuelwood respectively. However, no more than half these residues can be harvested due to difficulties in their collection [5].

Wood industries produce residues, such as chips and particles, sawdust, slabs, edgings and shavings. These residues can either be used in particleboards or pulp production or used for energy purposes in industrial boilers and for densified wood fuels production (pellets and briquettes). Bark is also included in industrial residues, since industrial wood is mainly debarked at the sawmills. However, in order to estimate the produced residues one needs to know the products output.

Wood industry production figures were not available on a regional level and therefore information on a national level from the Industrial Bulletin for FBiH and RS was used [1,2]. In 2008, almost 1 million m^3 sawmill products, 40.733 m^3 plywood and veneer sheets products and 2.428 m^3 particleboards were produced on a country level. Furniture and secondary wood industry products, such as doors, windows and parquets, were not included in this study's calculations, since they are given in different units (pieces, m^2, etc.).

Feedstock was calculated by employing FAO conversion factors for each wood processing industry. Sawmill residues (excluding bark) were assumed to comprise 40% of sawmill feedstock, while plywood and veneer sheets industry residues were assumed to comprise 45% of feedstock. Bark was separately calculated as 7% of sawmill feedstock [5]. These factors depend on a number of assumptions with regards type and modernization level of each process, the production capacity of each industry, the tree species processed, etc. The factors were found to be in good agreement with literature values for the Western Balkans region. Furthermore, the availability of wood industry residues is restricted by various technical factors and was assumed equal on average to 80% for all types of wood industry residues with the exception of bark for which availability was assumed to be 60%.

Black liquor is a byproduct of the chemical wood pulp production process. According to the Industrial Bulletin for FBiH and RS Statistics, 32.809 t of unbleached coniferous chemical wood pulp (90% dry substance) were produced in 2008 in FBiH, which in terms of energy is equivalent to 74.476 m^3 of fuelwood Moreover, 98.041 t of paper and paperboard were produced on a country level in 2008 [1,2]. Paper production is not a significant source of woody biomass in BiH, since the solid waste produced is very heterogeneous and contains non paper components, such as sand, metal, and glass, which cannot be used as a fuel [5].

Forest timber (fuel wood and forest residue) and wood waste from wood processing industry represent the major source of biomass for energy production in Bosnia and Herzegovina. Biomass residues from agricultural production have a significant energy potential in parts of northern and north-eastern Bosnia. Forests are one of the most important natural resources of Bosnia and Herzegovina. Bosnia and Herzegovina is one of the richest countries in Europe by the criteria of the forest coverage and diversity considering the total size of the State territory. The largest areas are covered by forests of broadleaf or deciduous trees, while about 10% of the country is covered by barren soils (i.e. one fifth of the forest soils). The total growing wood stocks in the forests of Bosnia and Herzegovina amount to 317,565,740 m^3 or 203.6 m^3/ha (62% broadleaf trees and 38% conifers). The annual volume increment of forests in Bosnia and Herzegovina is 9,500,600 million m^3 or 6.1 m^3/ha, the annual allowable level of wood cutting is 7,451,450 million m^3 or 4.75 m^3/ha [3].

The energy potentials of the natural wood residue resources in Bosnia and Herzegovina are presented in Table 1.

The production, harvesting and processing of timber is one of the oldest economic activities in the Country, and has a strategic importance for the country's economic development. Some statistical estimations shows that the wood export value within the total Bosnia and Herzegovina export value is probably in order of *15%*. It is further estimated that *15%* of the total population receives its livelihood through the activities in forestry and forest industry.

			quantities	quantities	minimum inferior calorific value	energy potential
			m^3/a	t/a	GJ/t	TJ/a
wood residue		broadleaf	295.529	212.781	10,28	2187
		conifers	202.866	91.290	10,28	938
sawmill wood waste	sawdust	broadleaf	283.300	203.976	10,28	2097
		conifers	145.227	65.352	10,28	672
	wood chops	broadleaf	212.475	152.982	10,28	1573
		conifers	145.227	65.352	10,28	672
Total			1.284.624	791.733		8139

Table 1. Quantities, types, structure and energy-related potential of wood residue in Bosnia and Herzegovina (based on an average volume of cutting in the period of 2007 – 2010.) [5].

Fuel wood is considered to have high value for local, small scale energy use, i.e. stoves, open fires and ovens. While this is clearly neither efficient nor perhaps environmentally optimal use of resource, it is nevertheless an essential, low cost resource for large numbers of rural people. From the 18,45 PJ estimated by this study, it is assumed that *20%* will be available for new, efficient small scale wood fired boilers, stoves, etc. This would account for 820 GWh heat production annually.

Saw mill waste production is generally high due to a low process efficiency of sawmills: the net end product (lumber) represents an estimated 40 - 45% of the log (a well managed mill in Europe runs at up to *50%* efficiency). The waste produced consists of wet sawdust, slabs, and the trimmings from cutting to length and width. Based on this ratio, waste from the primary and secondary wood processing industries would amount to approximately $1.14 \times 10^6\ m^3$.[6]

3.2. Agricultural biomass

The technical potential of straw production is limited by competing uses (e.g. animal feed and bedding), the need to leave material on the ground for nutrient replenishment etc, and is estimated to be *6,63 PJ*. Moreover, this resource is highly dispersed. Modern, straw-fire power stations require a considerable scale to be financially viable. Hence, it is assumed that one third of this resource could be exploited via local small scale straw fired baled fired boilers or straw pellet boilers supplying residential properties with heat. This would account for 491 GWh of heat annually [5].

Based upon livestock data (pigs, chickens, cattle), the amount of slurries and manures produced has been estimated. This could be exploited via anaerobic digestion (AD). The Theoretical Potential is 6,50 PJ biogas production. However, it is assumed that much of this resource could not be aggregated between farming units to provide sufficient feedstock that a typical AD unit may require. It is assumed that *20%* of theoretical potential could be realized, or *1,30*PJ. The installed capacity would be 18 MWe and annual output would be 126 GWh of electricity. Given both the remote, rural location of AD units, it is assumed that the amount of heat used would be negligible [5].

3.2.1. Agricultural sector overview

Out of the total Bosnia and Herzegovina territory, amounting to 5,112,879 ha, FBiH takes up 2,607,579 ha, while RS takes up 2,505,300 ha. Farmland covers approximately 2,600,000 ha (around 52%) of that territory, and the remaining 2,400,000 ha are woodlands (around 48%).

Fragmentation of farmland in BiH constitutes an additional problem, 54% of property is under 2 ha in size, 13.5% is between 2 and 3 ha, 16% of property is between 3 and 5 ha, 10% of property is between 5 and 8 ha, about 3% of property is between 8 and 10 ha in size, and only 2.9% or property is over 10 ha in size [7].

The crops structure of cultivated plants and their share in the total sowing structure constitute an important segment of the BiH plant production. According to statistics, in the RS, harvest areas amounted to 443,300 ha in 1990, to 285,731 ha in 1996, and to 356,548 ha in 1997. In the period between 2000 and 2006, about 67.17% of total area in crops was sowed with cereals, and 26.66% with fodder crops. The situation in Federation of BiH is not much different as the total sowing area is considerably smaller and it amounted to about 206,000 ha in 2001, and 197,000 ha in 2006. [1,2].

It is clear that the sowing structure is not favourable as it is not satisfactory in terms of the size of areas in crops and in terms of the yield per unit area, which are very small and low, respectively [1,2].

The crop structure is very unfavorable. The production of cereals in areas of 1-3 ha cannot be economically justified and a commercial livestock production cannot be built on it.

Another issue that brings us to the analysis of the technological level of agricultural production in BiH are average yields of the most common crops (over 80% of arable land in BiH). The comparison of yields with the same yields in the neighboring countries gives a clear picture of average yields of main agricultural crops, and it clearly shows that the agricultural production in BiH is completely behind--between 1.1 and 4.4 times less productive.

Thus, in addition to the unfavorable structure of agricultural crops, average yields in BiH are very low, which fully qualifies this production as extensive, unproductive and therefore barely sustainable. However, the natural conditions for agricultural production are favourable, and for some crops they are even optimal in comparison with some of the neighboring countries.

The analysis of production of main types of livestock in BiH clearly reflects the habits of autarchic village farms orientated towards satisfying their own needs and keeping their own livestock numbers at the biological minimum on one hand and the tardiness of the state and its institutions, i.e. agricultural experts, to launch development process on the other.

Based on the data from the RS Statistical Institute, in 1999, over 17% of total land in the RS – BiH were pastures. If we add 10% of natural meadows to this, we arrive at the fact that almost one third of the total land can be used for livestock production.

There are great possibilities for a quality livestock production on the territory of BiH, but the number of heads of cattle must be increased, the structure must be changed and the stock composition must be improved.

3.2.2. *Agriculture field crops and arboricultural residues*

Two large categories of field agricultural residues can be defined: field crop residues and arboricultural residues. Field crop residuals are the residues that remain in the field after the crops are harvested. Depending upon the crop, the harvesting method and other parameters, field agricultural residues may include various plant part such as stems, branches, leaves, chaff, pits, etc. varying in composition, moisture and energy potential. Arboricultural residues are the residues that remain in the field after farming activities performed during the cultivation of perennial crops (pruning vineyards and trees).

Total quantities of residues were estimated using recent statistical data for the production area for each crop as well as specific coefficients indicating the ratio of residues production to cultivated area.

For each crop i cultivated in region j, the annual energy theoretical potential $Erescrop_{i,j}$ is calculated by SYNENERGY Project, based on the following formula [5]:

$$Erescrop_{i,j} = r_i, A_{i,j}, H_i,$$

r_i country specific residue production per cultivated area [t/ha]

$A_{i,j,}$ cultivated area of crop i in region j [ha]

H_i country specific lower heating value of residue [GJ/t]

Data for crops production and harvested area in 2008 were obtained from the official statistical publications on the entity and state level. The coefficients used to estimate the quantities and the energy potential of agricultural field residues derived from local experts' estimations and references.

The estimation of the quantities of agricultural residues available for energy production is based on the degree of availability which is different for each crop, varies from year to year and depends on several factors such as:

- the harvesting method,
- the moisture content,
- the demand of agricultural residues for non-energy purposes (cereal straw, for example, is used for animal feeding, animal bedding, etc.),
- the need for some residues to remain on the soil to maintain the level of nutrients (sustainability reasons).

The availability factor for arable crop residues is estimated to be 30%. The same factor for arboricultural residues is estimated to be 80%, mainly due to technical difficulties in collection. Based on these factors, it is estimated that 527.765 t of field crop and arboricultural residues could be annually exploited for energy purposes (reference year of analysis 2008). This is equivalent to 7,47 PJ or 3,24 % of the total primary energy supply in 2008, which means that crop residues could contribute significantly to the energy supply of Bosnia and Herzegovina. Almost 90% of this potential comes from field crop residues, while arboricultural residues contribute the remainder.

Figures 1 and 2 present the technical potential of the most significant crop residues. Maize residues are the most abundant source of biomass contributing 75% to the field crop residues potential or 68% to the total crop residues potential. Wheat residues share in the field crop residues potential is also significant (17%), while barley, oilseeds, rye and oats residues contribute to a lesser extent. The major part of arboricultural residues comes from plum and apple tree prunings (73%). Other sources of arboricultural residues that should be taken into account are vineyards, pears, cherries, sour cherries and peaches prunings.

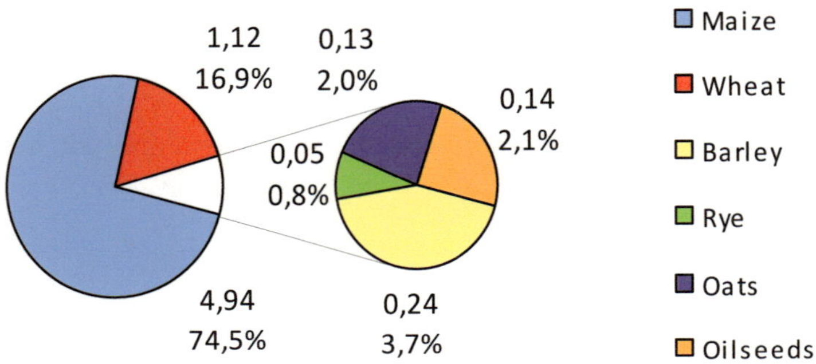

Figure 1. Arable crop residues technical potential in Bosnia and Herzegovina in PJ [5]

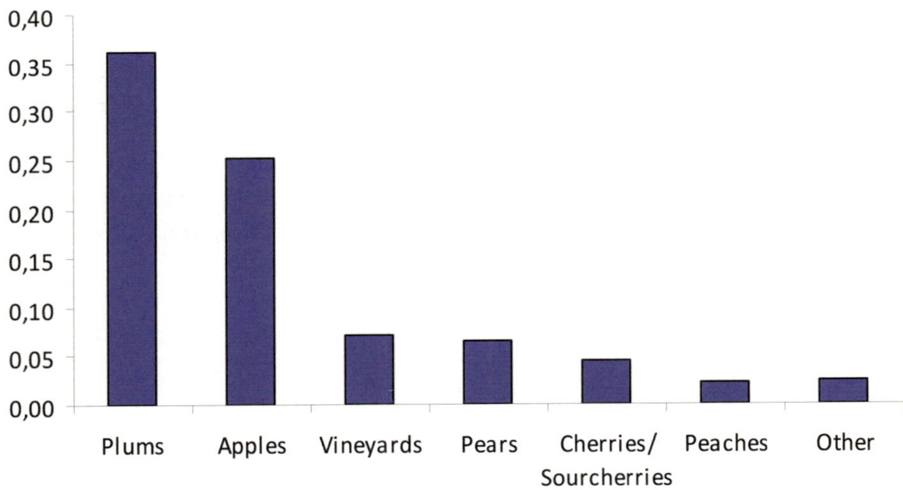

Figure 2. Pruning's technical potential in Bosnia and Herzegovina in PJ [5].

The crop residues potential in RS is more than twice that in FBiH and Brcko district and amounts to 5,20 PJ. In RS almost 90% of the potential comes from cereals, while this percentage is somewhat lower in FBiH (83%), where the contribution of arboricultural residues is higher (16%). Oilseed field residues have a minor contribution (1-2%) in both entities.

In the Federation of Bosnia and Herzegovina, 53% of the crop residues potential is found in the cantons of Tuzla (FBiH-K3) and Una-Sana (FBiH-K1). Another 30% of the potential is found in the cantons of Posavina (FBiH-K2) and Zenica-Doboj (FBiH-K4) as well as in the Brcko District.

3.2.3. Livestock manure

Energy can be derived from livestock manure as long as they are collected in lagoons or large tanks and can be considered feasible only in in-stall livestock systems, excluding therefore sheep and goats from such practices since their breeding is extensive making collection of manure impossible.

Since animal manure is of a high water content, it can be digested anaerobically for the production of biogas, which can be burnt for heat or/and electricity production.

Intensive livestock in Bosnia & Herzegovina consists of cattle, brood sows and poultry farming. According to official statistics there were 378.000 cattle (heads), 276.000 pigs and 11,26m poultry in 2008 [1,2]. The energy potential $Eresanim_{i,j}$ for animal species i in region j was evaluated based on the formula [5]:

$$Eresanim_{i,j} = p_i C_{i,j} Y_i H_i$$

$C_{i,j}$ number of animal species i nurtured in region j [heads]

p_i country specific manure generation factor for species i [t/head/yr]

Y_i country specific biogas yield [Nm^3/t manure]

H_i country specific lower heating value of biogas [GJ/Nm^3]

The manure generation factor, the biogas yield and the energy content of the produced biogas of the examined animal species depend on factors such as body size, kind of feed, physiological state (lactating, growing, etc.), and level of nutrition and coefficients regarding the residues produced on average per animal and the biogas yield per ton of produced residues were assumed according to the experts analysis in whole this region [5]. The amount of biogas that could be theoretically produced amounts to 292 million Nm^3, which is equivalent to 6,50 PJ In order to estimate the technically available livestock manure and since no further data regarding the regional distribution of animal farms that are of adequate size for biogas production were available, it was assumed that the technical potential of livestock manure would be 20% of its theoretical value, which is now the case for Croatia [5]. The available livestock manure for energy production amount to 1,30 PJ, or 0,56% of the total primary energy supply in the country in 2008 [5].

Residues from cows contribute the largest share to the total potential (50% in total), while poultry has a sizeable share (38%) and pig residues have the lowest share (12%).

Furthermore, in the same Figure it is shown that the potential is higher in the Federation of Bosnia and Herzegovina and Brcko District than in RS.

In FBiH the highest potential is found in the canton of Tuzla (FBiH-K3), which makes 40% of the total potential in FBiH and Brcko District. Furthermore, this canton exhibits the highest poultry residues potential, since 35% of the country's poultry is farmed there. Another 18% of the FBiH potential is found in the Zenica-Doboj canton (FBiH-K4) and therefore, 58% of the FBiH potential is concentrated in the north-east.

Exploitation of livestock manure for energy production via anaerobic digestion (AD) is considered to be feasible only for medium to large scale livestock units. A feasibility study called ANIWASTE financed by the EC in 2005 has sampled more than 300 farms in the wider region of Banja Luka and Lijevce polje, which is the region with the most intensive cattle raising activities. The average farm in this region has 100 pigs, 10-20 cows and 5.000-10.000 poultry. In general, the sector has passed through a post-war transition period in Bosnia and Herzegovina, which has resulted in small family farms [8].

4. Energy crops

In common with other resource assessments, the potential for energy crops is, in theory, large. It is also highly dependent on which crops are deemed to be most likely to be grown, what type of land is converted to their cultivation, and the areas of land used.

The estimations were based on two reasonably scenarios [5]:

a. *10%* of land currently used for grazing/pasture plus *5%* of fallow land are used to grow perennial grasses, and
b. *10%* of land currently used for grazing/pasture plus *25%* of fallow land are used to grow perennial grasses.

Total available land is 95.791 ha and 147.118 ha under scenarios A and B, respectively. 72% of the available land is found in the Federation of Bosnia and Herzegovina. Then the potential of energy crop *i* in the region *j* was calculated according to the following equation [5]:

$$Enercrop_{i,j}=Aencr_j\ CY_i\ BY_i\ H_i$$

$Aencr_j$ available land in region j [ha],

CY_i country specific yield of crop *i* [t/ha],

BY_i biofuel yield of crop *i* [t biofuel/ t crop],

H_i biofuel energy content of crop *i* [GJ/t].

Table 2 presents the energy crops considered in the two scenarios for Bosnia and Herzegovina, main energy markets and the energy potentials under the two scenarios.

The calculations are made for the whole land available in each case, e.g. if all the available land in Scenario A was used for biodiesel production with oilseeds the total potential would amount to 2,12 PJ, while if it was used for second generation bioethanol from Short Rotation Coppice (SRC) it would reach 6,21 PJ. These figures summarize potentials based on conversion efficiencies. In all cases the potential in the Federation of Bosnia and Herzegovina makes 72% of the total potential.

The respective Technical Potentials are estimated to be 15,33PJ and 23,54 PJ resource. The half of the resource would support local small scale energy crop fired baled fired boilers or energy crop pellet boilers supplying residential properties with heat. This would equate to 1.703 GWh of useful heat production per year.

Crop	End use	Energy potential (PJ)	
		Scenario A	Scenario B
Oilcrops	1st gen Biodiesel	2,12	3,26
Wheat	1st gen Bioethanol	2,13	3,26
Maize	1st gen Bioethanol	2,88	4,42
Perennial grasses	2nd gen Bioethanol	7,76	11,92
	Heat & Electricity	15,33	23,54
SRC (Short Rotation Coppice)	2nd gen Bioethanol	6,21	9,53
	Heat & Electricity	12,26	18,83

Table 2. Energy crops potential for biofuels (1st & 2nd generation) and bioenergy in BiH (2008) [5].

5. Municipal solid waste

Municipal solid waste (MSW) refers to waste collected by or on behalf of municipalities; this mainly originates from households but waste from commerce and trade, offices, institutions and small businesses is also included.

According to the EU legislation (Directive 2001/77/EC) energy produced from the biodegradable fraction of MSW is considered as renewable and therefore organic waste, waste paper and cardboard and textiles are a source of biomass. Due to lack of data regarding the share of the biodegradable part to the total quantities of MSW in BiH, the biodegradable fraction of 50% found in neighboring Serbia was employed. Furthermore, a lower heating value of 7,2 GJ/t for the biodegradable part was assumed [5].

Landfill gas. Municipal Solid Waste (MSW) production expected to reach 0,5 t/person/year (the EU 15 average). It is disposed and methane is captured and used to generate power. This assumes that, due to the location of the landfills, there are no local uses for heat. The theoretical biogas potential estimated in this study is 4,28 PJ.

In 2008, 1.367.097 t MSW was generated in Bosnia and Herzegovina, 86% of which (1.181.887 t) was collected [1,2]. This is equivalent to 308 kg of collected waste per capita per year. Other sources report a higher value of waste generation at around 500 kg/ per capita/ per year [4]. Nevertheless, it was decided to accept the number reported by the Agency for Statistics of Bosnia and Herzegovina, since it is in good agreement with waste generation rates found in other Western Balkan countries.

Table 3 shows estimated total MSW and household waste (HHW) amounts, in accordance with the methodology recommended in the SWMS , and population statistic [1,2,9].

	MSW generated in 1999 [Gg MSW]	MSW generated in 2010 [Gg MSW]	MSW generated in 2020 [Gg MSW]	MSW generated in 2030 [Gg MSW]
MSW in RS	724,269	1002,558	1347,354	1810,731
HHW in RS	362,134	501,278	673,676	905,364
MSW in FB&H	1138,0	1575,258	2117,015	2845,091
HHW in FB&H	569,0	787,629	1058,508	1422,546
Summary MSW	1862,269	2577,812	3469,369	4655,822
Summary HHW	931,134	1288,907	1732,183	2327,911

Table 3. Estimated Annual amounts of MSW and HHW at entity and country level [10].

Taking the above into account the theoretical potential of biomass from MSW can be estimated according to the following equation [5]:

$$Emsw = PpCoHo \quad (F.5)$$

P population,

p per capita waste generation [t/yr],

Co biodegradable waste fraction in MSW [%],

Ho biodegradable waste lower heating value [GJ/t].

The estimated theoretical potential amounts to 4,28 PJ or 1,9% of the country's total primary energy supply in 2008.

Currently, the main option for disposal of municipal waste is still landfilling, while most of the landfills are not sanitary. Furthermore, it is estimated that there are more than 2.000 open dumps, many located near to small municipalities in rural areas.

Implementation of SWMS commenced with WB/IDA credit for Project "Solid Waste Management Project" (Ex. Environmental Infrastructure Protection Project) in 2002. An analysis of the current situation in this sector has shown that the objectives concerning the construction of regional sanitary landfills defined in the SWSM are unrealistic. The plan is to have 16 regional landfills by December 2009, but until now, only 2 landfills have been constructed. Two regional sanitary landfills are anticipated in FBiH for 2010: "Smiljevac"- Sarajevo and "Mošćanica" - Zenica, where 10% and 8% of the total MSW collected in the FBiH would be disposed respectively. For RS, one regional sanitary landfill for MSW disposal "Ramići"- Banja Luka, is anticipated, where 16,7% of the total MSW collected in RS would be disposed. At the sanitary landfill in Sarajevo, the collected landfill gas is used for electricity generation, while at the Zenica landfill a flare system for the combustion of landfill gas has been constructed. The combustion of landfill gas by flare is also envisaged at the future sanitary landfill in Banja Luka.

In addition to landfills, according to the initial national communication of BiH under the UN framework convention on climate change (UNFCC), incineration of 20% of MSW with energy recovery is anticipated by 2030 [4]. It is further foreseen that recycling rates will be 10% of the total household waste (HHW) in 2020 and 20% for 2030. Moreover, 50% of the recycled HHW is foreseen to be biodegradable waste.[5].

6. Biomass as sustainable development driver

As already mentioned above bioenergy interest has been greatly increased in last period. Thus, at present factors may influence the prospects for bioenergy:

- increases in crude oil prices,
- concerns for enhancing energy security matters, by creating de-centralized solutions for energy generation,
- concerns for climate change and global warming, but also to
- preserve non-renewable resources,
- promotion of regional development and rural diversification by creating jobs and income in usually underdeveloped rural areas,

For the developing and transition countries as Bosnia and Herzegovina, the increased deployment of modern biomass based systems, as a reliable and affordable source of energy could be part of the solution to overcoming their current constraints concerning GDP growth. In any case, production and use of biomass should be sustainable in terms of the social, environmental and economic perspectives.

Success of biomass based projects depends on the understanding of the stakeholders on the all levels which have to understand biomass resource base, its purposes and potential use in some other competitive branches, benefits and disadvantages of use of such material for energy purposes on sustainable manner. All these aspects point strongly to the importance of coordination and coherence of policies directing the supply and use of biomass for different purposes [11]. Only with policy support, established promotional mechanisms and adequate investments environment it is possible to achieve certain level of the bioenergy involvement in energy balance of certain region or country.

An appropriate political and economic strategy of the biomass utilisation for bioenergy (including biomass price policy, subsidies) within the country would evidently encourage the creation of new jobs not only in forestry, agriculture, and wood processing industry but also in other industry branches. Today, it is obviously that issue of biomass utilisation for bioenergy has political, economic and environmental dimension. Thus, governmental regulations are indispensable to provide and secure stable economic and ecologic framework conditions [3].

According to findings from the book "European Energy Payhways" (2011), there are two pathways to sustainable Energy systems in Europe [12]:

- Policy pathway,
- Market pathway.

Policy pathway takes its departure from the EU Energy and Climate Package and has a strong focus on the targeted policies that promote energy efficiency and energy from renewable energy sources (RES). The Market Pathway leaves more of the responsibility for transforming the energy systems to the market, where is cost to emit GHG is dominating policy measure.

Both pathways require significant changes in the infrastructure of the energy system and related power plants, transmission networks, fuel infrastructures, buildings and transportation systems, which is not simple, particularly for transition countries like Bosnia and Herzegovina.

Chosen policies and their applications have direct and indirect impacts on the competitiveness of bioenergy compared with other sources. It is important to increase the knowledge about the design of different tax and support regimes to get the desired effect. The implementation of bioenergy is not solely influenced by financial instruments that support the construction and operation of bioenergy plant, but also depends on policies for agriculture, forestry and the environment as well as public support.

Taking into consideration variety of biomass types coming from different sectors, as agriculture, forestry, wood processing industry, food industry, municipal waste, crucial aspect is an adequate assessment of the resources. Obtaining as much as possible accurate data about available biomass resources is demanding job because potential variations in quantities from year to year. Only theoretical estimated biomass potential for biomass resources in certain area is still not indicative data for the project development, because technical availability depends on a lot of other factors as terrain configuration, equipment selection and type. From the other side economic and market potential depends on a lot of various factors which can be transportation fuel prices, or some other market related issues.

Due to that a lot of tools have been developed in order to give accurate and clear picture about available biomass resources. Tools as GIS (Geographical Information System) are in use today in order to identify biomass resources and their availability for technical exploitation taking into consideration roads infrastructure in certain area, as well as identification of the location for biomass energy plant or some other production plants, taking into consideration access to heat or power supply networks, etc.

Achieving a secure fuel stream that satisfies the business drivers of economy, efficiency and effectiveness whilst remaining within acceptable parameters for environmental impact, quality and future sustainability will be essential to future project development.[13]

Mentzer et al. coined commonly used and well-adopted definitions of supply chains. They define the supply chain as "a set of three or more entities (organizations or individuals) directly involved in the upstream and downstream flows of products, services, finances, and/or information from a source to a customer" [13].

Sustainability of the biomass based energy projects, strongly depends on the establishment of the whole energy supply chain, from the row material at the beginning until final product in a form of energy, synthetic fuel. Biomass is also utilized for food, feed, materials and chemicals, and bioenergy interacts with these areas; in many instances such interactions are synergistic, but they may also be in conflict.

Biomass use for bioenergy can take place immediately in the place of production or in the user's or intermediate producer's /processing firm's site. It must be economically acceptable, and it depends on the available quantities, the transport volume before and after processing, and the required technical equipment, including operation expenses. In all discussions about that, the following should be essential:

- the purpose for which the generated energy is used;
- the availability of biofuels in the close vicinity, including quantity, calorific value, processing and supply costs;
- the efficient use of that energy from biofuels, and the chances for its continuous sale to others.

On this basis, an economic model, including the selection of the equipment, has to be drawn up, and its feasibility should be verified in consideration of potential subsidies and earnings arising from selling energy to others.

The overall purpose of biomass supply chains for energy use is basically twofold: (1) Feedstock costs are to be kept competitive and (2) Continuous feedstock supply has to be ensured [14].

Future renewable energy projects will have to meet much more stringent regulations and guidelines on all areas of operations, from environmental emissions, feed stock materials, process residue disposal or recycling through to employment conditions.[13]

A significant barrier to the use of biomass in some regions is the public concern that its production is non-sustainable. In some instances, such as if harvesting native forests at a rate greater than their rate of natural regeneration, this view is clearly correct. There are simply some sources of biomass that for a variety of reasons (such as their aesthetic, recreational, biodiversity, water cycle management and carbon stock qualities) should never be used for energy purposes [14]

A key constraint to the expansion of biofuel production is the limited amount of land available to meet the needs for fuel, feed, and food in the coming decades. Large-scale biofuel production raises concerns about food versus fuel trade-offs, demands for natural resources such as water, and its potential impacts on environmental quality, biodiversity and soil erosion,.

There are also a number of economic and ecologic problems that could be solved before the economic and environmental effects will be visible in a community. The problems include:

- insufficient sensitisation of companies;

- insufficient sensitisation of the communities and their limited influence;
- insufficient knowledge of the decision-makers in the economy about the assets and opportunities that arise from biomass use for bioenergy;
- insufficient financial resources for changing the way of handling biomass resources;
- insufficient macro-economic incentives.

Biomass energy in Bosnia and Herzegovina has an important role mostly in terms of fuel wood for production of heat energy. This holds particularly true in the areas where the rural sector has a prominent role in the population structure, since historically the rural population in all areas was using the biomass for heating and/or cooking. Biomass in the form of fuel wood and charcoal is currently an ever increasing source of energy in BiH, whose average consumption is estimated at 1,323,286 m^3 per annum. However, the degree of efficiency of the energy conversion devices is very low. Unlike in households, biomass consumption is low in other sectors such as, for example, agriculture, trade and industry. Fuel wood is important mostly in the rural areas and small towns where no public heating network is available. In some areas of Bosnia and Herzegovina, the share of biomass in household heating reaches the level of up to 60% (parts of East Bosnia). As in many cases for development countries, the fuel security and rural development potential of bio fuels that tends to be of most interest. At this micro scale sustainable development drivers are more social-economic. Strategic approach for the rural areas has to offer new opportunities, in a sense that modern village is not only as food producer, with all difficulties related to competitiveness of its products, but also competitive energy producer, or supplier, which gives new dimension of its sustainability.

Most of the cities and rural households have its own heat supply systems, mainly low efficient boilers, which gives a chance to local producers of the biomass boilers and HVAC equipment as well as pellet and wood chips producers. This aspects will be analyzed, particularly in the context of the situation when the most large municipalities in Bosnia and Herzegovina has been signed "Covenant of Mayors" taking the real obligation for local GHG emission reduction.

There are some of district heating systems which have problem with sustainability because of low efficiency and use of expensive liquid fossil fuels. The analysis were shown that is possible to reconstruct some of them and switch the fuel to biomass, issuing lower prices of the heat produced as well as CER (Certified Emission Reduction) because such projects can be attractive as CDM.

There are a lot of small municipalities in Bosnia and Herzegovina with large physical potential of biomass and developed forestry and wood processing industry. It is easy to show that small municipalities in Bosnia and Herzegovina (with 10.000 to 20.000 inhabitants) with centralized wood processing industry can satisfy their all energy needs from its own wood waste, but also start some new business activities based on the available biomass..

Some estimations has shown that 50% of forest biomass this resource could supply medium scale CHP installations (*5 MWe +*) delivering power to grid and heat to residential/

commercial/ industrial users. The installed capacity would be around *21 MWe* and annual output would be *149 GWh* and *213 GWh* of electricity and heat respectively. If half of the scenario where potential of *7,66 PJ* would be available for bio-energy industry, or medium-scale CHP installations, delivering power to grid and heat to residential / commercial / industrial users, *106 MWe* installed capacity that would generate *745 GWh* electricity and *1.065 GWh* heat annually would be supported. Technical and economy aspects of the potential use of some technologies as steam turbines, steam engines, Stirling Engines, Organic Rankine Cycle and gasification technologies in the circumstances of Bosnia and Herzegovina will be analyzed.

Modern market opportunities offers many promotional mechanisms for bioenergy based projects. Some of them which are of the high importance has been analyzed: ESCO (Energy Service COmpanies) and Feed-in tariffs, because they already exists in Bosnia and Herzegovina. Due to that some of the aspects related to promotional mechanisms will be analyzed.

There are no any co-firing biomass based technologies in Bosnia and Herzegovina (except of a small demonstration unit at the Mechanical Engineering Faculty of Sarajevo), but it can became interesting because some analysis shows that use of *50%* of estimated forest residues would result in the production of *149 'green' GWh* within existing solid fuel power facilities, which are mainly from the seventies and use low rank lignite coal.

Biomass from the wood processing industry and forestry, together with agricultural and other forms of biomass is a significant energy source and due to that deserves careful planning and estimation because it can became one of the important economy drivers.

6.1. Heat and electricity opportunities from biomass in Bosnia and Herzegovina

This chapter assesses how the biomass resources - that have been identified and quantified within the previous chapters - could actually be exploited. It is obviously that the resources represent varied, sizeable and replicable opportunities for investment in modern power and heat generation technologies. Use of indigenous, renewable resources would contribute to energy-independence and give environmental benefits notably - but not only - carbon reduction.

Ways of using biomass resources include co-firing with fossil fuels; combustion in new build combined heat and power (CHP) units; anaerobic digestion; combustion at smaller scale ranging from individual stoves and ovens in households to larger, modern boilers for heat provision to buildings etc. The main options of biomass exploitation in the BiH heat, electricity and CHP market sectors are presented below. Based on the estimates on biomass technical potential the options considered for heat & electricity generation include:

6.1.1. Co-firing

Total power generation capacity in BiH is around 4 GW, 2 GW of which are hydropower plants (HPP), 600 MW lignite-fired plants, and the rest coal-fired units. There are 17 district heating (DH) systems operating in BiH. Solid fuels account for nearly 41% of total heat production [15]. It is estimated that around 180 MW_{th} of DH systems operate with brown coal and lignite [5]

Both literature and experts opinion suggests that cofiring 5 - 10% biomass feedstock with fossil fuel (on a weight basis) require relatively minor changes to the technology that is already in place, such as fuel feed systems, storage facilities, emissions controls etc and hence relatively low capital investment. Higher proportions of biomass fuel require more profound technical issues to be addressed and therefore higher investment. Power generation and district heating plant is typically at an advanced age and is unlikely to merit substantial investment, therefore it was considered that the most likely approach would be to co-fire at lower percentages.

Waste wood, forest and industrial residues as well as agriculture residues such as prunings and straw could be used for co-firing although wood chips are the preferable fuel. Forest residues, as estimated by the study, accounts for 3,07 PJ or 380.000 tonnes. If 50% of this could be used for co-firing this would result in the production of 149 'green' GWh within existing solid fuel power facilities [5]). Coal-fired boilers in the government sector could be co-fired or fully fired on biomass fuel. It could especially be realized within government sectors (schools, health organizations etc.) in rural areas of Bosnia and Herzegovina.

6.1.2. CHP generation using woody biomass

Technical potential for forest residues reaches 3,07 PJ or 380.000 tones per year. If the remaining 50% (from the abovementioned co-firing scenario) could supply medium scale CHP installations the total installed capacity would be around 21 MWe and annual output would be 149 GWh and 213 GWh of electricity and heat respectively (SYNENERGY, 2010).

6.1.3. Decentralised bio-gas units

The available livestock manure derived bio-gas can be utilized in small to medium bio-gas CHP units installed near the breeding farms. Nearly 18 MW of such installations may be fuelled by the 1,30 PJ of available bio-gas. These units could produce 126 GWh of electricity [5].

6.1.4. Small scale modern heating appliances

Currently biomass consumption comprises individual, traditional small stoves, ovens, boilers etc., with low efficiencies. The significant use of fuelwood indicates that there could be opportunities for the development of the market for modern biomass heating appliances. Over 54% of the total energy consumption in Bosnia and Herzegovina is in the household sector and 70% of this is fuel wood. Improved stoves and alternative fuels, while outside the scope of this study, are highly relevant in this context [6]).

The study estimated that if 20% of the 18,45 PJ of available fuel wood could be exploited for this purpose this would result in generating 820 GWh of heat annually [5].

6.1.5. Straw fired units

Agricultural residues in Bosnia and Herzegovina consisting primarily of straw account for some 6,63 PJ [5]. Straw may be directly used either in decentralized small, mainly farm based units producing heat for various purposes or in centralized CHP units. Large scale central straw fired units usually require strong economies of scale (capacities in EU are around 100 MW) and are coupled with an alternative fuel, usually conventional one. Considering the significant geographical spread of straw supply and the fact that logistics play a critical role to the economics of such plants it is unrealistic to expect new straw alone fired units to be built. In this respect straw could be merely used for heating purposes either in straw bale fired units or as straw pellet in pellet stoves and boilers. If one third of the technical available straw could be directed to this use it could produce nearly 491 GWh of useful heat [5]. One potential model for utilization of some agricultural residues is in the formation of rural agricultural processing companies. The company supplies seeds, fertilizers, equipment, and training to small rural farmers, and collects harvests (including residues) for centralized, high-tech processing. Sale for process heat and electricity generation converts residues into a valuable marketable product for local and international markets (and ash potentially used for fertilizers). This business model (excluding the energy components) is currently successfully used in some developing countries [6].

6.1.6. CHP using energy crops

The SYNENERGY study makes a reasonably conservative assumption that 10% of land currently used for grazing/pasture and 5% of the fallow land (low scenario) is used to grow perennial grasses. The Technical Potential, if perennial crops are used is estimated to be 15,33 PJ resource. It is assumed that half of this resource would be available for bio-energy industry, or medium-scale CHP installations (individual capacity 5 MWe plus) delivering power to grid and heat to residential / commercial / industrial users. This would support 106 MWe installed capacity that would generate 745 GWh electricity and 1.065 GWh heat annually [5].

6.1.7. Small scale heat with energy crops

In the study it is estimated that the other 50% of the 15,33 PJ for energy crops would support local small scale energy crop fired baled fired boilers or energy crop pellet boilers supplying residential properties with heat. This would equate to 1.703 GWh of useful heat production per year. [5])

6.1.8. Municipal Solid Waste

Effective treatment of municipal solid waste (MSW) represents challenge in the protection of the environment and natural resources, especially for countries in transition such as Bosnia and Herzegovina.

Cellulose or lignin derived materials, polymer based materials (plastic waste) together with inorganic material present the main components of MSW. Significant portion of plastic waste is disposed of on landfill, while only small part is recycled applying mechanical technology. Regarding the plastic waste, mechanical recycling can be recommended as a desirable technology because this makes no more pollution problems. But it is very difficult to separate various waste plastics with dust and metals into one-component raw material which can be recycled without any problems. So thermal recycle technologies are the objects of interest as alternatives for the mechanical recycle technologies. Also, the growing awareness in environmental concerns and the reducing landfill space have further prompted research in alternative methods of plastic recycling such as thermochemical conversion, particularly pyrolysis. In these technologies, pyrolysis may be favorably used for oil and monomer recovery from waste plastics. Also, this technology has more advantages than combustion technology in the view of discharging less pollutants. The resulting products of pyrolysis are solid char, liquid pyrolytic oil and gases. Each of the products formed has potential usage as energy carriers and chemical feed stocks for further processing.

Co-pyrolysis techniques have received much attention in recent years because they provide an alternative way to dispose and convert plastic polymer and cellulose (or lignin) derived materials into high value feedstock and the specific benefits of this method potentially include: the reduction of the volume of waste; the recovery of chemicals and the replacement of fossil fuels. Since MSW consist both wastes, plastic and cellulose or lignin derived materials, Co- pyrolysis techniques may be very attractive method of treating mixed MSW.

6.2. The possibility of using biomass in district heating systems (DHS) in Bosnia and Herzegovina as a way to achieve their sustainability

Maybe most obvious example of unsustainable energy systems in Bosnia and Herzegovina are district heating systems, there are several reasons for that: most of them are old, built in seventies, and requires reconstructions and technical improvements, a lot of them running on expensive liquid of gaseous fossil fuels, and tariff system is more socially oriented than market oriented. There is also one important issue which makes whole concept unsustainable and requires urgent solutions, mayor shareholders of those systems are local communities, and functioning of DHS is directly affecting on their annual budgets. Due to that bioenergy can became solution for some of them, particularly with approach which consider use of clean development mechanisms of Kyoto as the one of the approaches which can make those projects sustainable.

The Clean Development Mechanism (CDM) is the one of the three flexible mechanisms (the other two are l Emission Trading - ET, and Joint Implementation - JI) which allows entities from Annex I (developed) parties to develop emission-reducing projects in non-Annex I (developing) countries, and generate trade able credits – CER credits (CER - Certified

Emission Reduction, one CER is equivalent to one tonne of CO_2 emission reduction) corresponding to the volume of emission reductions achieved by that project.

Depending on the scale of the projects CDM projects can be classified into large-scale or small- scale projects.

There are three types of small-scale project activities; Type I: renewable energy project activities with a maximum output capacity of 15 megawatts (or an appropriate equivalent); Type II: project activities relating to improvements in energy efficiency which reduce energy consumption, on the supply and/or demand side, by up to 60 GWh hours per year (or an appropriate equivalent); Type III: other project activities that result in emission reductions of less than or equal to 60 kilotonnes of carbon dioxide equivalent annually.

Any CDM project activity not possessing the above mentioned characteristics is considered a large-scale CDM project activity.

Several options proposed under the CDM rules allow the development of CDM programmes, among them being bundles, PoAs, and several stand-alone CDM activities.

By definition, a CDM PoA is considered »a voluntary coordinated action by a private or public entity which coordinates and implements any policy/measure or stated goal (i.e. incentive schemes and voluntary programmes), which leads to GHG emission reductions or increases net GHG removals by sinks that are additional to any that would occur in the absence of the PoA, via an unlimited number of CDM programme activities (CPAs)« Bundling is a modality allowing the validation and registration of several project activities (small or large scale ones) within one CDM entry. Just like PoAs, bundles allow significant economy of scale while developing several CDM activities together.

In Bosnia and Herzegovina District Heating Systems are generally concentrated in larger cities. According to available data, currently in Bosnia and Herzegovina exists 25 District Heating Companies (12 in Republic of Srpska and 13 in Federation BiH).

District Heating Companies in Republic of Srpska mainly relies on its own boiler facilities, which mainly use fossil fuels (fuel oil, coal, gas). The exception is the District Heating Plant in Pale which as addition to coal use also biomass (waste wood) and Sokolac (only biomass). Estimated consumption of biomass in district heating sector in Republika Srpska in 2012 amounts about 1218,00 tonnes (0,1219 PJ) [16].

According to data listed in [17] the installed capacity of boilers in District Heating Companies in Republic of Srpska is 483.5 MW, the district heating sector is heated about 40 000 flats with a total area of about 2.3 million m^2 and about 460 000 m^2 of office space. According to available data, during the 2010 District heating companies in Republic of Srpska delivered to consumers about 1483 TJ of heat energy [1].

In Federation of BiH, the largest number of district heating systems also use fossil fuels (coal, fuel oil, gas). A certain number of district heating companies do not have their own thermal aggregates such as boiler units, but are connected to local heat production facilities

– thermal power plant on coal (Tuzla, Lukavac, Kakanj) or Ironworks in which is also the primary fuel coal (eg, Zenica). The largest district heating system is in Sarajevo (installed capacity of boilers is 488.694 MW and the connected heat load is about 333.162 MW) that uses mainly gas as a fuel [18].

In Federation of BiH, also two district heating systems (in Gradačac and in Livno) use biomass as primary fuel. According to available data, consumption of biomass in these two companies during the heating season 2010/11 amounted to 14 980 m^3 (Gračanica 12880 m^3, Livno 2100 m^3) and to consumers has delivered around 25.6 TJ (Gračanica 20,218 TJ, Livno 5,381 TJ) of heat energy.

According to available data, during the 2010 District heating companies in Federation of BiH delivered to consumers about 3913 TJ of heat energy [2].

One way to improve the current situation in the district heating systems which using fossil fuels is the partial or complete replacement with biomass fuels where it is possible. Those projects can be attractive as CDM project.

The analysis conducted in the District Heating Companies in Gradiškaand Prijedor] which use heavy fuel oil as fuel has shown that realisation of proposed CDM Programme of Activities (PoA) would led to lower heat prices, openning of the new jobs, reduction of fossil fuels dependency of Bosnia and Herzegovina and reduction of CO_2 emission. In addition, by selling CERs District Heating Companies would provide additional revenues that could invest partially in the modernization of existing systems.

The District Heating Company in Gradiška provides heating for about 1740 buildings (residential buildings, public buildings such as kindergarten, schools etc. and other facilities). The vast majority of these, about 50%, are residential apartment buildings. Heated floor area in residential buildings is about 75 000 m^2. It produces heat in a central boiler house, consisting of two 11.8 MW boilers with a combined capacity of 23.6 MW. The boilers are fired by heavy fuel oil, and the total connected heat load in the town is about 16.8 MW.

Average annual fuel consumption during the heating seasons (2008-2010) is about 1516 tonnes of heavy fuel oil, and heat supplied to the district heating network is about 13,35 GWh/yr. Consumption of heavy fuel oil has been increasing each year because of connection of new customers to the existing district heating network.

The District Heating Company in Gradiška intends to install a new 6 MW wood biomass boiler for production of thermal energy for heating residential and commercial facilities in Gradiška. The new biomass boiler will be installed within the existing boiler house of the company. During the heating season, the biomass boiler will provide the base heat load. In that way, the Public Communal Company "Toplana" A.D. Gradiška has estimated less consumption of heavy fuel oil (which is currently the only fuel for production of thermal energy) by approximately 1080 tonnes annually.

As part of the project, the wood biomass boiler will be connected with the existing boilers in a parallel function enabling the use of both heavy fuel oil boilers for covering peak heat load during the coldest winter days. As a result of implementation of this project, the new installed heat capacity in production will be 29,6 MW.

Biomass fuel should be transported by a truck from the local Forestry Company or local biomass factory, about 30 km to a storage area, which will be built close to the existing boiler house. The amount of transported biomass will be supported by invoices. Calculation shows that payback period with estimated investment of 2.87 million EURO and CDM is about 6 years and 5 months.

Toplana A.D. Prijedor, the district heating company (DHC) is a main producer of heat for the town of Prijedor and it covers nearly 320 000 m2 of building surface for heating. Installed heat power is 2x30 MW via two boilers. Total connected heat load in the town is about 30 MW (the second boiler is technical reserve). Annual heat energy production is approximately 50 GWh.

Today DHC uses heavy fuel oil for combustion. One of the existing boilers of 30 MW will be reconstructed in order to use wood pellets. This boiler will be the base load boiler, while the other existing boiler will be reserve and peak load boiler. The needed wood pellets will be produced by the DHC and it is a part of the project.

Production of wooden pellets includes a complete introduction of the new technological line for production of wood pellets (Figure 1). The wood pellets will have the following parameters: 6 mm diameter, 10% of moisture, 1 % of ash, and 5 kWh/kg calorific value.

The capacity of the technological line for production of wood pellets will enable production of 4 t of pellets per hour. This capacity will be sufficient for the continuous production of heat during the heating season. In addition, pellets will be produced outside the heating season and all production surpluses will be sold on the market. Raw materials used for the production are wooden sawdust, waste wood and wooden logs that are categorized as firewood. Warehouses for the reception of raw materials are located near of the boiler house and have the capacity of 20 000 m3. Energy from wood pellets will replace energy from 4901 t of heavy fuel oil. The total amount of pellets needed per heating season is about 11272 t. To produce this amount of pellets, DHC in Prijedor should provide at least 25362 m3 spatial raw wood with 50 % humidity.

Calculation shows that payback period with estimated investment of 4.4 million EURO and CDM is about 5 years and 8 months, which is one year shorter than project without CDM. Reduction of CO_2 emission from the project will be 14 381 t/yr.

7. Overview of all existing barriers to harnessing the biomass energy potential

When considering further developments in Bosnia and Herzegovina's energy sector, conventional energy wisdom has to be adapted to fit the specific context. Although hydropower will remain the mainstay of the renewable energy sector in the near future, biomass as an energy carrier does have potential on the Bosnia and Herzegovina market.

While the size of the Bosnia and Herzegovina's market place allows for some economies of scale, its capitalization, the purchasing power and even the monetization of Bosnia and Herzegovina remains low. In the rural areas, the private sector is still underdeveloped, but the human resource base is not limited, and the electricity grid is developed at a sufficient level.

Key barriers that were identified can be summarized (: the development of large-scale bioenergy plantations that can supply sustainable amounts of low-cost biomass feedstocks; the risks involved in designing, building and operating large integrated biomass conversion systems capable of producing bioenergy and biofuels at competitive prices with fossil fuels; and the development of nationwide biomass-to-bioenergy distribution systems that readily allow for consumer access and ease of use [19].

Decentralized renewable energy technologies and markets offer opportunities; but they need support, including targeted policies, capacity building, adequate financial resources to meet high up-front costs, and special effort to link-up with income generation activities. Specific barriers include [20]

7.1. Financial barriers

- The high capital cost of biomass energy systems is a major barrier to the increased use of these systems, despite such technologies being among the cheapest renewable energy technologies;
- The capacity to assess biomass energy proposals/loan applications is limited or non-existent;
- There are significant other priorities for public and private funds for reconstruction, food security, poverty alleviation, following the war, and local financial resources are consequently scarce;
- Since there are virtually no biomass energy projects there are no economies of scale;
- A large fraction of the energy economy (fuel wood) operates outside the formal economy;

In order to avoid financial barriers, some promotional mechanisms are usually used in realization of bioenergy projects [21]:

- **Feed-in tariffs and fixed premium;**. These systems exist in various European countries (including Bosnia and Herzegovina) and are characterized by a specific premium or total price, normally set for a period of several years, that domestic producers of green electricity receive. The additional costs of these schemes are either paid by suppliers in proportion to their total sales volume and are passed through to the power consumers, charged directly to buyers of green electricity or paid by national governments using environmental taxes on conventional electricity. Fixed feed-in systems are used, for example, in Austria and Germany. Fixed-premium systems are used in Denmark, the Netherlands and Spain.

- **Green Certificate Systems;** A system of green certificate systems currently exists in five EU Member States, as well as Australia. In this case, renewable electricity is sold at conventional power-market prices, but with the right to sell government-issued certificates that guarantee the renewable character of electricity to consumers or producers that are obliged to purchase a certain number of green certificates from renewable electricity producers according to a fixed percentage, or quota, of their total electricity consumption/production. Since producers/consumers wish to buy these certificates as cheaply as possible, a secondary market of certificates develops where renewable electricity producers compete with one another to sell green certificates.
- **Tendering;** Under a tendering procedure, the state places a series of tenders for the supply of renewable electricity, which is then supplied on a contract basis at the price resulting from the tender. The additional costs generated by the purchase of renewable electricity are passed on to the end-consumer of electricity through a specific energy tax. Pure tendering procedures existed until recently in Ireland and France.
- **Investment subsidies**; In some countries, direct investment subsidies apply for biomass combustion systems. This is the case, for example, in Germany for domestic wood pellet stoves.
- **Tax deduction;** Support systems based only on tax deduction are often applied as an additional policy tool to support renewable energy. In the Netherlands for example, a company investing in a biomass combustion system may deduct an additional 44 per cent of the investment cost from their taxable income.

7.2. Policy barriers

- Absence of an integrated policy and regulatory framework within Bosnia and Herzegovina that would otherwise encourage the use of biomass residues for energy generation;
- Suitable policies and regulations are yet to be enacted to provide a level playing field for renewable sources, including the biomass energy;
- Policies and governmental linkages between biomass energy use and income generation activities are weak and/or non-existent;

7.3. Information barriers

- There is limited availability and access to existing renewable energy resource information. Data frequently does not exist, and a central information point is lacking – information is scattered between sectors; e.g. public sector, private sector (including consultancy firms), development assistance, R&D centres and academia;
- There is a limited knowledge of the biomass energy potential due to lack of detailed market surveys;
- Where information on economics, market development, marketing, and technical issues exist, it is distributed between organizations that do not co-operate;

7.4. Awareness and perception barriers

- There is a lack of awareness of modern options for biomass energy. Knowledge on, for example, the competitiveness of life cycle costs of the biomass energy technologies (which can be the lowest cost option) is mostly absent.
- There is a perception that the traditional use of wood and charcoal must be reduced, so biomass energy is seen as something to be discouraged;
- There is little knowledge and no experience of the costs and benefits of the range of technologies available for modern biomass energy;
- Limited in-country capacity for renewable energy data collection and analysis is an important barrier for renewable energy project development;

7.5. Institutional barriers

- Modern biomass energy services are dealt with by various ministries, agencies and institutions, on different levels, making good coordination between them a necessity if efficient use of limited human and financial resources in an area is to be achieved;
- Generally speaking, government decision-makers (who have access to and control the budget) have little interaction with those at operational level. Operational lines of communication between operation and decision-making levels need to improve within government agencies;
- Limited geographic distribution of suppliers limits access to renewable energy technologies (hardware);

7.6. Technical barriers

- Bulk procurement of renewable energy technologies is limited due to the current small market for renewable energy based modern energy services. Hence the (technical) infrastructure to support renewable energy development does not exist;
- Local manufacturing and/or assembly of renewable energy technology components are currently mostly lacking;
- There is only limited technical capacity to design, install, operate, manage and maintain renewable energy based modern energy services, mainly as a result of lack of past activities in this field;
- The technical skills, including conclusive data comparing energy technologies for equivalent energy services, is limited;
- Norms and standards in terms of renewable energy performance, manufacture, installation and maintenance are weak and/or non-existing.

It is clear that without addressing the above barriers, it will be difficult to promote sustainable energy alternatives to increase biomass use in Bosnia and Herzegovina. At the same time, Governments at the entity and state level as well as the other institutions in Bosnia and Herzegovina have little capacity – financial, technical or institutional – to address these barriers [6].

8. Sector-wide technical assistance

In order to achieve a more significant application of biomass in BiH, first of all, it is necessary to carry out the following research:

- defining target areas in BiH where detailed research of economically and ecologically sustainable use of biomass should be performed,
- quantification of different flows of non-used biomass in target areas,
- estimation of biomass costs as a fuel in the future and a comparative analysis with the costs of other fuels,
- identification of the possibility for suitable, financially competitive solutions of biomass application,
- identification of the most suitable technologies, investment methods and incentive measures for selected solutions of biomass application,
- identification of obstacles in legislation and regulations that influence the selection of technologies for biomass application in the target areas in a most efficient way,
- identification of institutional obstacles for accepting the most efficient solutions for the construction of a biomass-fueled system for production of thermal and/or electrical energy

Implementation of the above mentioned steps would clearly show the real economical and ecological potential and solutions for the application of biomass-fueled facilities in the target areas in BiH, and it would help the competent authorities to plan the construction of such facilities. The identified activities greatly depend on the agriculture and forestry development strategy and the ministry of energy should plan and implement them together with the competent ministries for these areas [6].

Despite the high dependence of Bosnia and Herzegovina's rural population on biomass energy and the apparent large biomass energy resources, information related to the biomass energy sector is difficult to find and is frequently out of date. Modern biomass energy systems are virtually unknown, and consequently, there is a significant need for technical assistance that will benefit the sector as a whole. This should run parallel to targeted activities to develop specific biomass energy markets. Technical assistance, in the first instance, is required to deal with:

- Information on resource availability: scepticism about the availability of biomass for fuel was frequently expressed, and this manifested itself in a belief that biomass energy is something to be discouraged. Clear and unambiguous information about fuel availability (going well beyond the estimates based on production that have been used in this report) are needed;
- Awareness on options and benefits: knowledge of the options and benefits for biomass energy appears to be severely limited, and there is a need for targeted awareness raising (through, for example, study tours and training courses) and for information dissemination, including promotion campaigns and the like. This is closely tied to a need to training, human and institutional capacity building on supply, energy and heat generation, and demand.

Progress in these areas of technical assistance would provide the basis upon which further assistance could build. Assuming that a case can indeed be made for biomass energy (that sufficient bio-resources are available and cost effective technical options could be adopted), technical assistance would be required in a number of related areas including:

- Legal and contractual issues (standard biomass fuel supply contracts for example);
- Tariff adjustment, including issues related to installed capacity payments;
- Policy issues (including biomass based electricity in energy policy/plan), and the development of more coherent energy strategies for rural / remote areas;
- Government, NGO and private sector institutional requirements to support a modern biomass energy market
- Project financing and project development.

Technical assistance should go hand in hand with one or two targeted demonstration projects, characterized by relatively low risk, and with sufficient potential for the development of a viable market.

Since relevant data appears to be held by a wide variety of government, NGO and private sector stakeholders, in some cases with unhelpful competition, an institutional structure which is transparent and non-partisan is needed to make available basic and common information as envisioned through sector-wide technical assistance.

9. Conclusions

In contrast to many other energy technologies biomass and bioenergy production is connected to many policy areas, such as climate, energy, agriculture and waste policies, which means that definition of bioenergy policy requires more wider and inter sector approach in order to achieve sustainability of the whole system, An adequate policy concept is a key factor in the establishment of the sustainable bioenergy systems in any country. The availability and use of biomass resources are often intertwined with various major sectors of the economy: agriculture, forestry, food processing, building materials, traffic, etc, but from the other, positive side, this gives bioenergy many opportunities to generate multiple benefits apart from energy generation

It is obvious that only integral approach in establishment of biomass conversion systems, bioenergy and biofuels from biomass can find place on a market and became competetive to fossil fuels.. Bioenergy projects must be economically viable for the different actors in the value chain. Biomass used for energy purposes must be able to compete with other uses of the biomass, and at the same time the energy produced from biomass must be as cheap as or cheaper than energy produced from competing energy systems. Bosnia and Herzegovina is a country with significant potential in different biomass resources, and it is obviously that biomass and bioenergy in a forthcoming period can play more important role in the economy of the country. In order to achieve sustainable system of biomass use and exploitation, it is necessary to define an adequate policy and legal framework which leads to that goal. .

Author details

Petar Gvero and Milovan Kotur
University of Banja Luka, Faculty of Mechanical Engineering, Banja Luka, Republic of Srpska, Bosnia and Herzegovina

Semin Petrovic
IGT – Research and Development Centre of Gas Technology, Sarajevo, Bosnia and Herzegovina

Sasa Papuga
University of Banja Luka, Faculty of Technology, Banja Luka, Republic of Srpska, Bosnia and Herzegovina

10. References

[1] Republic of Srpska Institute of Statistics http://www.rzs.rs.ba/Publikacije/RSuBrojkama/RSuBrojkama2008.zip
[2] Federal Office of Statistics http://www.fzs.ba/Dem/Vital/VitalnaBos.htm
[3] ***. Cost/Benefit Analysis in the B&H Biomass Sector – With a Special Focus on Srebrenica Region. UNDP Study, Sarajevo 2010
[4] ***, Initial National Communication of Bosnia and Herzegovina under the United Nations Framework Convention on Climate Change (UNFCCC). Banja Luka, 2009.
[5] ***, SYNENERGY – Biomass in Bosnia and Herzegovina, Action 1: Renewable Energy Assessment, USAID HELLENIC AID ENERGY COOPERATION, CRES Greece, 2010.
[6] Tremeer, G.B., Gvero, P., Petrovic, S., Pilot Study on Using Biomass-fired Boilers in Rural Buildings of for Rural Productive Uses Local Wood Waste in Bosnia and Herzegovina, *UNDP Study*, Bosnia and Herzegovina, 2006.
[7] ***, National Action Environmental Plan (NEAP), 2003.
[8] ***. ANIWASTE – Integral Solving of Waste Problem form Farms and Slaughterhouses in NW B&H Region, Feasibility Study, EU CARDS Project, 2005Kautto, N., Towards More Coherent and Sustainable Biomass Policy. Examing European Biomass-to-Energy Planning. Ph.D.Thesis, IIIEE, Lund University, Sweden, 2011.
[9] ***, SWMS-Solid Waste Management Strategy for BiH, EU PHARE program, AEA Technology plc, 2000.
[10] Gvero, P.M., et.al., Renewable Energy Sources and Their Potential Role in Mitigation of Climate Changes and asa sustainable Development Driver in Bosnia and Herzegovina. Thermal Science. Vol.14, Issue 3 (pp.641-654), 2010.
[11] Kautto, N., Towards More Coherent and Sustainable Biomass Policy. Examing European Biomass-to-Energy Planning. Ph.D.Thesis, IIIEE, Lund University, Sweden, 2011
[12] ***, European Energy Pathways, Pathways to Sustaiable European Energy Systems. Alliance for Global Sustaiability, Goteborg, Sweden, 2011.

[13] Simpson, P. Supply chain development for biomass fuel -challenges and opportunities. IMechE Conference Transactions - Renewable Bioenergy - Technologies, Risks, Rewards, (pp.3-8), London, 2003.

[14] Gold, S., Seuring, S., Supply Chain and Logistics Issues of Bio-nergy Production.Journal of Cleaner Production, 19 (2011) (pp.32-42).

[15] International Energy Agency (IEA) 2010. http://www.iea.org/stats/index.asp,

[16] ***, Energy balance Republic of Srpska, the plan for 2012, Government of Republic Srpska, 2012.

[17] ***, Republika Srpska Energy Sector Development Planuntil 2030. Energetski institu Hrvoje Pozar, Ekonomski institut Banja Luka, 2010

[18] http://www.toplane-sa.co.ba/onama.php ,2012.

[19] ***, Cost/benefit analysis in the B&H biomass sector – with a special focus on Srebrenica region, UNDP Study, Bosnia and Herzegovina, 2010.

[20] Champagne, P., Biomas. Future Energy -Improved, Sustainable and Clean Options for Our Planet. In edition of Letcher, M.T., Elsevier, 2008.

[21] Van Lo., Koppejan, J., The Handbook of Biomass Combustion and Co-firing. Earthscan, 2008.

Chapter 2

Cereal Canopy Structure – Its Assessment and Use in Efficient Crop Management

Jan Křen, Tamara Dryšlová, Lubomír Neudert and Vojtěch Lukas

Additional information is available at the end of the chapter

http://dx.doi.org/10.5772/54528

1. Introduction

Cereals are economically the most important group of field crops. Detailed knowledge of stand structure and its development including interrelationships among individual plants in the stand (inter- and intra-plant competition) is a significant precondition for effective cropping treatments during the growing season, including agrochemicals application.

To attain higher effectiveness of crop management practices, extensive research on cereal stand structure was conducted in the 1980s and 1990s [1-6]. The stand state and structure reflect variability in soil conditions as well as cropping treatments. Results of individual methods used for modification of cropping treatments depend on a level of stand organization which is observed – a stand (plant population), plant, plant part (leaf, tiller) [7].

Stems and spikes are the most often assessed units of stand structure as for final and resulting expression of all factors affecting stand development. However, they are also reproductive units and basic units of an important cereal adaptation system - tillering [8]. Therefore, their appropriate assessment allows obtaining information of great biological and economic importance.

The chapter gives a review of evolution of approaches used for the assessment of cereal stand structure. Weaknesses and strengths of these approaches are discussed.

1.1. Approaches based on dividing of cereal yield and growth analysis

The growth in biology is usually described by observing temporal development of average measurement values. Similarly, for analyses of yield formation a number of formed and reduced yield elements per unit area (tillers, reproductive organs and grains) is observed and their average values are determined.

The assessment of cereal stands and yield formation is usually based on the classical concept as reported by Heuser [9] and later on by a number of other authors who divided grain yield into spike number per unit area, grain number per spike and grain weight (1000-grain weight) – Figure 1.

At present, this concept based on the plant number and numbers of formed and reduced tillers per stand unit area prevails in both applied research and practice. It uses advantages of plant as well as stand description on the basis of changes in the tiller number when no destructive analyses and higher labour intensity are needed. Recently, however, the concept has been often criticized because it does not provide enough precise quantification of differences in stands. Hunt [10] drew an attention to changes in the number and size of plant parts (modules) during plant growth and development. It is evident that the procedures based on the growth analysis are rather labour intensive; their simplification for practical use results in lower accuracy and does not allow to record spatial heterogeneity of the stand. Therefore, some authors expressed a need of available innovated criteria for stand assessment [11].

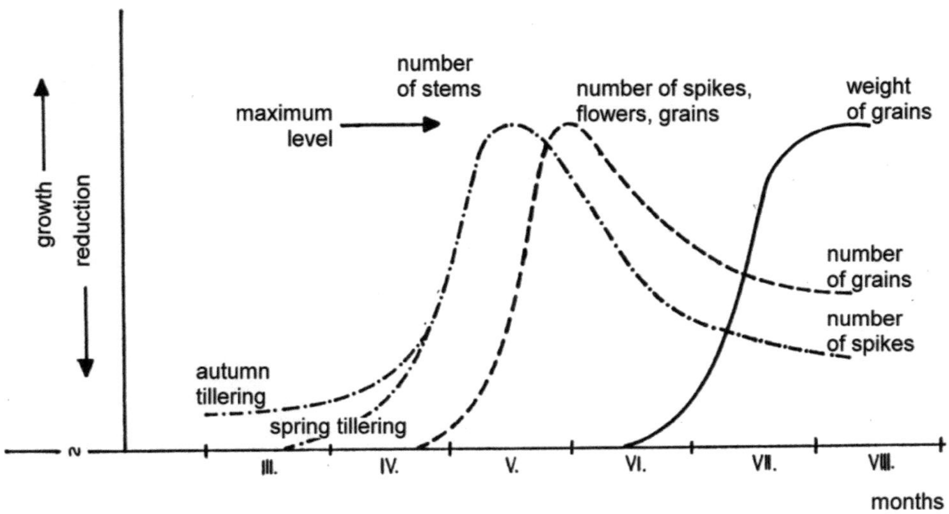

Figure 1. Schematic illustration of dynamics of yield elements formation and reduction in tillering cereals through the growing season [12]

1.2. Plant modular growth and population concept of stem system of the stand

Dividing of yield into individual yield components allowed cereal research and practice to get closer to so-called modular concept of plant and plant growth demographical analyses [13]. White [14] and Porter [15,16] report that plants can be studied as developing modular systems and their growth can be described similarly to processes of population type.

The growth and development of cereal plants consist of a number of growth and development stages of modules (leaves, shoots, stems and grains) that ovelap one another. Therefore, the growth and development of individual leaves and shoots are more determined than those of entire plant. The growth of entire plant does not stop unless the growth and development of the last module is finished, whereas the first formed modules finished their growth and development earlier. The size and properties of leaves and stems in the stand depend not only on their position on the plant, however, on the position of plants in the stand, i.e. on micro-conditions influencing the growth of individual plants [5]. Thus, in cereal stands the variability of site conditions is reflected in changes of inter- and intra-plant relationships, which is expressed by changes in variability of plant modular parts [6]. This concept enables to explain compensatory and autoregulatory processes in cereal stands by modification of both the number and size of plant parts. The stand structure can be described by density distribution (histogram, polygon) of their weight. For effective stand management of small-grain cereals it is important to assess the amount of biomass of productive stems per stand unit area or proportion of this "productive" biomass of the total amount of aboveground biomass.

1.3. Autoregulation and compensation in the stand

The cereal stand is usually interpreted as plants growing per unit area rather than individuals assembled in the population, and their interrelationships are most frequently measured as an average or average reaction [17]. However, some authors have earlier referred to relationships between the average and other statistical parameters – variance and skewness [18,19]. Křen [5,6] documented that these parameters enabled to evaluate intra-plant relations.

The canopy closure can be established both through an increased number of stems and their increased size (weight). Stand productivity, as a result of compensatory and autoregulatory processes, depends on total productive stem weight per unit area and is limited by site productivity (carrying capacity that represents a summary of all sources which are available to plants in given space and time).

The identical yield of aboveground biomass can be obtained by lower plant density and a longer period of their growth or vice versa. This logically results in mutual compensation of plant density and size. In practice it means that at the formation of biomass amount corresponding to site carrying capacity self-thinning takes place during the further growth.

Changes in the number and size of shoots in cereal stand during the growing season are analogical to changes in natural plant populations and can be illustrated using frequency curves. A large potential number of shoots capable to reproduce are formed by tillering. This amount, however, reduces to a final spike number, which usually corresponds to 1/2 to 1/4 of the tiller number at the beginning of stem elongation (BBCH 31), by the period of anthesis (BBCH 61), when the organization of stand structure is finished.

The presented rules reveal that the stand structure is always a result of a response of the plant population to site conditions. Their good knowledge should be a basis for assessment of stand structure, which will enable more effective utilization of vegetation factors of the location, cropping treatments and properties of varieties.

From this point of view, the development of root system is of a great significance. Strong root system is important for nutrients and water uptake, and leads directly to increasing the site carrying capacity. By the lower density of plants, they form more tillers and more roots. The rooted tillers exhibit better tolerance to unfavorable conditions. It leads to a general conclusion that to ensure high yield it is necessary to obtain as high number of productive stems and their biomass amount as possible by as the lowest plant number per unit area as possible. These are requirements for optimum development of plants in the stand and canopy closure. However, this general standpoint can be hardly implemented in practice since a large number of factors influencing the plant growth and development are impossible to control completely. Growers should take into consideration that the site productivity and duration of the growing season do not exhibit stable values that would enable to determine exactly sowing time, appropriate seeding rate and stand development during the vegetation. In particular cases, ability of knowing how to respond to the weather course in individual years using a way of stand establishment and consecutive cropping treatments is important. Thus, effective methods for the assessment of the stand during the growing season can be a valuable tool for farmers.

1.4. Possibilities of innovations in stand structure assessment

The stand structure depends on initial plant number, available sources and their change during the growing season. The value of obtained information should be adequate to consumed labour. In this respect, a sample size is of great importance. In general, it governs that with the increasing size and number of samples the exactness of results increases, however, labour intensity is also higher. The two problems (labour intensity as well as the value of obtained information) are to be solved, i.e. what information is provided by plant and stem analysis and how to use it.

Classical methods for the assessment of stand structure based on counting plants and stems (spikes) per unit area of the stand are labour consuming and interpretation of results is often difficult. They provide information on plant and stem numbers and/or their size (weight), however, they do not allow assessing the relationships in the stand (inter- and intra-plant competition).

Using a current level of knowledge and novel technologies could enable to make diagnostics of stand state and structure (to assess the amount of produced biomass and its structure) more effective. Based on data published over the last years [20-23], it can be assumed that spectral characteristics and area sensing of stands can be used for this purposes.

Based on the character of processes influencing the stand structure, the growing season of cereals was divided into the three parts:

1. vegetative, including the period from emergence till the end of tillering (BBCH 10-29),
2. generative, including the period of stem elongation and heading (BBCH 30-59),
3. reproductive, including anthesis, grain formation and maturation (BBCH 60-99).

2. Canopy structure development – problem statement

2.1. Vegetative period - from emergence till the end of tillering (BBCH 10-29)

Vegetative period lasts from emergence to the end of tillering (BBCH 10-29), it is time limited by the transition to the double ridge stage. It includes the growth of the first leaves, formation of buds in leaf axils, and tillering (Figure 2). A number of code systems have been proposed for describing tillering pattern, of which the Rawson's system [24] has become the most universally accepted. Tillers are designated by letter T and the index which is determined by the order of leaves in whose axils the tillers emerge (Figure 3). Tiller buds in axils of the fifth and sixth leaves only sporadically grow more than 5-6 mm, and in the axils of other leaves they remain small, clearly visible only after magnification. No buds can usually be observed in the axils of the eighth leaf and other leaves of the main stem [25].

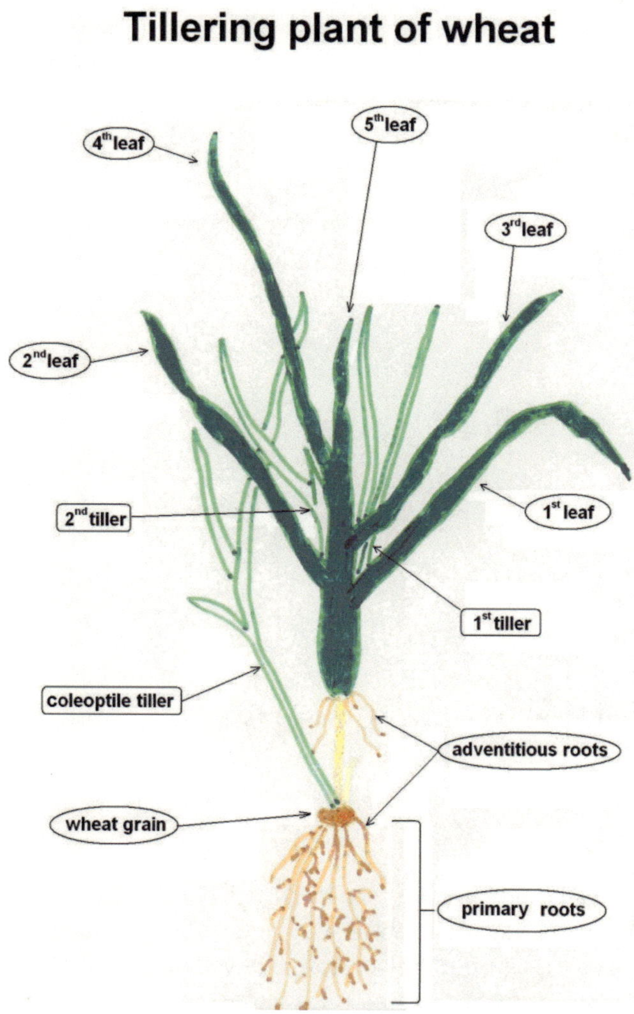

Figure 2. Illustration of tillering wheat plant [27]

After emergence of seeds, number of plants per area unit is determined, i.e. space limitation of their growth, and pool of tillers for further selection is thus formed. By sampling it is possible to identify individual plants and their tillers in the stand. Weight distribution of plants is left-sided skewed. Plant variability indicates stand establishment quality (Graph 1). The values of the CV range between 30 and 60 % in common stands [30].

Weight variability of tillers is the highest (CV = 50 to 80 %) in this period. It is caused by their gradual formation. The number of tillers per square meter in winter wheat and spring barley usually ranges from 1600 to 2500, in some cases exceeds 3000. Distribution of their weight is also strongly left-sided skewed, under favorable conditions is continuous and unimodal (Graph 1), which indicates good conditions for stem formation and stand structure development. Bimodal distribution indicates the effect of unfavorable conditions. Growth reduction occurs, tillers are smaller and apical dominance is strongly expressed. The segregating distribution therefore corresponds with the distribution of main shoots (Graph 1). As individual plants and tillers can be identified, the intra-plant relationships can directly be assessed, e.g. by regression analysis of tillers weight dependence on their sequence in the plants, however, it is very laborious.

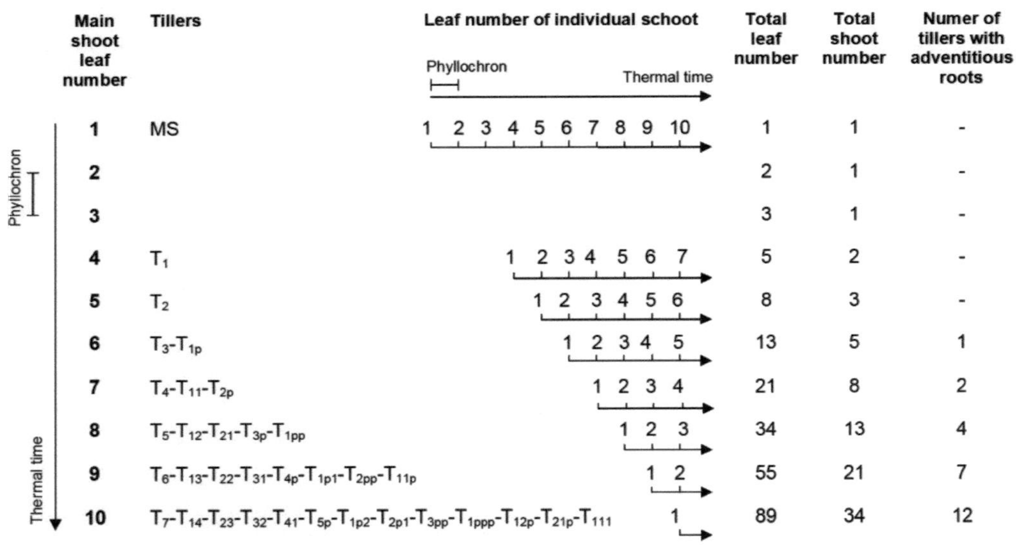

Figure 3. Schematic illustration of tillering wheat plant [28] – see Figure 2

Shoot growth is practically measurable since the first leaf emergence above ground (in the main shoot), and since bud emergence in leaf axils (in tillers). Tillers can also be formed, under certain circumstances, during the generative development (BBCH 30-59). Tillering at that time is undesirable with regard to efficient use of biomass for grain production as most of the late tillers are not fertile and those which are fertile increase grain variability.

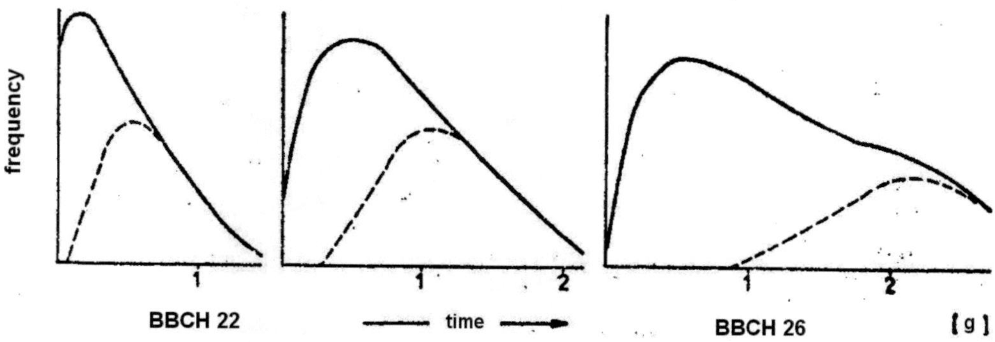

Graph 1. Schematic illustration of changes in density of tiller weight distribution of cereals during tillering [29]

Determination of plant and tiller density is in this stage important for assessment of the production potential of the stand and modification of cultivation measures, especially timing and dosing of N fertilizers, growth regulators and pesticides.

2.2. Generative period – from stem elongation and heading (BBCH 30-59)

Generative period is delimited by the stages of 'double ridge' (BBCH 30) and anthesis (BBCH 60) [3]. This period involves stages of stem elongation and heading, i.e. stages of an intensive growth and differentiation of tillers, their dying-off and formation of stems from the most robust ones; this is the result of selection resulting from competition among plants and tillers [8]. This results in a stabilization of numbers of productive stems per unit area at the end of canopy establishment. Depending on the variety and growing conditions, the number of productive stems per m^2 ranges from 500 to 800 and from 700 to 1000 in stands of winter wheat and spring barley, respectively [30].

Individual plants can be credibly identified within the stand only till the beginning of heading [29]. This limits the evaluation of their variability and intraplant competition.

The distribution of the weight of tillers has two peaks and this is the reason why the variability evaluated by means of the variation coefficient does not give an exact picture of differentiation processes [29]. As shown in Graph 2, the distribution of non-productive (V) and productive tillers (G) overlaps in the zone of local minimum. Critical weight of winter wheat tillers for transition to generative stage in analyses performed by [29] was about 2 g. This illustration of stem differentiation shows stochastic character of the development of stand structure. This means that tillers belonging to weight categories corresponding with the local minimum may either be transformed into productive stems or die off; their fate is dependent on the course of weather and on the efficiency of applied growing measures. It is quite logical that with running time and intensifying differentiation the numbers of these tillers (and thus the possibility of a modification of stand productive density) decrease.

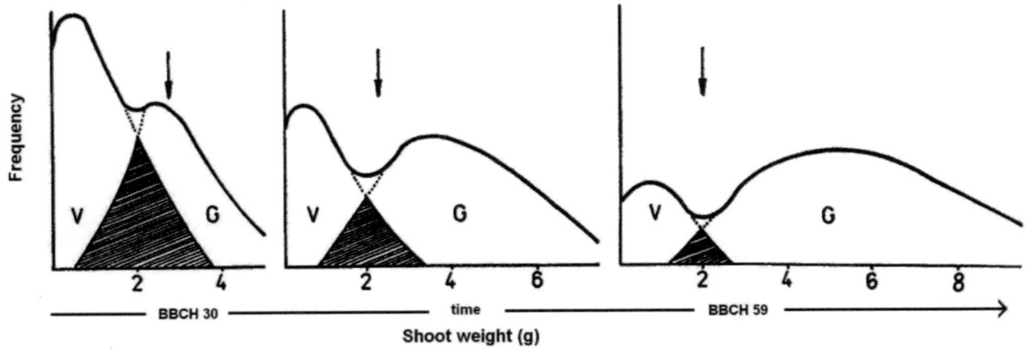

Graph 2. A schematic presentation of changes in the distribution of tiller weights during the period of generative development (BBCH 30-59). V – vegetative tillers, G – generative tillers (stems); the dark area represents tillers that can become, depending on availability of resources, either vegetative or generative [29].

An identification of productive and non-productive tillers is important for the estimation of the stand production potential and for the yield prognosis. This means that main objective should be the formation of a maximum possible number of productive tillers, i.e. of the maximum possible share of the so-called productive biomass in the total above-ground biomass. The development of the stand structure should be optimized in such a way that the produced biomass would maximally participate in the formation of grain yield.

2.3. Reproductive period - including anthesis, grain formation and maturation (BBCH 60-99)

The reproductive period includes pollination, grain filling and ripening (BBCH 61-91). Significant changes in the number of productive shoots occur only under very adverse conditions. However, their uniformity and variability can still be changed by a number of factors.

Individual plants cannot be identified in a stand therefore, the analyses carried out are limited to shoot variability and indirect assessment of intra-plant relationships. Weight distribution is at the beginning of this period bimodal with a pronounced right side representing the productive shoots, and increasing left side representing the rest of non-reduced unfertile tillers which usually die during grain filling (Graph 3). Weight distribution of the ears is in this period usually unimodal and nearly normal shape.

Shoot weight can be considered as basic information used for indirect estimation of the reproductive value and productivity of its ear. Rawson and Evans [31] reported narrow linear relationship (r = 0.94) between shoot weight at heading and a final number of grains in an ear. Our investigations also revealed highly significant linear correlation between shoot weight and number of embryos after flowering (r = 0.91-0.94). Therefore, proportion of shoot weight per one embryo can be considered as a basic factor determining the number of grain embryos in ear [32]. Narrow correlations between the amount of the aboveground biomass formed until flowering (BBCH 60) and number of grains per area unit of a stand and also yield are reported by a number of authors [33-36].

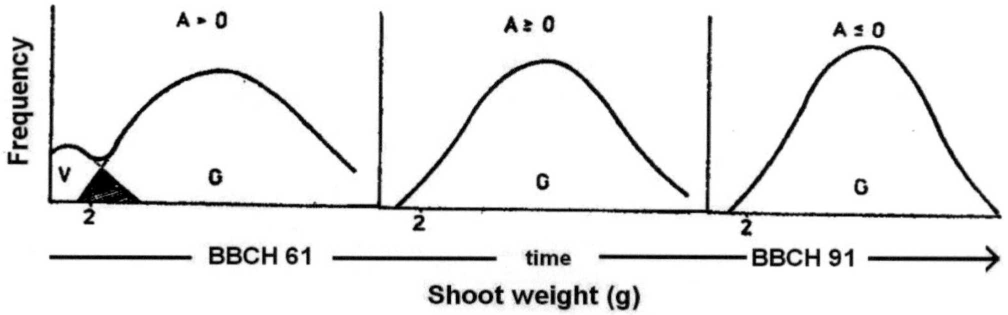

Graph 3. Schema of changes in tiller weight distribution during reproductive period (BBCH 60-91). V- vegetative tillers, G – generative tillers (stems), black area represents tillers which may be either vegetative or generative depending on source availability [29].

The amount of the biomass of productive tillers at the flowering (anthesis) can thus be used for prediction of the final stand yield. Duration of leaf area and redistribution of assimilates from stem to kernels is very important with respect to yield formation. Longer duration of the active green area supports higher yields. Many papers have been devoted to physiological processes of grain formation, most of which being based of the source x sink approach. Great attention has been paid to supplying of embryos and filling grain with carbohydrates and nitrogen substances, and to the possibilities of better availability of these sources for grain filling by breeding and growing management [37-40].

If grains are comprehended as modules, similar rules of the population biology of plants can be applied to their formation as to tiller formation [41]. Weight distribution of grains can have different shapes (both bimodal and asymmetric unimodal) in dependence on stand structure and growth conditions. One of the most important quality parameters is grain uniformity. It has an effect on grain yield (proportion of grain above 2.5 mm mesh) and bulk density. Both parameters are important at processing of food wheat and malting barley. Variability in weight of grains is influenced by relationships on individual levels of the hierarchic structure of a cereal plant - tillering node and ear structure [42].

Only a limited number of cultivation measures have been performed during grain filling. Qualitative N rate is usually applied to winter wheat at heading. Later application of N fertilizers is exceptional. Sometimes, N is applied in a tank-mix of urea together with fungicides against ear diseases. Cultivation measures are, therefore, predominantly oriented to maintenance and performance of production potential of stand which has been formed at previous stages of growth and development. Following factors are important for yield formation and grain quality during the reproductive period:

- amount of the aboveground biomass formed and its structure,
- active green area and duration of its functionality,
- course of translocation and assimilate redistribution processes.

3. Experiments and methods

3.1. Characterization of locations and field experiments

Evaluation of the canopy development of winter wheat and spring barley was carried out in small-plot field experiments established at two locations in Central and South Moravia (Table 1) within the period of three years (2005-2007). Experiments were conducted as contrast variants (Table 2) which took into account differences in the stand density and in the nutritional status of plants.

Parameter	Location	
	Žabčice	Kroměříž
Geographical location	49°01′20″ N; 16°37′55″ E	49°17′12″ N; 17°21′50″E
Soil type	Gleyey fluvisol (FMG)	Luvic chernozem (CMl)
Texture class	Clay loam	Silt loam
Altitude (m)	177	235
Annual mean temperature (°C)	9.2	8.7
Average annual sum of precipitations (mm)	480	599

Table 1. Characteristics of experimental locations

Crop and variety	Fore-crop	Variant	Seeding rate (seeds per m²)	Nitrogen fertilization (kg N.ha⁻¹)	Location	Year
Winter wheat, variety Cubus	Spring barley	A	350	40 prior to sowing	Žabčice	2006, 07
					Kroměříž	2005, 06, 07
		B	500	40 prior to sowing 40 during regeneration 40 at the beginning of stem elongation	Žabčice	2006, 07
					Kroměříž	2005, 06, 07
Spring barley, variety Malz	Maize	C	300	0	Žabčice	2005, 07
					Kroměříž	2005, 06, 07
		D	300	60/05 and 50/07 in the stage of the 3rd leaf	Žabčice	2005, 07
				50/05, 60/06 and 07 prior to sowing	Kroměříž	2005, 06, 07
		E	500	0	Žabčice	2005, 07
					Kroměříž	2005, 06, 07
		F	500	60/05 and 50/07 in the stage of the 3rd leaf	Žabčice	2005, 07
				50/05, 60/06 a 07 prior to sowing	Kroměříž	2005, 06, 07

Table 2. Characteristics of experimental variants

Each experimental variant was established in six replications: three of them were harvested, two were used for sampling, which enabled to analyse the structure and nutritional status of the stand, and one served for multispectral imaging of a demarcated area of 0.25 m^2 (0.5 m x 0.5 m) as well as for measuring of LAI by device SunScan System-SS1-R3-BF3 (manufacturer Delta-T Devices Ltd., U.K.). In sampling plots, squares of the size 0.25 m^2 (0.5 m x 0.5 m) were also demarcated to obtain plants samples used for analyses of stand structure and samples of soil used for the estimation of the content of mineral nitrogen (N_{min}) in depths of 0-30 and 30-60 cm ($N\text{-}NO_3$ and $N\text{-}NH_4$).

Multispectral images and samples of soil and plant material were obtained at the agronomically important developmental stages BBCH 25, 31, 37, 55, 65, 87 and 91. Analyses of stand structure and nutritional status involved:

- estimation of numbers and weights of individual tillers and plants,
- estimation of dry matter (DM) weight of the above-ground part of plants,
- estimation of chlorophyll content in leaves,
- analysis of DM of the above-ground part of plants (for contents of N, P, K, Ca, and Mg).

The segregation of tillers to productive and non-productive ones was performed as follows: When performing analyses, tillers were ordered according to their decreasing weight. The number of fully ripe ears per plot was taken as the number of productive tillers. In 2005, the total number of weighed tillers sampled at the growth stage BBCH 31 was lower than the number of fully ripe ears (only tillers heavier than 1 g were weighed so that their number was lower than that of ears). Due to this fact, the analysis was not performed. Tillers with the highest weight at the given developmental stage were rated as productive ones and the sum of their weights represented the so-called productive biomass. This value was separated from the total weight of fresh above-ground biomass per unit area of the stand.

3.2. Estimation of spectral characteristics of the stand

The imaging set consisted of a multispectral camera DuncanTech MS-3100 (manufacturer Geospatial Systems, Inc., USA) with the objective Sigma 14 mm F2.8 Aspherical HSM, a notebook Acer Aspire 1362 (AMD Sempron 2800+, 512MB RAM, 60GB HD), and a framegrabber (videograbber) National Instruments NI-1428, which was connected with the notebook via PCMCIA and communicated with the camera via a CameraLink interface. Power for the whole set was supplied from a portable 42 Ah accumulator with a 12V/220V changer (transformer). The camera was controlled via a COM interface and the recording of images was performed in DTcontrol software. Because of a problematic connectivity with the framegrabber the notebook was later on replaced by a desktop PC (Intel Pentium II 400 Mhz, 640 MB RAM, 40 GB HDD) with a framegrabber placed in the PCI slot and with a 17" LCD monitor.

The imaging part of the camera consisted of three CCD elements with filters of the size 7.6 x 6.2 mm, which recorded in three bands of electromagnetic radiation – green (500–600 nm, peak 550 nm), red (600–700 nm, peak 650 nm) and near infrared (750–900 nm, peak 830 nm).

By composing 8 bit images from CCD the so-called CIR (colour infrared) the image was obtained in false colours with the size of 1,392 x 1,039 pixels, and this was subsequently stored in a wireless TIFF format. The use of a wide (14 mm) objective enabled to reach (in combination with the physical size of the CCD element) a wide angle of recording (ca 26° x 20°).

For imaging from the height of 5.5 m, a mobile aluminium scaffold was used (Figure 4); this corresponded with the image area 2.54 x 1.9 m (4.83 m^2) and the spatial resolution of less than 2 mm per pixel. Since 2005, a set of optic etalons SphereOptics Zenith® was a part of each imaged scene to normalize the reflectance of radiation in accordance with existing light conditions. This set consisted of four polytetrafluorethylene (PTFE) etalons with exactly calibrated reflectances of 10 %, 25 %, 50 %, and 70 %. Used material had the so-called Lambert perfectly diffusing surface for which the intensity of reflected radiation was independent of direction of radiation incidence. From the viewpoint of field measurements resistance and washability of polytetrafluorethylene were also advantageous.

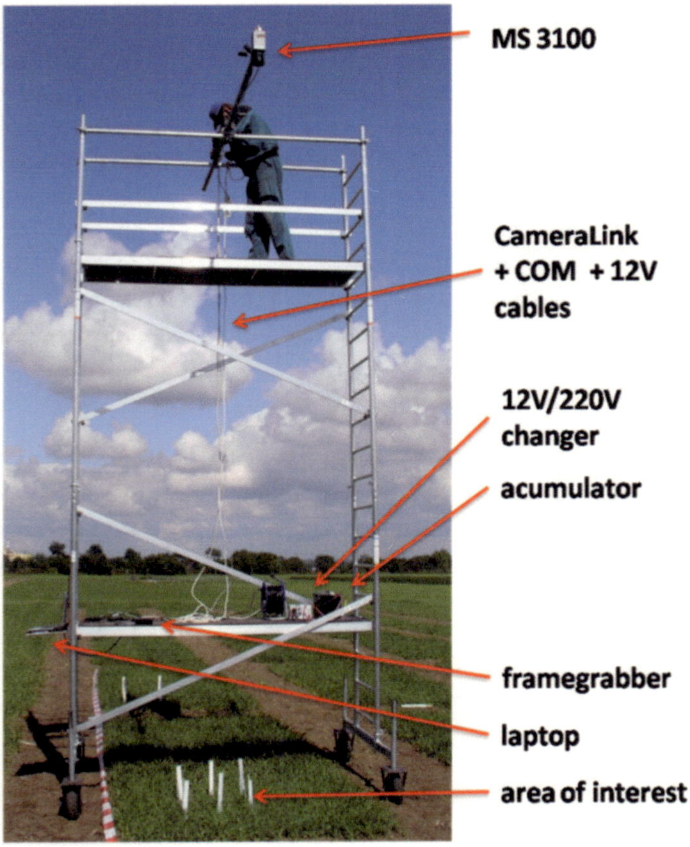

Figure 4. A view of imaging equipment

When processing recorded images, a radiometric correction was performed at first to eliminate differences in light conditions of imaging (sunny – overcast), i.e. to normalize reflectance. The normalization of images to constant light conditions was performed on the basis of a conversion of digital numbers (DN) of image pixels according to the measured and calibrated reflectance of a 25 % optic etalon. Using images processed in this way it was thereafter possible to calculate the Normalized Difference Vegetation Index (NDVI) according to the equation NDVI = (NIR – R)/(NIR + R). Simultaneously, average values of DN and NDVI of the demarcated part of the canopy were determined on the basis of CIR and NDVI images (NIR = Near Infrared, R = Red, CIR = Colour Infrared).

3.3. Methods of data processing

NDVI values were calculated from multispectral images by means of Erdas Imagine software. The obtained results were processed with Statgrafic and Statistica software using elementary statistical characteristics, analysis of variance (ANOVA) and correlation analysis.

4. Results and discussion

4.1. Vegetation period

The correlation values (Table 3 and 4) confirmed the possibility to compensate for higher stand density by available sources, which is in accordance with the law of final constant yield [43]. Its validity was later confirmed for tillering cereals by several authors [26,44,45,]. Source availability during vegetative growth and development makes plant density less important. What becomes more important is the number of tillers per plant and their performance. The increasing autonomy of tillers changes the character of cereal stand. Population of plants is gradually changing to metapopulation of tillers and in compensation relationships increases the importance of intra-plant competition [46,1,2]. Thus the relatively high correlations between stand height and production parameters, observed during tillering, can be explained, which is in accordance with our previous results [47].

Two types of relationships can be observed among shoots in the stand: relationships among plants which are more or less random, and intra-plant relationships which are controlled by the hierarchic structure of plants [18,26]. Reactions of plants in a stand to certain conditions are performed in the changes of intra-plant relationships which are then reflected in tiller variability. The structural conception [9], therefore, should not be only used for dividing yield to its components, which is difficult to interpret, but also for the assessment of intra-plant relationships which reflect plant adaptation to specific conditions [26, 48]. It is evident from the results that plant weight is increasing with increased N supply. In contrast, increased plant density reduces increase in their weight and variability.

It was possible to compensate low plant density by N fertilization. The effect of N fertilization on production parameters appears in spring barley as early as at the tillering stage due to earlier N application (prior to seeding or at the third leaf stage). At BBCH 31, the effect of N fertilization was obvious in both crops.

The results confirmed that the production potential is established by the amount of the above-ground biomass per area unit and its structural composition is being formed during the vegetative period. Regarding the possibilities to innovate the canopy management using its spectral characteristics and remote sensing, it is important to answer the question whether the information on the amount of the above-ground biomass per area unit or on the number of plants and tillers is more important at the tillering stage. The answer is a controversial requirement to create the largest possible amount of biomass of productive tillers using the smallest possible density of plants [49]. Thus the conditions for optimum plant growth and development within the stand are defined (inter- and intra-plant competition should arise as late as possible) as well as for sufficient tillering and formation of strong adventitious root system, which provide plants with nutrients and water at subsequent growth stages. This is also confirmed by practical experience showing that less dense stands can be managed better than over-dense ones. From this stand point, even distribution of plants (optimum size and shape of nutritive area) providing the least local variability at crop canopy and thus even distribution of competition relations is important for an effective canopy management [50].

	Parameter	BBCH	n-2	2	3	4	5	6	7	8
1	Number of plants per m^2	25	8	-0.222	0.049	0.521	-0.305	0.225	0.194	0.014
		31	8	-0.170	-0.353	0.606	0.236	0.714*	0.599	0.517
2	Average plant weight (g)	25	8	1	0.727*	0.480	0.451	0.887**	0.898**	0.795
		31	8	1	0.382	0.143	0.640*	0.554	0.612	0.560
3	Average number of tillers per plant	25	8		1	0.866**	-0.245	0.703*	0.714*	-0.911
		31	8		1	-0.330	-0.438	0.028	-0.132	0.212
4	Number of tillers per m^2	25	8			1	-0.383	0.692*	0.685*	-0.596
		31	8			1	-0.192	0.665*	0.446	0.621
5	Average tiller weight (g)	25	8				1	0.354	0.390	0.913
		31	8				1	0.587	0.728*	0.798
6	Weight of above-ground biomass ($g.m^{-2}$)	25	8					1	0.988**	0.449
		31	8					1	0.941**	0.928
7	Weight of above-ground dry matter ($g.m^{-2}$)	25	8						1	0.737
		31	8						1	0.892
8	Stand height (cm)	25/31	2							1/1

Table 3. Correlation between the assessed parameters in winter wheat at BBCH 25 and 31 (*statistical significance, **high statistical significance)

	Parameter	BBCH	n-2	2	3	4	5	6	7	8
1	Number of plants per m²	22	20	-0.212	-0.135	0.486*	-0.040	0.304	0.358	0.048
		31	14	-0.257	-0.312	0.658**	-0.061	0.562*	0.468	-0.014
2	Average plant weight (g)	22	20	1	0.600**	0.394	0.028	0.845**	0.786**	0.866**
		31	14	1	0.839**	0.410	0.610**	0.775**	0.812**	0.802*
3	Average number of tillers per plant	22	20		1	0.781**	-0.485*	0.465*	0.421*	0.318
		31	14		1	0.468	0.097	0.631**	0.586*	0.357
4	Number of tillers per m²	22	20			1	-0.466*	0.616**	0.619**	0.239
		31	14			1	0.068	0.888**	0.779**	0.373
5	Average tiller weight (g)	22	20				1	0.005	0.058	-0.188
		31	14				1	0.510*	0.625**	0.847**
6	Weight of above-ground biomass (g.m⁻²)	22	20					1	0.974**	0.845**
		31	13					1	0.959**	0.885**
7	Weight of above-ground dry matter (g.m⁻²)	22	20						1	0.811**
		31	13						1	0.907**
8	Stand height (cm)	22/31	12							1/1

Table 4. Correlation between the assessed parameters in spring barley at BBCH 22 and 31 (*statistical significance, ** high statistical significance)

Current practice requires rapid and effective methods for stand assessment. The conventional canopy management is very laborious. Regarding a large number of tillers per assessed area (usually 0.25 m²) and needed number of replications, it is practically impossible to carry out their accurate identification in plants as described by Rawson [24] or Klepper et al. [51]. Based on the obtained results, the following parameters should be considered by canopy management during the vegetative period:

- density of plants after emergence,
- tillering intensity, variability of plant size and stand height during tillering,
- tiller size or number of strong tillers, their uniformity at the beginning of stem elongation,
- the amount of total above-ground biomass per unit area of a stand.

4.2. Generative period

The process of tiller differentiation was evaluated by means of histograms illustrating the frequency distribution of their weights. To respect given size limits of this chapter, only data about the differentiation of tillers evaluated in experimental variants in Kroměříž in 2007 are presented: Graph 4 contains data about winter wheat and Graph 5 informs about spring barley. As shown in these histograms, the process of tiller differentiation was influenced both by the stand density and by an N dose. A higher density of plants and a lack of nitrogen accelerated the process of differentiation and made it also more intensive. The separation of tillers into subgroups of productive and non-productive ones could be identified on the basis of a local minimum. In variants without application of nitrogen (variant A in the experiment with winter wheat and variants C and E in the experiment with spring barley), symptoms of tillers separation were manifested in both crops as early as at the beginning of the stage of stem elongation (BBCH 31). On the other hand, however, the fertilised variants showed in this period a marked shift of values to the left, i.e. the proportion of lighter tillers was increased due to a more intensive tillering, which was supported by nitrogen. The differentiation of tillers appeared in histograms under conditions of a lack of resources (i.e. nitrogen). The lack of resources occurred in stands with a high density of plants (Graph 5, variant E in the experiment with spring barley). Application of nitrogen prolonged processes of differentiation till the stage of heading. From the viewpoint of yield formation, a gradual differentiation of tillers is beneficial because potentially productive tillers can be preserved for a longer time interval. On the other hand, however, too dense stands can suffer from a lack of resources (e.g. during dry periods).

To estimate the production potential and predict the yield, it is an advantage to know the numbers of productive tillers and their critical weight for the transition from the vegetative to the generative stage of growth and development. In our analyses this was done in two different ways:

- basing on the position of local minima in histograms (Graphs 4 and 5),
- by deduction of the strongest tillers (evaluated on the basis of the number of ears at BBCH 91) from their total number.

The observed critical weights of winter wheat and spring barley individual tillers were approximately 2 g and 1.5 g, respectively.

The separation of tillers into two groups, i.e. vegetative and generative (potentially productive) ones enabled us to determine the share of productive and non-productive biomass in the total above-ground biomass of the stand. This value can be an important indicator of the effectiveness of farming inputs into the crop cultivation. Thereafter an analysis of metapopulations of potentially productive tillers was performed.

Further analyses were focused on the evaluation of relationships between populations of productive and non-productive tillers. Correlations between total above-ground biomass and biomass of potentially productive tillers were positive and statistically highly significant while correlations between the total above-ground biomass and the share of the biomass of potentially productive tillers were very variable. Similar values and a similar character of correlations were also between the content of nitrogen in the total above-ground biomass and biomass of potentially productive tillers and also their share in the total above-ground biomass on the other hand (Table 5). Analysis of variability (Table 6) revealed low

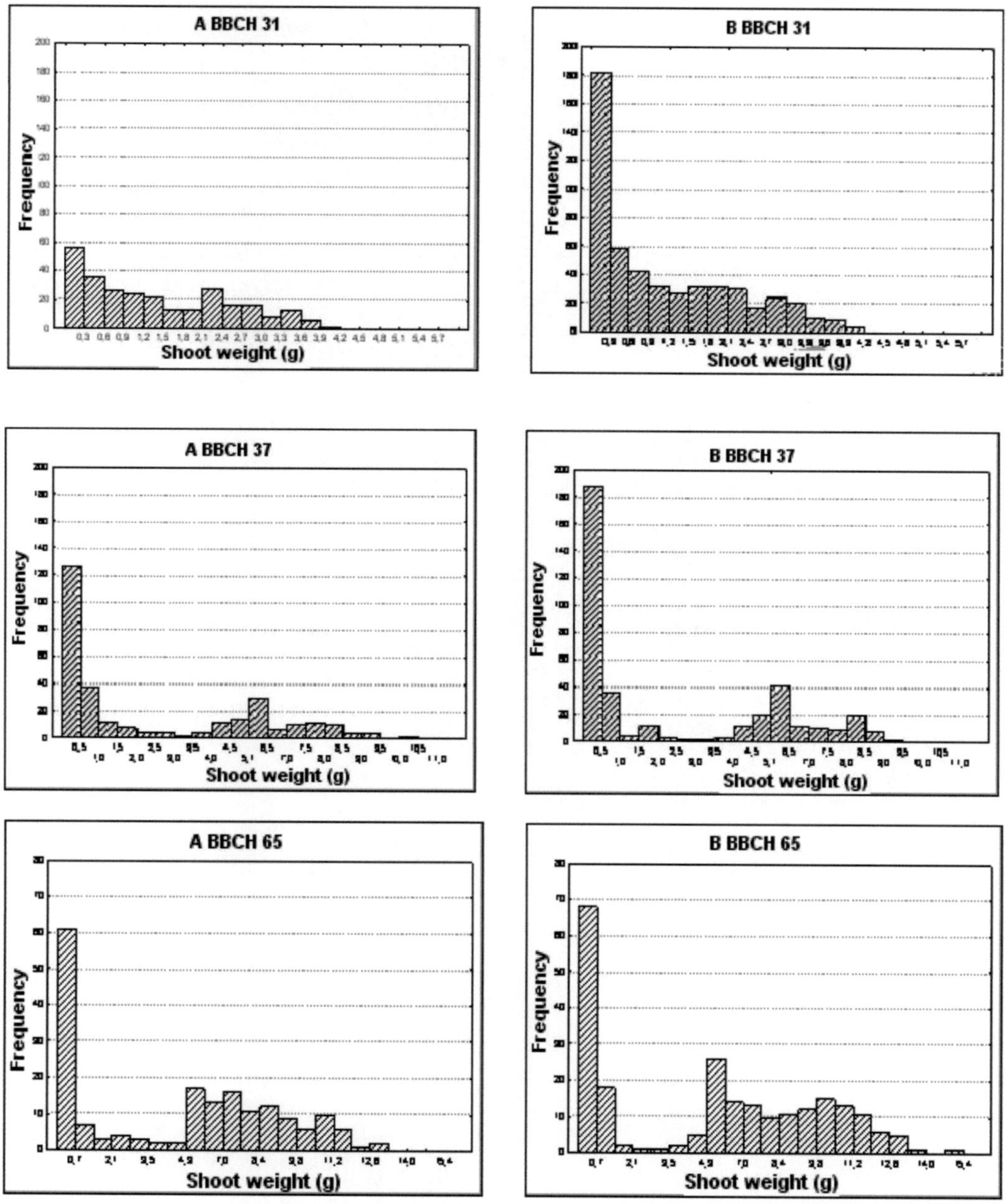

Graph 4. Development of tillers differentiation in winter wheat variety 'Cubus' in contrast variants of stand structure at the location of Kroměříž in 2007 (left side Variant A, right side Variant B)

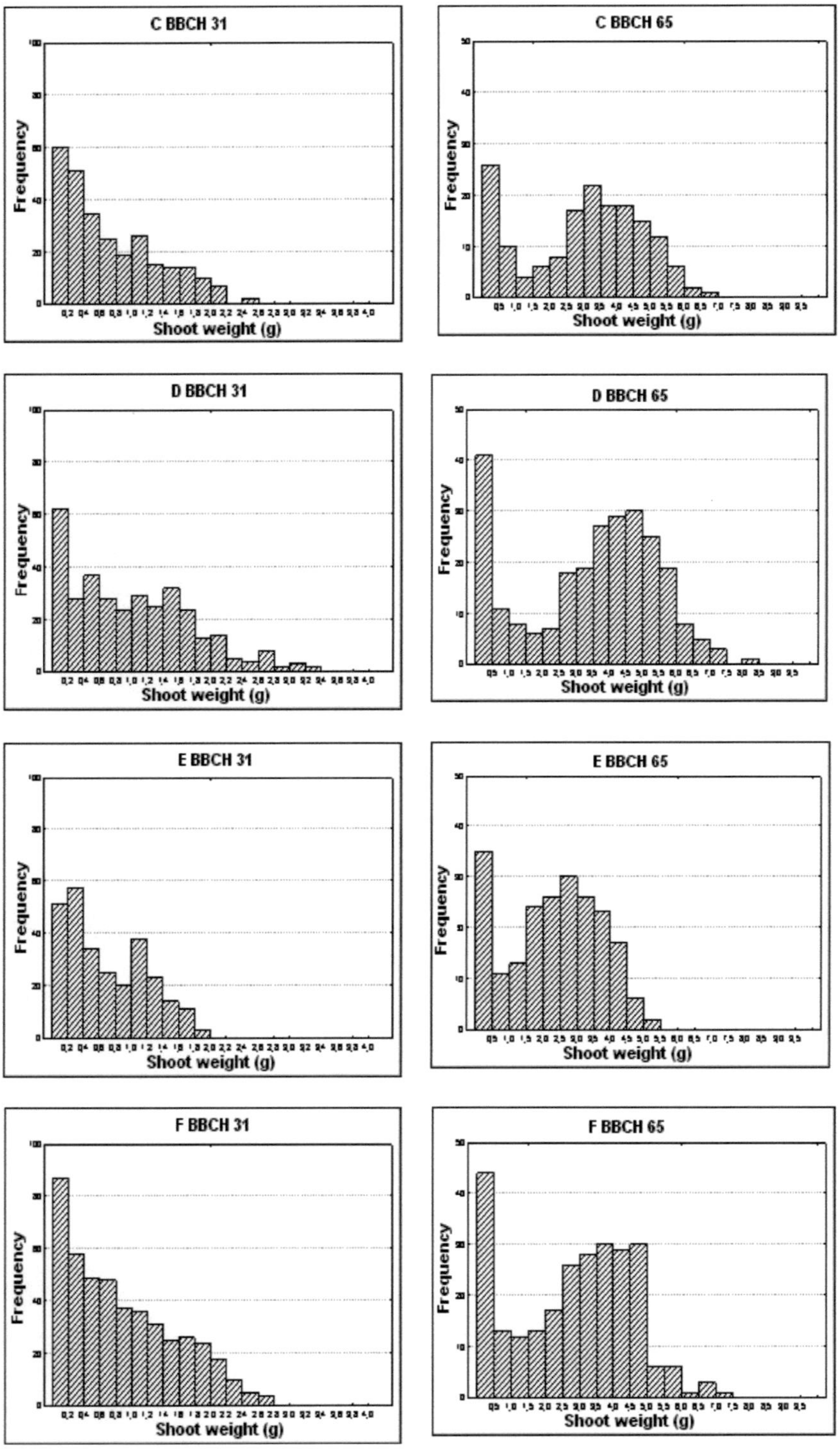

Graph 5. Development of tillers differentiation in spring barley variety 'Malz' in contrast variants of stand structure at the location of Kroměříž in 2007 (from top to bottom Variants C, D, E, F)

values (i.e. less than a half) of the CV for the proportion of biomass of productive tillers in the total above-ground biomass (ranging from 7.06 to 15.79 % and from 7.37 to 17.46 % for winter wheat and spring barley, respectively) while values of CV for other traits under study ranged from 29.81 to 54.43 %. Analogical correlations calculated on the basis of all and of productive tillers showed a similar character but they were lower (Table 7). Values of CV for correlated traits (Table 8) were mostly higher, including the proportion of productive tillers in their total number (22.85 to 26.11 % and 12.67 to 32.12 % for winter wheat and spring barley, respectively). Basing on this observation, it can be concluded that there is a close relationship between the total above-ground biomass and that of potentially productive tillers already at the stage of stem elongation. Because of a low variability the proportion of potentially productive tillers in the total above-ground biomass can be used for the estimation of a productive potential of the crop. This finding can be useful for more effective methods of canopy control based on spectral characteristics and on indirect estimation of above-ground biomass using the NDVI (Normalized Difference Vegetation Index).

Crop	Develop -mental stage BBCH	n-2	Correlation coefficient between			
			Total and productive biomass	Total biomass and share of productive biomass	N-content in bio-mass and productive biomass	N-content in biomass and share of productive biomass
Winter wheat	31	8	0.9111**	0.0850	0.8657**	0.1841
	37	6	0.9659**	-0.3515	0.9409**	-0.1133
	65	8	0.9720**	0.2182	0.9130**	-0.2676
Spring barley	31	14	0.9501**	-0.7810**	0.8847**	-0.7603**
	37	6	0.9995**	0.9657**	0.9664**	0.8949**
	55/65	18	0.9835**	0.1695	0.8743**	0.3365

Table 5. Relationships in stands of winter wheat and spring barley (*statistical significance, ** high statistical significance)

Crop	Develop-mental stage BBCH	n-2	Trait							
			Total biomass		N-content in biomass		Productive biomass		Share of productive tillers	
			Mean (g.m^{-2})	CV (%)	Mean (g.m^{-2})	CV (%)	Mean (g.m^{-2})	CV (%)	Mean (g.m^{-2})	CV (%)
Winter wheat	31	10	1,586	33.12	8.78	39.86	1031	38.74	0.65	15.79
	37	8	3032	30.76	12.65	46.64	2667	29.81	0.89	7.06
	65	10	3775	35.00	16.14	45.13	3533	34.31	0.94	7.57
Spring barley	31	16	1376	39.58	6.55	36.24	1060	30.99	0.80	12.80
	37	8	2430	54.43	8.03	49.95	2573	50.36	0.84	17.46
	55/65	20	2777	38.19	9.83	34.46	2485	40.44	0.89	7.37

Table 6. Mean values and coefficients of variation for traits in stands of winter wheat and spring barley

Crop	Developmental stage BBCH	n-2	Correlation coefficient between			
			Total number of tillers and number of productive tillers	Total number of tillers and share of productive tillers	N-content in biomass and number of productive tillers	N-content in biomass and share of productive tillers
Winter wheat	31	8	0.5262	-0.5621	0.7987**	0.0123
	37	6	0.8140**	-0.7203**	0.7729*	-0.6106
	65	8	0.8260**	-0.6578*	0.8114**	0.5930*
Spring barley	31	14	0.7107**	- 0.7505**	0.5803*	-0.7458**
	37	6	0.3323	0.0008	0.9156**	0.8332*
	55/65	18	0.9007**	-0.0612	0.7669**	0.1316

Table 7. Relationships in stands of winter wheat and spring barley (*statistical significance, ** high statistical significance)

Crop	Developmental stage BBCH	n-2	Trait							
			Total number of tillers		N-content in biomass		Number of productive tillers		Share of productive tillers	
			Mean (g.m^{-2})	CV (%)	Mean (g.m^{-2})	CV (%)	Mean (g.m^{-2})	CV (%)	Mean (g.m^{-2})	CV (%)
Winter wheat	31	10	1583	26.98	8.78	39.86	483	22.10	0.3163	26.11
	37	8	1072	36.83	12.65	46.64	520	27.49	0.5120	22.85
	65	10	718	34.21	16.14	45.13	510	24.80	0.7495	25.79
Spring barley	31	16	1324	33.68	6.55	36.24	676	21.09	0.5400	24.76
	37	8	1172	11.62	8.03	49.95	609	35.06	0.5200	32.12
	55/65	20	932	25.15	9.83	34.46	625	26.52	0.6700	12.67

Table 8. Average values and coefficients of variation for traits in stands of winter wheat and spring barley

The stand structure is a result of a simultaneous growth of individual plants within the framework of a population and changes in dependence on dynamics of growth processes, which are limited both spatially and temporally, in the course of the growing season. The temporal limitation results from developmental changes taking place during the life cycle of plants and while the spatial limitation is given by the size of their nutritive area and by available resources. These limitations are mutually interlinked and can stand for each other; this results from the law of "final constant yield" [43]. In practice this means that a shortening of the growing season, and especially of the period of generative growth and development, which is characterized by intensive growth, can be compensated by higher sowing rates and *vice versa*. An adequate supply of water and nutrients reduces the decrease in numbers of tillers both in dense and thin stands. Under conditions of abundance of resources the highest yields can be obtained in stands with a high productive density, which can be obtained on the basis of a lower number of plants. This can be explained by the fact that the competition is postponed till the end of the period of stem elongation. This observation corresponds with results of our earlier experiments, in which it was found out that the highest grain yield could be reached on the basis of maximum amount of above-ground biomass and density of productive stems at the minimum density of plants [49].

Corroborated were also conclusions drawn by Muravjev [8] that under favorable conditions the uniformity of stems and ears can be increased by competition induced in the period of stem elongation. Such a competition results in the selection of the strongest and the most vigorous tillers. However, the developmental biology of cereals permits only a relative synchronization of growth and development. Allometry and temporal sequence of the establishment and formation of identical organs (modules) support and work in favor of their increasing variability [3]).

An uneven growth of plants and tillers as well as their competition are the major factors which influence the stand structure. It is very difficult to evaluate directly the intensity of mutual interactions and competition of plants growing together only on the basis of the depletion of available resources [52]. For that reason the ecologists use the size distribution of plants as a suitable indicator of these relationships and of changes taking place in the structure of plant populations. The use of this indicator enabled to characterize the differentiation of tillers and to define relationships between potentially productive tillers on the one hand and non-productive ones on the other.

An effective utilization of growing factors and farming inputs is dependent on the fulfillment of two controversial requirements:

- formation of a maximum possible amount of above-ground biomass,
- establishment of a maximum possible proportion of productive tillers (stems) in the total above-ground biomass.

Basing on our results, it can be concluded that there is a close relationship between the total above-ground biomass and N-content in the total above-ground biomass on the one hand and the biomass of potentially productive tillers already in the period of stem elongation on the other. Because of a low degree of variability it is possible to use the share of potentially productive tillers in the total above-ground biomass for the estimation of overall productive capacity of a stand. This creates preconditions for a more efficient use of indirect methods of canopy assessment on the basis of its spectral characteristics, e.g. when using of the NDVI for optimization of the canopy management in dependence on the intensity of inputs. The obtained results also indicated that the relationships calculated at the level of biomass were more exact than those obtained by counting of the number of tillers. This indicates that values of the NDVI (and possibly also of other spectral canopy characteristics) will enable to obtain more exact estimation of the amount of above-ground biomass (both total and productive) than of its structure (numbers of plants and tillers).

The results also indicate that there is a possibility of the occurrence of various, dynamically changing situations in cereal crops. The population concept applied in studies concerning modular units (tillers) enabled to create a unifying base for these relatively chaotic phenomena. They can therefore be used when studying and testing new methods of efficient and areal screening of the condition of cereal stands by means of spectral characteristic and technologies of remote and terrestrial sensing [23,53,54].

4.3. Reproductive period

Although the differentiation and abortion of vegetative tillers is still proceeding, the process of yield structure formation is practically terminated. Shoot system has no more the adaptation function, which is now provided by ear structures and reproductive organs. Adaptation proceeds by grain filling; the importance of physiological processes taking part in it is increasing. These are usually explained by the theory of source x sink [37,55,56], which has been long applied in scientific papers oriented to yield formation in cereal crops. It is generally accepted that biomass of productive stems at anthesis correlates with the number of grains per m^2, and the duration of leaf area correlates with the of 1000-grain weight [36].

Analysis of stand structure was followed by analyses of yield structure. It was logical that in winter wheat (Table 9) higher yield was obtained in variant B compared with variant A by 15.7 % in Žabčice and by 14.8 % in Kroměříž, due to higher number of ears and kernels per m^2. Yield differences among variants were just under the limits of statistical significance (Table 11), on the other hand, the effects of year and locations on yield were highly significant. Difference was found between the variants in 1000-grain weight. In spring barley, the highest yield at both locations was reported in variant D and the lowest in variant E (Table 10), which was in accordance with the results of stand structure analyses. Yield differences among the variants were highly significant (Table 12). The effect of year on the yield was also statistically highly significant but the effect of location was only statistically significant. Similarly to winter wheat, the highest number of ears and grains per m^2 (Table 10) was formed in the best performing variant D. However, the highest number of grains in ear (21.14) was found in Kroměříž, and high values of 1000-grain weight (44.79 g in Žabčice and 44.34 g in Kroměříž) were found at both locations unlike of winter wheat. Variant C in Žabčice and E in Kroměříž were the least-yielding. They were characterized by the lowest number of grains per m^2 and the lowest 1000-grain weight. The results indicate an apparent effect of different pedological-climatic conditions of the locations on grain yield formation. Under drier climatic conditions in Žabčice, higher seeding rate had a positive effect on yield, while at the more productive location of Kroměříž with more balanced ratio of temperatures and precipitation, lower seeding rate in combination with N fertilization was beneficial. High density of plants in variant E in Kroměříž caused obviously a long-term N deficiency, which resulted in higher variability of weight in productive tillers at ripening.

The results also confirmed high yield determination by the number of grains per m^2, which corresponds with the data from literature. Further, negative correlation between 1000-grain weight and both ear number per m^2 and yield was confirmed in winter wheat. This can be explained by modular structure of cereal plants. High yields are usually obtained by higher

Parameter / Variant	Žabčice		Kroměříž	
	A	B	A	B
Number of ears per 1 m²	374	456	543	665
Number of grains per ear	34.85	31.79	31.51	31.82
Number of grains per m²	12 928	14 516	16 875	21 149
1000-grain weight (g)	38.17	41.26	42.78	39.76
Grain yield (g.m⁻²)	516	597	718	824

Table 9. Winter wheat yield components in Žabčice (two-year averages; 2006 and 2007), and in Kroměříž (three-year averages; 2005-2007)

Parameter / Variant	Žabčice				Kroměříž			
	C	D	E	F	C	D	E	F
Number of ears per m²	560	634	630	610	545	715	588	679
Number of grains per ear	19.88	19.47	19.78	20.33	18.55	21.14	17.17	19.84
Number of grains per m²	11 284	12 650	12 656	12 316	10 117	15 044	10 179	13 439
1000-grain weight (g)	43.92	44.79	44.00	42.70	44.55	44.34	41.41	43.01
Grain yield (g.m⁻²)	507	583	575	540	451	668	420	576

Table 10. Spring barley yield components in Žabčice (two-year averages; 2005 and 2007), and in Kroměříž (three-year averages; 2005-2007)

Source	Sum of squares	Df	Mean square	F-ratio	P-value
Location	25.557	1	25.557	42.37	0.000
Year	56.968	2	28.484	47.22	0.000
Variant	1.935	1	1.935	3.21	0.083
Replication	1.894	3	0.631	1.05	0.386
Residual	18.097	30	0.603		
Total	105.014	37			

Table 11. Analysis of variance of winter wheat yield

Source	Sum of squares	Df	Mean square	F-ratio	P-value
Location	3.965	1	3.965	4.12	0.046
Year	293.027	2	146.513	152.19	0.000
Variant	17.826	3	5.942	6.17	0.001
Replication	0.424	3	0.141	0.15	0.932
Residual	65.466	68	0.963		
Total	378.274	77			

Table 12. Analysis of variance of spring barley yield

number of ears formed in the plants of a certain stand. Ears in later formed tillers have usually smaller kernels. Then, it is clear that high yields are characterized by high number of grains per m^2 and lower 1000-grain weight. Higher 1000-grain weight in variant B in Žabčice (Table 9) can be explained by the fact that the stand was created predominantly by main stems. Number of ears at harvest was lower than the number of germinating seeds sown. In spring barley (Table 10), manifestation of these relationships was not unambiguous either. It is likely due to the fact that compensation processes were not been fully employed as it is evident from low numbers of grains per m^2 and grain yields at both locations.

Recently, an interesting discussion on this subject has been reported between Sinclar and Jamieson [57] and Fischer [58]. Both parts consider accumulation of sources till anthesis as important to grain yield determination. However, they differ in their opinion to the hypothesis that in the post-anthesis period yield is predominantly determined by kernel number. Sinclar and Jamieson [57] stated that yield is fundamentally driven by carbon and nitrogen resource accumulation, essentially independent of grain number. Fischer [58] considers the number of grains formed during anthesis, under optimal conditions, as essential for yield formation. Our current and former results concerning grain formation [41] are rather in accordance with the conception of Sinclar and Jamieson [57]. Changes in metapopulations of grains during their formation can be explained based on trophic approach and the rules of plant population biology. The sources for grain formation can be considered analogically to carrying capacity of the environment used to explain processes in plant populations [42].

During the reproductive growth and development, stand structure practically cannot be further influenced. Canopy control should, therefore, be focused on the assessment of the active green area and its heath status. Consequent crop management should predominantly be oriented to maintaining the functionality of the active green area as long as possible with emphasis on the flag leaf and ear supplying assimilates to the forming grains. The following stand parameters can be considered important during the reproductive period:

- number of productive shoots,
- canopy closure,
- total aboveground biomass,
- biomass of productive shoots,
- active green area, its duration and health status,
- resistance to lodging.

4.4. Canopy spectral characteristic - NDVI

NDVI, as the most frequently used spectral characteristic of the vegetation cover, is used above all for the assessment of the total amount of aboveground biomass per unit area. It is also known that NDVI is well correlated with the total biomass within the period of intensive growth but during canopy senescence the correlation gradually decreases [59]. This was also confirmed in our cereal canopy investigations. Correlations between NDVI and different canopy characteristic are presented in tables 13,14,15 for winter wheat and 16, 17, 18 for spring barley (see Appendix part).

Higher values of NDVI indicated:

- a greater amount of biomass and its dry matter per m^2, this corresponded with higher values of LAI, above all in the period of tillering (in spring barley) and in the period of stem elongation (winter wheat)
- a greater average weight of plants, a higher number of tillers per plant, and a higher number of plants per m^2 in the period of tillering,
- a greater average weight of tillers and a higher number of tillers per m^2 at the beginning of stem elongation,
- a more intensive green colour of the stand, which indicated a better nutritional status of the stand as far as nitrogen supply was concerned,
- a higher content of chlorophyll, above all in biomass of productive tillers.

This means that NDVI is positively correlated with the amount of aboveground biomass and its colour. Similar values of NDVI may obviously indicate either a greater amount of aboveground biomass with a nitrogen deficiency or a smaller amount of biomass in a good nutritional condition. It is also difficult to estimate on the basis of NDVI whether a given amount of aboveground biomass was produced by a higher number of less tillering plants or, on the contrary, by a lower number of plants with a higher number of tillers. This means that correlations between morphological (structural) and the physiological parameters of the stand condition and values of NDVI require further investigations. This is obviously one of the reasons of different results presented in papers dealing with possibilities of application of NDVI for the assessment of nutritional condition of crops and for the prediction of yields and quality of cereal grains.

Freeman et al. [60] found out weak correlations between NDVI and grain yield and grain protein content. These authors also mentioned that in the course of growing season there were no consistent relationships between NDVI and content of nitrogen in grain or in straw in different locations and years. Aparicio et al. [61] mentioned that, within the period from the stem elongation to ripening of grains, it was possible to explain 52 % and 39 % of variability in yields of durum wheat grown under and without irrigation, respectively, by means of NDVI values. Fetch et al. [62] reported a low efficiency of NDVI when applied for the determination of agronomic factors in barley (5–77 %). In spite of this, however, they concluded that the estimation of canopy reflectance might be a potential tool for the assessment of agronomic factors. At the same time they also pointed out that the effect of cultivars and developmental stage of on obtained results may be also important.

On the other hand, Zhao et al. [63] reported that vegetation indexes characterizing the canopy reflectance in green and red bands of electromagnetic spectrum were correlated highly significantly with the content of nitrogen in leaves at the stage of anthesis and with the protein content in wheat grains. They also mentioned a possibility of the use of correlations between spectral indexes and water content in leaves for the estimation of the protein content in grains. Jorgensen et al. [64] confirmed the possibility of the assessment of the stand nutritional condition by imaging in three 2-nm wide bands (450, 700, and 810 nm) which indicate well the lack of nutrients. Zhang et al. [65] mention 82–94 % and 55–70 % of determination using an NDVI-based regression model and validating experimental data from different locations, respectively. They concluded that NDVI can be used for remote sensing of nitrogen supply and nutritional status of stands. Similarly, Reyniers et al. [23] report close correlations between NDVI values and yield and N-content in grain at heading. Alvaro et al. [66] found out strong correlations between NDVI and growth characteristics and mentioned that the reliability of spectral reflectance measurement and non-destructive nature convert this method into a promising tool for the assessment of growth traits in spaced individual plants.

Negative correlations between NDVI and CV of plant weights and tillers identified in our investigations also indicate the importance of a spatial distribution of plants. This effect of stand heterogeneity should be taken into account when interpreting values of NDVI especially within the period before a canopy closure, i.e. usually till the beginning of the stem elongation (BBCH 31). The assessment of NDVI values in this period could be used for the evaluation of quality of stand establishment (i.e. uniformity of spatial distribution of individual plants). Nevertheless, Flowers et al. [20] mention the existence of close correlations between NIR digital counts and tiller density at BBCH 25 ($r = 0.67$-0.87) and also Philips et al. [22] recommended to use a high determination of the correlation between NDVI and density of tillers at BBCH 25 ($r^2 = 0.67$-0.99) for a variable application of nitrogen.

The contemporary level of knowledge enables a practical application of NDVI, especially for the evaluation of cereal crops heterogeneity in precision agriculture [23]. A great advantage is a possibility of quick and areal evaluation of canopy enabling variable cropping measures. In spring barley, this can be used within the period from stem elongation till the beginning of grain formation. Later on, when the canopy is already senescent, the correlation between NDVI and production traits is not so strong, which is mentioned by many authors [59,67,68]. This was confirmed partly by low and insignificant values of correlation coefficients between NDVI and productive traits and partly by results of variance analysis, which revealed lower values of NDVI during grain filling (BBCH 87).

Regarding the fact that NDVI correlates the most with the amount of aboveground biomass, it can be expected that it is potentially usable also for an indirect assessment of local differences in the stand microclimate from the viewpoint of the spread of fungal diseases. Relationships between NDVI and productive traits of the stand are the most significant within the period of stem elongation it can be concluded that for winter wheat the NDVI could be used for a variable application of nitrogen production doses or for growth regulators to protect stands against lodging. As usual, these measures are taken at the beginning of stem elongation. At later stages of growth and development (i.e. after anthesis) the relationships between NDVI and productive characteristics of the stand are not so significant. This is confirmed partly by low and insignificant correlations between NDVI and productive traits and partly by results of ANOVA which revealed lower values of NDVI at anthesis (BBCH 65).

5. Conclusions

5.1. Vegetative period

The changing plant density and availability of sources results, in cereals during the vegetation period and development, in changes of tillers number and size. Both the factors influence the organization of canopy structure. The tillering intensity affects the formation of adventitious roots, thus creating conditions for water and nutrients uptake.

Due to different fertilization scheme, these processes are different in winter wheat and spring barley. In winter wheat, the assessed production parameters were first influenced by different seed rate and later by different N fertilization in the regenerative doses at BBCH 23. In spring barley, N was applied prior to seeding or at the third leaf stage and the first assessment was only carried out at tillering (BBCH 22). At this time, all fertilized variants manifested higher values in all production parameters.

During tillering, the influence of plant density on the total amount of the above-ground biomass and dry matter per m^2 decreased and the influence of tillers number increased in both crops. Increasing plant density resulted in increase in stand height and decrease of the average weight and number of tillers per plant.

Due to higher density of stand caused by higher seed rate or higher N dose, the competition in the stand increased, which influenced the variability in plant and tiller size. Higher inter-plant competition was expressed by lower values of the CV for plant weight and number of tiller per plant. On the other hand, intra-plant competition increased the values of CV for tiller weight. These effects were most expressed at BBCH 31 in variants with higher seed rate and N fertilization in both crops.

Canopy management during the vegetative growth should predominantly be focused on the following parameters:

- density of emerged plants,
- intensity of tillering, variability in plant weight (size) and stand height during tillering,
- strong tillers and their uniformity at the beginning of stem elongation.

5.2. Generative period

A higher density of plants and a lack of nitrogen accelerated the differentiation of tillers and made it more intensive. A segregation of tillers into two subgroups (i.e. productive and non-productive ones) was possible due to an identification of the local minimum in histogram of their weight distribution. In variants without the nitrogen application, symptoms of the minimum appeared already at the beginning of the stage of stem elongation.

Variants fertilized by nitrogen showed in this period a marked shift of values to the left, i.e. the proportion of lighter tillers was increased due to a more intensive tillering. The differentiation of tillers occurred under conditions of a lack of nitrogen. Under conditions of a high density of plants this situation occurred earlier. Application of nitrogen prolonged processes of differentiation till the end of the heading. From the viewpoint of yield formation, a gradual differentiation of tillers is beneficial because potentially productive tillers can be preserved for a longer time interval. On the other hand, however, too dense stands can suffer under conditions of lacking resources (e.g. during dry periods).

The observed critical weights of winter wheat and spring barley individual tillers for transition from vegetative to generative growth and development were about 2 g and 1.5 g, respectively.

The separation of tillers into two groups, i.e. vegetative and generative (potentially productive) ones enabled to determine the share of productive and non-productive biomass in the total above-ground biomass. This value can be an important indicator of the effectiveness of farming inputs into the crop cultivation.

The application of nitrogen was manifested in:

- a higher weight of an average tiller even under conditions of an increased stand density;
- lower values of CV for tiller weight.

As compared with potentially productive tillers, the values of CV for the weight of all tillers were mostly doubled in the majority of cases due to their two-peak distribution. Decreased variability of productive tillers was associated with their higher density and with a higher amount of productive biomass per unit area of the stand.

Correlations between total above-ground biomass and that of potentially productive tillers were positive and statistically highly significant while correlations existing between the total above-ground biomass and the share of biomass of potentially productive tillers were very variable. Similar values and a similar character of correlations were found also between the content of nitrogen in the total above-ground biomass on the one hand and biomass of potentially productive tillers and their share in the total above-ground biomass on the other. As compared with other traits under study, values of the CV for the proportion of productive tillers biomass in the total above-ground biomass were low (i.e. less than a half). This enables to conclude that there is a close relationship between the total above-ground biomass and that of potentially productive tillers already at the stage of stem elongation. Because of a low variability it is possible to use the proportion of potentially productive tillers in the total above-ground biomass for the estimation of a productive potential of the crop. This finding can be useful for more effective methods of canopy control based on spectral characteristics and on indirect estimation of above-ground biomass using the NDVI.

5.3. Reproductive period

Although the differentiation and abortion of vegetative tillers is still proceeding, the process of yield structure formation is practically terminated. Shoot system has no more the adaptation function, which is now provided by ear structures and reproductive organs. Adaptation proceeds by grain filling; the importance of physiological processes taking part in it is increasing. These are usually explained by the theory of source x sink, which has been long applied in scientific papers oriented to yield formation in cereal crops. It is generally accepted that biomass of productive stems at anthesis correlates with the number of grains per m^2, and the duration of leaf area correlates with the of 1000-grain weight.

The following stand parameters can be considered important during the reproductive period:

- number of productive shoots,
- canopy closure,
- total aboveground biomass,
- biomass of productive shoots,
- active green area, its duration and health status,
- resistance to lodging.

5.4. Canopy spectral characteristic - NDVI

The obtained results confirmed the certain possibility of NDVI application for the assessment of the condition of cereal stands. Higher values of NDVI indicated:

- a greater amount of biomass and its dry matter per m^2 and higher values of LAI, above all in the period of stem elongation,
- a greater average weight of plants, a higher number of tillers per plant, and a higher number of plants per m^2 within the stage of tillering,
- a greater average weight of tillers and a higher number of tillers per m^2 at the beginning of stem elongation,
- a more intensive green colour of the stand, which indicated a better nutritional status of the stand as far as nitrogen supply was concerned.

At later phenological stages, these relationships between NDVI and productive traits of the stand were not so strong.

Similar values of NDVI may indicate either a greater amount of aboveground biomass with a deficit of nitrogen or a smaller amount of it in a good nutritional status. It is also difficult to decide whether the given amount of aboveground biomass was produced by a greater number of less tillering plants or, on the contrary, of a smaller number of plants with more tillers. The effect of canopy morphological (i.e. structural) and physiological characteristics on values of NDVI, therefore, requires further investigations.

Traits characterizing heterogeneity of the stand (i.e. CV of the weights of plants and tillers) correlated negatively with the NDVI. This finding may be important for the interpretation of NDVI values, above all in the period when the stand canopy is not fully developed, i.e. usually till the beginning of stem elongation (BBCH 31) when NDVI values are influenced by radiation reflected by plants and soil surface. In this period, the assessment of NDVI could be used to evaluate quality of stand establishment (i.e. of the uniformity of plant distribution).

In spite of a difficult interpretation of structural stand parameters the present level of knowledge enables to use NDVI in practice. A great advantage of NDVI is a possibility of a quick areal evaluation of heterogeneity of cereal canopy within the fields which can be used in precision agriculture.

Appendix

Correlations between NDVI and different canopy characteristic - tables 13,14,15 for winter wheat.

Correlations between NDVI and different canopy characteristic - tables 16, 17, 18 for spring barley.

	Trait	BBCH	n-2	2	3	4	5	6	7	8
1	Number of plants per m^2	25	8	0.521	0.225	0.194	1			0.031
		31	8	0.606	0.714*	0.599	1	0.697*	0.663*	0.609
		37	2	0.211	0.378	-0.001	-0.074	0.913	0.297	0.436
2	Number of tillers per m^2	25	8	1	0.692*	0.685*	1			0.553
		31	8	1	0.665*	0.446	1	0.596	0.422	0.702
		37	4	1	0.947**	0.860*	0.665	0.223	0.888*	0.295
		65	8	1	0.979**	0.923**	0.485	0.835**	0.942**	0.739*
3	Weight of total above-ground biomass ($g.m^{-2}$)	25	8		1	0.988**	1			0.800**
		31	8		1	0.941**	1	0.601	0.888**	0.960**
		37	4		1	0.827*	0.532	0.245	0.866*	0.206
		65	8		1	0.954**	0.549	0.829**	0.943**	0.709*
4	Total dry matter weight ($g.m^{-2}$)	25	8			1	-1			0.868**
		31	8			1	1	0.518	0.937**	0.834**
		37	4			1	0.863*	-0.134	0.879*	0.514
		65	8			1	0.628	0.884**	0.952**	0.595
5	LAI ($m^{-2}.m^{-2}$)	25	0				1			-1
		31	0				1	1	1	
		37	4				1	-0.377	0.660	0.824*
		65	10				1	0.474	0.560	0.510
6	Number of productive tillers per m^2	25	0					1		
		31	8					1	0.707*	0.243
		37	4					1	0.345	-0.391
		65	10					1	0.904**	0.552
7	Weight of productive tillers ($g.m^{-2}$)	25	0						1	
		31	8						1	0.714*
		37	4						1	0.365
		65	10						1	0.639*
8	NDVI	25/31	7/6							1/1
		37/65	4/10							1/1

Table 13. Correlations between NDVI and morphological traits at the level of winter wheat stand (* statistically significant, ** statistically highly significant)

	Trait	BBCH	n-2	2	3	4	5	6	7	8	9
1	Average plant weight (g)	25	8	-0.606	0.727*	-0.738*	0.451				0.804**
		31	8	-0.468	0.382	0.182	0.640*	0.047	0.631*	-0.533	0.513
		37	2	-0.975*	0.934	-0.814	0.616		0.996**	-0.870	0.710
2	CV of plant weight (%)	25	8	1	-0.869**	0.886**	0.301				-0.437
		31	8	1	-0.015	0.201	-0.479	-0.464	-0.708*	0.658*	-0.977**
		37	2	1	-0.835	0.886	-0.773		-0.991**	0.949	-0.736
3	Average number of tillers per plant	25	8		1	-0.905**	-0.245				0.590
		31	8		1	0.603	-0.438	-0.688*	-0.313	-0.568	0.088
		37	2		1	-0.582	0.297		0.901	-0.669	0.652
4	CV of the average number of tillers per plant	25	8			1	0.143				-0.557
		31	8			1	-0.344	-0.197	-0.297	-0.069	-0.258
		37	2			1	-0.859		-0.839	0.827	-0.388
5	Average tiller weight (g)	25	8				1				0.622
		31	8				1	0.590	0.871**	-0.122	0.493
		37	4				1		0.570	-0.866*	0.048
		65	8				1	-0.966*	0.400	0.120	0.288
6	CV of the weight of tillers (%)	25	0					1			
		31	8					1	0.624	0.290	0.439
		37	0					1			
		65	2					1	0.468	-0.522	0.851
7	Average weight of a productive tiller (g)	25	0						1		
		31	8						1	-0.331	0.746*
		37	4						1	-0.866*	0.568
		65	10						1	-0.753**	0.491
8	CV of the weight of productive tillers (%)	25	0							1	
		31	8							1	-0.866**
		37	4							1	-0.384
		65	10							1	- 0.241
9	NDVI	25	7								1
		31	6								1
		37	4								1
		65	10								1

Table 14. Correlations between NDVI and morphological traits under study at the level of individual plants and tillers of winter wheat (* statistically significant, ** statistically highly significant)

	Trait	BBCH	n-2	2	3	4	5
1	Uptake of N in above-ground biomass (g)	25	4				-0.465
		31	4	0.834*	0.649	0.860*	0.946**
		37	4	0.944**	0.532	-0.128	0.506
		65	10	0.957**	0.337	0.547	0.802**
2	Uptake of N in biomass of productive tillers (g)	31	8	1	0.561	0.879*	0.767*
		37	4	1	0.278	-0.149	0.408
		65	8	1	0.307	0.527	0.546
3	Content of chlorophyll in above-ground biomass	25	4		1		-0.749
		31	4		1	0.722	0.781
		37	4		1	-0.511	0.670
		65	4		1	0.760	0.682
4	Content of chlorophyll in biomass of productive tillers	31	4			1	0.889*
		37	4			1	-0.486
		65	4			1	0.819*
5	NDVI	31	6				1
		37	4				1
		65	10				1

Table 15. Correlations between NDVI and uptake of nitrogen and chlorophyll content in aboveground biomass of winter wheat (* statistically significant, ** statistically highly significant)

	Trait	BBCH	n-2	2	3	4	5	6	7	8
1	Number of plants per m^2	22	16	0.568*	0.324	0.377	0.514			0.233
		31	12	0.653*	0.540	0.452	0.285	0.264	0.405	0.167
2	Number of tillers per m^2	22	16	1	0.692**	0.715**	0.961*			0.438
		31	12	1	0.891**	0.783**	0.827**	0.736**	0.845**	0.609*
		37	4	1	0.058	0.106	-0.509	-0.121	-0.040	0.216
		55	6	1	0.953**	0.915**	0.491	0.933**	0.953**	0.878**
		65	8	1	0.930**	0.853**	0.632	0.862**	0.905**	0.701
		87	2	1	0.837	0.862	0.102			0.547

	Trait	BBCH	n-2	2	3	4	5	6	7	8
3	Weight of total above-ground biomass (g.m^{-2})	22	16		1	0.979**	0.966*			0.801**
		31	11		1	0.959**	0.824**	0.605*	0.979**	0.793**
		37	4		1	0.992**	0.994**	0.946**	0.993**	0.783
		55	6		1	0.934**	0.490	0.889**	0.979**	0.898**
		65	8		1	0.870**	0.641	0.903**	0.992**	0.836**
		87	2		1	0.858	0.662	0.893	0.981*	0.843
4	Total dry matter weight (g.m^{-2})	22	16			1	0.989*			0.811**
		31	11			1	0.697*	0.441	0.944**	0.880**
		37	4			1	0.974**	0.913*	0.976**	0.814*
		55	6			1	0.170	0.817*	0.893**	0.973**
		65	8			1	0.665	0.845**	0.871**	0.840**
		87	2			1	-0.231			0.709
5	LAI (m^2.m^{-2})	22	2				1			0.920**
		31	9				1	0.719*	0.833**	0.501
		37	3				1	0.980**	0.995**	0.695
		55	6				1	0.530	0.531	0.124
		65	6				1	0.629	0.629	0.484
		87	14				1			0.827
6	Number of productive tillers per m^2	31	12					1	0.635	0.354
		37	4					1	0.974**	0.643
		55	6					1	0.937**	0.857**
		65	8					1	0.942**	0.857**
7	Weight of productive tillers (g.m^{-2})	31	12						1	0.840**
		37	4						1	0.745
		55	6						1	0.886**
		65	8						1	0.853**
8	NDVI	22	16							1
		31	11							1
		37	4							1
		55	6							1
		65	6							1
		87	4							1

Table 16. Correlations between NDVI and morphological traits at the level of spring barley stand (* statistically significant, ** statistically highly significant)

	Trait	BBCH	n-2	**2**	**3**	**4**	**5**	**6**	**7**	**8**	**9**
1	Average plant weight (g)	22	16	-0.231	0.658**	-0.330	0.050				0.696**
		31	12	0.084	0.843**	-0.559	0.628*	-0.056	0.617*	-0.289	0.831**
2	CV of plant weight (%)	22	10	1	-0.157	0.749**	-0.247				-0.382
		31	10	1	0.224	0.527	-0.243	0.343	-0.157	0.226	-0.140
3	Average number of tillers per plant	22	16		1	-0.323	-0.287				0.260
		31	12		1	-0.531	0.129	0.081	0.194	0.013	0.647*
4	CV of the average number of tillers per plant	22	10			1	-0.485				-0.360
		31	10			1	-0.279	0.326	-0.251	0.134	-0.534
5	Average tiller weight (g)	22	16				1				0.207
		31	12				1	-0.204	0.874**	-0.528	0.626*
		37	4				1	-0.798	0.981**	0.399	0.756
		55	6				1	0.099	0.944**	0.628	0.869**
		65	8				1	0.100	0.927**	0.064	0.793*
		87	2				1	-0.566			0.064
6	CV of the weight of tillers (%)	31	12					1	0.176	-0.062*	0.331
		37	4					1	-0.698	-0.505	-0.311
		55	6					1	0.333	0.058	0.036
		65	8					1	0.240	0.150	-0.376
		87	2					1			0.651
7	Average weight of a productive tiller (g)	31	12						1	-0.760**	0.811**
		37	4						1	0.232	0.821*
		55	6						1	0.543	0.799*
		65	8						1	-0.239	0.715*
8	CV of the weight of productive tillers (%)	31	12							1	-0.490
		37	4							1	0.112
		55	6							1	0.322
		65	8							1	-0.134
9	NDVI	22/31	16/11								1/1
		37/55	4/6								1/1
		65/87	6/4								1/1

Table 17. Correlations between NDVI and morphological traits at the level of individual plants and tillers of spring barley (* statistically significant, ** statistically highly significant)

Trait		BBCH	n-2	2	3	4	5
1	Content of N in above-ground biomass (g)	22	10		0.143		0.852**
		31	12	0.969**	0.461**	0.921**	0.899**
		55	6	0.970**	0.235	0.659	0.751*
		65	8	0.995**	0.800	0.983**	0.918**
		87	2	0.960*			0.725
2	Content of N in biomass of productive tillers (g)	31	11	1	0.637*	0.969**	0.889**
		37	4	1			0.739
		55	6	1	0.209	0.733*	0.799*
		65	8	1	0.780	0.980*	0.829*
3	Content of chlorophyll in above-ground biomass	22	10		1		0.047
		31	10		1	0.657*	0.440
		55	6		1	0.645	0.370
		65	2		1	0.889	0.969
4	Content of chlorophyll in biomass of productive tillers	31	9			1	0.925**
		55	6			1	0.912**
		65	2			1	0.974**
5	NDVI	22/31	16/11				1/1
		37/55	4/6				1/1
		65/87	6/4				1/1

Table 18. Correlations existing between NDVI and physiological traits under study (* statistically significant, ** statistically highly significant)

Author details

Jan Křen, Tamara Dryšlová, Lubomír Neudert and Vojtěch Lukas
Department of Agrosystems and Bioclimatology, Faculty of Agronomy, Mendel University in Brno, Brno, Czech Republic

Acknowledgement

This study was supported by the Ministry of Agriculture of the Czech Republic, projects N°s: QI111A133 and QJ1210008.

6. References

[1] Masle-Meynard J, Sebillotte M (1981) Etude de l'hétérogénéité d'un peuplement de blé d'hiver. I. Notion de structure du peuplement (Study on the heterogeneity of a winter-wheat stand. 1. Notion of stand structure). Agronomie 1(3):207–215.

[2] Masle-Meynard J, Sebillotte M (1981) Etude de l'hétérogénéité d'un peuplement de blé d'hiver. II. Origine des différentes categories d'individus du peuplement: éléments de

description de se structure. (Study on the heterogeneity of a winter-wheat stand. 2. Study on the different sorts of individuals of the stand – factors allowing the description of its structure). Agronomie 1(3):216–224.

[3] Porter J. R (1984) A model of canopy development in winter wheat. Journal of agricultural Science 102:383–392.

[4] Křen J (1990) Autoregulace a zákonitosti populační biologie rostlin v porostu pšenice (Autoregulation and the regulatories of plant population biology in the wheat stand). Rostlinná Výroba 36 (8):839–845.

[5] Křen J (1990) Rozdíly mezi rostlinami a stébly v porostu pšenice (Differences between the plants and stems in stand of winter wheat). Rostlinná výroba 36 (11):1121–1127.

[6] Křen J (1990) Změny variability rostlin a stébel v porostu pšenice (The changes in variability of plants and shoots in the wheat stand). Rostlinná výroba 36 (12):1251–1259.

[7] Kren J, Smoldas R, Lukas V, Misa P (2007) Within-crop variability affecting variable N application to winter wheat. 6th European Conference on Precision Agriculture (6ECPA), 3-6 June 2007, Skiathos, Greece, Wageningen Academic Publisher, pp. 697-703.

[8] Muravjev S.A (1973) Stebleotbor v zlakovom fitocenoze (Stem selection in the cereal phytocenosys). Riga, Zinatne.

[9] Heuser W (1927/28) Die Ertragsanalyse von Getreide-züchtungen (Yield analyses in cereals breeding). Pflanzenbau 4, 6.3:353–357.

[10] Hunt R (1978) Demography versus plant growth analysis. New Phytologist 80:269–272.

[11] Petr J, Hnilica P, Hradecká D (1983) Studium optimální úrovně výnosových prvků a modelování tvorby výnosu (Optimum level of yield components research and modelling of yield formation). [Research report] Praha, VŠZ.

[12] Petr J et al (1983) Intenzívní obilnářství (Intensive grain production). SZN Praha, 377 pp.

[13] Watkinson A.R, White J (1985) Some life history consequences of modular constructions in plants. Philosophical Transactions of the Royal Society (London), B 313, pp. 31-51.

[14] White J (1979) The plant as a metapopulation. Annual Review of Ecology and Systematics 10:109–145.

[15] Porter J.R (1983) A modular approach to analysis of plant growth. I. Theory and principles. New Phytologist 94:183–190.

[16] Porter J.R (1983) A modular approach to analysis of plant growth. II. Methods and results. New Phytologist 94:191–200.

[17] De Wit C.T (1960) On competition. Versl. Landbanka-Onderz. 66:1–82.

[18] Koyama H, Kira T (1956) Intraspecific competition among higher plants. VIII. Frequency distribution of individual plant weight as affected by the interaction between plants. J. Inst. Polytechn. Osaka Univ. D 7:73–94.

[19] Malet P (1979) Liaison statistique entre le development et la croissance chez les plantes: application d'un modele simplific a l'analyse de la croissance (Statistical relationship

between development and growth of plants: application of simplified model to growth analyses). Annales Agronomiques 30 (5) :415–430.

[20] Flowers M, Weisz R, Heiniger R (2001) Remote sensing of winter wheat tiller density for early nitrogen application decisions. Agronomy Journal 93:783–789.

[21] Flowers M, Weisz R, Heiniger R, Tarleton B, Meijer A (2003) Field validation of a remote sensing technique for early nitrogen application decisions in wheat. Agronomy Journal 95: 95–167.

[22] Phillips S.B, Keahey D.A, Warren J.G, Mullins G.L (2004) Estimating winter wheat tiller density using spectral reflectance sensor for early-spring, variable-rate nitrogen applications. Agronomy Journal 96:591–600.

[23] Reyniers M, Vrindts E, De Baerdemaeker J (2006) Comparison of an aerial-based system and an on the ground continuous measuring device to predict yield of winter wheat. European Journal of Agronomy 24:87–94.

[24] Rawson H.M (1971) Tillering patterns in wheat with special reference to the shoot at the colleoptille node. Austral. J. biol. Sci., 24: 829-841.

[25] Baker C. K, Gallagher J.N (1983) The development of winter wheat in the field. I. Relation between apical development and plant morphology within and between seasons. J. agric. Sci., Camb., 101: 327-335.

[26] Křen J. (1987) Výzkum variability odnožování a tvorby stébel u ozimé pšenice [Research of tillering and stalk formation variability of winter wheat]. Academic dissertation, Oseva – Cereal Research Institute Kroměříž, 134 s.

[27] Klepper B, Rickman R.W, Peterson C.M (1982) Quantitative characterization of vegetative development in small cereal-grains. Agron. J., 24: 789-792.

[28] Kirby E.J.M, Appleyard M, Fellowes G (1985) Effect of sowing date and variety on main shoot leaf emergence and number of leaves of barley and wheat. Agronomie, 5: 117-126.

[29] Křen J (1991) Využití poznatků o variabilitě rostlin a stébel v porostu pšenice. [The use of knowledge in variability of plants and stems in the wheat stand]. Rostlinná výroba 37 (4):297-305.

[30] Křen J, Souček A (1990) Jak posuzovat strukturu porostu pšenice. [How to assess wheat stand structure]. Úroda 38 (1):13-15.

[31] Rawson H.M, Evans L.T (1971) The contribution of stem reserves to grain development in a range of wheat cultivars of different height. Aust. J. agric. Res. 22:851-863.

[32] Křen J, Vlach M (1984) Některé zákonitosti tvorby výnosu stébel ozimé pšenice (Regulatories of stem yield formation in winter wheat). Rostlinná Výroba 30:95–102.

[33] Fischer R.A, Stockman Y.M (1980) Kernel number per spike in wheat (*Triticum aestivum* L.): responses to preanthesis shading. Aust. J. Plant Physiol. 7:169-180.

[34] Diekmann F, Fischbeck G (2005) Differences in wheat cultivar response to N supply. I: Differences in grain yield formation. Journal of Agronomy and Crop Science 191:351-361.

[35] Ugarte C, Calerdini D.F, Slafer G.A (2007) Grain weight and grain number responsiveness to pre-anthesis temperature in wheat, barley and triticale. Field Crops Research 100:240-248.

[36] Bancal P (2008) Positive contribution of stem growth to grain number per spike in wheat. Field Crops Research 105:27-39.

[37] Borras L, Slafer G.A, Otegui M.E (2004) Seed dry weight response to source-sink manipulations in wheat, maize and soybean: a quantitative reappraisal. Field Crops Research 86:131-146.

[38] Barbottin A, Lecomte C, Bouchard C, Jeuffroy M.H (2005) Nitrogen remobilization during grain filling in wheat: Genotypic and environmental effects. Crop Science 45:1141-1150.

[39] Gooding M.J, Gregory P.J, Ford K.E, Ruske R.E (2007) Recovery of nitrogen from different sources following applications to winter wheat at and after anthesis. Field Crops Research 100:143-154.

[40] Ehdaie B, Alloush G.A, Waines J.G (2008) Genotypic variation in linear rate of grain growth and contribution of stem reserves to grain yield in wheat. Field Crops Research 106:34-43.

[41] Kren J, van Keulen H, Grashoff C (1992) Grain growth in oats: experimentation and modeling. Research report, Centre for Agrobiological Research (CABO-DLO), Wageningen, The Netherlands, October, 132 pp.

[42] Kren J (1996) Possibilities of using modular growth and plant hierarchical structure to evaluate resource use in cereal growing. 4th Congress of the European Society for Agronomy, 7-11 July, Veldhoven – Wageningen, The Netherlands, Book of Abstracts, II: 560–561.

[43] Kira T, Ogawa H, Shinozaki K (1953) Intraspecific competition among higher plants. I. Competition – density – yield interrelatioships in regularly disperse populations. J. Inst. Polytech., Osaka Univ., D 4: 1-16.

[44] Darwinkel A (1978) Patterns of tillering and grain production of winter wheat at a wide range of plant densities. Neth. J. agric. Sci., 26: 383-398.

[45] Drezner W (1984) Zdolnost krzewienia sie pszenicy ozimej zwyczajnej jako cynik plonotworczy oraz element kszaltowania struktury plonu [Tillering ability of winter wheat as an indicator of yield formation and the component of yield structure]. Monograph, D 203.

[46] Slavíková J (1986) Ekologie rostlin [Plant ecology]. SNP, Praha, 366 s.

[47] Vlach M, Křen J (1984) Organizace porostu ozimé pšenice v závislosti na počtu rostlin na jednotce plochy [Organization of winter wheat stand in relation to number of plantes per unit of area]. Rostlinná výroba, 30: 287-294.

[48] Gustafssson A, Dormling I, Ekman G (1982) Barley and wheat under controlled climatic conditions: a model experiment of vigour and variability. Hereditas, 96: 1-11.

[49] Křen J (1993) Zásady zakládání porostů ozimé pšenice [Principles of winter wheat stand establishment]. Obilnářské listy, 1: 10-11.

[50] Křen, J., Benada, J. 1990. Možnosti snižování výsevků obilnin [Possibilities of reducing seeding amount of cereals]. Úroda, 38: 343-345.

[51] Klepper B, Rickman R.W, Belford R.K (1983) Leaf and tiller identification on wheat plants. Crop Science, 23: 1002-1004.

[52] Harper J.L (1977) Population biology of plants. Academic press, London – New York – San Francisco.

[53] Hammen V.C (2001) On-line sensing of cereal crop biomass. Dissertation, Humboldt Universität Berlin, pp. 171 [On-line] (http://edoc.hu-berlin.de/dissertationen/hammen-volker-carsten-2001-07-17)

[54] Scotford I.M, Miller P.CH (2004) Estimating tiller density and leaf area index of winter wheat using spectral reflectance and ultrasonic sensing techniques. Biosystems engineering 89:395-408.

[55] Aggarwal P.K, Fischer R.A, Lipoon S.P (1990) Source sink relations and effects of post-anthesis canopy defoliation in wheat at low latitudes. J. Agric. Sci. 114:93-99.

[56] Arakawa A, Nakagawa H, Horie T (1999) Analysis of grain filling in two-rowed barley based on sink source relation. Japanese Journal of Crop Science 68:561-569.

[57] Sinclair T.R, Jamieson P.D (2008) Yield and grain number of wheat: A correlation or causal relationship? Authors' response to "The importance of grain or kernel number in wheat: A reply to Sinclair and Jamieson" by R.A. Fischer. 2008. Field Crops Research 105:22-26.

[58] Fischer R.A (2008) The importance of grain or kernel number in wheat: A reply to Sinclair and Jamieson. Field Crops Research 105:15-21.

[59] Calera A, Gonzalez-Piqueras J, Melia J (2004) Monitoring barley and corn growth from remote sensing data at field scale. International Journal of Remote Sensing 25:97-109.

[60] Freeman K.W, Raun W.R, Johnson G.V, Mullen R.W, Stone M.L, Solie J.B (2003) Late-season prediction of wheat grain yield and grain protein. Communications in Soil Science and Plant Analysis 34:1837-1852.

[61] Aparicio N, Villegas D, Casadesus J, Araus J.L, Royo C (2000) Spectral vegetation indices as nondestructive tools for determining durum wheat yield. Agronomy Journal 92:83-91.

[62] Fetch T.G. Steffenson, B.J. Pederson, V.D (2004) Predicting agronomic performance of barley using canopy reflectance data. Canadian Journal of Plant Science 84:1-9.

[63] Zhao C.J, Liu L.Y, Wang J.H, Huang W.J, Song X.Y, Li C.J (2005) Predicting grain protein content of winter wheat using remote sensing data based on nitrogen status and water stress. International Journal of Applied Earth Observation and Geoinformation 7:1-9.

[64] Jorgensen R.N, Christensen L.K, Bro R (2007) Spectral reflectance at sub-leaf scale including the spatial distribution discriminating NPK stress characteristics in barley using multiway partial least squares regression. International Journal of Remote Sensing 28:943-962.

[65] Zhang, J.H, Wang K, Bailey J.S, Wang R.C (2006) Predicting nitrogen status of rice using multispectral data at canopy scale. Pedosphere 16:108-117.

[66] Alvaro F, Del Moral L.F.G, Royo C (2007) Usefulness of remote sensing for the assessment of growth traits in individual cereal plants grown in the field. International Journal of Remote Sensing 28:2497-2512.

[67] Lorenzen B, Jensen A (1991) Spectral properties of a barley canopy in relation to the spectral properties of single leaves and the soil. Remote Sensing of Environment 37:23-34.

[68] Behrens T, Müller J, Diepenbrock W (2006) Utilization of canopy reflectance to predict properties of oilseed rape (*Brassica napus* L.) and barley (*Hordeum vulgare* L.) during ontogenesis. European Journal of Agronomy 25:345-355.

Chapter 3

The Oil Palm Wastes in Malaysia

N. Abdullah and F. Sulaiman

Additional information is available at the end of the chapter

http://dx.doi.org/10.5772/55302

1. Introduction

Oil palm is the most important product from Malaysia that has helped to change the scenario of it's agriculture and economy. Lignocellulosic biomass which is produced from the oil palm industries include oil palm trunks (OPT), oil palm fronds (OPF), empty fruit bunches (EFB) and palm pressed fibres (PPF), palm shells and palm oil mill effluent palm (POME). However, the presence of these oil palm wastes has created a major disposal problem. The fundamental principles of waste management are to minimise and recycle the waste, recover the energy and finally dispose the waste. These principals apply to agro-industrial wastes such as palm oil residues as they do to municipal waste. We can simply no longer afford to dispose the residues when there is an economically useful alternative. We must first consider the current uses and disposal of mill residues in order to address the potential for recovery of energy in the palm oil industry. One of the unique aspects of Malaysian renewable energy sources is that the palm oil mill is self-sufficient in energy, using PPF, EFB and shell as fuel to generate steam in waste-fuel boilers for processing, and power-generation with steam turbines as described in Section 2.2.

World palm oil production in 1990 doubled to 11.0 million tonnes from 5.0 million tonnes in 1980, and by the year 2000, the production doubled to 21.8 million tonnes. Malaysia produced about half of the world palm oil production (10.8 million tonnes), thus, making Malaysia as world's largest producer and exporter of palm oil during this period [1]. In 2008, even though Malaysia had produced 17.7 million tonnes of palm oil based on 4,500,000 hectares of land used for its plantation, Indonesia became the world's largest producer and exporter of palm oil, replacing Malaysia as a chief producer [2,3] Palm oil has made impressive and sustained growth in the global market over the past four decades, and it is projected in the period 2016 – 2020, the average annual production of palm oil in Malaysia will reach 15.4 million tonnes [4]. In 1999, the land area under oil palm plantation is about 3.31 million hectares, and it has been projected that Sarawak will have about one million tonnes hectares of oil palm by the year 2010 [5].

The oil palm industry has always been linked to the environment because it is a land intensive industry. Any unplanned development will lead to the degradation of the forest systems, loss of habitats including plants and animals, extreme land degradation and pollution (water and airborne) due to the use of large quantities of pesticides and herbicides required to maintain the plantation. The Roundtable for Sustainable Palm Oil (RSPO) was established in recent years with the support from the government and Malaysia Palm Oil Council (MPOC). RSPO consists of palm oil producers, processors, traders, consumer goods manufacturers, retailers and non-governmental organizations (NGOs), and they will develop the principles and criteria of a sustainable palm oil industry, and facilitate the development of sustainable palm oil production. The proposed guidelines include commitment to transparency, compliance with all applicable local, national and ratified international regulations, adoption of sustainable cultivation practices (including water management, pesticide control and soil erosion), conservation of resources and biodiversity and community development [6]. The oil palm industry has long avoided the openings of virgin forest land, which thus minimize environment degradation and enhance the sustainability of oil palm growing. As initiatives, the Ministry of Plantation Industries and Commodities (MPIC) had announced in 2006, the RM20 million Malaysian Palm Oil Conservation Fund (MPOCF) with aims to help protect affected wildlife (including orang utan and other protected species) and to sustain biodiversity conservation programmes that are expected to be beneficial to both the industry and society.

Oil palm is the most important product of Malaysia that has helped to change the scenario of its agriculture and economy. Despite the obvious benefits, oil palm mill also significantly contributes to environmental degradation, both at the input and the output sides of its activities. On the input side, crude palm oil mills use large quantities of water and energy in the production processes, and on the output side, manufacturing processes generate large quantities of solid waste, wastewater and air pollution. The solid wastes may consist of empty fruit bunches (EFB), mesocarp fruit fibers (MF) and palm kernel shells (PKS). The liquid waste is generated from an extraction of palm oil of a wet process in a decanter. This liquid waste combined with the wastes from cooling water and sterilizer is called palm oil mill effluent (POME). During POME digestion, odor released into surrounding air, thus, reduces air quality in the surrounding lagoons area. Disposal of EFB into oil palm plantation without recovering remnant oil in the EFB contributes to oil spills. Incineration of EFB means wasting renewable energy source and heat which actually could be provided for boiler in palm oil mill. At present, PKS and MF wastes are used extensively as fuel for steam production in palm-oil mills. EFB is a resource which has huge potential to be used for power generation, currently not being utilized. The application of shells for road hardening has no impact to the environment, however, current practice is actually wasting potential renewable energy source. Methane gas is one among other green house gases which can cause ozone depletion. However, at present, methane in biogas generates during POME digestion is not being utilized or captured and it just escapes into the atmosphere. Palm oil mill residues are currently underutilised; therefore, maximizing energy recovery from the wastes is desirable for both economic and environmental reasons.

All economic activity begins with physical materials and energy carriers such as fuels and electric power. Without materials, there might be no food and shelter technology; without energy, there might be no work, thus, no economic activity. The reliable sustainable resource is important to fulfill the need of energy. Oil palm waste is a reliable resource because of its availability, continuity and capacity for renewable energy solution. Furthermore, in current situation the presence of oil palm wastes has created a major disposal problem, thus, affect the environmental. The technological, economic, energy balance, and environmental considerations must be kept at a balance to meet the best solution of utilization oil palm wastes. There is abundance of raw materials available of the palm tree consisting of around 90% of biomass wastes and only around 10% of oil. About 90 million tonnes of oil palm fruit production was recorded in 1998; however, 43-45% of this was mill residues in the form of EFB, shell and fibre. Palm fronds and stems are currently underutilised, and the presence of these oil palm wastes has created a major disposal problem. Therefore, maximising energy recovery from the wastes is desirable for both the environmental and economic reasons. Direct combustion, gasification, pyrolysis, liquefaction, fermentation and anaerobic digestion are alternate conversion technologies available to maximise energy recovery. Therefore, sustainable development can be promoted by encouraging energy projects for the long term, utilising local skills and creating employment.

2. Oil palm industry

Traditionally the oil palm (Elaeis guineensis) was grown in semi-wild groves in tropical Africa. It was first introduced to Malaysia for planting in the Botanical Gardens in Singapore in 1870 [7]. Germination takes around 3 months, after which the seedlings are planted in small plastic bags where they are left in a so-called pre-nursery for several months. They are transplanted into bigger plastic bags and grow in a nursery for several more months to a size of about 1 meter, before they are transplanted into a field at an age of around 1 year.

The new improved crosses begin to flower after less than one year of transplantation and produce their first bunches of fruit after less than 2 years. At this age, their leaves have a size of over 2 meters in height and diameter. During its young age, the trunk grows at a rate of about 35 to 75 cm per year and produces alternate rows of leaves, depending on its gene [8]. The base of the old leaves surround the stem and begin falling off at the age of 12 to 15 years [9]. By this time, growth and production have slowed down.

The number of leaves in an oil palm plant increase from 30 to 40 in a year at the age of 5 to 6 years. After that, the generation of leaves decreases to about 20 to 25 per year [9]. The average economic life-span of the oil palm is 25 years to 30 years [10]. A marked increase in the cultivation of oil palm began in 1960 [11], for which by the year 1990 onwards there was a peak in replanting. This provided a good opportunity to harness the by-products of the oil palm. During the re-plantation, the heights of the oil palm tree are in the range of 7 m to 13 m, with a width of between 45 cm to 65 cm, measuring 1.5 m from the surface of the soil. There are about 41 leaves in each frond of the mature oil palm tree. It is estimated that in the year 2000, the process of re-plantation would generate about 8.36 million tonnes dried biomass, consisting of 7.02 million tonnes of trunk and 1.34 million tonnes of leaves [5]. Due to the high moisture of about 70% fresh weight, the newly chopped tree trunk cannot be burnt in the plantation. To leave the old trunk for natural decomposition not only obstructs the re-plantation process but harbours insects that would harm the new trees as well. The tree trunk usually takes between five to six years to decompose [12].

Most crude palm oil mills harness the energy from the fibre and shell in their own low pressure boilers and normally, the EFB's are burnt causing air pollution or returned to the plantation. A 60 tonnes of fresh fruit bunches (FFB) per hour mill based within a 10,000 hectare plantation, can generate enough energy to be self sustaining and supply surplus electricity to the grid if it utilises all of its wastes. In order to provide a better understanding of the palm oil industry in Malaysia, the following sections give an overview of the oil palm industry in Malaysia including oil palm plantation and the mass balance of the oil palm industry as it is self-sufficient in energy.

2.1. Malaysian palm oil scenario

The first commercial oil palm estate in Malaysia was set up in 1917 at Tennamaran estate, Selangor. Palm oil is one of the seventeen major oils and fats in the world market. The government encouraged crop diversification from rubber to oil palm in the late 1950s. The area utilised for oil palm plantations in Malaysia has increased to 3.31 million hectares by the year 1999; where 62% of the total area is located in Peninsular Malaysia while Sabah and Sarawak 28% and 10%, respectively [4].

The oil palm fruit produces two distinct oils which are palm oil and palm kernel oil. Palm oil is obtained from the mesocarp while palm kernel oil is obtained from the seed or kernel. Palm oil is used mainly for the production of margarine and compounds in cooking fats and oils and also for the production of candles, detergents, soap and cosmetic products. Production of palm kernel oil is about 12% of the production of its palm oil.

The success of the Malaysian palm oil industry is the result of the ideal climatic conditions, efficient milling and refining technologies and facilities, research and development, and efficient and adequate management skills. Practically all palm oil mills generate their own heat and power through the co-generation system [13]. The Malaysian government is fully committed to the expansion of the industry and encourages global expansion of palm oil production. Palm oil is now readily accepted globally and Malaysia has exported palm oil to more than 140 countries in the world.

Most palm oil is currently produced in South East Asia, even though the oil palm is originally an African crop, which was introduced to South East Asia in the 19th century. The two largest producers are Malaysia and Indonesia, who together account for roughly 85% of the world palm oil production [14]. In 2004 Malaysian production exceeded Indonesian production. However, the US Department of Agriculture notes that mature palm area in Indonesia is being expanded from 5 to 8 million hectares, which should easily overtake

Malaysia in the near future [15]. There are plans for expansion of palm area in South America [16] and Africa [17], both of which in principle offer large tracts of suitable tropical land. Compared to the potential expansion, however, these plans are embryonic and current production is low and largely for domestic consumption.

Palm oil and related products represented the second largest export of Malaysia in the first nine months of 2005, after electronics, but just ahead of crude oil [18]. In 2005, Malaysian palm oil production is projected to reach approximately 15 million tonnes (301,000 barrels per day), which is very close to the actual value of 14.96 million metric tones recorded by Malaysian Palm Oil Board (MPOB) [19,20]. By comparison, Malaysian petroleum production in 2004 is estimated at 43 million tonnes (855,000 barrels per day), of which 16 million tonnes (321,000 barrels per day) were exported. Domestic petroleum demand of 26 million tonnes represented 44% of the total energy demand of 60 million tonnes of oil equivalent [21].

The total oil palm planted area in Malaysia increased by 2.8% to 4.17 million hectares in 2006. The area expansion occurred mainly in Sabah and Sarawak with a combined growth of 4.5% compared to 1.6% in Peninsular Malaysia [22]. Sabah remained the largest oil palm planted state with 1.24 million hectares or 30% of the total planted area. Table 2.2 shows the oil palm planted areas by state in Malaysia for 2005 until 2008 (in hectares) [22,23].

State	2005	2006	2007	2008
Johor	667,872	671,425	670,641	na
Kedah	75,472	76,329	75,096	na
Kelantan	89,886	94,542	99,763	na
Melaka	52,015	52,232	49,113	na
N.Sembilan	155,164	161,072	170,843	na
Pahang	606.821	623,290	641,452	na
Perak	340,959	348,000	350,983	na
Perlis	278	258	260	na
P.Pinang	14,074	14,119	13,304	na
Selangor	132,100	128,915	129,315	na
Terengganu	163,967	164,065	161,287	na
Peninsular Malaysia	**2,298,608**	**2,334,247**	**2,362,057**	**-**
Sabah	1,209,368	1,239,497	1,278,244	na
Sarawak	543,398	591,471	664,612	na
Sabah & Sarawak	**1,752,766**	**1,830,968**	**1,942,856**	**-**
Malaysia	**4,051,374**	**4,165,215**	**4,304,913**	**4,487,957**

Table 1. Oil Palm Planted Area 2005 - 2008 (Hectares) [22,23]

The production of crude palm oil increased by a further 6.1% to 15.9 million tonnes in 2006 from 15.0 million tonnes the previous year as shown in Figure 2.3. Figure 2.4 shows that the increase was mainly attributed to the expansion in matured areas by 2.0% and rise in the average fresh fruit bunches yield per hectare by 3.8% to 19.6 tonnes due to better management and agricultural inputs. Figure 2.4 also shows that the oil yield per hectare

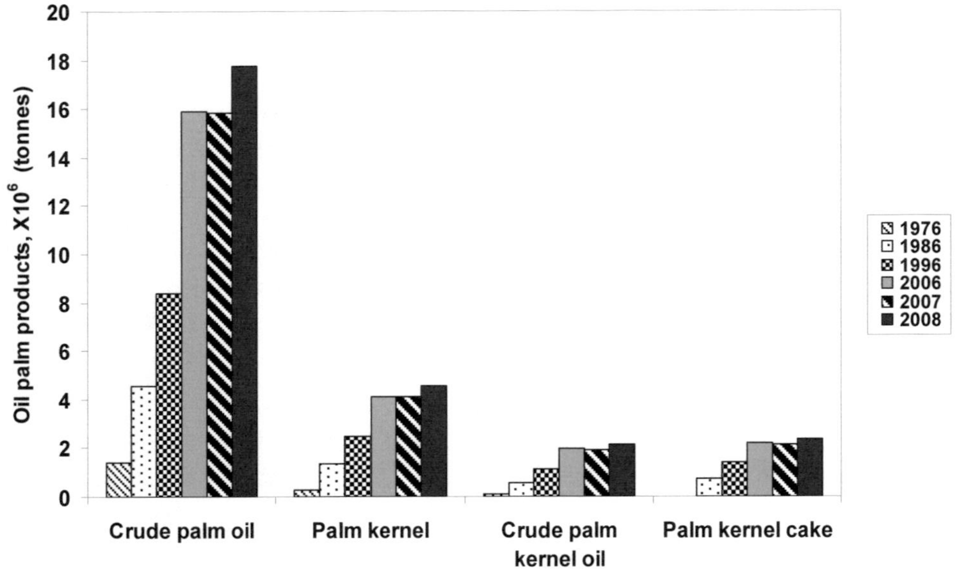

Figure 1. Production of Palm Oil Products every Ten Years from 1976 – 2006, 2007 and 2008 [18,23]

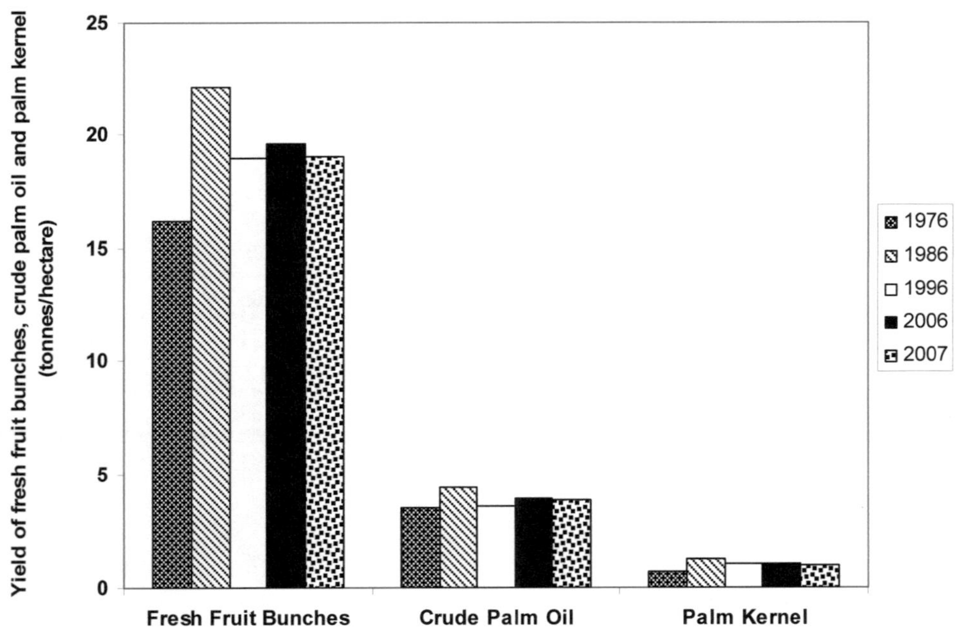

Figure 2. Yield of Fresh Fruit Bunches, Crude Palm Oil and Palm Kernel every Ten Years from 1976 – 2006 and 2007 [18].

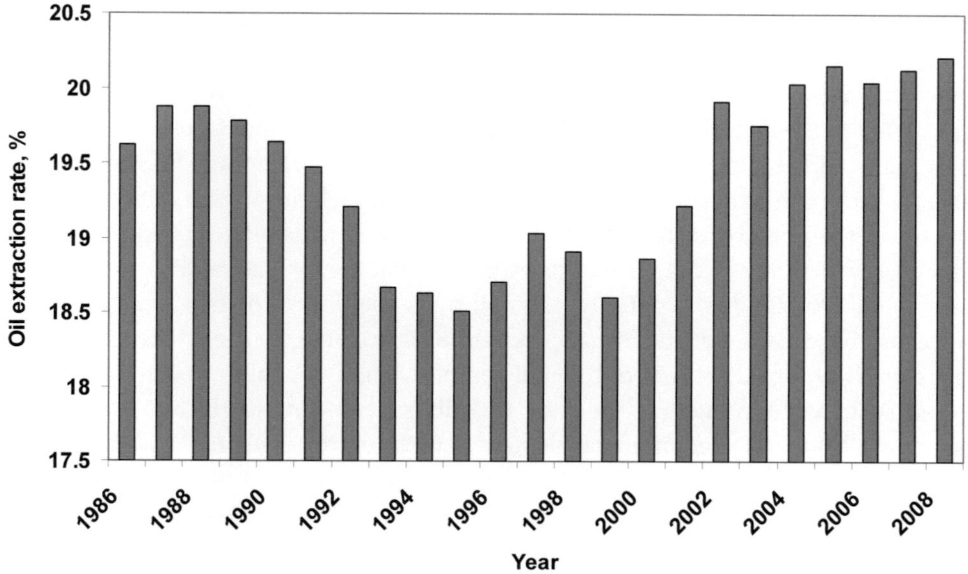

Figure 3. Annual Oil Extraction Rate (OER) for 1986-2008 (%) [22,23]

had increased by 3.4% to 3.9 tonnes, despite the oil extraction rate (OER) declining marginally by 0.5% to 20.04% as shown in Figure 2.5. The decrease in OER in the years 1993 to 2001 which is significant is due to the global recession accounting for a lower demand of export market. However, despite a weak global economy, there is a significant recovery in 2002 as the government implemented prudent policies to assist the Malaysian oil palm industry. These include the expansion of oil palm in matured areas and the campaign on improved productivity in the oil palm industry, coupled with providing competitive prices of oil palm, liberalization of export duties and the encouragement of counter-trades for higher exports [24]. Crude palm kernel oil production rose by 6.1% to 1.96 million tonnes in tandem with a 4.1% growth in palm kernel production as shown in Figure 2.3 [18,23].

The rapid expansion of oil palm cultivation has raised concerns about the sustainability and environmental impact of oil palm plantations, in particular with regard to biodiversity, destruction of old growth rainforest and air pollution [25,26]. To illustrate the potential impact, it is worthy to reflect on the fact that with a palm oil yield of 4 tonnes per hectare tropical forest of roughly the size of the United States would be required to satisfy current world crude oil demand. Increased yields are one avenue for reducing the area imprint for oil palm plantations. It is estimated, based on fundamental factors and actual yields achieved on experimental plots that yields as high as approximately 10 tonnes per hectare may eventually be achievable [27]. At these yields, current world oil demand could be met on roughly 4 million square kilometres, which is 40% of the area of the United States, or over half the land mass of Brazil.

2.2. The mass balance of the oil palm industry

The palm oil mill is self-sufficient in energy, using waste fibre and shell as fuel to generate steam in waste-fuel boilers for processing, and power-generation with steam turbines. As an example, The Federal Land Development Authority (FELDA) palm oil mill in Sungai Tengi, Selangor, Malaysia, employs the standard oil extracting process [28]. In the standard milling process, used in the factories with a milling capacity of over 10 tonnes of raw material per hour, water is added into a digester [29]. More than 19.7 million tonnes FFB were processed in 2000 [28]. The standard sized mills processing 60 tonnes/hour of fruit bunches normally produce 40 tonnes/hour of steam. Part of the steam is used to generate 800 kW of electricity and the rest is used as process steam. It is estimated that the total generating capacity of the mills is about 200 MW [28]. Typically palm oil mills use fibre and shell as a boiler fuel to produce process steam for sterilisation, etc and also possibly for electricity generation to supply electricity for other parts of the mill complex. These oil palm wastes make oil palm mills self sustainable in energy. The shell and fibre alone can supply more than enough energy to meet the mill's requirements using low pressure relatively inefficient boilers. The EFB have traditionally been burnt in simple incinerators, as a means of disposal and the ash recycled onto the plantation as fertiliser. However, this process causes air pollution and has now been banned in Malaysia, furthermore, under this route of disposal, no energy is recovered. Alternatively EFB can be composted and returned to the plantation, or returned directly as mulch. Figure 2.1 shows a proposed plan for the operational process and product of the palm oil industry if EFB is used as fuel beside palm shell and fibre.

Referring to Figure 2.1, as the fresh fruit bunches reach the processing plant, the sterilisation process begins with the steam temperature at 140°C, pressure at 2.5 to 3.2 kg/cm^2 for 50 minutes [28]. After this process, the stripping process will take over. In the stripping process, a rotating divesting machine is used to separate the sterilized oil palm fruit from the sterilized bunch stalks. The empty fruit bunches (EFB) will fall in the collector and are brought to the burning place as a fuel. After the bunches have been stripped, the sterilised fruits are fed into a digester where water at 80°C is added. This is performed in steam-heated vessels with stirring arms, known as digesters or kettles. The most usual method of extracting oil from the digested palm fruit is by pressing. The type of press used in this palm oil is the screw type press.

The crude oil extracted from the digested palm fruit by pressing contains varying amounts of water, together with impurities consisting of vegetable matter, some of which is dissolved in the water. Centrifugal and vacuum driers are used to further purify the oil before pumping it into a storage tank. When the digested fruit is pressed to extract the oil, a cake made up of nuts and fibre is produced. The composition of this cake varies considerably, being dependent on the type of fruit. The cake is given a preliminary breaking treatment before being fed into the nut/fibre separator called depericarper. When the fibre has been separated from the nuts, the latter can then be prepared for cracking. Any uncracked nuts must be removed and recycled and the shell separated from the kernels. The waste fibre and shell are also transported to the burning place as a fuel. The kernels are packed and sold to kernel oil mills.

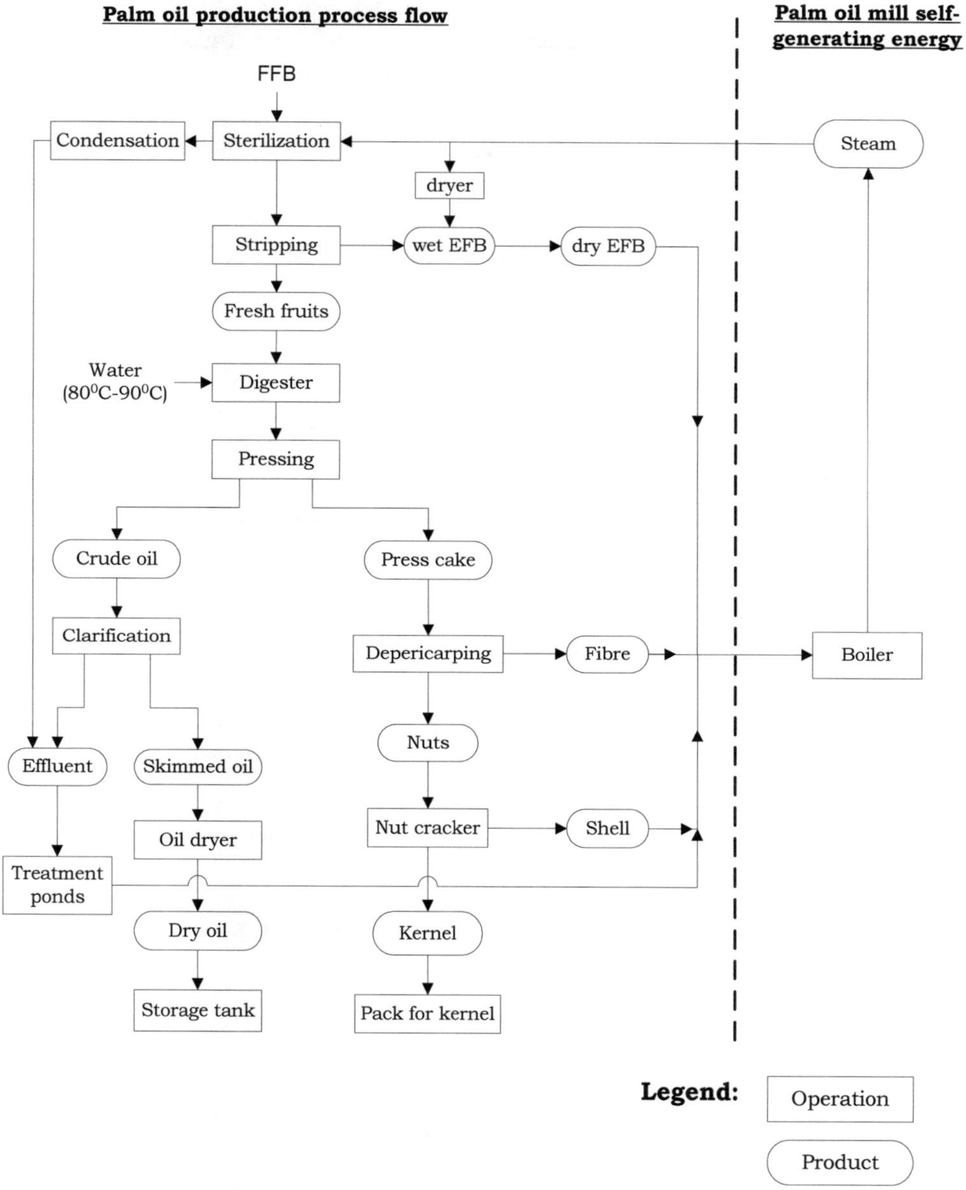

Figure 4. Proposed plan for operation of a Palm Oil Mill (adapted from [28])

Palm oil mills in Malaysia typically meet most of their electricity and process steam requirements by burning some of the wastes, with energy for start-up generally being provided by back-up diesel [13,28,30]. Not all of the wastes are burnt. For each kg of palm oil, electricity consumption is around 0.075-0.1 kWh and steam demand around 2.5 kg. This represents a steam to electricity ratio of around 20 to 1 and could be met by burning 0.3-0.4 kg of waste. As the boiler efficiency is only around 70%, actual consumption is correspondingly higher [31]. Little effort was made in the past to optimise process steam consumption or boiler or turbine efficiency, as the fuel was substantially treated as a waste that was incinerated to be disposed of. The electricity co-generated in Malaysian palm oil mills therefore only amounts to roughly 1-1.5 billion kWh or less than 2% of 2003 generation of over 82 billion kWh. To illustrate the kinds of waste available, the process flow of a palm oil mill is summarised in Figure 2.2 (simplified from [28]) and a typical product stream distribution is shown in Table 2.1 (adapted from [30]). The total product stream distribution in oil palm mills is greater than 100% in wet basis as extra water is added during the process, for example during sterilization with steam. Most of this water ends up in POME.

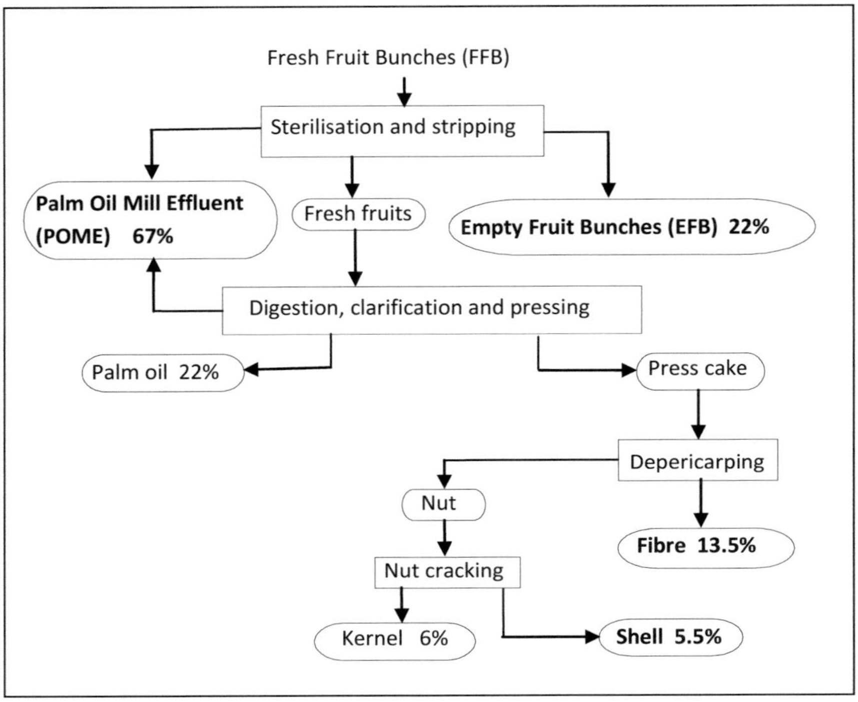

Note: main waste streams in bold, all percentages on wet FFB basis

Figure 5. Simplified process flow diagram of an oil palm mill

As can be seen in Table 2.1, the moisture content of fresh EFB is very high. Typically it is over 60% on a wet EFB basis. Consequently, it is a poor fuel without drying and presents considerable emissions problem that its burning is discouraged by the Malaysian government. Palm oil mills therefore typically use shell and the drier part of the fibre product stream, rather than EFB, to fuel their boilers [31]. Palm Oil Mill Effluent (POME) is so wet that it is usually treated by anaerobic digestion before the discharge of the effluents [32].

For each kg of palm oil, roughly a kg of wet EFB is produced. As over 60% of the wet EFB consists of water, and the heating value of the dry EFB is roughly half that of palm oil, the energy obtainable from the EFB product stream amounts to roughly 0.2 kg of oil equivalent per kg of palm oil. Based on Malaysia's 2005 palm oil production of 15 million tonnes, the energy value of the EFB waste is therefore around 3 million tonnes of oil equivalent, which would amount to $1.2 billion for an assumed $400 per tonne ($55 per barrel).

	Wet FFB basis (tonnes per hectare)	% FFB	**Dry FFB basis** (tonnes per hectare)	% FFB
FFB	20.08	100	10.6	100
Palm oil	4.42	22.0	4.42	41.7
Palm kernel	1.20	6.0	1.20	11.4
EFB	4.42	22.0	1.55	14.6
POME	13.45	67.0	0.67	6.3
Shell	1.10	5.5	1.10	10.4
Fibre	2.71	13.5	1.63	15.4
Total	27.3	136.0	10.6	99.8

Table 2. Typical product stream distribution in oil palm mills [30]

As mentioned before, most crude palm oil mills harness the energy from the fibre and shell in steam boilers. However, the introduction of advanced cogeneration (combined heat and power) also can play a role in combatting climate change, as well as introducing significant economic benefits. Through cogeneration, the costs of energy will be cut because it uses fuels at high conversion efficiencies can reduce the emissions of carbon dioxide and other pollutants. However, it is only worth doing if one can sell the additional surplus energy (electricity) to customers at an economical rate. Today, the ability to sell electricity into the local grid provides an opportunity to turn waste into a valuable commodity.

2.3. Options for the disposal of oil palm wastes

The total land area in Malaysia amounts to 32.90 million hectares. According to Hoi and Koh [32], the major agricultural crops grown in Malaysia are rubber (39.67 %), oil palm (34.56 %), rice (12.68 %), cocoa (6.75 %) and coconut (6.34 %) which indicated that major production of the agricultural sector had been rubber derived products including wood residues, however, by 1995 oil palm products became more significant [34].

Lignocellulosic biomass which is produced from the oil palm industries include oil palm trunks (OPT), empty fruit bunches (EFB), fronds, palm pressed fibres (PPF) and shells. Table 2.3 shows the breakdown of wastes from palm oil production in 2007 [35].

Wastes	Quantity (ktonnes)
Fronds	46,837
Empty fruit bunches (EFB)	18,022
Palm pressed fibres (PPF)	11,059
Oil palm trunks (OPT)	10,827
Shell	4,506

Table 3. Wastes from palm oil production [35]

One of the major characteristics of the forestry and agricultural sector is the production of large quantities of processing residues that have no economic value other than energy generation. Their presence in recent years has created a major disposal problem due to the fact that open burning is being discouraged by the Department of Environment in Malaysia. Other than biomass from the plantations, the palm oil industry also produces other types of waste in large quantities mainly EFB, PPF, shell and palm oil mill effluent (POME). Table 2.4 shows the breakdown of product or waste from each bunch of fresh fruit (FFB) [36].

Products/Wastes	Percentage by weight to FFB (dry basis)
Palm oil	21
Palm kernel	7
Fibre	15
Shell	6
Empty fruit bunches	23
POME	28
Total	100

Table 4. Products/wastes from each bunch of FFB [36]

The EFB are usually air dried until the moisture content reaches about 40% when it is ready to be used as fuel in the palm oil processing plant [37]. The burnt waste is then used as fertiliser in plantations [38]. Other than that, EFB were also used in the plantations as a mulch, thus, can reduce the applied fertiliser cost and is a step towards environmental conservation by reducing dependence on fossil fuel required for the manufacture of inorganic fertilizer [39]. It is claimed that using the EFB as mulch has several advantages for the nutritional sustainability of the plantation. Some plantation owners claimed that the benefits of EFB as a fertiliser and as a soil conditioning agent are significant, because it releases nutrients slowly to the soil via microorganisms therefore effectively recycling the plant nutrients. It improves the soil structure due to better aeration, increases the water holding capacity and increases the soil pH, whilst other mill owners welcomed alternative methods of disposal. This is due to the inconvenience of handling and transporting, as well as the costs and problems concerning disposal of the waste on the plantation. However, open burning is no longer allowed by the authority because this process causes air pollution and by this means of disposal no energy is recovered [40].

Oil-palm fronds have been successfully used as a substitute for tropical grasses by ruminant producers in Malaysia [41]. Nowadays, the PPF is usually burnt in the palm oil processing plant as fuel and the excess is disposed of in the plantations [42]. The PPF are burnt in a boiler with some palm shells to produce the power for running the mill (self-sufficient). The boilers used are normally of grate-type beds which are manufactured locally [13]. Most of the crude palm oil mills harness the energy from the shell and fibre in their own low-pressure boilers and normally the oil palm trunk would be left to decompose naturally at the plantation [37]. This practice not only disturbs the process of plantation due to the low decomposition rate, it also encourages the spread of diseases and insects like *rhinoces beetles* and *ganoderma* that are harmful to the plantation [37]. Moreover, most of the plantations have to adopt the push-felling technique and trunk-shredding which leads to burning [43].

The utilization and generation of oil palm biomass is widely accepted and offers benefits for rural areas related to employment, rural infrastructure, the conservation of cultivated areas and hence the attractiveness of rural regions. The new markets for Malaysia can be developed, especially for developing countries, where oil palm biomass has a higher contribution to the overall energy supply. Also the establishment of an industry related to 'oil palm biomass for energy'' technology could be supported.

3. Utilization of oil palm wastes

Another route to obtain more energy from oil palm plantations is the more efficient use of oil palm biomass other than the palm oil. There are no detailed statistics for oil palm dry matter production. Such statistics are only compiled for palm oil, palm kernel and fresh fruit bunches (FFB). Rough extrapolations, however, can be made based on estimates of the ratio of palm oil to other dry matter. For each kg of palm oil roughly another 4 kg of dry biomass are produced; approximately a third of which is found in FFB derived wastes and the other two thirds is represented by trunk and frond material [27,30,44]. On an energy basis, the palm oil represents roughly a third of the biomass yield, as it has roughly twice the heating value of the other oil palm dry matter, which therefore amounts to approximately 2 kg on a palm oil equivalent basis. Based on 2005 production, around 30 million metric tonnes of oil equivalent of non palm oil dry biomass matter were available for energy production from Malaysian palm oil plantations, or in other words approximately half of 2004 total primary energy demand. Only a small fraction of this potential was used, and that vary inefficiently. Open burning is still too common and responsible for substantial air pollution problems in South East Asia, indicating that other solutions urgently need to be found. Some of the biomass is used for mulching and as fertiliser, though this use is limited by labour and logistical limitations and concerns about encouraging oil palm pests [45].

Generally, oil palm mills generate a numbers of oil palm wastes. The oil palm wastes contribute about RM6379 million of energy annually [46]. However, there is much to be done to optimise the utilization of oil palm wastes for cogeneration in Malaysia. Various studies conducted in Malaysia have indicated that the used of biomass as a source of energy is one of the most promising ways of effectively using the residues. Some of the commercial projects and research activities are include treatment of palm oil mill effluent [47,48], pyrolysis of oil palm shell [49], chars from oil palm waste [50], solid biofuels from biowastes [51], briquetting of palm fibre and shell [36], palm oil effluent as a source of bioenergy [52] ethanol fermentation from oil palm trunk [53] and converting oil palm trunks and cocoa wood to liquid fuels [37]. In the following sections, potential uses of oil palm wastes are presented.

3.1. Potential uses of Pome

POME is the effluent from the final stages of palm oil production in the mill. It is a colloidal suspension containing 95-96% water, 0.6-0.7% oil and 4-5% total solids including 2-4%

suspended solids [54]. Most palm oil mills and refineries have their own treatment systems for POME, which due to its high organic content is easily amenable to biodegradation. The treatment system usually consists of anaerobic and aerobic ponds. However, because of silting and short circuiting many do not reach discharge standards to water courses. This situation can be significantly improved by introducing enclosed anaerobic digestion systems which reduce the biological oxygen demand (BOD) of the effluent and capture methane, one of the more potent greenhouse gases. The energy in the methane can then be recovered, either as a supplementary boiler fuel, or in a biogas engine generator. For each tonne of crude palm oil (CPO) produced, about an average of 0.9-1.5m^3 POME is generated. The biological oxygen demand (BOD), chemical oxygen demand, oil and grease, total solids and suspended solids of POME ranges from 25000 to 35000 mg/L, 53630 mg/L, 8370 mg/L, 43635 mg/L and 19020 mg/L respectively [55]. Therefore, this had created environmental problem because the palm oil mill industry in Malaysia produces the largest pollution load into the rivers throughout the country [56]. However, POME contains high concentrations of protein, nitrogenous compounds, carbohydrate, lipids and minerals that could be converted into useful material using microbial process [57,58]. As example, bio-gas can be produced by processing POME through anaerobic treating system. Anaerobic digestion is a series of processes in which microorganism break down biodegradable material in the absence of oxygen. About 400m^3 of bio-gas produced from 100 tonnes of POME, of which this amount of POME had been released during processing of 20 tonnes of fresh fruit bunches [59-61].

Currently, fertilizers is also derived from POME and used in the farms and vegetation areas [62]. It is also found that the gas composition contained hydrogen (66-68%) and carbon dioxide (32-34%) that can be produced from POME using anaerobic micro flora and this generated gas is free from methane [63]. At present, a renewable energy power plant developer in Malaysia, known as Bumibiopower is in the progress of setting up a plant from methane extraction and power generation using POME near Pantai Remis at the west coast of Peninsular Malaysia. A closed anaerobic system is installed to produce and collect consistently high quality of methane-rich biogas from POME. The installation of a generator of size between 1 and 1.5 MW is also included in this project [64].

3.2. Potential uses of bio oil derived from oil palm wastes

Bio-oil is a renewable, which is produced from biomass through a process known as fast pyrolysis. Fast pyrolysis represents a potential route to upgrade the biomass to value added fuels and renewable chemicals. There is an urgent need to develop a sustainable energy supply as the impact of burning fossil fuels on our climate is becoming ever more obvious and the availability of fossil fuels is decreasing. Bio-oil contributes to the reduction of greenhouse gas emissions and it offers several advantages, as it is easy to use, to store, and to transport. Bio-oil that can be extracted from dried biomass including dried oil palm wastes is currently under investigation as a substitute for petroleum [65]. Bio-oil contains fragments of cellulose, hemicelluloses, lignin, and extractives and they are typically brown liquids with a pungent odor. For woody feedstocks, temperatures around 500ºC together with short vapour residence times are used to obtain bio-oil yields of around 70%, and char and gas yields of around 15% each [66]. Bio-oil is a high density oxygenated liquid, which can be burned in diesel engines, turbines or boilers, though further work is still required to demonstrate long term reliability [67]. It is also used for the production of speciality chemicals, currently mainly flavourings. Renewable resins and slow release fertilisers are other potential applications, which have been the subject of research [68]. At this stage, fast pyrolysis is a novel and relatively untested technology. There are several pilot plants in North America and Europe, but no consistent track record yet outside of the manufacture of flavourings.

To date, fast pyrolysis of biomass has received very limited attention by researchers in Malaysia. Normally, fibre and shells are burnt in the palm oil processing plants to generate fuel to produce power for running the mill (self-sufficient) [13,69]. So far, research involving fast pyrolysis has been carried out by Universiti Teknologi Malaysia and Universiti Malaya on oil palm shell, rubber waste and rice husk waste, scrapped tyres and tubes [70-74]. One of the authors of this book and the research group from MPOB investigated on the fast pyrolysis of empty fruit bunches (EFB) [75,76].

The utilisation of bio-oil derived from pyrolysis process of oil palm wastes to substitute for synthetic phenol and formaldehyde in phenol formaldehyde resins is possible. Phenol can be used to manufacture moulding products for automotive parts, household appliances, and electrical components; in bonding and adhesive resins for laminating, plywood, protective coating, insulation materials, abrasive coating; in foundry industries for sand moulds and cores. However, producing resins from bio-oil has received very limited attention by researchers in Malaysia and still in research stage. A group of researchers from Universiti Teknologi Malaysia had studied the extraction of phenol from oil palm shell bio-oil [77]. They found that the quantity of phenol in the extracted oil was 24.2 wt% of total extracted oil.

In 2005, with the co-operation between Malaysian based Genting Sanyen Bhd and BTG Biomass Technology Group BV from The Netherlands, the first commercial bio-oil plant has already started production in Malaysia on a scale of 2 t/hr [78,79]. The main achievements of this project are more than 1,000 tonnes of bio-oil have been produced, the bio-oil is co-fired, replacing conventional diesel in a waste disposal system located 300 km from site, maximum capacity of the plant so far is about 1.7 t/hr on a daily continuous basis, the bio-oil quality can be controlled by the operating conditions, the drying of EFB to 5 wt.% moisture is possible using the excess heat from the pyrolysis process, the energy recovered from the process can be used effectively for drying the wet EFB, and potentially to generate the electricity required. Indeed, this is a breakthrough step in Malaysia for the utilisation of oil palm wastes as a source of bio-oil [80].

3.3. Potential uses of dry residues from oil palm wastes

The main products produced by the palm oil mills are crude palm oil and palm kernels. However, it also produces huge quantities of residues such as fibre, shell and empty fruit bunches as shown in Figure 2.2. Dry residues from oil palm wastes can be utilised to produce various types of products. EFB had been studied to convert into paper-making pulp by the researches from MPOB because EFB can be categorized as fibrous crop residues know as lignocellulosic residues. The high number of fibres/unit weight indicates the paper from EFB would have good printing properties and a good formation within paper making. EFB could produce thin, high quality printing paper, speciality papers for example for cigarette and photographic papers and security papers. The total chlorine-free methods had been used to bleach the pulp for producing paper [59,81]. Products such as paper and pulp that are obtained by processing the oil palm wastes can be used in many ways such as cigarette paper and bond papers for writing [82]. Normally, the excess shell are used to cover the surface of the roads in the plantation area.

Various types of wood such as saw-wood and ply-wood or lumber had been produced from oil palm trunk. Oil palm trunks have been chipped and waxed with resin to produce pre formed desk tops and chair seats for schools. The furniture is characterised for resistance against knocks, scratches, ink, termites and fungus The ply-wood or lumber can be utilised as core in producing blackboard. The saw-wood is used for furniture but it is not suitable as building material due to its low specific density. It was found that the strength of the ply-wood made from oil palm trunk was comparable with the commercial ply-wood. The particle board with chemical binders also can be produced from oil palm trunk. Some of the oil palm trunks are mixed with EFB and palm fibres to be combusted to produce energy [81,83,84]. Besides this, the palm shell and palm fibres have been convert of into briquettes in a study [36].

Medium density fibre-boards and blackboards can be produced from EFB and palm fibre [84,85]. Currently, the MDF industry has 14 plants with a total annual installed capacity of 2.9 million. The total export of MDF was RM1.2 billion in 2008. The industry has started utilising *acacia mangium* and mixed hardwood to produce MDF as alternatives to rubber wood. At present, Malaysia is the world's third largest exporter of MDF, after Germany and France. MDF from Malaysia has attained international standards such as British (BS), European (EN), Asia-Pacific: Japan Australia and New Zealand (JANS) standards [86]. High-density fiberboard (HDF), also called hardboard, is a type of fiberboard, which is an engineered wood product. It is similar to MDF, but is denser and much harder and stronger because it is made out of exploded wood fibers that have been highly compressed. Agro-Bio Fibre Sdn Bhd in Malaysia holds the patent for the EFB-based MDF over the last 10 years, has invested RM30 million to develop the technology to produce MDF and other products from the oil palm wastes. This company had signed a MoU with the Forest Research Institute of Malaysia (FRIM) to develop HDF used mainly for the production of floorboards that would use 100% EFB as its raw material [87].

Oil palm fibre is non-hazardous biodegradable material extracted from empty fruit bunch that are considered as waste after the extraction oil palm fruits. The fibres are clean, non-carcinogenic, and free from pesticides and soft parenchyma cells. Palm fibres are versatile and stable and can be processed into various dimensional grades to suit specific applications such as erosion control, mattress cushion production, soil stabilization, horticulture and landscaping, ceramic and brick manufacturing, paper production, acoustics control, livestock care, compost, fertilizer and animal feed. Palm fibres can also be used as fillers in thermoplastics and thermoset composites which have wide applications in furniture and automobile components. Production of thermoplastic and thermostat composites has reached commercialization stage when PROTON (Malaysian national car maker) entered into agreement with PORIM (Palm Oil Research Institute of Malaysia) [88,89].

Similar to EFB, according to a study fronds from oil palm trees can be converted into pulp [90]. Oil palm fronds also can be processed as roughage source for ruminants such as cattle and goats [91]. A new product known as oil palm frond based ruminant pellet can be used as balanced diet for fattening beef cattle which is developed by the Malaysian Agricultural Research and Development Institute (MARDI) [91].

Oil palm ash (OPA) can be utilised as an absorbent for removing pollutant gases such as nitrogen oxide and sulphur oxide. The combustion of oil palm fibre and shell as boiler fuel to generate steam in palm oil mill will produce OPA. It was found that OPA contains high amount of calcium, silica, potassium and alumina which can be utilised to synthesize active compounds to absorb the pollutant gases into absorbent [92,93]. The presence of some functional groups such as hydroxyl, lactone and carboxylic in oil palm shell have a high affinity towards metal ions. Thus, the charcoal derived from oil palm shell can be coated with chitosan to use as a remover of heavy metal especially chromium from wastewater industry; however, it is still at research stage [94].

Processing the oil palm wastes such as EFB, fibre, shell and palm kernel cake into a uniform and solid fuel through briquetting process will be an attractive option. Palm kernel cake is a by-product of crushing and expelling oil from palm kernel. Briquetting is a process of compacting loose material to form a homogeneous and densified product. The material can be densified into briquettes at high temperature and pressure using screw of extrusion techniques either with or without binder addition. Oil palm briquettes are often favoured for household and industrial heating unit operation such as boiler because of their enhanced physical properties, as well as being easy to handle and feed. According to a study, the equilibrium moisture content for the briquettes made of palm fibre and palm shell is about 12 mf wt.% [36]. It was found that briquettes made from 100% pulverised EFB exhibited good burning properties. It is recommended to blend with sawdust in order to produce better quality briquettes from EFB and palm kernel cake [95]. Oil palm briquettes can be used as fuel in producing steam, district heating and electricity generation for larger commercial scale. The local sawdust briquettes or charcoal briquettes are rarely used in the local market because it could not compete with the availability of cheap fuels such as charcoal and wood which are widely used in the rural areas and restaurants [96]. Therefore, the products are exported for oversea markets [97].

One of the promising technologies which utilise the oil palm wastes or plant matter involves the production of carbon molecular sieve (CMS) from lignocellulosic materials. Production of CMS from oil palm wastes which are cheap and abundant carbon source will enhance the economical feasibility of adsorption process. A CMS is a material containing tiny pores of a precise and uniform size that is used as an adsorbent for gases and liquids, and normally it is used to separate nitrogen from the other gases contained in air. A survey of literature indicated that palm shell have been used the most as the substrate for CMS production by many researcher in Malaysia [97-101]. Basically, there are three steps involve to prepare the CMS from oil palm wastes which are carbonisation of the wastes, activation of the chars produced and pore modification of the activated carbons to obtain CMS. Activated carbon is produced from carbonaceous source materials such as nutshells, oil palm wastes, peat, wood, coir and lignite. Activated carbon also called activated charcoal is a form of carbon that has been processed to make it extremely porous and have a very large surface area, thus available for adsorption or chemical reactions. Activated carbon can be produced by either physical reactivation or chemical activation. In physical reactivation, the precursor is developed into activated carbons using gases by carbonization and/or oxidation process. For chemical activation, prior to carbonization, the raw material is impregnated with certain chemicals such acid, base or salt [102]. According to a study, the optimum conditions for preparing activated carbon from EFB for adsorption of 2,4,6-TCP were found as follows : activation temperature of 814°C, CO_2 activation time of 1.9h and IR of 2.8, which resulted in 168.89 mg/g of 2,4,6-TCP uptake and 17.96% of activated carbon yield [103].

Biochar is commonly defined as charred organic matter, produced to abate the enhanced greenhouse effect by sequestering carbon in soils and improve soil properties. Biochar is a stable carbon compound that can be kept in the ground for a long time, until thousands of years. Biochar is created when biomass is heated to temperatures between 300 and 1000°C, under low or zero oxygen concentrations. Universiti Putra Malaysia (UPM) with the collaboration of Nasmech Technology Sdn Bhd have successfully built a plant producing biochar from EFP and also the first large-scale biochar production plant in the region. They have constructed a carbonator - driven plant to produce the biochar from residue materials including the EFB about 20 tonnes daily [104].

Besides converting dried oil palm wastes into various value added products, it also have potential as a source of renewable energy. Utilization of oil palm wastes as a source of energy will bring other environmental benefit like reduction in CO_2 .emissions. The greenhouse gases that are present in the atmosphere include water vapor, CO_2, methane and ozone, and the increase of greenhouse gases primarily CO_2 is the major cause for global warming. Oil palm wastes such as fiber, shell and EFB can be used to produce steam for processing activities and for generating electricity [105]. At present, there are more than 300 palm oil mills operating with self-generated electricity from oil palm wastes. The electricity generated is for their internal consumption and also sufficient for surrounding remote areas [106].

A cement company in Malaysia had used palm shell as fuel in the boiler and they found they the emissions of CO_2 can be reduced by 366.26 thousand metric tonnes in the year 2006 alone [107]. Hence, the emission of CO_2 in Malaysia can be decreased significantly if all industries in Malaysia can replace or partially replace fossil fuel with oil palm wastes to generate energy without degrading the environment.

Hydrogen is a synthetic fuel, which can be obtained from fossil fuels, nuclear energy and renewable energy sources such as oil palm wastes. In almost any application replacing fossil fuels, hydrogen may be used as fuel especially as feedstock for synthesis of clean transportation fuels or as a gaseous fuel for power generation [108,109]. Gasification is one of the technologies for producing hydrogen. Oil palm wastes such as EFB, fiber, shell, trunks and fronds can be used for gasification [109,110]. The benefits of using hydrogen as transportation fuel are higher engine efficiencies and zero emissions [111]. However, production of hydrogen from oil palm wastes is still at the early stage of research in Malaysia.

4. Conclusions

Malaysia is one of the world's primary palm oil producers and has been taking steps to promote the use of renewable energy. The utilization of renewable energy resources, in particular oil palm wastes is strategically viable as it can contribute to the country's sustainability of energy supply while minimizing the negative impacts of energy generation on the environment. It will help the government to achieve its obligation to prolong the fossil fuel reserves. The efficient use of oil palm biomass other than the palm oil itself for food consumption is a promising route to obtain more energy from oil palm plantations. It will also solve the agriculture disposal problem in an environmental friendly manner while recovering energy and higher value chemicals for commercial applications like bio-fuel, coal replacement, building products and many others. The current principle adopted in Malaysia is a cost pass-through mechanism for electricity generation which is the same principle adopted for renewable power generation. This method would result in a small increase in the price of electricity paid by electricity consumers, but at the same time, the consumers may benefit from revenues derived from renewable energy generation. Although this effort pales in comparison to other countries which had become leaders in renewable energy growth, the acceptance of this form of renewable energy contribution calls for a paradigm shift among the people in the realm of sustainable energy. In general, the maturity of the country is marked by an acceptance of the need for the country to wean reliance on a depleting and environmentally damaging fuel source.

Author details

N. Abdullah* and F. Sulaiman

School of Physics, Universiti Sains Malaysia, 11800 Minden, Penang, Malaysia

Acknowledgement

The authors would like to acknowledge the three research grants provided by Universiti Sains Malaysia, Penang (1001/PFIZIK/814087, 304/PFIZIK/6310087, 304/PFIZIK/6310073) that has made this research possible.

* Corresponding Author

5. References

[1] Yusof Basiron. Palm Oil and Its Global Supply and Demand Prospects. Oil Palm Industry Economic Journal 2002; 2:1, 1-10.

[2] Malaysian Palm Oil Industry Performance 2008. Global Oils & Fats Business Magazine Vol.6 Issue 1 (Jan-March), 2009 Available on line at <http://theoilpalm.org/wp-content/themes/oilpalm/pdf/GOFB%206-1%20Malaysian%20Palm%20Oil%20Industry%20Performance%202006.pdf>, (accessed 25 June 2012).

[3] World Growth Palm Oil Green Development Campaign: Palm Oil — The Sustainable Oil a Report by World Growth September 2009. Available on line at <http://www.worldgrowth.org/assets/files/Palm_Oil.pdf>, (accessed 25 June 2012).

[4] Teoh Cheng Hai. Land Use and the Oil Palm Industry in Malaysia, A bridged report produced for the WWF Forest Information System Database under project MY 0057 – Policy Assessment of Malaysia Conservation Issues, Kinabatangan, November 2000.

[5] Lim KO. The Future Energy Potential Of Replanting Oil Palm Tree In Malaysia. Renewable Energy 1986; 8: 76 – 85.

[6] United Nations Development Programme. Malaysia Generating Renewable Energy from Oil Palm Wastes, UNDP, Kuala Lumpur. 2007.

[7] Government of Malaysia. The Oil Palm in Malaysia. Report published by the Ministry of Agriculture and Cooperatives, Kuala Lumpur, Government of Malaysia. 1966.

[8] Corley RHV, Hardon JJ, Wood BJ. Developments in Crop Science 1, Oil Palm Research, Amsterdasm:New York:Elsevier Scientific Publishing Company. 1976.

[9] Hartley CWS. Oil Palm (Elaeis Guineensis Jacq.) 2nd ed., London:Longman. 1977.

[10] Lim KO, Zainal Alauddin ZA, Gulam Abdul Qadir, Abdullah MZ. Energy Productivity of Some Plantation Crops in Malaysian and the Status of Bio Energy Utilization. To appear in Proceeding of WREC. 1999.

[11] Kamaruddin H, Abdul Halim H, Riazuan R, Mohd. Zain J. Improvement of Oil Palm Fronds CTM-Pulp Brightness-A Preliminary Study, in Proceeding of Seminar on Oil Palm Trunks and Other Palmwood Utilization (Oil Palm Tree Utilization Committee of Malaysia), Kuala Lumpur, Malaysia. 1991.

[12] Mohamad Husin, Abd. Halim Hj. Hassan, Ahmad Tarmizi Mohammed. Availability and Potential Utilisation of Oil Palm Trunks and Fronds up to the year 2000, PORIM occasional paper. 20, 1-7. 1986.

[13] Noor, M.A., Ani, F.N. and Kannan, K.S. Renewable Energy Scenario In Malaysia, 2nd International Seminar on Renewable Energy for Poverty Alleviation at IEB, Dhaka, Bangladesh, 1999.

[14] Foreign Agricultural Service, USDA, Oil Seeds Circular, Growing Industrial Use of Vegetable Oil Expected to Impact EU Oilseeds and Products Trade. October 2005.

[15] USDA, Oil Crops Outlook, Indonesian Production and Exports of Palm Oil Are Expanding Rapidly. September 2005.

[16] da Costa RC. Potential for Producing Bio-Fuel in the Amazon Deforested Areas. Biomass and Bioenergy 2004, 26:405-15.

[17] Baryeh EA. Effects of Palm Oil Processing Parameters on Yield. Journal of Food Engineering 2001, 48:1-6.

[18] Department of Statistics Malaysia, Malaysian External Trade Statistics, September 2005.

[19] Foreign Agricultural Service, USDA, Oil Seeds Circular, Table 9. October 2005.

[20] Review of Annual Production Of Oil Palm Products : 1975 - 2005. Malaysian Oil Palm Board (MPOB). <http://econ.mpob.gov.my/economy/Production1_05.htm>, (accessed 5 July 2009).

[21] Energy Information Administration (EIA), US Department of Energy, Country Analysis Brief Malaysia, Update January 2005.

[22] Review of the Malaysian Oil Palm Industry 2006. Malaysian Oil Palm Board (MPOB).

[23] A Summary On The Performance if The Malaysian Oil Palm Industry .(2008). Report by Malaysian Oil Palm Board. <http://econ.mpob.gov.my/economy/Performance-130109.html>, (accessed 1 August 2009).

[24] Overview of the Malaysian Oil Palm Industry. (2002). Report by Malaysian Oil Palm Board. <http://econ.mpob.gov.my/review 2002.html>, (accessed 6 May 2009).

[25] Radzi Abas M. Oros DR, Simoneit BRT. Biomass Burning as The Main Source of Organic Aerosol Particulate Matter in Malaysia During Haze Episodes. Chemosphere 2004; 55:1089–95.

[26] Basiron, Y. and Weng, C.K. The Oil Palm and Its Sustainability. Journal of Oil Palm Research 2004; 16, 1:1-10.

[27] Wahid MB, Abdullah SNA, Henson IE. Oil Palm – Achievements and Potential. In: New Directions for a Diverse Planet, Proceedings of the 4th International Crop Science Congress, 26 Sep – 1 Oct 2004, Brisbane, Australia. <www.cropscience.org.au>, (accessed 21 September 2009).

[28] Mahlia TMI, Abdulmuin MZ, Alamsyah TMI, Mukhlishien D. An Alternative Energy Source from Palm Waste Industry For Malaysia and Indonesia, Energy Conversion and Management 2001; 42, 2109 – 18.

[29] Prasertsan, S. and Prasertsan, P. Biomass Residues from Palm Oil Mills in Thailand: An Overview on Quantity and Potential Usage, Biomass and Bioenergy 1996; 11, 5:387 – 95.

[30] Yusoff S. Renewable Energy from Palm Oil - Innovation on Effective Utilization of Waste. Journal of Cleaner Production 2006; 14:87-93.

[31] Husain Z, Zainal ZA, Abdullah MZ. Analysis of Biomass-Residue-Based Cogeneration System in Palm Oil Mills. Biomass and Bioenergy 2003; 24:117–24.

[32] Ahmad AL, Ismail S, Bhatia S. Water Recycling from Palm Oil Mill Effluent (POME) using Membrane Technology. Desalination 2003; 157:87-95.

[33] Hoi Why Kong, Koh Mok Poh. Newsletter of Wood and Non-Wood Products Utilization Branch (FOPW) of the FAO Forest Products Division, Issue No 9 December 2001. <http://www.fao.org/DOCREP/003/y3198E/y3198e00.htm> (accessed 19 September 2002).

[34] Kong HY. Current status of biomass utilization in Malaysia. Forest Research Institute Malaysia; 2000. p. 1–15.

[35] Goh CS, Tan KT, Lee KT, Bhatia S. Bio-Ethanol From Lignocellulose: Status, Perspectives and Challenges in Malaysia. Bioresource Technology 2010;101(July (13)):4834–41.

[36] Hussain Z, Zainac Z, Abdullah Z. () 'Briquetting of Palm Fibre and Shell from The Processing of Palm Nuts to Palm Oil. Biomass and Bioenergy 2002; 22, 6:505-509.

[37] Yeoh HH. The Potential Of Producing Liquid Fuel From Oil Palm Wastes. M.Sc. thesis. Universiti Sains Malaysia, Penang, Malaysia; 1999.

[38] Lim B. The New Straits Times 2000. Thursday, December 28, 2000.

[39] Hamdan AB, Tarmizi AM, Mohad D, Tayeb. Empty Fruit Bunch Mulching and Nitrogen Fertilizer Amendment: The Result Effect On Oil Palm Performance And Soil Properties, PORIM Bulletin 1998, Palm Oil Resesrch Institute of Malaysia.

[40] Zain AZ, Koh Mok Poh, Mohd. Yusoff MN, Khoo Kean Choon, Mohd. Nasir N. Commercial Utilization of Oil Palm Empty Fruit Bunches, in Processing of the Third National Seminar on Utilisation of Oil Palm Tree and Other Palms, Forest Research Institute Malaysia, Kuala Lumpur; 1994.

[41] Abu Hassan O, Ishida M. Status of Utilization of Selected Fibrous Crop Residues and Animal Performance with Emphasis on Processing of Oil Palm Fronds (OPF) For Ruminant Feed in Malaysia. Tropical Agriculture Research Center TARS No.25. Ministry of Agriculture, Forestry and Fisheries, Tsukuba, Japan; 1992.

[42] Takigawa A, Oshio S, Mohd. Jaafar Daud, Abu Hassan Osman. Processing and Utilization of Palm By-product for Ruminant, MARDI-TARC Collaborative Study; 1990.

[43] Khozirah Shaari, Khoo Kean Choon () Introduction. In Khozirah Shaari, Khoo K.C, Abdul Razak Mohd. Ali (eds.) Oil Palm Stem Utilisation 1991: Review of Research, FRIM Research Pamphlet No. 107, Kepong, Forest Research Institute Malaysia; 1991.

[44] Husin M, Ramli R, Mokhtar A, Hassan WHW, Hassan K, Mamat R, Aziz AA. Research and Development of Oil Palm Biomass Utilization in Wood-based Industries. Palm Oil Developments 2002, 36:1-5.

[45] Menon, N.R., Rahman, Z.A. and Bakar, N.A. Empty Fruit Bunches Evaluation: Mulch in Plantation vs. Fuel for Electricity Generation. Oil Palm Industry Economic Journal 2003; 3(2):15-20.

[46] Jaafar MZ, Kheng WH, Kamaruddin N. Greener Energy Solutions for A Sustainable Future: Issues and Challenges for Malaysia, Energy Policy 2003; 31(11):1061-72.

[47] Fakhru'l-Razi A, Noor MJMM. Treatment of Palm Oil Mill Effluent (POME) with the Membrane Anaerobic System (MAS), Water Science and Technology 1999; 39, 10-11:159-63.

[48] Chin KK, Lee SW, Mohammad HH. A Study of Palm Oil Mill Effluent Treatment Using a Pond System, Water Science and Treatment 1996; 34, 11:119-23.

[49] Ani FN, Islam MN. Liquid Oil from Fluidised Bed Pyrolysis of Rice Husk Waste and Its Characterisation, RERIC International Energy Journal 1998; 20, 1:55-65.

[50] Lua AC, Guo J. Preparation And Characterisation of Chars from Oil Palm Waste, Carbon 1998; 36, 11:1663-70.

[51] Lim KO. Production of Solid Biofuels from Biowastes, in Proceedings of The Second International Energy Conference, Kuala Lumpur, Energy from Biomass Residues, Forest Research Institute Malaysia; 1992.

[52] Chew TY, Yeoh BG. Palm Oil Effluent As A Source of Bioenergy, in Proceedings National Symposium on Oil Palm By-Products for Agro-Based Industries, Kuala Lumpur; 1985.

[53] Akmar PF, Elham P. Ethanol Fermentation from Oil Palm Trunk Hydrolyzate, in Oil Palm Trunk and Other Palmwood Utilization, Proceeding of a National Seminar, Kuala Lumpur, Malaysia; 1991

[54] Mohammad AW, Yeong WT, Md Jahim J, Anuar,N. Palm Oil Mill Effluent (POME) Treatment and Bioresource Recovery Using Ultrafiltration Membrane: Effect of Pressure One Membrane Fouling, Biochemical Engineering Journal 2008; 35(3):309-17.

[55] Ma AN. A Novel Treatment for Palm Oil Mill Effluent, Palm Oil Research Institute of Malaysia (PORIM) 1995; 29:201-12.

[56] Hwang TK, Ong SM, Seow CC, Tan HK. Chemical Composition of Palm Oil Mill Effluent, Planter 1978;54:749-56.

[57] Habib MAB, Yusoff FM, Phang SM, Ang KJ, Mohamed S. Nutritional Values of Chironomid Larvae Grown in Palm Oil Mill Effluent and Alga Culture. Aquaculture 1997; 158:95-105.

[58] Agamuthu P, Tan EL. Digestion of Dried Palm Oil Mill Effluent by Cellulomonas Species. Microbios Letters 1985; 30: 109-13.

[59] Gurmit S, Lim Kim H, Teo L, David Lee, K. Oil Palm and the Environment a Malaysian Perspective, Malaysian Oil Palm Growers Council 1999; vol.1, p. 253.

[60] 'Palm biomass power plants in Malaysia'. Biopact article. Sunday, August 06, 2006. <http://biopact.com>, (accessed 23 September 2009).

[61] 'The world's most productive ethanol crop is oil palm'. Biopact article. June 21st, 2006. <http://biopact.com>, (accessed 23 September 2009).

[62] Esther OU. Anaerobic Digestion of Palm Oil Mill Effluent and Its Utilization as Fertilizer for Environmental Protection. Renewable Energy 1997; 10(213): 291-4.

[63] Fakhru'l-Razi, Yassin AAA, Lyuke SE, Ngan MA, Morimoto M. Bio-hydrogen Synthesis from Wastewater by Anaerobic Fermentation Using Microflora. International Journal of Green Energy 2005; 2: 387-96.

[64] 'Bumibiopower Methane Extraction and Power Generation'. Project report by Mitsubishi Securities Clean Energy Finance Committee. (January 2004). <http://dnv.com/certification/.../PDD_BumibiopowerBiomass_2004-03-19.pdf> (accessed 5 August 2009).

[65] Wikipedia. The free encyclopedia: pyrolysis oil. <http://en.wikipedia.org/wiki/Bio-oil> (accessed 23 April 2011).

[66] Bridgwater AV, Peacocke GVC. Fast Pyrolysis Processes for Biomass. Sustainable and Renewable Energy Reviews 2000; 4(1):1 – 73.

[67] Chiaramonti D, Oasmaa A, Solantausta Y. Power Generation using Fast Pyrolysis Liquids from Biomass. Renewable and Sustainable Energy Reviews 2007; 11(6): 1056-86.

[68] Bridgwater AV, Meier D, Radlein D. An Overview of Fast Pyrolysis of Biomass. Organic Geochemistry 1999; 30(12):1479-93.

[69] 'Bumibiopower Biomass Power Plant Project Malaysia'. Project Design Document for Small-Scale SDM Activity, Version 02, March 2006. <http://www.dnv.com/certification/climatechange/Upload/PDD_Bumibiopower%20Biomass_2004-03-19.pdf > (accessed 23 December 2007).

[70] Nurul Islam M, Zailani R, Ani FN. Pyrolytic Oil from Fluidised Bed Pyrolysis of Oil Palm Shell and Its Characterisation, Renewable Energy 1999; 17, 1:73-84.

[71] Kawser MDJ, Farid NA. Thermochemical Processing of Rubber Waste to Liquid Fuel, Journal of Plastics, Rubber and Composites Processing and Applications 2000; 29(2):100-4.

[72] Ani FN, Islam MN. Liquid Oil from Fluidised Bed Pyrolysis of Rice Husk Waste and its Characterisation, RERIC International Energy Journal 1998; 20(1):55-65.

[73] Ani FN, Islam MN. Pyrolytic Recycling of Agro-Industrial Solid Wastes in Malaysia. Journal of the Institute of Energy 1998; 71: 55-8.

[74] Abnisa F, Wan Daud WMA, Husin WNW, Sahu JN. Utilisation Possibilities of Palm Shell as a Source Of Biomass Energy In Malaysia By Producing Bio-Oil In Pyrolysis Process. Biomass and Bioenergy 2011; 35:1863-72.

[75] Abdullah N. An Assessment of Pyrolysis for Processing Empty Fruit Bunches. PhD thesis, Aston University, Birmingham, UK; 2005.

[76] Sukiran MA, Chin CM, Abu Bakar NK. Bio-oils from Pyrolysis of Oil Palm Empty Fruit Bunches. American Journal of Applied Sciences 2009; 6(5): 869-75.

[77] Kawser MDJ, Farid Nasir A. Oil Palm Shell as a Source of Phenol. Journal of Oil Palm Research 2000; 12(1): 86-94.

[78] 'Genting Group Unveils Malaysia's First Commercially Produced Bio-oil Using Breakthrough Technology'. Report by Genting Group, Press Release, 21st August, 2005.

[79] Venderbosch RH, Gansekoele E, Florijn JF, Assink D. Pyrolysis of Oil Palm Residue in Malaysia. BTG Biomass Technology Group BV, Ng HY, Genting Bio-oil BHD. <http://www.pyne.co.uk/docs/PyNe%20Section1.pdf;2005>, (accessed 3 April 2011).

[80] BTG-BTL website. Biomass To Liquid:Pyrolysis Oil-The Sustainable Alternative. <http://www.btg-btl.com/index2.php> (accessed 2 April 2011).

[81] Poster A. Malaysia Palm Oil Council. Palm Oil Industry- a Learning Experient. 2006. <http://www.mpoc.org.,y/>, (accessed January 2009).

[82] 'Long Term Benefit from Oil Palm Biomass'. Bernama, The Malaysian National News Agency. Press Release 16th August, 2001.

[83] Husin M, Ramli R, Mokhtar A, Wan Hassan WH, Hassan K, Ropandi, M, Abdul Aziz A. Research and Development of Oil Palm Biomass in Wood Based Industries. Malaysian Palm Oil Board Bulletin. Palm Oil Development 36. 2005.

[84] Sumanthi, S., Chai, S.P. and Mohamed, A.R. Utilization of Oil Palm as a Source of Renewable Energy in Malaysia. Renewable and Sustainable Energy Reviews 2008; 12(9): 2404-21.

[85] Malaysian Timber Council (MTC). Kuala Lumpur, Malaysia (1992-2004). <http://www.mtc.com.my/news/pr33.htm>, (accessed 7 January 2009).

[86] Wood Based Industry. Report by Malaysia Industrial Development Authority (MIDA). http://www.mida.gov.my/en_v2/index.php?page=wood-based-industry; (accessed 7 April 2011).

[87] MIDA e-News: Agro-Bio Fibre expands production, ties up with FRIM on HDF. Report by Malaysia Industrial Development Authority (MIDA). <http://www.mida.gov.my/en_v2/index.php?mact=News,cntnt01,detail,0&cntnt01articleid=234&cntnt01returnid=388>, (accessed 7 April 2011).

[88] Ministry of Plantation Industries and Commodities. Speech by Y.B. Dato' Seri Lim Keng Yaik in the Opening Ceremony of the Seminar on Oil Palm Biomass: Opportunities for Commercialization (2001). <www.kppk.gov.my/index .php?option=com_content&task=view&id=86&Itemid=46-53k>, (accessed 9 September 2009).

[89] Shuit SH, Tan KT, Lee KT, Kamaruddin AH. Oil Palm Biomass as a Sustainable Energy Source: Malaysian Case Study, Energy 2009; 34(9): 1225-35.

[90] Hassan OA, Ishida M, Shukri Mohd I, Tajuddin ZA. Oil Palm Fronds as a Roughage Feed Source For Ruminants in Malaysia FFTC for the Asia and Pacific Region 1994. <http://www.agnet.org/library/eb/420/;1994>, (accessed 6 September 2009).

[91] 'Oil Palm Frond Based Ruminant Pellet'. (2008). Report by MARDI, <http://www.mardi.my/main.php?Content=home&FolderID=157&CurLocation=131>, (accessed 7 September 2009).

[92] Zainudin NF, Lee KT, Kamaruddin AH, Bhatia S, Mohamed AR. Study of Adsorbent Prepared From Oil Palm Ash (OPA) for Flue Gas Desulphurization. Separation and Purification Technology 2005; 45:50-60.

[93] Mohamed AR, Zainudin NF, Lee KT, Kamaruddin AH. Reactivity of Adsorbent Prepared From Oil Palm Ash (OPA) for Flue Gas Desulphurization : Effect of SO_2 Concentration and Reaction Temperature. Studies in Surface Science and Catalysis 2006; 159:449-52.

[94] Nomanbhay SM, Palanisamy K. Removal of Heavy Metal from Industrial Wastewater Using Chitosan Coated Oil Palm Shell Charcoal. Electronic Journal of Biotechnology 2005; 8(April 15):44-53.

[95] Nasrin AB, Ma AN, Choo YM, Mohamad S, Rohaya MH, Azali A, et al. Oil Palm Biomass as Potential Substitution Raw Materials for Commercial Biomass Briquettes Production. American Journal of Applied Science 2008; 5(3):179–83.

[96] Hoi Why Kong. Wood and Charcoal Briquetting in Malaysia. In: Proceedings of International Workshop on Biomass Briquetting, (eds P.D.Grover and S.K Mishra). 1995. pp 68-74. FAO Regional Wood Energy Development Programme. 1995.

[97] Ahmad M, Wan Daud W, Aroua M. Adsorption Kinetics of Various Gases In Carbon Molecular Sieves (CMS) Produced From Palm Shell. Colloids and Surfaces A 2008; 312:131-5.

[98] Tan J. Ani F, Carbon Molecular Sieves Produced From Oil Palm Shell for Air Separation. Separation and Purification Technology 2004; 35: 47-54.

[99] Ahmad M, Wan Daud W. Aroua M. Synthesis of Carbon Molecular Sieves from Palm Shell by Carbon Vapor Deposition. Journal of Porous Matter 2007; 14: 393-9.

[100] Adinata D, Wan Daud W, Aroua M. Production of Carbon Molecular Sieves From Palm Shell Based Activated Carbon by Pore Sizes Modification With Benzene For Methane Selective Separation. Fuel Process Technology 2007; 88: 599-605.

[101] Daud W, Ahmad M, Aroua M. Carbon Molecular Sieves from Palm Shell: Effect of the Benzene Deposition Times On Gas Separation Properties. Separation and Purification Technology 2007;57: 289-93.

[102] Wikipedia. The free encyclopedia: Activated Carbon. <http://en.wikipedia.org/wiki/Activated_carbon#cite_note-0>, (accessed 3 April 2011).

[103] Hameed BH, Tan IAW, Ahmad AL. Preparation of Oil Palm Empty Fruit Bunch-Based Activated Carbon For Removal Of 2,4,6-Trichlorophenol: Optimization using Response Surface Methodology. Journal of Hazardous Materials 2009, 164; 2-3:1316-24.

[104] Bernama. Biochar Malaysia: UPM-NASMECH Effect of Producing EFB Biochar : World's First. 7 January 2010. <http://biocharmalaysia.blogspot.com/2010/01/upm-nasmech-effort-of-producing-efb.html>, (accessed 4 April 2011).

[105] Keck Seng (M) Berhad. Biogas utilization. 2001. <http://www.biogen.org.my/bris/BioGen/Tech/(b)Project/technology(b)2.pdf>, (accessed 4 April 2011).

[106] Ludin NA, MBakri MA, Hashim M, Sawilla B, Menon NR, Mokhtar H. Palm Oil Biomass For Electricity Generation In Malaysia. Pusat Tenaga Malaysia, Malaysia Palm Oil Board, SIRIM Berhad. 2004. <http://www.biogen.org.my/bris/BioGen/Tech/(d)Documents/technology(d)7.pdf>, (accessed 5 April 2011).

[107] Dit M. Palm Kernel Shell (PKS) Is More than Biomass for Alternative Fuel after 2005. Proceedings of Chemistry and Technology Conference. 2007. <cdm.unfccc.int/Projects/DB/DNV-CUK1182238337.59/ReviewInitialComments/6MHGPZIM56I19DT6NGL30G9OGNTIOK>, (accessed 3 April 2011).

[108] Barbir F. Transition to Renewable Energy Systems with Hydrogen as an Energy Carrier. Energy 2009;34:308–12.

[109] Azali A, Nasrin AB, Choo YM, Adam NM, Sapuan SM. Development of Gasification System Fuelled With Oil Palm Fibres and Shells. American Journal of Applied Science 2005:72–5.

[110] Kelly-Yong TL, Lee KT, Mohamed AR, Bhatia S. Potential of Hydrogen From Oil Palm Biomass as a Source Of Renewable Energy Worldwide. Energy Policy 2007; 35:5692–701.

[111] Neef HJ. International Overview of Hydrogen and Fuel Cell Research. Energy 2009; 34: 327–33.

Chapter 4

Biosorption of Lanthanides Using Select Marine Biomass

Naoki Kano

Additional information is available at the end of the chapter

http://dx.doi.org/10.5772/51164

1. Introduction

Contamination of toxic metals in the aquatic environment is one of the most debated problems in the world with industrial development. Thus, the minimization and recovery of harmful pollutants such as heavy metals in natural environment is very significant [1]. Various treatment technologies such as ion exchange, precipitation, ultrafiltration, reverse osmosis and electrodialysis have been used for the removal of heavy metal ions from aqueous solution [2]. However, these processes have some disadvantages, such as high consumption of reagent and energy, low selectivity, high operational cost, and difficult further treatment due to generation of toxic sludge [3].

Among environmentally friendly technologies for the removal of heavy metals from aquatic effluent, biosorption has attracted increasing research interest recently [4-5]. The major advantages of biosorption are its high effectiveness in reducing the heavy metals and the use of inexpensive biosorbents [6]. Biosorption studies using various low cost biomass as adsorbents have been currently performed widely for the removal of heavy metals from aquatic effluent [7-18].

Among many biosorbents, marine seaweed can be an excellent biosorbent for metals because it is well known to concentrate metals [19-20]. Seaweeds are reported to accumulate hydrocarbons (as well as metals); and they are exposed to the ubiquitous presence of organic micropollutants and can work as suitable biomonitors [21]. Furthermore, it is considered that the shell (usually treated as waste material) can be also an promising adsorbent. The shell has an internal structure comprised of three distinct layers. The innermost layer (i.e., hypostracum) consists of aragonite; the middle layer (i.e., ostracum), which is the thickest of the three, consists of various orientations interbedded with protein molecules (conchiolin); and the outermost layer (i.e., periostracum) consists of chitin, which is represented as $(C_8H_{13}NO_5)_n$ [22]. Particularly, it is considered that protein (called "conchiolin") including amino acid group play an important role for collecting trace metal in shell [23].

Biosorption studies have been mainly focused on toxic elements such as Cd, Pb, Cu, As and Cr for subject elements [24]. In our research, the objective elements are mainly rare earth elements (REEs) from the viewpoint of resources recovery, although REEs do not represent a common toxic threat.

Rare earth elements (REEs) find wide range of applications as functional materials in agriculture and as other industrial products, then the demand of REEs in modern technology has increased remarkably over the past years [25-26]. These elements and their compounds have found a variety of applications especially in metallurgy, ceramic industry and nuclear fuel control [27]. For example, current applications of lanthanum as a pure element or in association with other compounds are in super alloys, catalysts, special ceramics, and in organic synthesis [28]. However, the shortage of trace metals including REEs (and the problem of stable supply for these metals) has been a concern in recent years. Therefore, the establishment of the removal or recovery method for trace metals is important from the viewpoint of resources recovery.

It is known that alginate is an exopolymer extracted mainly from brown algae (and various bacteria) that has been used both as immobilization material and as biosorbent of several heavy metals [29]. Then, biosorption studies using seaweed have been generally concentrated on brown algae so far [30-31]. Green and red algae as well as brown algae were also used for biosorbent of REEs in the present work.

Considering the above-mentioned, laboratory model experiments for confirming the efficiency of marine biomass (seaweed and shell) as sorbent for REEs was designed in present work. Furthermore, the surface morphology of the marine biomass used in this work was determined by SEM (Scanning Electron Microscope) before and after metal adsorption.

The crystal structure, and the specific surface area of the shell biomass were also determined by by XRD (X-ray powder diffraction), and BET (Brunaeur, Emmet and Teller) and Langmuir method, respectively.

2. Experimental work

2.1. Samples

The seaweed biomass

Many kinds of seaweeds samples (10 species of green algae, 21 species of brown algae and 21 species of red algae) were taken along several coasts in Niigata Prefecture (referred to the figure in our previous paper [32]) since April, 2004. Among seaweed species, the seaweeds for biosorbent used in this work were *Sargassum hemiphyllum* (brown algae), *Ulva pertusa* (green algae) and *Schizymenia dubyi* (red algae). Each seaweed sample was washed in the surrounding seawater to remove attachment at sampling place. After transport back to the laboratory, the seaweed was first washed with tap water and ultrapure water thoroughly and then air-dried for 2-3 days. Afterwards, it was dried overnight in an electric drying oven (Advantec DRA 430DA) at maximum temperature of 55 °C to avoid degradation of the binding sites, the biomass was ground. Sizes of biomass ranging from 0.5 mm to 1 mm were obtained by passing through sieves (SANPO Test Sieves).

Based on Diniz and Volesky's study [31], each sieved biomass sample was loaded with Ca^{2+} in a solution of 50 mmol • dm^{-3} $Ca(NO_3)_2$ (biomass concentration of 10 g • dm^{-3}) for 24 h under gentle agitation in order to remove the original cations on seaweed. Later, the biomass was washed with ultrapure water to remove excess Ca^{2+} until the mixture was reached approximately pH 5. Finally, the washed biomass was dried again overnight at 50°C in an electric drying oven, and stored in desiccators (containing silica gel as a desiccant) before use.

The shell biomass

Buccinum tenuissimum shellfish used for shell biomass were collected at fishermen's cooperative association. After being separated from the meat by boiling, organism shells were washed thoroughly with ultra-pure water after washed with tap water repeatedly. After drying, the shells were ground and sieved through a sieve (SANPO Test Sieves) to remove particles having size more than 500μm. Sieved material was used for adsorption experiments. Afterwards, a part of this sieved materials was heated for 6 h at 480°C or 950°C in an electric furnace (ISUZU Muffle Furnace STR-14K, Japan). Moreover, adequate ultrapure water was added to a part of heat-treatment (950°C, 6h) samples, and heated at 100 °C on a hotplate for evaporation to near dryness (removing water), and finally dried in an electric drying oven at 60 °C.

2.2. Sorption experiment for lanthanides using seaweed and shell biomass

The following sorption experiments were performed using the above-mentioned marine biomass. Experimental conditions (i.e., pH, contact time and biosorbent dose rate) in this work were optimized and determined based on our preliminary experiments [e.g., 21] and other literatures [24, 31]. The pH of each solution was adjusted by using 0.1 mol·dm^{-3} NH_3aq / 0.1 mol·dm^{-3} HNO_3.

The seaweed biomass

Samples of 0.4 g of the biomass were contacted with 200 cm^3 of solution containing known initial each lanthanide (La, Eu or Yb) concentration ranging from 0.1 to 4 mmol • dm^{-3}. Afterwards, the suspensions were shaken for 24 h in a water bath at ambient temperature (~25 °C) at pH 4.

The shell biomass

Each sample of 0.2 g was contacted with 100 cm^3 of multi-element standard solution (prepared by XSTC-1) including known initial lanthanide concentration (10 to 500 μg·dm^{-3})

in a 200 ml conical flask. Afterwards, the suspensions were shaken for 30 min in a water bath at room temperature at pH 5.

Following with each sorption experiment, the suspension containing biomass and lanthanides standard solution was filtered through a 0.10 μm membrane filter (Advantec Mixed Cellulose Ester, 47mm) to remove lanthanides that have been adsorbed into the biomass, and the concentration of these metals in the filtrate was determined with ICP-MS or ICP-AES.

The metal uptake by the marine biomass was calculated using the following mass balance equation [33]:

$$q = (C_i - C_f)V / W \ [mg \times g^{-1}] \tag{1}$$

where q = metal uptake (mmol·g^{-1}); C_i = initial metal concentration (mmol·dm^{-3}); C_f = equilibrium metal concentration (mmol·dm^{-3}); V = volume of the solution (dm^3); and W = dry mass of seaweed (g).

The removal efficiency (RE, %) of the biosorbent on the metal in the solution was determined by the following equation [24]:

$$RE = (C_i - C_f) \times 100 \ / C_i \tag{2}$$

2.3. Langmuir and Freundlich isotherm model

Langmuir adsorption isotherm model was applied based on Tsui et al. [24] in this study, and the model assumes monolayer sorption onto a surface and is given as below.

$$q = (q_e \times C_f) \ / (A^{-1} + C_f) \tag{3}$$

where q_e = maximum metal uptake (mmol·g^{-1}) (i.e., the maximum attainable binding capacity); and A = affinity constant (1 mmol^{-1}) (i.e., the affinity of the metal ion toward the biomass).

The Freundlich equation is widely used in the field of environmental engineering, and was applied based on based Dahiya et al. [10-11]. Freundlich isotherm can also be used to explain adsorption phenomenon as given below.

$$log_{10} q_e = log_{10} K_F + (1/n) log_{10} C_f \tag{4}$$

where K_F and n are constants incorporating all factors affecting the adsorption capacity and an indication of the favorability of metal ion adsorption onto biosorbent, respectively. It is shown that $1/n$ values between 0.1 and 1.0 correspond to beneficial adsorption. That is, q_e versus C_f in log scale can be plotted to determine values of $1/n$ and K_F.

3. Results and discussion

3.1. The seaweed samples

3.1.1. Metal sorption capacity at different species of seaweed

The equilibrium sorption isotherms of La, Eu and Yb by three kinds of Ca-loaded seaweed biomass are shown in Fig. 1. Sorption experiment of Eu using *U.p.* could not be conducted in this work due to the lack of sample. The adsorption data obtained in this work were analyzed using Langmuir and Freundlich equations. The correlation coefficients (R^2) of Langmuir and Freundlich isotherms for La, Eu and Yb using three kinds of seaweed biomass are shown in Table 1 along with other parameters.

From this table, it is found that R^2 value for each datum is comparatively large for both isotherms. The value of $1/n$ less than unity indicates better adsorption and formation of relatively stronger bonds between adsorbent and adsorbate [10]. That is to say, favorable adsorption for La, Eu and Yb by these seaweed biomass used in this work is presented. Furthermore, it is noted that R^2 values for these data are particularly large for Langmuir isotherm than for Freundlich isotherm. This result suggests that the adsorption on these samples mainly occurred by monolayer reaction. Therefore, the curves obtained from non-linear regression of the data by Langmuir isotherm are also shown in Fig. 1.

	Sargassum hemiphyllum			*Schizymenia dubyi*			*Ulva pertusa*		
	La	Eu	Yb	La	Eu	Yb	La	Eu	Yb
Langmuir									
q_e / mmol g^{-1}	0.700	0.781	0.769	0.651	0.980	0.926	0.930	---	0.719
A / mmol^{-1}	5.19	4.33	3.03	2.53	1.05	1.08	6.03	---	1.59
R^2	0.992	0.987	0.999	0.982	0.969	0.983	0.996	---	0.991
Freundlich									
K_F / g^{-1}	0.575	0.648	0.561	0.450	0.475	0.443	0.816	---	0.409
$1/n$	0.440	0.462	0.500	0.531	0.683	0.646	0.461	---	0.595
R^2	0.806	0.731	0.937	0.839	0.969	0.923	0.878	---	0.923

Table 1. Langmuir and Freundlich parameters for biosorption of lanthanides on three kinds of Ca-loaded biomass at pH 4.0 (--- represents the missing data due to the lack of sample)

From Table 1, it is found that the data of q_e for Eu and Yb by *Sargassum hemiphyllu* (brown algae) obtained in this study are similar to the result of species of *Sargassum* conducted by Diniz and Volesky [31], although q_e for La is slightly small. Moreover, it is noteworthy that both Langmuir parameters: q_e and A for La by *Ulva pertusa* (green algae) is large. Then, the comparison of sorption isotherms of La among three kinds of seaweed biomasses is shown in Fig. 2. From this figure, the sorption capacity of La by *U. p.* (green algae) is considerably large compared to that by other algae: *S. h.* (brown algae) and *S. d.* (red algae). It is generally known that brown seaweed is superior to red and green seaweed in metal sorption capacity for heavy metals such as Cd, Pb, Cu [34-35]; and such a high value

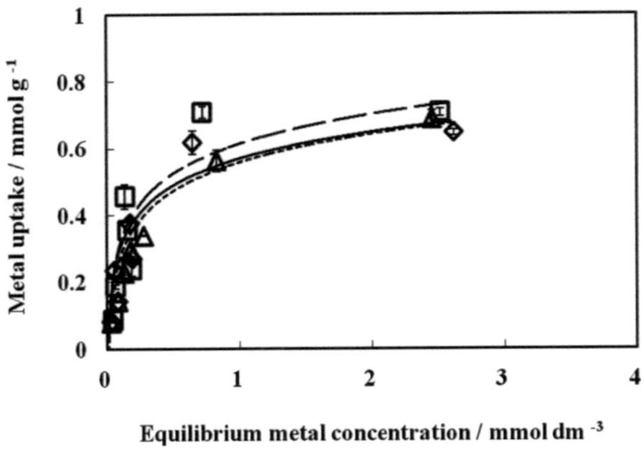

(a)

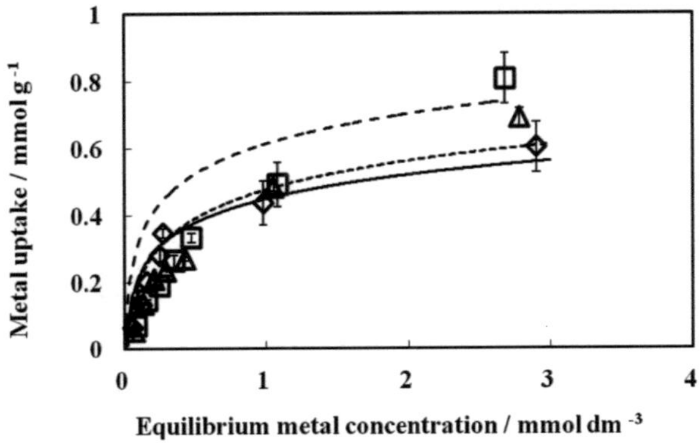

(b)

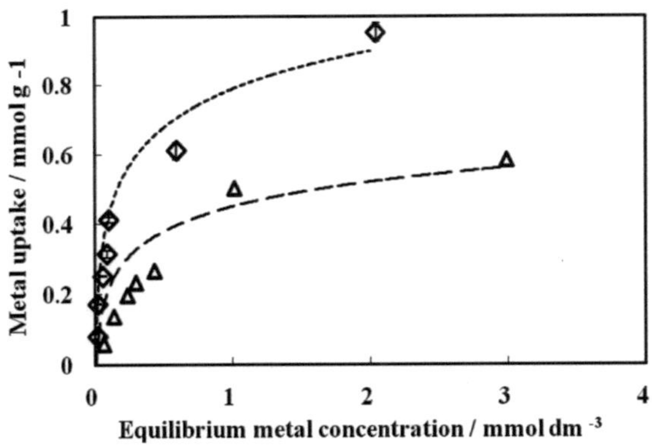

(c)

Figure 1. Sorption isotherms of La (◇), Eu (□) and Yb (△) by three kinds of Ca-loaded seaweed biomass at pH 4.0, (a): *Sargassum hemiphyllum*, (b): *Schizymenia duby* and (c): *Ulva pertusa*. The curves obtained from non-linear regression of the data by Langmuir isotherm are also shown (La: solid curve, Eu: broken curve and Yb: dotted curve). Data are mean±standard deviation (n=3).

of q_e (as observed in this work) using green algae have not been reported so far. In other words, it is significant outcome to find that *Ulva pertusa* (green algae) can be a promising biosorbent for removing La.

According to our previous work [32], in case of U, the mean concentration is the highest in brown algae and is the lowest in green algae among phyla (i.e., green, red and brown algae). However, as for the mean concentration of light REE (LREE) such as La, a slightly higher concentration is found in green algae; whereas the concentration of heavy REE (HREE) such as Yb or Lu in green algae is smaller than that in brown algae (as shown in the figure in our previous paper [32]).

Then, large sorption capacity of La by *Ulva pertusa* may be related to the character (or the constituent) of green algae; and it is possible that "La adsorption on *Ulva pertusa*" is due to "metal-specific", although further studies is needed to confirm the peculiarity by investigating the other combinations of metals and seaweed species.

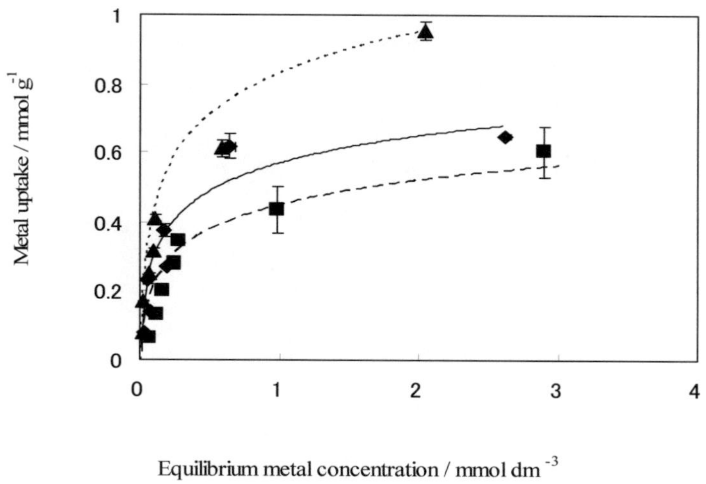

◆ : *Sargassum hemiphyllum*, ■ : *Schizymenia duby*, ▲ *Ulva pertusa*. The curves obtained from non-linear regression of the data by Langmuir isotherm are also shown (*S. h.*; solid curve, *S. d.*; broken curve and *U. p.*; dotted curve). Data are mean±standard deviation (n=3).

Figure 2. Comparison of sorption isotherms for La among three kinds of Ca-loaded seaweed biomass at pH 4.0

SEM pictures of three kinds of Ca-loaded seaweed biomass before and after adsorption of lanthanum are shown in Fig. 3 and Fig. 4, respectively.

According to SEM observation, the surface of *S. d.* (red algae) seems to be relatively flat, whereas *S. h.* (brown algae) and *U. p.* (green algae) have more extensive surface area, although the specific surface area of the seaweed biomass could not be measured due to their small specific surface area. Furthermore, by comparing SEM pictures in Fig. 3 with that in Fig. 4, it is found that the morphology of *S. h.* and *U. p.* surface has hardly changed even after exposing lanthanum. On the other hand, the distinct change of the surface morphology on *S. d.* was observed after adsorption of metals.

(a)

(b)

(c)

Figure 3. SEM pictures of seaweed biomass before adsorption of lanthanum. (a) : *Sargassum hemiphyllum*. (b) *Schizymenia duby* and (c) : *Ulva pertusa*.

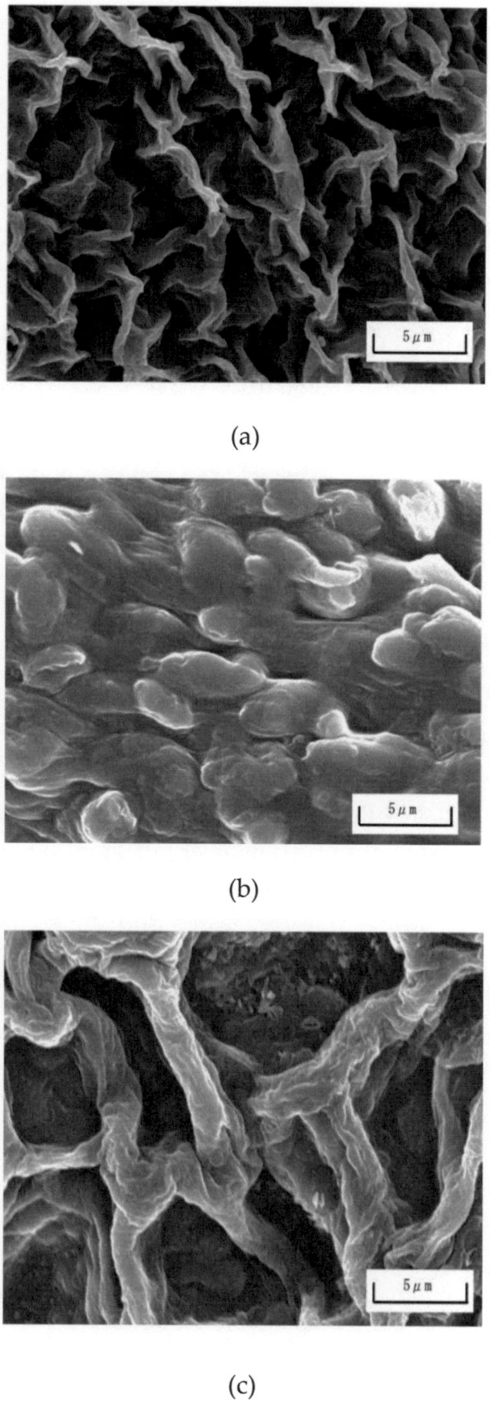

Figure 4. SEM pictures of seaweed biomass after adsorption of lanthanum. (a) : *Sargassum hemiphyllum*. (b) *Schizymenia duby* and (c) : *Ulva pertusa*.

From the above observation, two kinds of biomass: *Sargassum hemiphyllum* and *Ulva pertusa* should be predicted to withstand the repeated use; and hence it can be a good adsorbent for lanthanides.

3.1.2. Removal efficiency and binding mechanism of seaweed biomass

The removal efficiency (RE) of 3 kinds of seaweed biomass as a function of initial metal concentrations (*Ci*) for 3 lanthanides is shown in Fig. 5((a): La, (b): Eu, (c): Yb). With increasing *Ci*, the RE generally decreased exponentially; and at high *Ci*, similar RE (i.e., about 40%) occurred for each lanthanide even with any biomass. These data are well fitted into an exponential function (R^2 ranging from 0.866 to 0.994) shown in Fig. 5; and the equations and R^2 for each lanthanide in each biomass are shown in Table 2.

From the viewpoint of recovering trace metals from aqueous environment such as seawater, the removal efficiency at low concentration of metal is particularly important. The coefficient before exponential function in each equation in Table 2 represents the value of RE at low *Ci* near approximately zero $mmol \cdot dm^{-3}$. From Table 3, the coefficient for each lanthanide in *Sargassum hemiphyllum* and especially that for La in *Ulva pertusa* is large. This implies that *U. p.* could be an efficient adsorbent for La as well as *S. h.* for lanthanides in aqueous environment such as seawater.

The amount (mmol·g) of adsorbed lanthanide and released Ca from three kinds of Ca-loaded seaweed biomass is shown in Tables 3-5. Based on the data in these tables, relationship between the uptake of each lanthanide ion and calcium ion released from each biomass is shown both in terms of mill equivalent per gram ($meq \cdot g^{-1}$) in Fig. 6. Good and linear relationship is generally found for these samples between the uptake of each lanthanide and Ca released from these biomasses into the solution as shown in Fig. 6. Particularly, in case of *S. h.* and *U. p.*, the slope of the line is about one with the y-intercept of the graphs almost passes through the origin. It indicates that ion-exchange process is found to be the main mechanism responsible for the sorption of lanthanide ion onto the seaweed as Tsui et al. [24] and Diniz & Volesky [31] also pointed out.

	Sargassum hemiphyllum		*Schizymenia dubyi*		*Ulva pertusa*	
	Equation	R^2	Equation	R^2	Equation	R^2
La	RE=91.4exp(-0.239*Ci*)	0.939	RE=78.7exp(-0.241*Ci*)	0.939	RE=98.5exp(-0.191*Ci*)	0.993
Eu	RE=94.3exp(-0.223*Ci*)	0.866	RE=66.2exp(-0.136*Ci*)	0.973	---------------	---
Yb	RE=88.9exp(-0.234*Ci*)	0.994	RE=68.2exp(-0.173*Ci*)	0.944	RE=70.3exp(-0.213*Ci*)	0.975

--- represents the defect of data due to the lack of sample

Table 2. Equations and correlation coefficients (R^2) to describe the relationships between removal efficiency (RE) and initial concentrations (*Ci*) of different lanthanides in the sorption system

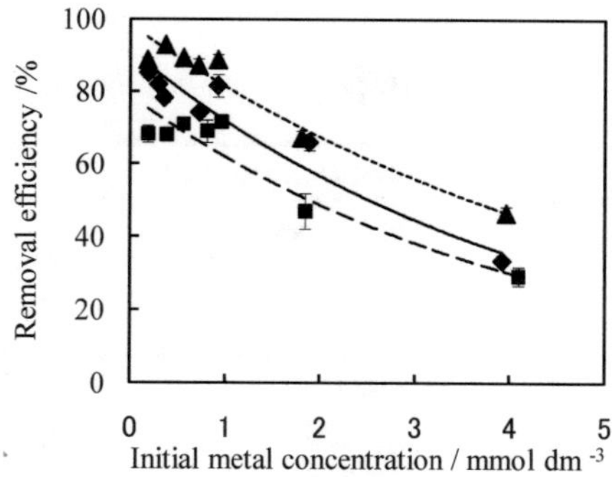

(a)

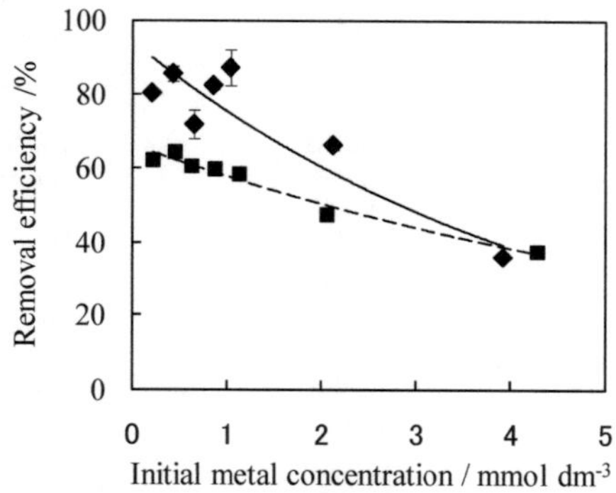

(b)

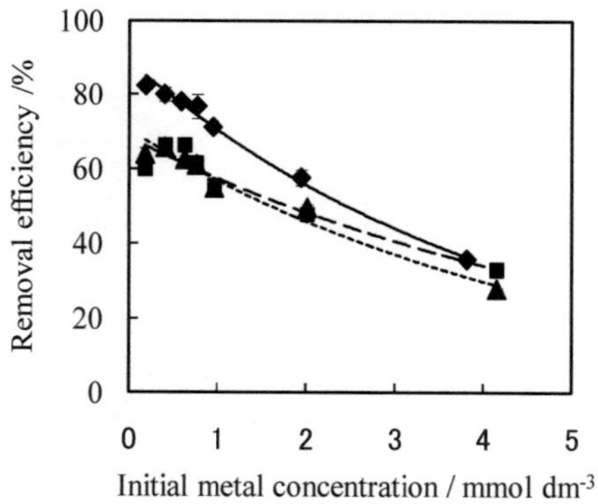

(c)

Figure 5. Removal efficiency of lanthanides ((a) :La, (b): Eu, (c) :Yb) by Ca-loaded seaweed biomass at different initial concentrations,◆: *Sargassum hemiphyllum*,■ : *Schizymenia duby*, ▲:*Ulva pertusa*. Each exponential function is also shown (*S. h.*; solid curve, *S. d.* : broken curve and *U. p.* : dotted curve). Data are mean±standard deviation (n=3).

3.2. The shell samples

3.2.1. Characteristics of Buccinum tenuissimum shell biomass

	Sargassum hemiphyllum			
	Adsorbed lanthanide / mmol g^{-1}	Total Released Ca / mmol g^{-1}	Ca excess (blank) / mmol g^{-1}	Net Released Ca / mmol g^{-1}
La	0.081	0.208	0.082	0.126
	0.121	0.226	0.082	0.144
	0.150	0.276	0.082	0.194
	0.153	0.305	0.082	0.224
	0.234	0.507	0.082	0.426
	0.375	0.595	0.082	0.514
	0.617	0.995	0.082	0.913
Eu	0.085	0.155	0.080	0.075
	0.186	0.298	0.080	0.218
	0.237	0.377	0.080	0.297
	0.355	0.576	0.080	0.496
	0.456	0.791	0.080	0.711
	0.707	1.021	0.080	0.941
	0.708	1.507	0.080	1.427
Yb	0.079	0.194	0.077	0.117
	0.160	0.288	0.077	0.211
	0.227	0.403	0.077	0.326
	0.296	0.494	0.077	0.417
	0.339	0.579	0.077	0.502
	0.563	0.838	0.077	0.761
	0.688	0.958	0.077	0.881

Table 3. Amount of adsorbed lanthanide and released Ca by Ca-loaded *Sargassum hemiphyllum* biomass

X-ray powder diffraction (XRD) patterns of the four kinds of *Buccinum tenuissimum* shell biomass samples are shown in Fig. 7. The crystal structure of the shell biomass was transformed from aragonite (CaCO3) into calcite (CaCO3) phase by heat-treatment (480°C, 6h). Moreover, the crystal structure of the shell biomass was mainly transformed into calcium oxide (CaO) by heat-treatment (950°C, 6h); and was mainly into calcium hydroxide ($Ca(OH)_2$) by adding water after heat-treatment (950°C, 6h). SEM pictures of the four kinds of sieved shell biomass samples are shown in Fig. 8. Comparing Fig. 8(b) with Fig. 8(a), comparatively clear crystal with a lot of big particles may be observed by heat-traetment (480°C, 6h). It is suggested that ground original sample contains a lot of organic materials such as protein, and most of organic matter seem to disappear by heat-traetment. Moreover, fine crystal particle is not observed in Fig. 8 (c). This may be attributable to the phenomena that many crystals were connected largely with each other due to high-temperature sintering. Meanwhile, relative clear crystal (sizes are mostly 1.0-4.0μm) is observed in Fig. 8 (d).

Schizymenia dubyi				
	Adsorbed lanthanide / mmol g^{-1}	Total Released Ca/ mmol g^{-1}	Ca excess (blank) / mmol g^{-1}	Net Released Ca / mmol g^{-1}
La	0.053	0.376	0.240	0.136
	0.132	0.469	0.240	0.229
	0.201	0.571	0.240	0.331
	0.281	0.657	0.240	0.417
	0.346	0.747	0.240	0.507
	0.436	0.757	0.240	0.517
	0.603	0.835	0.240	0.595
Eu	0.070	0.655	0.240	0.415
	0.146	0.778	0.240	0.538
	0.192	0.862	0.240	0.622
	0.263	1.025	0.240	0.785
	0.333	1.176	0.240	0.936
	0.493	2.102	0.240	1.862
	0.807	2.146	0.240	1.906
Yb	0.055	0.557	0.240	0.317
	0.136	0.594	0.240	0.354
	0.210	0.694	0.240	0.454
	0.236	0.796	0.240	0.556
	0.270	0.771	0.240	0.531
	0.483	0.950	0.240	0.710
	0.691	1.036	0.240	0.796

Table 4. Amount of adsorbed lanthanide and released Ca by Ca-loaded *Schizymenia dubyi* biomass

Ulva pertusa				
	Adsorbed lanthanide / mmol g^{-1}	Total Released Ca/ mmol g^{-1}	Ca excess (blank) / mmol g^{-1}	Net Released Ca / mmol g^{-1}
La	0.079	0.259	0.160	0.099
	0.172	0.314	0.160	0.154
	0.251	0.415	0.160	0.255
	0.315	0.658	0.160	0.498
	0.412	0.776	0.160	0.616
	0.610	1.546	0.160	1.386
	0.929	1.672	0.160	1.512
Yb	0.059	0.263	0.160	0.103
	0.134	0.349	0.160	0.189
	0.198	0.449	0.160	0.289
	0.234	0.644	0.160	0.484
	0.266	0.704	0.160	0.544
	0.503	1.132	0.160	0.972
	0.584	1.421	0.160	1.261

Table 5. Amount of adsorbed lanthanide and released Ca by Ca-loaded *Ulva pertusa* biomass

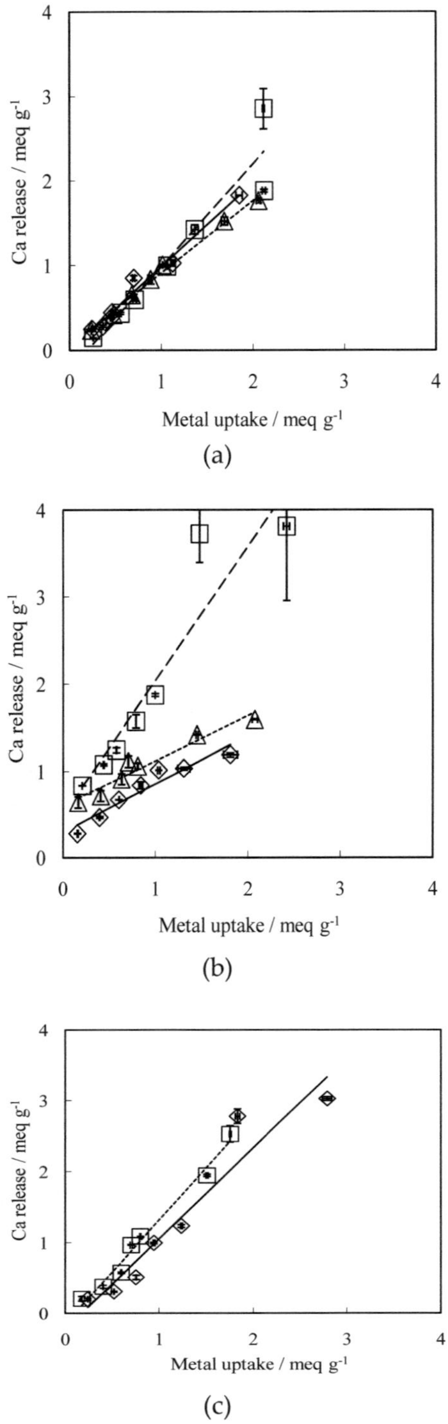

Figure 6. Relationship between metal uptake and Ca release from Ca-loaded seaweed biomass, (a): *Sargassum hemiphyllum*, (b): *Schizymenia duby* and (c):*Ulva pertusa*. ◇:La,□:Eu,△:Yb. Each regression line is also shown (La: solid line, Eu: broken line and Yb: dotted line). Data are mean±standard deviation (n=3).

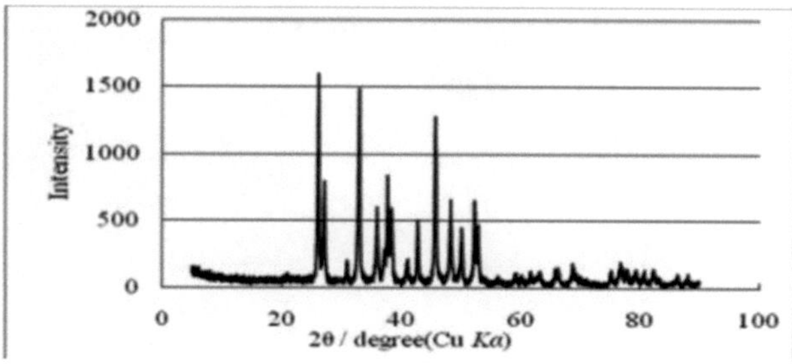

(a)

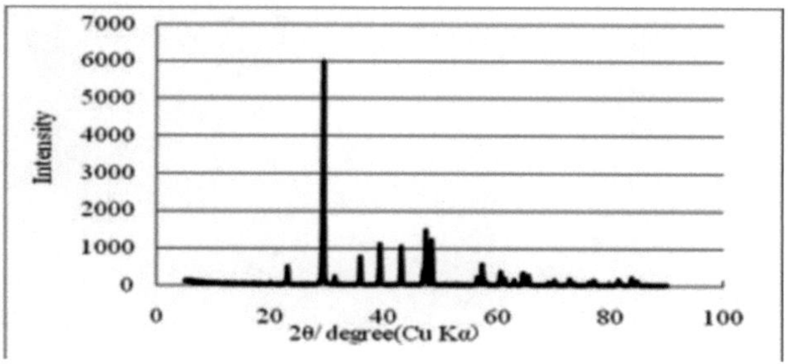

(b)

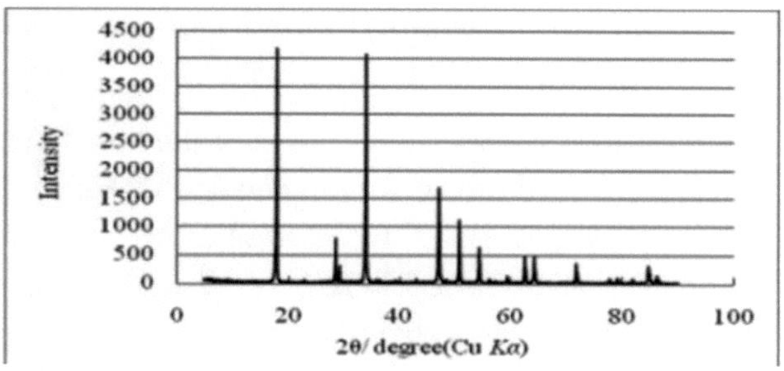

(c)

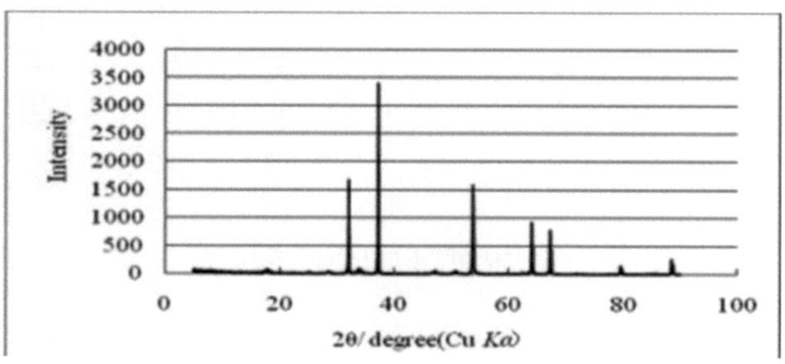

(d)

Figure 7. X-ray diffraction (XRD) patterns of *Buccinum tenussimum* shell biomass before adsorption of metals. (a) ground original sample, (b) heat-treatment (480 °C) sample, (c) heat-treatment (950 °C) sample, (d) heat-treatment (950 °C) and water added sample.

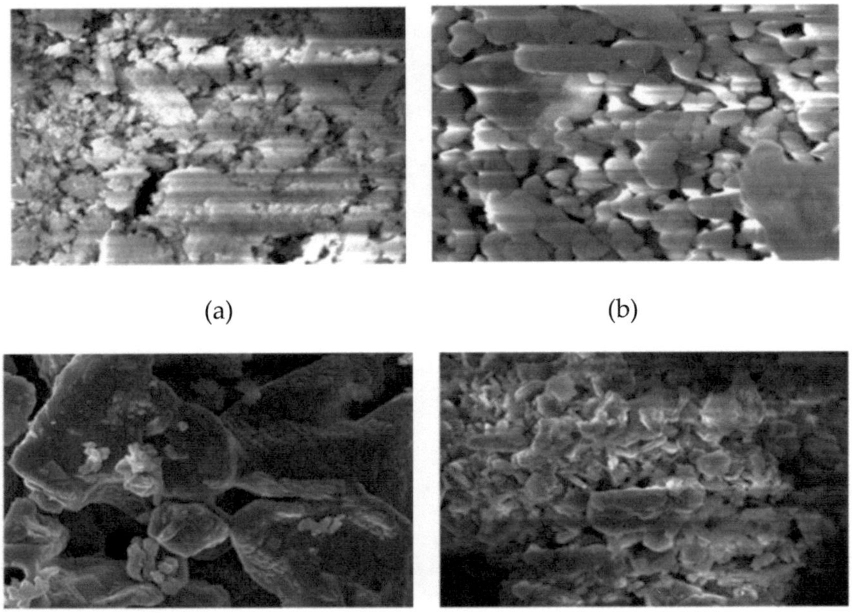

(a) (b) (c) (d)

Figure 8. SEM pictures of *Buccinum tenussimum* shell biomass before adsorption of metals. (a) ground original sample, (b) heat-treatment (480°C) sample, (c) heat-treatment (950°C) sample, (d) heat-treatment (950°C) and water added sample

Furthermore, the measurement of specific surface area of the four kinds of sieved samples was performed in this study; and the results are shown in Table 6 along with the main crystal structure of these samples. Remarkably decrease of specific surface area (i.e., from 3.32m^2/g to 0.390m^2/g for BET, or from 5.35 m^2/g to 0.612 m^2/g for Langmuir) was found after heat-treatment (480°C, 6h). It is suggested that the crystal structure transformation (i.e., from aragonite ($CaCO_3$) into calcite ($CaCO_3$) phase) and also the difference of the surface morphology can be closely related to the remarkable decrease of specific surface area of the shell biomass. On the other hand, the surface area of "heat-treatment (950°C, 6h) sample" was 1.88m^2/g for BET or 3.10m^2/g for Langmuir respectively; and that of "heat-treatment (950°C, 6h) and water added sample" was 6.37m^2/g for BET or 9.91m^2/g for Langmuir, respectively.

3.2.2. Comparison for sorption capacity of lanthanides by four kinds of sieved biomass

The comparison for sorption capacity of lanthanides by four kinds of sieved *Buccinum tenuissimum* shell samples is shown in Fig. 9. In this experiment, the initial lanthanides concentration was taken as 100μg·dm^{-3}. From this figure, it is found that all kinds of sieved samples showed excellent sorption capacity under this experimental condition. However, the sorption capacity in sample (b) (i.e., the main phase is calcite) decreases slightly relative to that of the original material (i.e., (a): the main phase is aragonite) and others. The decrease

Sample	Main crystal structure	Specific surface area
(a)Ground original sample	Aragonite ($CaCO_3$)	3.31m²/g (BET) 5.35m²/g (Langmuir)
(b) Heat-treatment (480°C) sample	Calcite ($CaCO_3$)	0.390m²/g (BET) 0.612m²/g (Langmuir)
(c) Heat-treatment (950°C) sample	Lime syn. (CaO)	1.88m²/g (BET) 3.10m²/g (Langmuir)
(d) Heat-treatment (950°C) and water added sample	Portlandite ($Ca(OH)_2$)	6.37m² /g (BET) 9.91m²/g (Langmuir)

Table 6. The crystal structures and the specific surface areas of four kinds of sieved *Buccinum tenuissimum* shell biomass

of sorption capacity in sample (b) may be attributable to the remarkable decrease (i.e., by a factor of less than one eighth) of specific surface area of the biomass.

Prieto et al. [36] pointed that the sorption capacity of calcite is considerably lower than that of aragonite for Cd. In case of lanthanides, similar tendency of sorption capacity were suggested from our work.

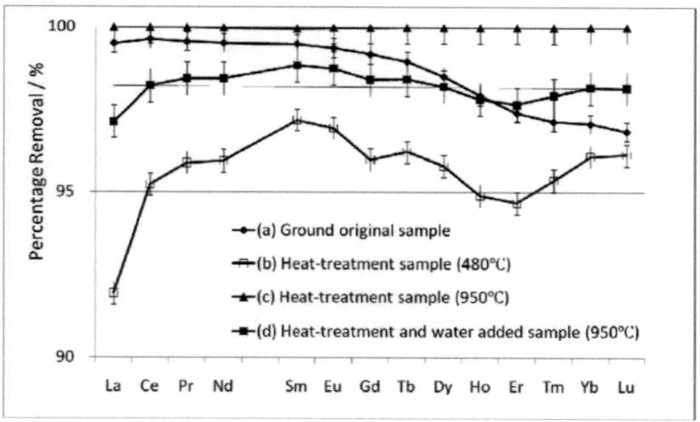

Figure 9. Comparison for sorption capacity of lanthanides by four kinds of sieved *Buccinum tenuissimum* shell samples.

3.2.3. Effect of competitive ions on the sorption of lanthanides

The percentage removal of lanthanides under the presence of common ions (Ca^{2+}, Mg^{2+}, Na^+ and K^+) at different concentrations 50, 100 and 200 mg·dm^{-3} is shown in Fig. 10. From this figure, the remarkable decrease of sorption capacity of lanthanides was not observed. Even when the concentrations of common ions are 200 mg·dm^{-3}, the percentage removal of light REE (LREE) such as La or Ce decreased slightly (2-3%), whereas the removal decreased about 5% for heavy REE (HREE) such as Yb or Lu. This implies that the shell biomass can be an efficient adsorbent for lanthanides in aqueous environment such as seawater, although it requires further investigations to apply the shell biomass to use as an adsorbent for lanthanides more practically.

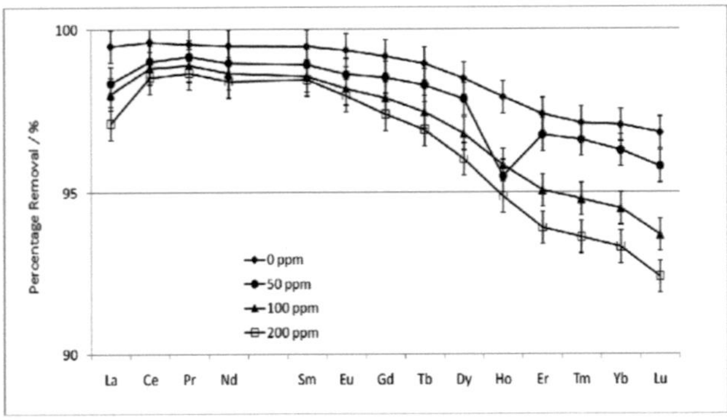

Figure 10. Effect of common ions (Ca^{2+}, Mg^{2+}, Na^{+} and K^{+}) on the removal efficiency of lanthanides using ground original sample.

3.2.4. Characteristics of Buccinum tenuissimum shell biomass after adsorption of metals

X-ray diffraction (XRD) patterns of four kinds of sieved samples after adsorption of metals are shown in Fig. 11. Similar to the XRD patterns before adsorption of metals, aragonite and calcite were found as the main crystal structure in (a): Ground original sample and (b): Heat-treatment (480°C, 6h) sample, respectively. However, the decrease of peak and increase of noise were also observed in both patterns, particularly in the ground original material as shown in Fig. 11(a). Bottcher [37] pointed out that the natural powdered aragonite was transformed to mixed rhombohedral carbonates by the reaction with (Ca, Mg)-chloride solutions. Therefore, there is the possibility that the transformation of aragonite occurred by the reaction with lanthanides in our experiment.

Moreover, according to XRD analysis, the main crystal structure of (c): "Heat-treatment (950°C) sample" was transformed from calcium oxide (CaO) to the mixture of calcium hydroxide ($Ca(OH)_2$) and calcite ($CaCO_3$) after exposing metals; and that of (d): "Heat-treatment (950°C) and water added sample" was transformed from calcium hydroxide($Ca(OH)_2$) to calcite ($CaCO_3$) after adsorption of metals. These changes may be due to the reaction with water or carbon dioxide in atmosphere.

SEM pictures of four kinds of shell biomass after adsorption of metals are shown in Fig. 12. By comparing SEM pictures in Fig. 8 with that in Fig. 12, it is found that the morphology of sample (a) and (b) has hardly changed even after exposing metals. From this observation, these sieved samples should be predicted to withstand the repeated use; and hence it can be a good adsorbent.

In contrast to sample (a), clear crystal structure (sizes are mostly 0.25-2.0μm) was observed in sample (b) even after adsorption of metal. In case of Cd conducted by Kohler et al. [38], the difference of procedure for reaction with metals between aragonite and calcite was suggested. According to their work, the precipitation of several distinct types of crystals was observed after exposing metals in the case of aragonite. Then, it is anticipated that similar phenomenon were occurred by adsorption of lanthanides in case of our samples. On the

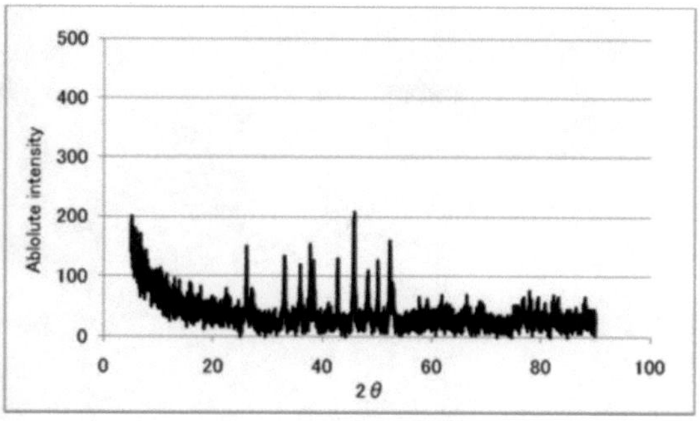

(a)

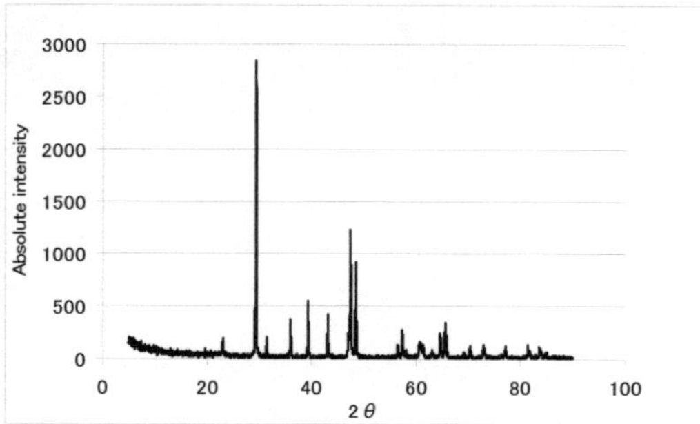

(b)

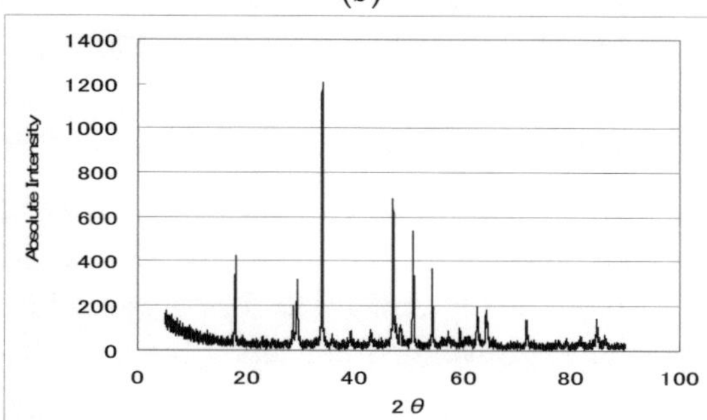

(c)

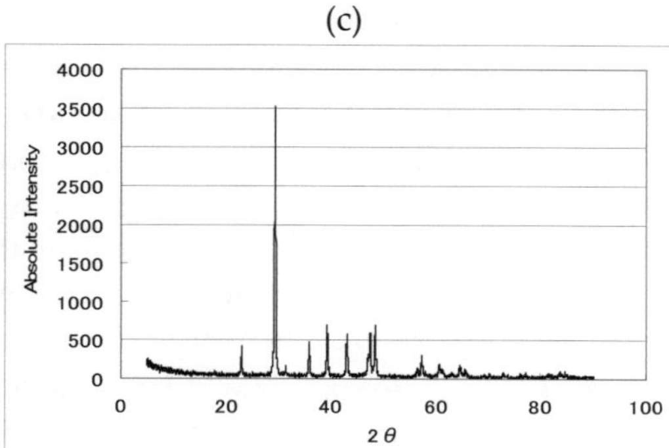

(d)

Figure 11. X-ray diffraction (XRD) patterns of *Buccinum tenussimum* shell biomass after adsorption of metals. (a) ground original sample, (b) heat-treatment (480°C) sample, (c) heat-treatment (950°C) sample, (d) heat-treatment (950°C) and water added sample

Figure 12. SEM pictures of *Buccinum tenussimum* shell biomass after adsorption of metals. (a) ground original sample, (b) heat-treatment (480°C) sample, (c) heat-treatment (950°C) sample, (d) heat-treatment (950°C) and water added sample

other hand, the surface of sample (c) and (d) after exposing metals have changed largely compared to that before adsorption of metals (Fig. 8). This is in good accord with the results of XRD patterns. Particularly, remarkable transformation was observed in the morphology of sample (d). The reaction of sample (d) with metal is supposed to proceed rapidly.

3.2.5 Adsorption isotherms of lanthanides by Buccinum tenuissimum shell biomass

The adsorption data obtained for lanthanides using *Buccinum tenuissimum* shell biomass were analyzed using Langmuir and Freundlich equations. The correlation coefficient (R^2) of Langmuir and Freundlich isotherms for lanthanides using ground original shell biomass is shown in Table 7 along with other relevant parameters.

From this table, it is found that R^2 value for lanthanides is comparatively high. It indicates the applicability of these adsorption isotherms satisfactorily for lanthanides in this sample. The dimensionless parameter Hall separation factor (R_L) for lanthanides is in the range of $0<R_L<1$, which means that the sorption for lanthanides by this shell biomass is favorable. Furthermore, the negative value of ΔG indicates that the sorption is spontaneous. The higher R^2 value for Freundlich model rather than for Langmuir isotherm (0.638-0.886 for Langmuir isotherm and 0.844-0.932 for Freundlich one) suggests that the adsorption on this sample is due to multilayer coverage of the adsorbate rather than monolayer coverage on the surface. It is noted that the value of $1/n$ less than unity indicates better adsorption mechanism and formation of relatively stronger bonds between adsorbent and adsorbate [10]. That is to say, favorable adsorption for lanthanides by this shell biomass is presented.

On the other hand, R^2 and other parameters of Langmuir and Freundlich isotherms for lanthanides using "heat-treatment (480°C) sample" is shown in Table 8. It is noteworthy that R^2 value for REEs in this sample is still more large (0.947-0.982 for Langmuir isotherm and 0.948-0.975 for Freundlich one), compared with the original ground sample (Table 7).

	Langmuir					Freundlich		
	a	b	R^2	ΔG_{ads}/ kJmol^{-1}	R_L	K_F	$1/n$	R^2
La	400	0.490	0.638	-15.3	0.0200	115	0.654	0.844
Ce	370	1.17	0.886	-17.5	0.0084	163	0.583	0.864
Pr	400	0.714	0.750	-16.3	0.0138	145	0.658	0.853
Nd	400	0.610	0.681	-15.9	0.0161	133	0.662	0.846
Sm	417	0.6.32	0.740	-16.0	0.0156	145	0.709	0.863
Eu	417	0.571	0.794	-15.7	0.0172	136	0.723	0.883
Gd	435	0.418	0.788	-15.0	0.0234	114	0.765	0.904
Tb	476	0.328	0.800	-14.4	0.0296	105	0.778	0.912
Dy	476	0.239	0.849	-13.8	0.0367	81.8	0.804	0.932
Ho	476	0.183	0.870	-12.9	0.0519	65.2	0.799	0.924
Er	476	0.148	0.878	-12.4	0.0633	55.6	0.787	0.920
Tm	476	0.135	0.842	-12.2	0.0687	52.6	0.785	0.910
Yb	476	0.136	0.818	-12.2	0.0683	53.3	0.762	0.887
Lu	500	0.119	0.786	-11.8	0.0775	51.3	0.759	0.877

Table 7. Coefficient of Langmuir and Freundlich isotherms for lanthanides using original *Buccinum tenuissimum* shell biomass

	Langmuir					Freundlich		
	a	b	R^2	ΔG_{ads} kJmol^{-1}	R_L	K_F	$1/n$	R^2
La	192	0.243	0.982	-13.6	0.0395	57.2	0.258	0.948
Ce	278	0.234	0.972	-13.5	0.0410	70.4	0.292	0.956
Pr	303	0.229	0.962	-13.5	0.0418	71.5	0.321	0.955
Nd	313	0.225	0.956	-13.4	0.0425	72.5	0.328	0.954
Sm	345	0.266	0.948	-13.8	0.0362	78.4	0.359	0.962
Eu	345	0.248	0.947	-13.7	0.0388	76.0	0.364	0.963
Gd	303	0.231	0.955	-13.5	0.0415	75.0	0.299	0.954
Tb	323	0.221	0.961	-13.4	0.0432	69.8	0.354	0.968
Dy	323	0.195	0.957	-13.1	0.0488	66.3	0.358	0.964
Ho	294	0.178	0.961	-12.8	0.0532	61.0	0.346	0.960
Er	294	0.171	0.963	-12.7	0.0553	59.1	0.355	0.964
Tm	303	0.176	0.964	-12.8	0.0539	59.0	0.372	0.968
Yb	323	0.181	0.960	-12.9	0.0523	60.8	0.395	0.974
Lu	333	0.176	0.966	-12.8	0.0536	62.4	0.389	0.975

Table 8. Coefficient of Langmuir and Freundlich isotherms for lanthanides using *Buccinum tenuissimum* shell biomass after heat-treatment (480°C, 6h)

Futhermore, this result indicates the stronger the monolayer adsorption (the surface adsorption) on the heat-treatment sample relative to on the original sample (before heat-treatment). Judging from the value of R_L or $1/n$ in Table 4, the heat-treatment (480°C) sample also exhibits the favorable property for lanthanides adsorption.

The correlation coefficient (R^2) and other parameters of Langmuir and Freundlich isotherms for lanthanides using "heat-treatment (950°C) sample" is shown in Table 9. It is found that R^2 value for lanthanides in this sample is fairly small compared with the values of "ground original sample" or "heat-treatment (480°C) sample" (In case of La, Ce, Yb and Lu, R^2 can not be estimated due to the lack of sorption data at low initial concentration). The low correlation coefficient (R^2) in this "heat-treatment (950°C) sample" may indicate that the removal of lanthanides occurred not by adsorption mechanism,

	Langmuir					Freundlich		
	a	b	R^2	ΔG_{ads} kJmol^{-1}	R_L		K_F	$1/n$
La	—	—	—	—	—	—	—	—
Ce	—	—	—	—	—	—	—	—
Pr	—	—	0.0553	—	—	5690	0.999	0.418
Nd	—	—	0.0727	—	—	10100	1.27	0.375
Sm	—	—	0.00190	—	—	838	0.724	0.157
Eu	—	—	0.2107	—	—	41600	1.65	0.521
Gd	—	—	0.157	—	—	44300	1.69	0.526
Tb	—	—	0.0974	—	—	7980	1.08	0.599
Dy	—	—	0.108	—	—	13900	1.32	0.506
Ho	—	—	0.101	—	—	11100	1.26	0.529
Er	—	—	0.110	—	—	8290	1.14	0.625
Tm	—	—	0.0915	—	—	8830	1.15	0.588
Yb	—	—	—	—	—	—	—	—
Lu	—	—	—	—	—	—	—	—

Table 9. Coefficient of Langmuir and Freundlich isotherms for lanthanides using *Buccinum tenuissimum* shell biomass after heat-treatment (950°C, 6h)

Particularly R^2 value is remarkably small for Langmuir isotherm, and then other relevant parameters can not be estimated. As for Freundlich one, not only R^2 value is relatively small (0.157-0.625), but the value of $1/n$ for most lanthanide is more than unity. That is to say, the almost perfect removal of lanthanides for this sample may be due to other mechanism rather than the adsorption on the biomass. However, the cause or mechanism of lanthanides removal on this sample has yet to be sufficiently clarified in our work, and further investigation to survey the mechanism is needed.

Finally, R^2 and other parameters of Langmuir and Freundlich isotherms for lanthanides using "heat-treatment (950°C) and water added sample" is shown in Table 10. It is found that R^2 value for lanthanides in this sample is fairly large particularly for Langmuir isotherm (0.992-0.999 for Langmuir isotherm and 0.885-0.951 for Freundlich one). This result is similar to that for "heat-treatment (480°C) sample", and indicates the stronger the monolayer adsorption on this sample. Judging from the value of R_L or $1/n$ in Table10, this sample also exhibits the favorable conditions for lanthanides adsorption.

	Langmuir					Freundlich		
	a	*b*	R^2	ΔG_{ads}/ kJmol^{-1}	R_L	K_F	1/*n*	R^2
La	161	0.969	0.999	-17.0	0.0102	59.3	0.264	0.951
Ce	200	0.980	0.999	-17.1	0.0101	70.3	0.283	0.950
Pr	217	0.852	0.998	-16.7	0.0116	69.3	0.327	0.919
Nd	222	0.789	0.997	-16.5	0.0125	68.0	0.340	0.930
Sm	233	0.878	0.996	-16.8	0.0113	72.0	0.364	0.937
Eu	233	0.782	0.996	-16.5	0.0126	68.3	0.380	0.917
Gd	227	0.647	0.997	-16.0	0.0152	61.6	0.384	0.937
Tb	227	0.629	0.996	-16.0	0.0157	61.7	0.390	0.937
Dy	233	0.506	0.996	-15.4	0.0194	57.2	0.409	0.936
Ho	227	0.404	0.996	-14.9	0.0242	50.6	0.425	0.931
Er	222	0.372	0.996	-14.7	0.0262	47.5	0.433	0.928
Tm	233	0.352	0.996	-14.5	0.0276	48.0	0.450	0.922
Yb	244	0.398	0.994	-14.8	0.0245	55.3	0.417	0.934
Lu	217	0.495	0.992	-15.4	0.0198	53.2	0.409	0.885

Table 10. Coefficient of Langmuir and Freundlich isotherms for lanthanides using *Buccinum tenuissimum* shell biomass after heat-treatment (950°C, 6h) and adding water

As mentioned above, biosorption studies have been mainly focused on toxic metals elements such as Cd, Pb, As and Cr so far, and a few reports are focused on lanthanides. The sorption experiments using shell biomass in this work were carried out under low concentration of lanthanide (i.e., 100 cm^3 of multi-element standard solution including known initial lanthanide concentration (10 to 500 μg·dm^{-3})). Then, sorption experiment for three lanthanides (La, Eu and Yb) in single component system by this shell biomass is being planned using the solution individually prepared by each nitrate salt: $La(NO_3)_3{\cdot}6H_2O$, $Eu(NO_3)_3{\cdot}6H_2O$, or $Yb(NO_3)_3{\cdot}3H_2O$ as the case of seaweed biomass in our work.

4. Conclusion

From this work, it was first quantitatively clarified that seaweed biomass could be efficient sorbents for lanthanides, and exhibit high ability of chemical adsorption. Particularly, *Ulva pertusa* is found to be a promising biosorbent for removing La. It is also suggested that the adsorption on seaweed biomass is mainly due to monolayer sorption because of well-fitting for Langmuir model.

Biosorption characteristic of *Buccinum tenuissimum* shell biomass was also studied for lanthanides. Sorption isotherms were analyzed using Langmuir and Freundlich equations to confirm the efficiency of shell biomass as sorbent. The shell biomass samples showed excellent sorption capacity for lanthanides under our experimental condition, even the presence of diverse ions (Ca^{2+}, Mg^{2+}, Na^+ and K^+) up to the concentration of 200 mg·dm^{-3}.

From these results, it was quantitatively clarified to some extent that shell biomass can be an efficient sorbent for lanthanides. It is very significant information from the viewpoint of environmental protection that the shell (usually treated as waste material) can be converted into a biosorbent for lanthanides.

The data obtained and the method used in this work can be useful tool from the viewpoint of resource recovery in future work.

Author details

Naoki Kano
Department of Chemistry and Chemical Engineering, Faculty of Engineering, Niigata University, Japan

Acknowledgement

The present work was partially supported by a Grant-in-Aid for Scientific Research (Research Program (C), No. 22510084) of the Japan Society for the Promotion of Science.

The author wish to express his thanks to Dr. M. Baga of the Marine Ecology Research Institute and Dr. N. Sugai, Dr. H. Handa, Dr. O. Sato and Dr. R. Ishikawa of Niigata Prefectural Fisheries and Marine Research Institute for giving helpful advice concerning sampling, identification and pretreatment of seaweed and shellfish. The author is also grateful to Dr. K. Satoh of Fac. of Sci., Dr. K. Fujii and M. Ohizumi of Office for Environmental and Safety, Mr. N. Saito and Mr. T. Hatamachi of Fac. of Eng. in Niigata University for permitting the use of instrument (ICP-MS, ICP-AES XRD, SEM and Surface Area Analyzer) and facilities and for giving helpful advice in measurement.

5. References

[1] Saether O. M, Storroe G, Segar. D, Krog R (1997) Contamination of soil and groundwater at a former industrial site in Trondheim, Norway. Appl. Geochem. 12:327-332.
[2] Fu F., Wang Q. (2011) Removal of heavy metal ions from wastewater:A review. Journal of Environmental Management. 92:407-418.
[3] Rahmati M.M., Rabbani P., Abdolali A., Keshtkar A.R. (2011) Kinetics and equilibrium studies on biosorption of cadmium, lead, and nickel ions from aqueous solutions by intact and chemically modified brown algae. J. Hazard. Mater. 185: 401– 407.
[4] Chen J (1997) Batch and continuous adsorption of strontium by plant root tissues. Bioresource Technology. 60: 185-189.
[5] Mohantly K, Jha M, Meikap B C, Biswas M. N (2006) Biosorption of Lanthanides Using Three Kinds of Seaweed Biomasses.Chem. Eng. J. 117: 71-77.
[6] Sudhir D, Tripathi R.M., Hegde A.G. (2008) Biosorption of lead and copper from aqueous solutions by pre-treated crab and arca shell biomass. Bioresource Technology. 99: 179-187

[7] Acosta R I, Rodriguez X, Gutierrez C (2004)Biosorption of chromium (VI) from aqueous solutions onto fungal biomass. Bioinorganic Chemistry and Application. 2: 1-7.

[8] Conard K, Hansen H C B (2007) Sorption of zinc and lead on coir. Bioresource Technology. 98: 89-97.

[9] Sharma P, Kumari P, Srivastava M M, Srivastava. S (2007) Ternary biosorption studies of Cd(II), Cr(III) and Ni(II) on shelled Moringa oleifera seeds. Bioresource Technology, vol. 98:474-477.

[10] Dahiya S, Tripathi R M, Hegde A. G (2008) Biosorption of lead and copper from aqueous solutions by pre-treated crab and arca shell biomass. Bioresource Technology. 99: 179-187.

[11] Dahiya S, Tripathi R M, Hegde A G (2008) Biosorption of heavy metals and radionuclide from aqueous solutions by pre-treated arca shell biomass. Journal of Hazardous Materials.150:376-386.

[12] Sari A, Tuzen M (2009) Removal of mercury(II) from aqueous solution using moss (Drepanocladus revolvens) biomass: Equilibrium, thermodynamic and kinetic studies. Journal of Hazardous Materials. 171: 500-507.

[13] Sari A, Tuzen M, Mendil D, Soylak M (2009) Biosorptive removal of mercury(II) from aqueous solution using lichen (Xanthoparmelia conspersa) biomass: Kinetic and equilibrium studies. Journal of Hazardous Materials. 169: 263-270.

[14] Sari A, Tuzen M (2008) Biosorption of cadmium(II) from aqueous solution by red algae (Ceramium virgatum): Equilibrium, kinetic and thermodynamic studies. Journal of Hazardous Materials. 157: 448-454.

[15] Sari A, Tuzen M (2008) Biosorption of Pb(II) and Cd(II) from aqueous solution using green alga (Ulva lactuca) biomass. Journal of Hazardous Materials. 152:302-308.

[16] Sari A, Tuzen M (2010) Biosorption of selenium from aqueous solution by green algae (Cladophora hutchinsiae) biomass: Equilibrium, thermodynamic and kinetic studies. Chemical Engineering Journal. 158:200-206.

[17] Periasamy K, Namasivayam C (1996) Removal of copper(II) by adsorption onto peanut hull carbon from water and copper plating industry wastewater. Chemosphere. 32: 769-789.

[18] Mevra C. Y, Yusuf K, O. Algur F (2006) Response surface optimization of the removal of nickel from aqueous solution by cone biomass of Pinus sylvestris. Bioresource Technology. 97:1761-1765.

[19] Sakao S., Ogawa Y., Uchida H. (1997) Determination of trace elements in sea weed samples by inductively couple plasma mass spectrometry. Analytica Chimica Acta. 355: 121-127.

[20] Fu F., Akagi T., Yabuki S., Iwaki M., Ogura N. (2000) Distribution of rare earth elements in seaweed:implication of two different sources of rare earth elements and silicon seaweed. Journal of Phycology. 36: 62-70.

[21] Punin Crespo M.O., Lage Yusty M. A. (2006) Comparison of supercritical fluid extraction and Soxhlet extraction for the determination of aliphatic hydrocarbons in seaweed samples. Ecotoxicology and Environmental Safety. 64: 400-405.

[22] Cubillas P., Kohler S., Prieto M., Chaiirat C., Oelkers H. E. (2005) Experimental determination of the dissolution rates of calcite,aragonite, and bivalves. Chemical Geology. 216: 59-77.

[23] Sikes C. S., Wheeler A. P., Wierzbicki A., Dillaman R. M., De Luca L. (1998) Oyster Shell Protein and Atomic Force Microscopy of Oyster Shell Folia. Biol. Bull. 194: 304-316.

[24] Tsui M. T. K, Cheung, K.C, Tam N F Y, Wong M H (2006) A comparative study on metal sorption by brown seaweed. Chemosphere. 65:51-57.

[25] Nader B.,Yadollah Y. (2005) On-line preconcentration of some rare earth elements in water samples using C18-cartridge modified with l-(2-pyridylazo) 2-naphtol (PAN) prior to simultaneous determination by inductively coupled plasma optical emission spectrometry(ICP-OES). Analytica Chimica Acta. 540: 325–332.

[26] He M., Hu B., Zeng Y., Jiang Z.C. (2005) ICP-MS direct determination of trace amounts of rare earth impurities in various rare earth oxides with only one standard series. Journal of Alloys and Compounds. 390: 168–174.

[27] El-Dessouky S.I., El-Sofany E.A., Daoud J.A. (2007) Studies on the sorption of praseodymium (III), holmium (III) and cobalt (II) from nitrate medium using TVEX–PHOR resin. Journal of Hazardous Materials. 143:17-23.

[28] Sert S., Kutahyali C., Inan S., Talip Z., Cetinkaya B., Eral M. (2008) Biosorption of lanthanum and cerium from aqueous solutions by *Platanus orientalis* leaf powder. Hydrometallurgy. 90: 13-18.

[29] Seki H., Suzuki A. (1998) Biosorption of Heavy Metal Ions to Brown Algae,Macrocystis pyrifera,Kjellmaniella crassiforia, and Undaria pinnatifida. Journal of Colloid and Interface Science. 206: 297-301.

[30] Yang J., Volesky B. (1999) Cadmium Biosorption Rate in Protonated Sargassum Biomass. Environ. Sci. Technol. 33: 751-757.

[31] Diniz V., Volesky B. (2005) Biosorption of La, Eu and Yb using Sargassum biomass. Water Research, 39:239-247.

[32] Sakamoto N., Kano N., Imaizumi H. (2008) Determination of rare earth elements, thorium and uranium in seaweed samples on the coast in Niigata Prefecture by inductively coupled plasma mass spectrometry. Applied Geochemistry. 23: 2955-2960.

[33] Davis T.A., Volesky B., Vieira R.H.S.F (2000) Sargassum seaweed as biosorbent for heavy metals. Water Research. 34: 4270-4278.

[34] Hashim M.A, Chu K.H (2004) Biosorption of cadmium by brown, green, and red seaweeds. Chemical Engineering Journal. 97: 249–255.

[35] Sheng P.X., Ting Y.P., Chen J.P., Hong L. (2004) Sorption of lead, copper, cadmium, zinc, and nickel by marine algal biomass: characterization of biosorptive capacity and investigation of mechanisms. Journal of Colloid and Interface Science. 275: 131–141.

[36] Prieto M, Cubillas P, Fernandez-Gonzalez A (2003) Uptake of dissolved Cd by biogenic and abiogenic aragonite: a comparison with sorption onto calcite. Geochimica et Cosmochimica Acta. 67: 3859–3869.

[37] Bottcher M. E (1997) The transformation of aragonite to $Mn_xCa_{(1-x)}CO_3$ solid-solutions at 20 °C: An experimental study. Marine Chemistry. 57: 97-106.

[38] Kohler S J, Cubillas P, Rodrguez-Blanco J D, Bauer C, Prieto M (2007) Removal of Cadmium from Wastewater by Aragonite Shells and the Influence of Other Divalent Cations. Environ. Sci. Technol. 41: 112-118.

Section 2

Biomass Systems

Chapter 5

Diagnosis and Recommendation Integrated System (DRIS) to Assess the Nutritional State of Plants

Ademar Pereira Serra, Marlene Estevão Marchetti, Davi José Bungenstab,
Maria Anita Gonçalves da Silva, Rosilene Pereira Serra,
Franklyn Clawdy Nunes Guimarães, Vanessa Do Amaral Conrad
and Henrique Soares de Morais

Additional information is available at the end of the chapter

http://dx.doi.org/10.5772/54576

1. Introduction

Approximately 80-90% of fresh biomass composition of plants consists of water, and 10-20% of fresh biomass comprises the dry biomass.

The elemental composition of dry biomass of plants consists above 90% of carbon, hydrogen and oxygen, the remains of nutrition composition is made of other essential nutrients to plants, such as: nitrogen, phosphorus, potassium, calcium, magnesium, sulphur, boron, zinc, iron, manganese, nickel, silicon and other elements uptaken from the environment (Epstein & Bloom, 2006).

The nutritional state of plants influence the dry biomass production. The nutritional deficiency of some essential nutrient prevents the maximum potential productive of plants. According to Serra et al. (2011), the fresh and dry biomass production from medicinal plant *Pfaffia glomerata* Pedersen (Spreng.) was negatively influenced by nitrogen (N) and phosphorus (P) concentration into the plant, furthermore, the limitation of P in soil generated less growth on plant with less biomass yield and expressed visible N and P nutrition deficiency.

The nutritional diagnose of plants consists on determination of nutrients contents, this determination is made with the comparison of the nutrient content with standard values, and this procedure called by leave diagnose that uses information from chemical analyses of plant tissue. However, there is the visible diagnose that is made with visual observation of nutritional deficiency or excess symptoms.

The visual diagnose can be little practical, because, when the deficiency symptoms show in plants, the plant metabolism has been already damaged and the correction of deficiency can note be taken good benefits on increase of yield or better products quality, besides, the deficiency symptoms is shown in plant when the deficiency is acute (Marshner, 1995).

The tissue analyses has been considered the direct way to evaluation the nutritional state of plants, but, to do this evaluate it is necessary a well specific part from the plant to take this diagnose, this specific part is the leaf tissue that is the most used (Malavolta, 2006; Mourão Filho, 2003; Hallmark & Beverly, 1991; Beaufils, 1973).

The leaf tissue is considered the most important part of the plant where the physiologic activate happens and this tissue shows easily the nutritional disturb. To use the leaf tissue is necessary to have the chemical analyses. Furthermore, to assess the nutritional status there is the need to have leaf standard to sample, this leaf standard depend on the crop that intend to evaluate, but, nowadays there are many information about the most cultivated commercial crops.

The leave diagnose can be a useful tool to assess the nutritional status of plant, but, the procedure to analyse the data must be appropriate. Furthermore, because of natural dynamic of the leaf tissue composition that is strengly influenced by leaf age, maturation stage and interaction among nutrients on uptake and translocation into the plant, if all the damages criteria were not observed the leaf diagnose becomes very difficult to understand and used (Walworth & Sumner, 1987).

The interpretation of nutrients contents in leaf analyses can be made by several methods to assess plant nutritional status. To interpretate results of traditional chemical analyses of plant tissue for the assessment of the nutritional status of plants, the methods of critical level and sufficiency range are used more frequently (Beaufils, 1973; Walworth & Sumner, 1987; Mourão Filho, 2004; Serra et al., 2010a,b; Camacho et a., 2012; Serra et al., 2012).

There are other diagnose systems, such as: Compositional Nutrient Diagnosis (CND) (Parent & Dafir, 1992), plant analysis with standardized scores (PASS) (Baldock & Schulte, 1996), these two methods are less studied then critical level and sufficiency range, but there is CND standard published on Serra et al. (2010a,b) for the West region of Bahia, a state in Brazil and other authors (Parent, 2011; Wairegi and Asten, 2012).

The sufficiency range is the most used method of diagnose, and this method consists on optimum ranges of nutrients concentration to establish the nutritional state of crops, otherwise to use the sufficiency range it is necessary to develop regional calibration that is very expensive.

The Diagnosis and Recommendation Integrated System (DRIS) relate the nutrient contents in dual ratios (N/P, P/N, N/K, K/N...), because of the relation between two nutrients, the problem with the biomass accumulation and reduction of the nutrients concentration in plants with its age is solved (Beaufils, 1973; Walworth & Sumner, 1987; Singh et al., 2000). The use of DRIS on concept of nutritional balance of a plant is becoming an efficient method to assess the nutritional status of plants, this method puts the limitation of nutrients in order of plant demand, enabling the nutritional balance between the nutrient in leaf sample.

Because of several factors that can influence nutrient concentration in plants, Jones (1981) suggests that it is necessary to be critical in relation to reliability of DRIS standard, because in this way the use of leaf diagnose method can be well used.

2. Diagnosis and Recommendation Integrated System (DRIS)

The Diagnosis and Recommendation Integrated System (DRIS) was developed by Beaufils in 1973, this method consist in dual relation between a pair of nutrients (N/P, P/N, N/K, K/N...) instead of the use of sufficiency range or critical level that are called univariate methods, because only the individual concentration of the nutrients in leaf tissue is taken into consideration while no information about the nutritional balance is provided. DRIS enables the evaluation of the nutritional balance of a plant, ranking nutrient levels in relative order, from the most deficient to the most excessive.

With the use of dual relation on DRIS, the problem with the effect of concentration or dilution on the nutrients in plants is solved, because, according to Beaufils (1973); Walworth & Sumner (1987) with the growth of leaf tissue, on one hand the concentration of nitrogen, phosphorus, potassium and sulphur decrease in older plants and the concentration of calcium and magnesium increase in older plants on the other hand. When it is used the DRIS method, where the dual ratio is used, the values remain constant, minimizing the effect of biomass accumulation, that is one of the major problem with sufficiency range and critical level method.

It is feasible to find on literature some crops on which DRIS had already been used to assess the nutritional status of plants, such as; pineapple (Sema et al., 2010), cotton (Silva et al., 2009; Serra et al., 2010a,b; Serra et al., 2012), rice (Guindani et al., 2009), potato (Bailey et al., 2009; Ramakrishna et al., 2009), coffee (Nick, 1998), sugarcane (Elwali & Gascho, 1984; Reis Jr & Monnerat, 2002; Maccray et al., 2010), orange (Mourão Filho et al., 2004), apple (Natchigall et al., 2007a,b), mango (Hundal et al., 2005), corn (Reis Jr, 2002; Urricariet et al., 2004), soybean (Urano et al., 2006, 2007), Eucalyptus (Wadt et al., 1998), among other crops.

According to Baldock & Schulte (1996), there are four advantages of DRIS; (1) the scale of interpretation is continuous numeric scale, and easy to use, (2) put the nutrients in order of the most deficiency to the most excessive, (3) identify cases where the yield of plant is been limited by into factor as nutritional status and (4) the Nutritional Balance Index (NBI) give a result of combined effects of nutrients. Nevertheless, the disadvantage of this methodology is that the DRIS index is not independent, because one nutrient concentration can have hard influence on the other DRIS index for one nutrient but this problem can be corrected in parts with a hard selection of the nutrient that will compound the DRIS norms.

3. DRIS norms

To be feasible the use of DRIS to assess the nutritional status of plants, the first step is establish the DRIS norms or standard. The DRIS norms consist on average and standard deviation of dual ratio between nutrients (N/P, P/N, N/K, K/N, etc.) obtained from a crop reference population (Table 1), but, it is necessary that the crop reference shows high yield (Beaufils, 1973). This method has been followed along the years (Jones, 1981; Alvarez V. & Leite, 1999; Silva et al., 2009; Maccray et al., 2010; Serra et al., 2010a,b; Serra et al., 2012).

The data bank to compose the DRIS norms is formed by the crop yield and chemical analysis of leaf tissue, and this information can be obtained from commercial crop or experimental units. The size of the data bank is not a factor that is directly related to the quality of the DRIS norms (Walworth et al., 1988; Sumner, 1977).

Walworth et al. (1988) observed that, when they used 10 data to establish the DRIS norms, the results obtained were more accurate then the use of a large number of data. What is more important to improve efficiency on DRIS norms is the quality of the data, because it is not accepting the use of sick plants to compose the data bank to establish the DRIS norms.

To make part of the DRIS norms, the rations between nutrients can be selected by the direct form (N/P) or reverse (P/N), but, there is more than one way to change the ratio that is going to compose the DRIS norms. Bataglia et al. (1990) used the entire dual ratio without selecting the direct or reverse form, and other researchers used the transformation by natural log (Beverly, 1987; Urano et al., 2006, 2007; Serra et al., 2010a,b; Serra et al., 2012).

With many ways to select the ratio to compose the DRIS norms there is a necessity to establish the most efficiency way for each crop that results in a better efficiency of the system. Silva et al. (2009) tested the dual ratio selection using the "F" value (Jones, 1981; Letzsch, 1985; Walworth & Sumner, 1987) and "r" value (Nick, 1998) in cotton crop, on his turn, Silva et al. (2009) did not test the criterion of choice the ratio by log transformation or the use of all nutrient ratio as it were made by Alvarez V. & Leite (1999) and Serra et al. (2010a,b).

Results obtained by Serra et al. (2012) showed that the use of "F" value or log transformation in nutrient ratio to define the norms produced different DRIS index, furthermore, when the DRIS index is interpret by Beaufils ranges the difference observed among index was reduced, showed less difference between the two groups of norms.

Following the premises of DRIS proposed by Beaufils (1973), it is feasible to change the dual ratio (A/B or B/A) that is more important to compose the DRIS norms. This way it is expected that the dual ratio from crop with high-yielding (reference population), composed with healthy plants, shows less variation than the population of plants with low-yielding (non-reference population), thus, the relation between variance ratio method, the F value, was defined as the variance ratio of low-yielding (non-reference) and high-yielding population (reference), and the order of the ratio with the highest value was chosen among the variance ratios (Jones, 1981; Letzsch, 1985; Walworth & Sumner, 1987).

The utilization of the relationship between variance ratio method ("F" value) from low-yielding and high-yielding is the most used method to define the DRIS norms. The method "F" value is defined on the data bank divided into two groups (non-reference and reference), and the choice of ratio directly (A/B) or inverse (B/A) defined by relationship between variances from the two populations, in which the ratio chosen will result arises from the following analysis (Jones, 1981; Letzsch, 1985; Walworth & Sumner, 1987):

If:

$$\left[\frac{S^2\left(\frac{A}{B}\right) non-reference\ population}{S^2\left(\frac{A}{B}\right) reference\ population}\right] > \left[\frac{S^2\left(\frac{B}{A}\right) non-reference\ population}{S^2\left(\frac{B}{A}\right) reference\ population}\right]$$

Then: the dual ratio that will make part of the DRIS norms will be A/B, on the another it will be B/A. S^2 is the variance of the dual ratio of the reference population and non-reference.

Besides the selection of forward or reverse ratio to compose the DRIS norms, the same principle can be selected with regard to the significance of F value, which can be 1%, 5% or 10% (Wadt, 1999), and feasible to use all dual ratio, which was selected by the largest ratio of variances, without the rigour of significance (Beaufils, 1973; Jones, 1981; Walworth & Sumner, 1987; Serra, 2011).

One can observe on literature that there are not any consensus about which methodology is more efficient to use. Jones (1981) did not select for significance, but he selected by the biggest reason of variances, as well as Raghupathi et al. (2005); Guindani et al. (2009); Sema et al. (2010); Serra et al. (2012). However, Wadt (2005) used the "F" value for the selection of dual ratio with a significance of 10%, excluding from the norms the dual ratio that was with significance above this value.

When selecting the dual ratio by significance of the "F" value, the sum of DRIS indexes does not give a zero value, in this case some nutrients can remain with a larger number of dual ratio than those with fewer ratios. However, Wadt et al. (1999) concludes that the rigour of the selection by the significance of "F" value generates greater efficiency for the diagnosis, in studies made with coffee crop (*Coffea canephora* Pierre).

Variable	Average	s	Criteria			Variable	Average	s	Criteria		
			r	F	ADR				r	F	ADR
N/P	15,1416	1,8617	X	X	X	S/B	0,1970	0,1001	X		X
N/K	2,2317	0,4022			X	S/Zn	0,4561	0,2180			X
N/Ca	1,5598	0,3095			X	S/Cu	1,2012	1,0183			X
N/Mg	10,5226	1,8856	X		X	S/Mn	0,2917	0,1646	X		X
N/S	4,5765	2,2252	X	X	X	S/Fe	0,1299	0,0691	X		X
N/B	0,7503	0,2404	X	X	X	B/N	1,4916	0,5351			X
N/Zn	1,6754	0,3502		X	X	B/P	22,6312	8,8362	X		X
N/Cu	3,9724	2,1953	X	X	X	B/K	3,3058	1,2388			X
N/Mn	1,0587	0,4183		X	X	B/Ca	2,2463	0,6303	X	X	X
N/Fe	0,4701	0,1493		X	X	B/Mg	15,3008	4,6670	X		X
P/N	0,0671	0,0087			X	B/S	6,3151	3,1483		X	X
P/K	0,1500	0,0354			X	B/Zn	2,4320	0,8360		X	X
P/Ca	0,1048	0,0265	X		X	B/Cu	6,1969	4,6239	X		X
P/Mg	0,7003	0,1316			X	B/Mn	1,5206	0,6858	X		X
P/S	0,3080	0,1597	X	X	X	B/Fe	0,6875	0,2750	X	X	X
P/B	0,0504	0,0181		X	X	Zn/N	0,6224	0,1281	X		X

Variable	Average	s	Criteria			Variable	Average	s	Criteria		
			r	F	ADR				r	F	ADR
P/Zn	0,1117	0,0251			X	Zn/P	9,3606	1,9666	X	X	X
P/Cu	0,2613	0,1369	X	X	X	Zn/K	1,3733	0,3150	X		X
P/Mn	0,0694	0,0246		X	X	Zn/Ca	0,9508	0,1766	X		X
P/Fe	0,0316	0,0112	X		X	Zn/Mg	6,4829	1,4814			X
K/N	0,4611	0,0758	X	X	X	Zn/S	2,7703	1,4366	X	X	X
K/P	6,9883	1,4438	X	X	X	Zn/B	0,4551	0,1419	X		X
K/Ca	0,7088	0,1301			X	Zn/Cu	2,4792	1,5065	X		X
K/Mg	4,8383	1,1228			X	Zn/Mn	0,6445	0,2446			X
K/S	2,0914	1,0735	X	X	X	Zn/Fe	0,2898	0,1007	X		X
K/B	0,3446	0,1226	X	X	X	Cu/N	0,3491	0,2056			X
K/Zn	0,7636	0,1646		X	X	Cu/P	5,2285	3,1722			X
K/Cu	1,7900	1,0192	X		X	Cu/K	0,7498	0,3941		X	X
K/Mn	0,4861	0,2024	X	X	X	Cu/Ca	0,5274	0,2854			X
K/Fe	0,2175	0,0769			X	Cu/Mg	3,7280	2,4233			X
Ca/N	0,6615	0,1095	X	X	X	Cu/S	1,6651	1,2996	X	X	X
Ca/P	10,0208	2,0527		X	X	Cu/B	0,2730	0,2010		X	X
Ca/K	1,4559	0,2605	X	X	X	Cu/Zn	0,5869	0,3864		X	X
Ca/Mg	6,9029	1,3999			X	Cu/Mn	0,3843	0,3414		X	X
Ca/S	2,8917	1,2685	X	X	X	Cu/Fe	0,1679	0,1328			X
Ca/B	0,4830	0,1436			X	Mn/N	1,1626	0,6971	X		X
Ca/Zn	1,0873	0,2011		X	X	Mn/P	17,1314	9,3481	X		X
Ca/Cu	2,5864	1,5065	X	X	X	Mn/K	2,5865	1,6147			X
Ca/Mn	0,6904	0,2674	X		X	Mn/Ca	1,7850	1,0510		X	X
Ca/Fe	0,3071	0,0926	X	X	X	Mn/Mg	11,6390	5,9153			X
Mg/N	0,0980	0,0173		X	X	Mn/S	5,4646	5,1483		X	X
Mg/P	1,4705	0,2403	X	X	X	Mn/B	0,8287	0,4293		X	X
Mg/K	0,2179	0,0517	X	X	X	Mn/Zn	1,9123	1,1921	X	X	X
Mg/Ca	0,1516	0,0351	X	X	X	Mn/Cu	4,6606	4,0784	X		X
Mg/S	0,4404	0,2311		X	X	Mn/Fe	0,5444	0,3329	X	X	X
Mg/B	0,0723	0,0256		X	X	Fe/N	2,3817	0,8848	X		X
Mg/Zn	0,1626	0,0387	X	X	X	Fe/P	36,101	14,2700		X	X
Mg/Cu	0,3909	0,2223	X	X	X	Fe/K	5,3725	2,5203	X	X	X
Mg/Mn	0,0994	0,0321	X	X	X	Fe/Ca	3,7003	1,6352			X
Mg/Fe	0,0464	0,0165		X	X	Fe/Mg	25,7109	13,4116	X		X
S/N	0,2829	0,1438			X	Fe/S	10,9353	7,3733		X	X
S/P	4,2903	2,2582			X	Fe/B	1,7883	0,9809			X
S/K	0,6233	0,3228			X	Fe/Zn	3,9554	1,6004		X	X
S/Ca	0,4192	0,1857			X	Fe/Cu	9,4342	6,2703	X	X	X
S/Mg	2,8927	1,3862	X		X	Fe/Mn	2,6114	1,7679			X

Data obtained from doctorate thesis of Serra (2011).

Table 1. DRIS norms, following the methodology of "F" values (Jones, 1981; Letzsch, 1985; Walworth & Sumner, 1987), "r" value (Nick, 1998) and all dual ratio (ADR) (Serra, 2011)

4. DRIS index

Several changes in the methodology of DRIS indexes calculation were proposed in order to increase the accuracy in the nutritional diagnosis for several crops. The calculation of the functions or standard deviation units can be defined by the methodology originally developed by Beaufils (1973), Jones (1981) or Elwali & Gascho (1984), there are some conflicting results in the literature regarding the effectiveness of each method of calculation. According to Mourão Filho (2004), there is still no clear definition of what would be the best recommendation to calculate the functions or standard deviation units for the DRIS.

According to Serra (2011), the use of the methodology proposed by Jones (1981) when compared with Beaufils (1973) and Elwali & Gascho (1987) showed better efficacy on DRIS index for cotton crop (*Gossypium hirsutum* r latifolium). The measure of the efficacy used by Serra et al. (2011) was the relation between yield and nutritional balance index (NBI).

Beaufils (1973):

For A/B < a/b;

$$f(\left(\frac{A}{B}\right) = \left[1 - \frac{a/b}{A/B}\right] \cdot \frac{100 \cdot K}{CV\%}$$

f(A/B)=0, for A/B = a/b

For A/B > a/b;

$$f(\left(\frac{A}{B}\right) = \left[\frac{A/B}{a/b} - 1\right] \cdot \frac{100 \cdot K}{CV\%}$$

Jones (1981):

$$f(\left(\frac{A}{B}\right) = \left[\left(\frac{A}{B}\right) - \left(\frac{a}{b}\right)\right] \cdot \frac{c}{s}$$

Elwali & Gascho (1984):

For A/B < a/b-1s

$$f(\left(\frac{A}{B}\right) = \left[1 - \frac{a/b}{A/B}\right] \cdot \frac{100 \cdot K}{CV\%}$$

f(A/B)=0, to the range between a/b-1s to a/b+1s

For A/B > a/b+1s

$$f(\left(\frac{A}{B}\right) = \left[\frac{A/B}{a/b} - 1\right] \cdot \frac{100 \cdot K}{CV\%}$$

After defining the functions DRIS, the DRIS index is calculated and for each nutrient a DRIS index is determined, which may have positive or negative values, that represent the arithmetic average of functions in which the nutrient is involved, when the result is negative (below zero), this means deficiency and when the positive value indicates excess, as proposed by Beaufils (1973):

$$DRIS\ Index\ A = \frac{\Sigma f\left(\frac{A}{B}\right) - \Sigma f\left(\frac{B}{A}\right)}{n}$$

n=number of DRIS functions of each dual ratio defined by criteria of chosen of the norms, in that the A nutrient is involved.

The sum of DRIS index in module of the nutrients in a sample diagnosed, generates the nutritional balance index (NBI), in an increasing scale, the higher NBI the greater nutritional imbalance in the plant and consequently low productivity, and the correlation between NBI and yield is considered one measure of the effectiveness of the system DRIS (Beaufils, 1973; Nachtigall & Dechen, 2007b; Guindani et al., 2009).

Mourão Filho (2004) concludes that researches on DRIS are still incipient, therefore, many accurately factors must still be better studied, factors such as the criteria for choosing the reference populations, the combination of methods to be used, so there is a need to more refined studies on these aspects.

5. Nutritional Balance Index (NBI)

When assessing the nutritional status of plants, looking up the nutritional balance of the plant, however, this goal can not be reached when using the traditional methods of nutritional diagnosis, such as the sufficiency range and critical level, because, both of them lead into account only the individual concentrations of nutrients in the plant, with no relationship among these nutrients.

The Diagnosis and Recommendation Integrated System (DRIS) provides the relationship between nutrients through dual ratio (A/B and/or B/A). Thus it is feasible by calculating DRIS index to obtain the nutritional balance (Baldock & Schulte, 1996). In addition to the DRIS index, which may take positive and negative values, there is the nutritional balance index (NBI), which is the sum in modulus of the DRIS indices from a sample and thus the lower the value of the NBI would be more nutritionally balanced in the crop.

Despite of the diagnosis of nutritional status, the DRIS can be a useful tool to indicate situations where yield is limited by other factors than nutritional, however, it does not discriminate the factors that would be limiting the yield. In crops that have low yield and low NBI it is expected that other factors were limiting productivity, not being a limitation by the nutritional status of the plant (Beaufils, 1973).

The Nutritional Balance Index (NBI) was calculated by summing the value in module of the index generated in the sample. This NBI may be useful to indicate the nutritional status of the plant. The higher the NBI, the greater the nutritional imbalance (Beaufils, 1973; Mourão Filho, 2003). The average NBI generates NBIa (Nutritional Balance Index average), according to the formula below:

$$NBI = |I\ DRIS\ A| + |I\ DRIS\ B| + |I\ DRIS\ C| + \cdots + |I\ DRIS\ N|$$

$$NBIa = \frac{NBI}{n}$$

Where: n is the number of DRIS index involved in the analysis.

The NBI has been used to prove the effectiveness of the DRIS system in diagnosing the nutritional status of the plant, because the greater the relationship between NBI and yield better the diagnostic system response, to point out the nutritional status of plants (Silveira et al. 2005b) (Figure 1). Guindani et al. (2009) used the NBI to select the reference population to compose DRIS norms relating to the NBI tracks yield and the yield range that had the highest coefficient of determination (R^2) was selected as the reference population.

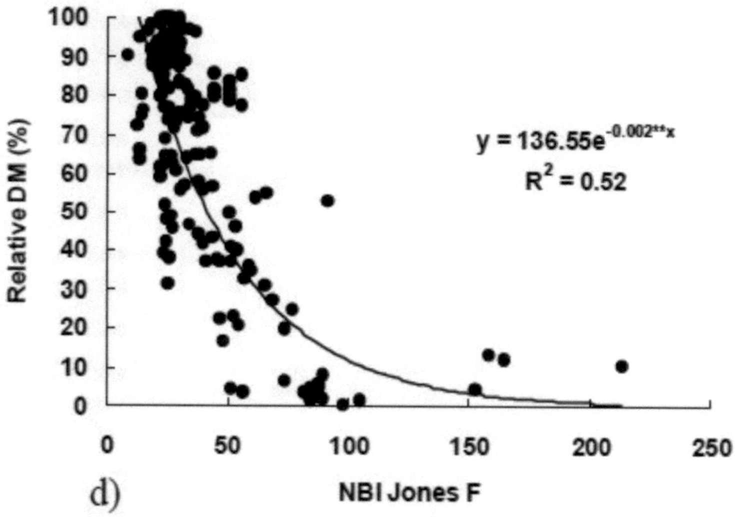

Figure 1. Relationship between the relative dry mass (DM) production of Signal grass and Nutritional Balance Index (NBI) obtained by the method of Jones for combinations of methods choice of ratio order among nutrients (F and R values) in the first growth using the norm of the first growth (a), in the first growth using the general norm (b), in the second growth using the norm of the second growth (c) and in the second growth using the general norm (d) (Silveira et al., 2005b).

6. Interpretation of DRIS index

The interpretation of DRIS index is the identification of nutrients that are limiting the crop yield from the presenting in nutritionally balanced or non-limiting. DRIS index can provide all null values or null values, positive and negative. However, the probability of having all zero values is small, therefore, it is necessary that all dual ratios show the same mean value of the standards. What happens under the conditions of analysis with the DRIS system is the presence of null values, positive and negative (Beaufils, 1973).

Null values mean that the average deviation for a given nutrient, are equidistant and cancelled in the expression of the final value for the DRIS index. It is said therefore that this nutrient is in a state of nutritional balance (Walworth & Sumner, 1984). After determination of DRIS index is necessary to interpretate these positive and negative values of a particular nutrient, it would be a situation in which the nutrient would be in excess (+) or deficiency (-).

6.1. Interpretation of DRIS index values by the order of the value

The usual method that is used for the interpretation of DRIS index is the ordering of the values of the indices, the ordering is more limiting disabilities by the most limiting excess.

By this method of ordering of the index establish that the lowest DRIS index and negative has been considered the most limiting, the second lowest, the second most limiting disability and until the most limiting excess, which would have the DRIS index greater and positive (Walworth & Sumner 1987; Bataglia & Santos, 1990). These criteria have been used both to evaluate the accuracy of the method (Jones, 1981) and for nutritional surveys, when there is the DRIS as a tool for identifying classes of farms and the distribution of nutritional status (Beaufils, 1973; Eymar et al., 2001; Hundal et al., 2005, Silva et al., 2009; Sema et al., 2010).

6.2. Interpretation of DRIS index by nutrient application potential response

The interpretation of DRIS index for the nutrient application potential response, was originated by Wadt (1996). This method of interpretation consists on grouping five categories of nutrient application potential response (NAPR), by comparing the rates of each nutrient DRIS with the nutrient balance index average (NBIa), which is the arithmetic average of the module of all DRIS index. The NBIa was chosen to be a value that reflects the average of the deviations of each dual ratio relative to the reference value (Wadt, 1996), as seen in Table 2.

The nutrient status of "highest deficiency" represents the situation where there is greater likelihood of positive response with the addition of the nutrient to soil. This positive response should be represented by higher crop yields, or by improving the quality of the agricultural product into a commercially desired degree. In turn, the status of "deficiency" also indicates that it is likely to increase in crop yield with the application of the nutrient, however, this probability is lower than the nutrient with the highest degree of deficiency ("highest deficiency") (Wadt, 1996).

The status "balanced" means that no crop response is expected in relation to the application of the nutrient in soil, there would be no response or a response of the crop "null". The nutrient status of "highest excess" represents the situation where the application of the nutrient may result in negative response on the crop yield, decreased productivity. Finally, the status of "excess" indicates that the addition of nutrients in soil may also result in negative response of the crop and its yield, but that this effect on yield can be controlled by higher nutrient excess (Wadt, 1996).

As recommended by Wadt (1996) the central concept for the addition of the nutrient to soil by the nutrient application potential response is that this increase should be considered as an adjustment in the fertilizer to soil. For example, when it is sure that the nutrient is in a

Nutricional	Criteria	Type of nutrient application potential response
Deficiency	I DRIS A < 0, | I DRIS A | > NBIa [††] and I_A is the index of lower value.	Positive, with higher probability (p)
Deficiency-prone	I DRIS A < 0 and | I DRIS A | > NBIa	Positive, with low probability (pz)
Sufficient	| I DRIS A | = NBIa	Null (z)
Excess-prone	I DRIS A > 0 and | I DRIS A | > NBIa	Negative, with a low probability (nz)
Excess	I DRIS A > 0, | I DRIS A | > NBIa and I_A is the index of higher value.	Negative, with a higher probability (n)

[†]the NAPR was calculated according to Wadt (1996).
[††]NBIa = Nutritional Balanced Index average.

Table 2. Criteria to interpret the DRIS index (I DRIS) by nutrient application potential response (NAPR[†]) (Wadt, 1996).

status of balance and adds it to the crop, it will not result in improved yield, yet, it does not mean that this nutrient should be excluded from the fertilizer recommendation, but that should be kept at fertilization at the same dosages that had been used.

For extraction of nitrogen in the soil, the extractants that have been used do not show a good correlation between the contents extracted by plants with the growth of plants or amount absorbed, and the fertilizer recommendations arising from fertilization of tables that are constructed by means of average curves response generated under field conditions, with data from multiple trials and different locations. Thus, it is expected that the diagnostic system allows adjustments to the amount of each nutrient to be applied, and the interpretation of DRIS index for the nutrient application potential response a useful tool for this purpose.

The nutritional diagnosis would be a complementary tool for the recommendation of the nutritional need of crops, however, it is not feasible to take off the use of soil analysis, because it is essential to check the evolution of soil fertility, and ability to supply nutrients (Wadt, 1996).

The use of nutrient application potential response (NAPR) for interpreting the DRIS index is well seen in Brazil, where, Wadt (1996), Dias et al. (2011), Serra et al. (2010a,b) and Serra (2011) (Figure 2) used to interpret the DRIS index in assessing the nutritional status of the cotton crop, Dias et al. (2011) used the NAPR in the cupuaçu crop (*Theobroma grandiflorum*).

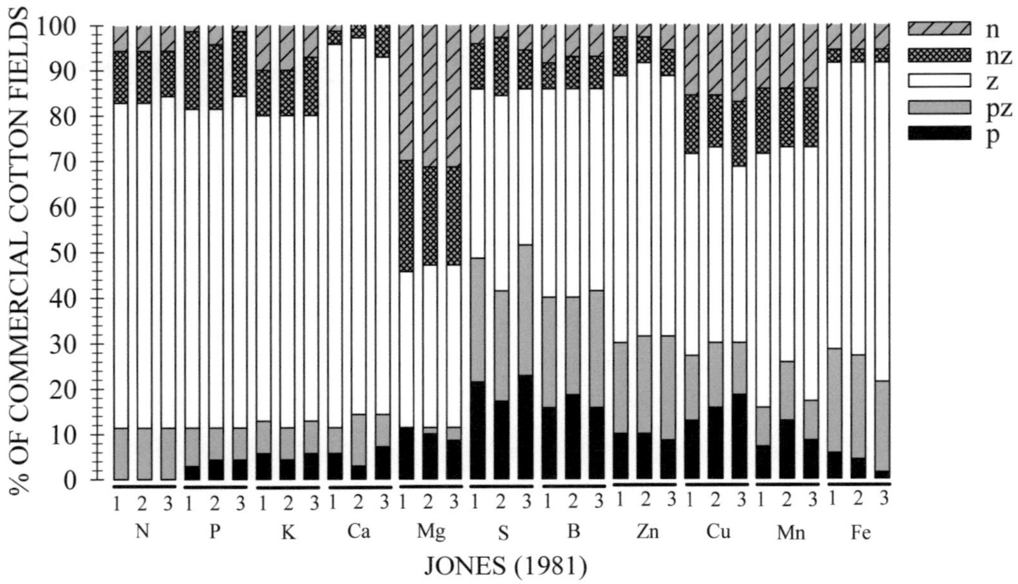

Figure 2. Percentage of plots diagnosed with the method of interpretation of DRIS index named nutrient application potential response (NAPR) (Wadt, 1996): (n) Negative response, with a higher probability; (nz) Negative response, with a low probability; (z) Nula response; (pz) Positive response, with low probability; (p) Positive response, with higher probability. (1) norms with all dual ratio; (2) norms with F value; (3) norms with r value (Serra, 2011).

7. Interpretation of leaf contents by Beaufils ranges

The determination of the Beaufils ranges consists in optimal ranges of nutrients for the assessment of leaf nutrients (Table 3 and 4). This method consists of determining the ranges by means of statistical models of the relationship between leaf concentrations and DRIS index, and, Beaufils (1973) found that from the optimal values of DRIS index were determined intervals of standard deviation of DRIS index for each range of nutritional assessment. Following this criterion, the range that would include the nutrients that would be deficiency was below - 4/3 standard deviation (s); deficiency-prone between - 4/3 to 2/3 s; sufficient between - 2/3 to 2/3 s and a excess-prone 2/3 to 4/3 s; excessive greater than 4/3 s (Table 3).

Thus, it creates the Beaufils ranges, which can be used to interpretate the nutrient concentration in chemical analysis of leaves (Table 3 and 4). As such use, recommended for the specific regions where they were certain, because if extrapolated to other region, it is expected that the results do not follow a favorable response.

Nutrient	Norm	Deficiency	Tendency to deficiency	Sufficient	Tendency to excess	Excess
				g kg^{-1}		
N	NL	<42.5	42.5 – 43.9	43.9 – 46.7	46.7 – 48.1	>48.1
	F value	<42.6	42.6 – 43.9	43.9 – 46.7	46.7 - 48.1	>48.1
P	NL	<2.5	2.5 – 2.8	2.8 - 3.3	3.3 - 3.5	>3.5
	F value	<2.5	2.5 – 2.8	2.8 – 3.3	3.3 - 3.5	>3.5
K	NL	<17.2	17.2 – 19.0	19.0 - 22.6	22.6 - 24.4	>24.4
	F value	<17.2	17.2 – 19.0	19.0 - 22.6	22.6 - 24.4	>24.4
Ca	NL	<25.2	25.2 – 27.5	27.5 - 32.1	32.1 - 34.4	>34.4
	F value	<25.3	25.3 – 27.5	27.5 - 32.0	32.0 - 34.3	>34.3
Mg	NL	<3.6	3.6 – 4.0	4.0 - 4.8	4.8 - 5.3	>5.3
	F value	<3.6	3.6 – 4.0	4.0 - 4.8	4.8 - 5.2	>5.2
S	NL	<4.8	4.8 – 8.7	8.7 - 16.5	16.5 - 20.4	>20.4
	F value	<5.2	5.2 – 8.9	8.9 - 16.3	16.3 – 20.0	>20.0
				mg kg^{-1}		
B	NL	<40.8	40.8 – 53.6	53.6 - 79.4	79.4 - 92.2	>92.2
	F value	<41.2	41.2 – 53.8	53.8 - 79.2	79.2 - 91.8	>91.8
Zn	NL	<21.7	21.7 – 24.9	24.9 - 31.2	31.2 - 34.4	>34.4
	F value	<21.8	21.8 – 24.9	24.9 - 31.2	31.2 - 34.3	>34.3
Cu	NL	<3.9	3.9 – 9.8	9.8 – 22.0	22.0 – 27.9	>27.9
	F value	<3.7	3.7 - 9.8	9.8 - 22.0	22.0 – 28.1	>28.1
Mn	NL	<14.9	14.9 – 33.4	33.4 - 70.8	70.8 - 89.3	>89.3
	F value	<14.9	14.9 – 33.4	33.4 - 70.8	70.8 - 89.4	>89.4
Fe	NL	<52.3	52.3 – 80.2	80.2 - 136.5	136.5 - 164.4	>164.4
	F value	<52.6	52.6 – 80.4	80.4 - 136.3	136.3 - 164.1	>164.1

Serra et al. (2012)

Table 3. Beaufils ranges determined for the nutritional diagnosis of cotton plants based on DRIS norms (NL transformation and F value) (Serra et al., 2012).

Nutrient	Norm	Deficiency	Deficiency-prone	Sufficient	Excess-prone	Excess
				% of plots		
N	NL	19.44	15.74	38.89	9.26	16.67
	F value	19.44	15.74	39.81	8.33	16.67
P	NL	7.41	19.44	45.37	12.04	15.74
	F value	7.41	19.44	45.37	12.04	15.74
K	NL	16.67	14.81	44.44	13.89	10.19
	F value	16.67	14.81	44.44	13.89	10.19
Ca	NL	19.44	17.59	46.30	7.41	9.26
	F value	19.44	17.59	43.52	10.19	9.26
Mg	NL	12.04	9.26	31.48	18.52	28.70
	F value	12.04	9.26	31.48	14.81	32.41
S	NL	2.78	54.63	21.30	11.11	10.19
	F value	7.41	50.93	20.37	2.78	18.52
				% of plots		
B	NL	13.89	33.33	32.41	9.26	11.11
	F value	14.81	32.41	32.41	9.26	11.11
Zn	NL	25.00	25.93	25.93	10.19	12.96
	F value	25.93	25.00	25.93	8.33	14.81
Cu	NL	0.00	36.11	42.59	11.11	10.19
	F value	0.00	37.96	40.74	11.11	10.19
Mn	NL	0.00	22.22	55.56	11.11	11.11
	F value	0.00	22.22	55.56	11.11	11.11
Fe	NL	0.93	37.04	50.00	2.78	9.26
	F value	0.93	37.04	50.00	2.78	9.26

Serra et al. (2012)

Table 4. Percentage of plots diagnosed by Beaufils ranges as deficient, deficiency-prone, sufficient, excess-prone or excess leaf nutrient contents of cotton, based on the criteria of natural log transformation (NL) and F value (Serra et al., 2012).

8. Conclusion

The DRIS developed by Beaufils (1973) had among its objectives, to correct the problem of correlation with the sampling time of the plant nutrients, and using dual ratio that promote the relationship among. Hence, improving efficacy of plant nutritional diagnosis allows the determination of the evaluation of the nutritional balance.

With the advent of Diagnose and Recommendation Integrated System (DRIS) by Beaufils (1973), researchers were setting to this system of nutritional diagnosis in order to increase their efficiency. However, evolution has brought a number of possibilities for calculation of DRIS' norms and functions, that are needed to be tested to determine the best combination of methodology.

The use of DRIS is still being widely disseminated in the world, DRIS brings results consistently good in assessing the nutritional status of plants, showing the nutritional balance, a fact which is not observed with traditional systems (sufficiency range and critical level).

Author details

Ademar Pereira Serra and Davi José Bungenstab
Brazilian Agricultural Research Corporation – Embrapa, Brazil

Marlene Estevão Marchetti, Franklyn Clawdy Nunes Guimarães, Vanessa Do Amaral Conrad and Henrique Soares de Morais
Agrarian Science Faculty, Universidade Federal da Grande Dourados – UFGD, Brazil

Maria Anita Gonçalves da Silva
Agrarian Science Faculty, Universidade Estadual de Maringá – UEM, Brazil

Rosilene Pereira Serra
Federal University of Mato Grosso, Brazil

9. References

Alvarez Venegas, V.H. & Leite, R. de A. (1999). Fundamentos estatísticos das fórmulas usadas para cálculo dos índices DRIS. Boletim Informativo da Sociedade Brasileira de Ciência do Solo. v.24, p.20-25.

Baldock, J.O. & Schulte, E.E. (1996). Plant analysis with standardized scores combines DRIS and sufficiency range approaches for corn. Agronomy Journal. v.88, n.3, p.448-456.

Bailey, J.S.; Ramakrishna, A. & Kirchhof, G. (2009). Na evaluation of nutritional constraints on sweetpotato (*Ipomoea batatas*) production in the central highlands of Papua New Guinea. Plant and Soil. v.316, p.97–105.

Bataglia, O.C. & Santos, W.R. (1990). Efeito do procedimento de cálculo e da população de referência nos índices do sistema integrado de diagnose e recomendação (DRIS). Revista Brasileira de Ciência do Solo, v.14, p.339-344.

Beaufils, E.R. (1973). Diagnosis and recommendation integrated system (DRIS). A general scheme of experimentation and calibration based on principles developed from research in plant nutrition. South Africa: University of Natal, Pietermaritzburg. 132p. (Soil Science Bulletin, 1).

Beverly, R.B. (1987) Modified DRIS method for simplified nutrient diagnosis of 'Valencia' oranges. Journal of plant Nutrition, v.10, p.1401-1408.

Camacho, M.A.; Silveira, M.V.; Camargo, R.A.; Natale, W (2012). Faixas normais de nutrientes pelos métodos ChM, DRIS e CND e nível crítico pelo método de distribuição normal reduzida para laranjeira-pera. Revista Brasileira de Ciência do Solo. v.36, n.1, p. 193-200.

Dias, J.R.M.; Wadt; P.G.S.; Folle, F.A.; Solino, A.J.S.; Delarmelinda, E.A. & Tavella, L.B. (2011) Potencial de resposta à adubação para N, P, K, Ca e Mg em cupuaçueiros avaliados por diferentes normas DRIS. VOL. 41(1), p.77 – 82.

Elwali, A. M. O. & Gascho, G. J. (1984). Soil testing, foliar analysis, and DRIS as guide for sugarcane fertilization. Agronomy Journal, v. 76, p. 466-470.

Epstein, E. & Bloom, A. (2006). Nutrição mineral de plantas: princípios e perspectivas. 2.ed. Londrina: Editora Planta. 401p.

Guindani, R.H.P.; Anghinoni, I.; Nachtigall, G.R. DRIS na avaliação do estado nutricional do arroz irrigado por inundação. Revista Brasileira de Ciências do Solo, v.33, p.109-118, 2009.

Eymar, E.; Cadahía, C. & Africa, S. (2001) Foliar nutrient reference levels obtained in hydroponic cultures as preliminary norms for dris to fertigate conifers', Communications in Soil Science and Plant Analysis, 32:1, p.267-282, 2001.

Hallmark, W.B. & Beverly, R.B. (1991). Review – An update in the use of the Diagnosis and Recommendation Integrated System. Journal of Fertilizer Issues. v.8, p.74-88.

Hundal, H.S.; Singh, D.; Brar, J.S. (2005). Diagnosis and recommendation integrated system for monitoring nutrient status of mango trees in submountainous area of Punjab, India. Communications in Soil Science and Plant Analysis. v.36 (15): 2085-2099.

Jones, C.A. (1981). Proposed modifications of the diagnosis and recommendation integrated system (DRIS) for interpreting plant analyses. Communications in Soil Science and Plant Analysis. v.12, p.785-794.

Letzsch, W.S. (1985). Computer program for selection of norms for use in the diagnosis and recommendation integrated system (DRIS). Communications in Soil Science and Plant Analysis. v.16, p.339–347.

Mccray, J.M.; JI, S., Powell, G.; Montes, G. & Perdomo, R. (2010) Sugarcane Response to DRIS-Based Fertilizer Supplements in Florida. Journal of Plant Nutrition. v.196, p.66–75.

Malavolta, E. Manual de nutrição mineral de plantas. São Paulo, Agronômica Ceres, 2006. 638p.

Marschner, H. (1995) Mineral nutrition of higher plants. 2.ed. San Diego: Academic Press. 889p.

Mourão Filho, F.A.A. & Azevedo, J.C. (2003). DRIS norms for 'Valencia' sweet orange on three rootstocks. Pesquisa Agropecuária Brasileira, v.38, p.85-93.

Mourão Filho, F.A.A. . DRIS: Concepts and applications on nutritional diagnosis in fruit crops. Scientia Agricola, Piracicaba, v. 61, n.5, p. 550-560, 2004.

Nachtigall, G.R. & Dechen, A.R. (2007a). DRIS norms for evaluating the nutritional state of apple tree. ScientiaAgricola, v.64, n.3, p.282-287, 2007a.

Nachtigall, G.R. & Dechen, A.R. (2007b). DRIS Use on apple orchard nutritional evaluation in response to potassium fertilization. Communications in Soil Science and Plant Analysis, v.38, n.17, p.2557-2566, 2007b.

Nick, J.A. DRIS para cafeeiros podados. Piracicaba:USP/ESALQ, 1998. 86p. (Dissertação - Mestrado).

Parent, L.E. & Dafir, M. (1992). A theoretical concept of compositional nutrient diagnosis. Journal of the American Society for Horticultural Science, v.117, p.239-242.

Parent, L. (2011) Diagnosis of the nutrient compositional Space of fruit crops. Revista Brasileira de Fruticultura, v. 33, n. 1, p. 321-334.

Raghupathi, H. B., Reddy, Y. T. N., Reju, M. Kurian & Bhargava, B. S. (2005). Diagnosis of Nutrient Imbalance in Mango by DRIS and PCA Approaches. Journal of Plant Nutrition, v.27, n.7, p.1131-1148.

Reis Junior, R.A.; Corrêa, J.B.; Carvalho, J.G. & Guimarães, P.T.G. (2002). Diagnose nutricional de cafeeiros da região sul de Minas Gerais: normas DRIS e teores foliares adequados. Revista Brasileira de Ciência Solo, v.26, n.3, p.801-808.

Ramakrishna, A.; Bailey, J.S.; Kirchhof, G. (2009). A preliminary diagnosis and recommendation integrated system (DRIS) model for diagnosing the nutrient status of sweet potato (Ipomoea batatas). Plant and Soil, v.316, p.107–116.

Sema, A.; Maiti, C.S.; Singh, A.K. & Bendangsengla, A. (2010). DRIS nutrient norms for pineapple on alfisols of india. Journal of Plant Nutrition, v.33, n.9, p.1384-1399.

Serra, A.P.; Marchetti, M.E.; Vitorino, A.C.T.; Novelino, J.O. & Camacho, M.A. (2010a). Desenvolvimento de normas DRIS e CND e avaliação do estado nutricional da cultura do algodoeiro. Revista Brasileira de Ciências do Solo, v.34, n.1, p.97-104.

Serra, A.P.; Marchetti, M.E.; Vitorino, A.C.T.; Novelino, J.O.; Camacho, M.A. (2010b). Determinação de faixas normais de nutrientes no algodoeiro pelos métodos CHM, CND e DRIS. Revista Brasileira de Ciência do Solo, v.34, n.1, p.105-113.

Serra, A.P. Normas e funções DRIS para avaliação do estado nutricional do algodoeiro (*Gossypium hirsutum* r latifolium). Tese de doutorado. Dourados, MS: UFGD, 2011. 200f.

Serra, A.P.; Marchetti, M.E.; Rojas, E.P.; Vitorino, A.C.T. (2012). Beaufils ranges to assess the cotton nutrient status in the southern region of Mato Grosso. Revista Brasileira de Ciência do Solo, v.36, n.1, p. 171-182.

Silva, M.A.C.; Natale, W.; Malheiros, E.B. & Pavinato, A. (2009). Estabelecimento e validação de normas DRIS para a cultura do algodão no centro-oeste do Brasil. Acta Scientiarum Agronomy, v.31, n.1, p.93-99.

Silveira, C.P.; Nachtigall, G.R. & Monteiro, F.A. (2005a). Norms for the diagnosis and recommendation integrated system for signal grass. Scientia Agricola, v.62, n.6, p.513-519.

Silveira, C.P.; Nachtigall, G.R. & Monteiro, F.A. (2005b). Testing and validation of methods for the diagnosis and recommendation integrated system for signal grass. Scientia Agricola, v.62, n.6, p.520-527.

Singh, N.P.; Awasthi, R.P. & Sud, A. (2000). Diagnosis and recommendation integrated system (DRIS) norms for apple (Malus x Domestica Borkh. L. CV. Starking Delicius) in Himachal Pradesh. Indian Journal of Hoticulturae, v.53, n.3, p.196-204.

Sumner, M. E. (1977). Aplication of Beaufils' diagnostic indices to corn data published in literature irrespective of age and conditions. Plant and Soil, v. 46, p.359-363.

Parent, L (2011). Diagnosis of the nutrient compositional space of fruit crops. Revista Brasileira de Fruticultura, Jaboticabal - SP, v. 33, n. 1, p.321-334.

Urano, E.O.M.; Kurihara, C.H.; Maeda, S.; Vitorino, A.C.T., Gonçalves, M.C.; Marchetti, M.E. (2006). Avaliação do estado nutricional da soja. Pesquisa Agropecuária Brasileira, v.41, n.9, p.1421-1428.

Urano, E.O.M.; Kurihara, C.H.; Maeda, S.; Vitorino, A.C.T., Gonçalves, M.C.; Marchetti, M.E. (2007). Determinação de teores ótimos de nutrientes em soja pelos métodos chance matemática, sistema integrado de diagnose e recomendação e diagnose da composição nutricional. Revista Brasileira de Ciência do Solo, v.31, p.63-72.

Urricariet, S.; Lavado, R.S. & Martín, L. (2004) Corn response to fertilization and SR, DRIS, and PASS interpretation of leaf and grain analysis. Communications in Soil Science and Plant Analysis, v.35, n.3, p.413-425.

Wadt, P.G.S. Os métodos da chance matemática e do sistema integrado de diagnose e recomendação (DRIS) na avaliação do estado nutricional de plantios de eucalipto. Viçosa, MG, Universidade Federal de Viçosa, 1996. 123p. (Tese de Doutorado)

Wadt, P.G.S.; Novais, R.F. De; Alvarez Venegas, V.H.; Fonseca, S.; Barros, N.F. De; Dias, L.E. (1998). Três métodos de cálculo do DRIS para avaliar o potencial de resposta à adubação de árvores de eucalipto. Revista Brasileira de Ciência do Solo, v.22, p.661-666.

Wadt, P.G.S.; Novais, R.F.; Alvarez V., V.H. & Bragança, S.M. (1999). Alternativas de aplicação do "DRIS" à cultura de café Conilon (*Coffea canephora* Pierre). Scientia Agricola, v.56 n.1. p.188-199.

Wadt, P.G.S. (2005). Relationships between soil class and nutritional status of coffee plantations. R. Bras. Ci. Solo, 29:227-234.

Walworth, J.L. & Sumner, M.E. (1987). The diagnosis and recommendation integrated system (DRIS). Advance in Soil Science, v.6, p.149-188.

Walworth, J. L.; Wooddard, H. J. & Sumner, M. E. (1988). Generation os corn tissue norms from a small, high-yield data base. Communications in Soil Science and Plant Analysis, v.19, n.5, p.563-577.

Waireg, L. W. I. & Van Asten, P. J. A. (2012). norms for multivariate diagnosis of nutrient imbalance in arabica and robusta coffee in the east African highlands. Experimental Agriculture,v.48, p.448-460.

Chapter 6

Bioprocess Modeling and Control

Mihai Caramihai and Irina Severin

Additional information is available at the end of the chapter

http://dx.doi.org/10.5772/55362

1. Introduction

The bioprocess advancement is determined by the living cells capabilities and characteristics, the bioreactor performance as well as by the cultivation media composition and the main parameters evolution. The high metabolic network complexity inside the cells often determine very sophisticated, non-linear growth and product formation kinetics, with further consequences on the bioprocess behavior, but at the same time on the product quality and yield.

The key issue of this rather complicated situation is the use of modeling and further on of computer assisted control as a powerful tool for bioprocess improving. The process models, as relationships of the input, output and inner variables, though incomplete and simplified, can be effective to describe the phenomena and the influences of great importance for control, optimization and better theoretical knowledge. The function of any biological model is to describe the metabolic reactions rates and their stoichiometry on the basis of bioreactor conditions, with the main difficulties-the identification of principal factors affecting cellular growth and bioproduct formation, and the building up of a suitable model structure for the intracellular processes.

Moreover the scheduling, supervision and automatic control in modern bioprocessing is done by advanced process control systems, where all the functions are implemented in software (in accordance with the Figure 1). The main bioprocess control attributes are: handling of off-line analyses; recipe and scheduling; high level overall control; state and parameters estimation; simulation; prediction; optimization.

For the industrial developments the central and manifold objective of the computer control is the realization of the economic interests in assuring high operational stability, process reproducibility and increased product yield together with the maintaining of rigorous safety and the implementation of the GMP or environmental regulations, important requests in modern biomanufacture imposed by the product quality improving needs.

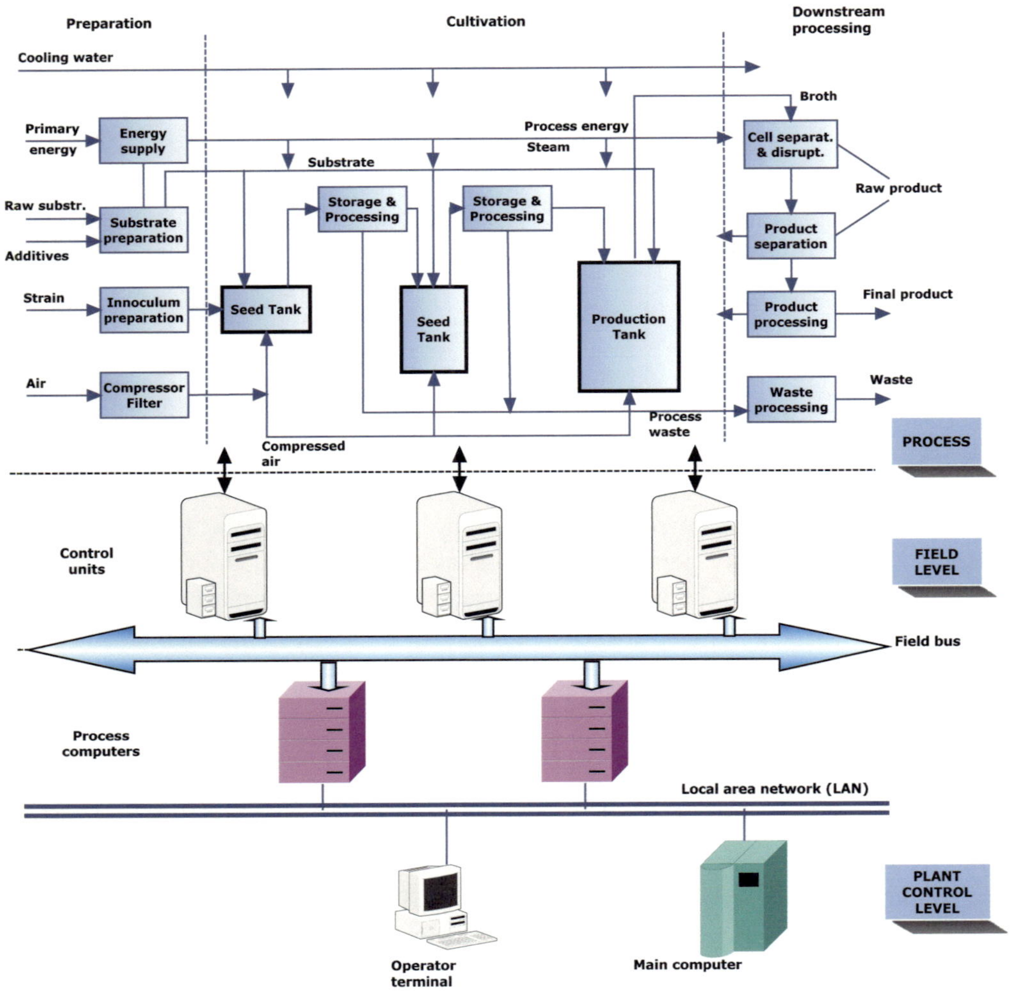

Figure 1. General presentation of the computerized bioprocess control [2]

2. Instruments and techniques for bioprocess variables determination and monitoring

To achieve the biological potential of cells, the optimal environmental conditions must be maintained in the bioreactor for cell growth / product formation, at least with regard to the key parameters. Generally speaking, biological systems are influenced by different process variables, which have a direct influence on cell metabolism. Sensors for these variables are (typically) inserted into specially designed ports on the bioreactor. As bioreactors increase in size (i.e. in the industry field), the mixing problems become usual and probe location becomes problematic. To accurately outline large fermenters, probes may be collected from several locations.

A. Direct physical determinations

The existence of defined and optimal environmental conditions for biomass and product formation means that different physical and chemical parameters require to be kept constant or conforming to an optimal evolution trend during the process, i.e. any deviation from a specified optimum might be corrected by a control system.

The standard direct physical determinations are[3]: (1) temperature; (2) pressure (over pressure); (3) agitator shaft power and rate of stirring; (4) foam; (5) gas and liquid flow; (6) weight.

Temperature determination is important for bioprocess evolution as well as other process operations (i.e. sterilization, concentration, and purification). The temperature measurement is made in the range +20°C to +130°C through mercury-in-glass thermometers, bimetallic thermometers, pressure bulb thermometers, thermocouples, metal-resistance thermometers or thermistors; all of them must be steam-sterilizable at 120°C. The most popular are the Pt100 resistance thermometers.

Pressure measurements may be needed for several reasons; the most important of them is the safety. Industrial and laboratory equipment is designed to withstand a specified working pressure plus a factor of safety. Also, the measurement of pressure is important in media sterilization. Moreover, the pressure will influence the solubility of gases and contribute to the maintenance of sterility, when a positive pressure is present. The standard measuring sensor is the membrane pressure gauge based on strain or capacitance measurements.

The formation of **foam** can create serious problems in no controlled situations: loss of broth, clogging of gas analyzers, infections, etc. It is a common practice to add an antifoam agent when the culture starts foaming above a certain predetermined level. A standard foam sensing consists in an electrical conductivity / capacitance / heat conductivity probe.

A number of mechanical antifoam devices have been made, including discs, propellers, brushes attached to the agitator shaft above the surface of the broth. Unfortunately, most of the mechanical devices have to be used in conjunction with an antifoam agent, without negative influence on the bioprocess behavior.

B. Direct chemical determinations

The regular chemical determinations are [3]: (1) pH; (2) redox potential; (3) dissolved oxygen concentration (pO_2); (4) exit-gas analysis; (5) on-line analysis of other chemical factors (ion-specific sensors, enzyme electrodes, microbial electrodes, mass spectrometers, fluorimeters).

In most processes there is a need for **pH** monitoring and control if maximum yield of a product is to be obtained. The pH may be further controlled by the addition of appropriate quantities of alkaline or acid solutions, depending of the characteristic pH trend evolution. Normally, the pH drift is only in one direction. pH measurement is carried out using a combined glass reference electrode that will withstand repeated sterilization at temperature of 120°C and pressures of $138 kN/m^2$.

In most aerobic fermentations it is essential that the **dissolved oxygen concentration** does not fall below a specified minimal level. If in small fermenter the most used electrodes are galvanic, the polarographic electrodes are more commonly used in pilot or production bioreactors. For an increase of precision, they are both pressure and temperature compensated.

C. On-line analysis of other chemical compounds

Ion-specific sensors have been developed to measure NH_3^+, Ca^{2+}, K^+, Mg^{2+}, PO_4^{3-}, etc. However, none of these probes are steam sterilizable.

The **mass spectrometer** [4] can be used for in-line analysis since it is very versatile, but unfortunately expensive. It does allow for monitoring of gas partial pressures (O_2, CO_2, CH_4, etc.), dissolved gases (O_2, CO_2, CH_4, etc.), concentrations of volatiles (methanol, ethanol, acetone).

The **fluorimetric measurements** [5,6] are very specific and rapid, but their use in bioprocess is quasi-limited nowadays. Hence, the measurement of NAD (provided it remains at a constant concentration in cells) would be an ideal method for continuous measurement of microbial biomass concentration.

The **biosensor** [7] is based on a biological receptor, which is coupled to an electronic transducer that converts the biological signal into an electrical signal by measuring voltage, current, light, temperature. Biosensors can be used to measure the concentration of different substrates / metabolites in the culture broth. In order to avoid the possible effects on growth / product formation (i.e. inhibition), the biological receptor can be immobilized on a separate membrane or on the transducer surface. **Enzyme electrodes** are the most applied, normally for in-line determinations (no steam sterilizable). The specific enzyme is immobilized on a membrane held in close contact to a pH or oxygen electrode. Also **microbial electrodes** using immobilized whole cells have been used for determination of sugars, acetic acid, ethyl alcohol, vitamin B, nicotinic acid, glutamic acid and cephalosporin. Generally speaking, the biosensors have been investigated with limited success. Hence, only the glucose biosensors have been fully applied. The main difficulties in developing an on-line biosensor are the thermal stability of the immobilized biomaterial, i.e. inactivation during sterilization, and the limited linear range inherent to biologic species.

It is to consider two most recent directions of development regarding the bioprocess variables monitoring and control [8]: (a) **realization of miniaturized sensors for the *in situ* measurement** of temperature, pH or dissolved oxygen; (b) **use of analyzers** not yet applied for on-line process monitoring in biotechnology, but of real interest: the continuous air-segmented flow analyzer (**CFA**); the flow injection analyzer (**FIA**); and the High Performance Liquid Chromatography (**HPLC**).

3. The mathematical modeling of the aerobic bioprocess

The aerobic bioprocess modeling is an useful tool to accomplish several important tasks [2]: (a) it can be the basis for adequate optimization and control technique applications; (b) it can provide the necessary information about the features of the chosen bioprocessing system; (c) it synthesizes the characteristics of the specified living cells' evolution and hence, it is the best technique to predict the process efficiency.

The models show the complex biosystems attributes; so they must be as possible as extensive and non-speculative. Moreover the models are an acceptable compromise between the presentation of processes in detail, with considerable number of parameters, and the use of few parameters, easy to apply and estimate.

Most important properties of a biological mathematical model were defined in the Edwards and Wilke' postulates [2]: (a) it is capable to represent all the culture phases; (b) it is flexible enough to approximate different data types without the insertion of significant distortions; (c) it must be continuously derivable; (d) it must be easy to operate, once the parameters evaluated; (e) each model parameter is to have a physic significance and must be easy to evaluate.

The attempts to realize high global models were not successful: firstly, due to the impossibility to measure on-line the great number of bioprocess parameters, and secondly, due to the high degree of complexity. Finally several types of models can represent the evolution of the aerobic bioprocess. The most important categories will be presented further on.

1. The unstructured global models are in use nowadays as the main tool for both the bioprocess modeling, but also for being applied in overall computer control [2]. Their limit is they are a simplified representation of the bioprocess behavior: conforming to this concept the bioprocess evolution depends directly and only on the macroscopic variables representing the working conditions in the bioreactor. Therefore the unstructured models are essentially kinetic equations that describe the variation of substrate or product concentrations and of a unique biological state variable-the cell concentration, and can also express the influences of some important process variables (pH, pO_2, temperature, and others), and only sometimes they are balance equations.

Generally speaking [9], one considers that the specific growth rate ($\mu = \frac{1}{X}\frac{dX}{dt}$) is the key variable for cell growth, substrate consumption and product formation. The specific growth rate is time dependent and dependent on different physical, chemical and/or biological parameters (substrate concentration-S, cell concentration-X, product concentration-P, pH, temperature-T, dissolved oxygen concentration-C, and different inhibitors-I).

Conforming to the literature assumptions [10], the specific growth rate dependence upon different process parameters can be considered as follows:

$$\mu = f\left(S, X, P, pH, C, I, ..., t\right) \tag{1}$$

a. $\mu=\mu(S)$ Kinetic models with growth limitation through substrate concentration (without inhibition) Main model equations [2, 11] are presented in Table 1.

Model equation	Constants	Authors	Comments
$\mu(S)=\frac{\mu_{max}S}{K_S+S}$ (2)	μmax=max specific growth rate [1/h] KS = saturation constant [g/L]	Monod equation (1942, 1949)	Empirically derived from the Michaelis & Menten equation
$\mu(S)=\frac{\mu_{max}S^n}{K_S+S^n}$ (3)		Moser equation (1988)	Analogy with a Hill kinetic (n>0)
$\mu(S)=\mu_{max}\frac{S}{K_S+K_D+S}$ (4)	KD=diffusion constant	Powell equation (1958)	Influence of cell permeability, substrate diffusion and cell dimensions through KD parameter

Table 1. Models $\mu=\mu(S)$

There are also some models, which utilize the substrate concentration in more complex structures. Nyholm (1976) introduces a dual function for substrate utilization: consumption (including assimilation and dissimilation in the liquid phase) and growth (substrate utilization for growth):

$$S \xrightarrow{k_{elim}} S_e \xrightarrow{k_{degrad}} S_a \tag{5}$$

S_e is the substrate for growth and S_a the substrate used for consumption. The growth rate is linked to the intracellular concentration of limiting substrate (S_{int}/X) and to *preserved* substrates (i.e. inorganic ions or vitamins, not decomposed through cell metabolism) with application in wastewater bio treatment:

$$\mu=\mu\frac{\frac{S_{int}}{X}}{\frac{dS_{int}}{dt}}=r_{S_{elim}}-r_{S_{degrad}} \tag{6}$$

b. μ=μ(X, S) The influence of cell and substrate concentrations upon the specific growth rate[2, 11]

Model equation	Constants	Authors	Comments
$\mu(X)=\mu_{max}(1-k_x X)$ (7)	kX=kinetic constant	Verhulst (1845)	It is known as growth logistic model
$\mu(X,S)=\mu_{max}\frac{S_0-\frac{X}{Y}}{K_S+S_0-\frac{X}{Y}}$ (8)	So=substrate initial concentration Y=substrate/cell yield.	Meyrath (1973)	It is based on Monod kinetics.

$N = N_0 \exp(\mu_{max} t) = \dfrac{N_0 \mu^0_{max} \exp(\mu^0_{max} t)}{\mu^0_{max} + m_x N_0 (\exp(\mu^0_{max} t) - 1)}$ (9)	N=population density m=limiting size of the population (the carrying capacity)	Verhulst – Pearl kinetics	Logistic growth: combination between the population trend to growth according to a geometric progression and the environment tendency to limit the excessively high densities of the population
$\mu = \mu_{max} \dfrac{S}{K_X X + S}$ (10)	KX=kinetic constant	Contois (Contois – Fujimoto) equation (1959):	If S = constant, the only dependence remains μ = f(X).

Table 2. Models$\mu = \mu(X, S)$

c. Growth kinetics with substrate inhibition

In most cases, the kinetic model equations are derived (like the Monod model) from the inhibition theory of enzymatic reactions. Consequently they are not generally valid and can be applied in connection with experimental acceptability [2, 11].

Model equation	Constants definition	Authors name	Comments
$\mu = \mu_{max} \dfrac{1}{1 + \dfrac{K_S}{S} + \dfrac{S}{K_i}} = \dfrac{S}{K_S + S} \dfrac{1}{1 + \dfrac{S}{K_i}}$ (11)	Ki = inhibition constant	Andrews model (1968)	Substrate inhibition in a chemostat
$\mu = \mu_{max} \dfrac{S(1 + \dfrac{S}{K^l_S})}{S + K_S \dfrac{S^2}{K^l_S}}$ (12)	Ksl= inhibition constant	Webb model (1963)	
$\mu = \mu_{max} \dfrac{1}{1 + \dfrac{K_S}{S} + \sum_j (\dfrac{S}{K_{i,S}})^j}$ (13)	Ki,S= inhibition constant	Yano model (1966)	
$\mu = \mu_{max} \dfrac{S}{K_S + S} e^{-\frac{S}{K_{i,S}}}$ (14)		Aiba model (1965)	

Table 3. Growth kinetics with substrate inhibition

d. μ = f(S, P) Growth kinetic with product inhibition [2, 11]

Hinshelwood (1946) detected product inhibition influences upon the specific growth rate: linear decrease, exponential decrease, growth sudden stop, and linear/exponential decrease in comparison with a threshold value of P. The first type (Hinshelwood - Dagley model):

$$\mu(S,P) = \mu_{max} \frac{S}{K_S + S}(1 - kP) \tag{15}$$

where: k = inhibition constant (considering the product concentration influence).

Model equation	Constants definition	Authors name
$\mu(P) = \mu_{max} - K_1(P - K_2)$ (16)	K1, K2 = constants (>0)	Holzberg model (1967)
$\mu(P) = \mu_{max}(1 - \frac{P}{P_{max}})$ (17)	Pmax = maximum product concentration.	Ghose and Tyagi model (1979)
$\mu(P) = \mu_{max} e^{-K_1 P(t)}$ (18)	K1 = constant	Aiba (1982):
$\mu(S,P) = \mu_{max} \frac{S}{K_S + S} e^{-KP}$ (19)		Aiba and Shoda model (1989)

Table 4. Models$\mu = f(S, P)$

e. The influence of dissolved oxygen (as a second substrate) upon the specific growth rate

In some cases it is needed to consider the dissolved oxygen as a second substrate. The most used equation is the kinetic model with double growth limitation, μ(S, C) [2, 11]

i. Olsson model:

$$\mu(S,C) = \mu_{max} \frac{S}{K_S + S} \frac{C}{K_C + C} \tag{20}$$

where: K_C = oxygen saturation constant.

ii. *Williams' model,* which also quantifies the P influence (K_P=P saturation constant; K_1, K_2, K_3, K_4=modeling constants):

$$\mu(S,C,P) = \left(\frac{K_1 S}{K_S + S} \cdot \frac{K_2 P}{K_P + P} \right) \cdot \left(\frac{C}{K_C + C} + K_3 C - K_4 \right) \tag{21}$$

f. $\mu(S_1, S_2)$ Kinetic models based on different substrates

Besides the case when the dissolved oxygen is considered as a second substrate, there are many cases when two or more carbon sources are taken into consideration. There are two typical situations: (1) the cells grow through the sequential (consecutive) substrate consumption (diauxic growth), where a simple Monod model can be applied; (2) the cells grow through the simultaneous consumption of substrates (e.g. wastewater treatment); in this case, the mathematical modeling is more complex.

g. Unstructured kinetic models for product formation

The product formation kinetic is taken into account in conjunction with the growth kinetic. Nowadays, the Gaden [3] classification is still useful. Based on this categorizing, four kinetic types can be defined:

Type 0: This production type occurs even in resting cells that use only a little substrate for their own metabolism. The microbial cells function only as enzyme carriers. Some examples are provided by steroid transformation and vitamin E synthesis by *Saccharomyces cerevisiae.*

Type 1: Type-1 situations include processes in which product accumulation is directly associated with growth; in this case the product formation is linked to the energy metabolism. Examples include fermentation to produce alcohol and gluconic acid and situations in biological wastewater treatment.

Type 2: Type-2 bioprocesses include fermentations in which there is no direct connection between growth and product formation (for example, penicillin and streptomycin synthesis).

Type 3: This production type includes those having a partial association with growth and thus, an indirect link to energy metabolism (e.g. citric acid and amino acid production)

Afterward there are now more advanced models, the structured and the segregated models.

2. In case of the structured models [12, 13] the biotic phase is not any more viewed as a homogenous component, but they provide information about the physiological state of the cells, their composition and regulatory adaptation to the environment. Conforming to this concept the cell mass is structured in several intracellular compounds and functional groups, which are connected to each other and to the environment by fluxes of material and information. The structured models can be: multi compartment models, genetically structured models, and biochemical structured models.

A case study of **the biochemical structured model is the modeling of Penicillin V biosynthesis:** The model of *Penicillin V* biosynthesis [2] is a tool for both: the understanding of the kinetic function of the precursors, the dissolved oxygen, enzymes activities, formation of metabolic intermediates and by-products (the determination of the metabolic step responsible for the global rate limitation can be a basis for the genetic engineering modification of the enzyme expression involved in this metabolic reaction); the bioprocess computer control.

First it is the metabolic pathway with the L-Cysteine, L-Valine and α-Aminoadipic Acid (AAA) as the initial substrates, which can form together Tripeptide ACV (α-α-aminoadipyl-L-cysteinyl-D valine). The further cyclisation reaction of Tripeptide ACV to Isopenicillin N (IPN) is oxygen dependent. The following reactions can be done directly in one step or in two steps. In this second case the intermediate is the 6-Aminopenicillanic Acid (6-APA), with the precursors Phenylacetic Acid (PAA) for Penicillin G or Phenoxyacetic Acid (POA) for Penicillin V, to be incorporated into the Penicillin molecule during the last step. It is also possible in parallel with Penicillin G and Penicillin V formation that 6-APA is alternatively carboxylated with CO_2 to form 8-HPA (8-hydroxy-Penicillinic Acid).The model for Penicillin V biosynthesis is presented in Table 5.

Metabolic step	Kinetic equation
ACV formation by ACV Synthetase	$r_1 = k_1 X_{ACVS} \cdot \dfrac{1}{\left(1+\dfrac{K_{AAA}}{C_{AAA}}+\dfrac{K_{CYS}}{C_{CYS}}+\dfrac{K_{VAL}}{C_{VAL}}\right)} \cdot \dfrac{1}{1+\dfrac{C_{ACV}}{K_{ACV}}}$ (22)
Isopenicillin N formation by IPN Synthetase	$r_2 = k_2 X_{IPNS} \cdot \dfrac{C_{ACV}}{C_{ACV} + K_o(1+\dfrac{C_{Glut}}{K_L})C_o}$ (23)
Formation of 6-APA from IPN by Isopenicillin N Amidohydrolase (IAH)	$r_3 = k_3 X_{IAH} \dfrac{C_{IPN}}{C_{IPN} + K_{IPN}}$ (24)
Formation of Penicillin V from activated side chain precursor and 6-APA by Acyl-CoA and 6-APA Acyltransferase (AT)	$r_4 = k_4 X_{AT} \cdot \dfrac{1}{1+\dfrac{K_{6APA-POA}}{C_{6APA}}+\dfrac{K_{POA}}{C_{POA-CoA}}}$ (25)
One step conversion of IPN to Penicillin V	$r_5 = k_5 X_{AT} \cdot \dfrac{1}{1+\dfrac{K_{IPN-POA}}{C_{IPN}}+\dfrac{K_{POA}}{C_{POA-CoA}}}$ (26)
Carboxylation of 6-APA to 8-HPA (first order kinetics if CO2 concentration is considered as constant)	$r_6 = k_6 X_{AT} . C_{6APA}$ (27)
Cleaving of Penicillin V to 6-APA and Phenoxyacetic Acid by Penicillin Amidase (PA) (reversible reaction of Penicillin formation)	$r_7 = k_7 X_{PA} \cdot \dfrac{C_{PenV}}{C_{PenV} + K_{PenV}}$ (28)

where X=the activity of the corresponding enzyme

Table 5. Model for Penicillin V biosynthesis

The parameters values from the above model were determined in a fed-batch bioprocess; it was found that the IPNS enzyme is metabolic flux limiting and further on the ACVS enzyme. As the IPN formation from Tripeptide ACV is dependent on the O_2 concentration, the dissolved oxygen concentration superior to 45% from the saturation can increase productivity.

3. The segregated models [12, 13] can describe more complex phenomena like: alterations or disturbances in the physiology and cell metabolism; cells 'morphological differentiation; genome mutations; spatial segregations of growth regions; cells aggregation; mixed cultures (including the competition between two or more species for the same substrate). On the contrary the unstructured and structured models have the limit to consider a homogenous population of cells and only one species in the bioreactor. The segregated models can be built by using ordinary differential equations to describe the behavior of several classes of independent/correlated cells. Each cell class behavior can be described by both unstructured and structured models.

4. Metabolic modeling

The most sophisticated modeling tool is that introduced by the metabolic engineering. This approach relies upon the concept of metabolic pathways as sequences of specific enzyme-catalyzed reaction steps converting substrates into cells' products. The manipulation of metabolic pathways to improve the cellular properties and especially the yield or the productivity of some important metabolites is of interest. So the metabolic engineering is recently developed with the purpose of generating information for the oriented modification of the enzymatic, regulatory or transport activities of the cells. The information will be used to build upgraded cells by the further application of the recombinant DNA technology.

The determination and the correct interpretation of the structure and the control mechanisms of metabolic networks are the first critical tasks of the metabolic engineering in order to fulfill the goal of rational pathway manipulation [14]. The main accent is towards considering the metabolic network as a whole and not the individual reactions. Due to the increased complexity of these networks and of the corresponding regulatory mechanisms the physiological state (metabolic steps characteristics at specific genetic and environmental conditions) of the cells is determined by the *in vivo* metabolic fluxes and their control.

The flux can be defined as the rate of material processing through a whole metabolic pathway. The value of the flux does not introduce information about the activity of the enzymes from the considered pathway, but it represents only their contribution regarding the substrate conversion into the final metabolite of this pathway.

The quantification of the metabolic fluxes is the principal objective realized by the techniques of **Metabolic Flux Analysis** (MFA) [15]. Metabolite balancing is the first operation in the determination of fluxes, done with the major hypothesis that the intracellular fluxes can be evaluated by measuring the extracellular fluxes. The metabolite balancing is performed by using a stoichiometric model for the intracellular reactions and by applying a mass balance around each intracellular metabolite, without any enzyme kinetic information. The general defining relationship is of matrix form:

$$S \cdot \underline{v} = \underline{r} \tag{29}$$

where: S=stoichiometric matrix of the metabolic network

$\underline{v}$=vector of unknown fluxes

$\underline{r}$=vector of measured metabolite extracellular concentrations, whereas the metabolite intracellular concentrations is 0.

The rows number is representing the number of metabolites in the pathway and the number of columns is equal to the unknown number of fluxes at steady state condition. The resulting system is normally underdetermined as the number of reactions is normally greater than the metabolite number. There are also various network structure characteristics (metabolic branch, reversible reactions, and metabolic cycles) that can increase the system degree of freedom.

So beside the metabolic balancing constraints additional constraints are needed to solve the equations system. If finally there are more constraints than the freedom degree, the system becomes over determined and redundant equations are to be used to test the consistency of the overall balances. The supplementary constraints can be obtained by using other information regarding the intracellular biochemistry and/or by applying others techniques [15].

So, another tool to perform MFA is the *Linear Programming* (LP) [14, 16]. Conforming to this method the metabolic fluxes are determined by simultaneously accomplishing 2 conditions: to be in line with the metabolic balances constraints and to optimize a certain objective function. So it is to formulate the mathematical problem:

$$\text{Minimize } \underline{c \cdot v} = \sum c_i v_i$$

$$\text{Subject to } S \cdot \underline{v} = \underline{r} \tag{30}$$

where: c= vector of the weight factors of fluxes in the objective function.

The objective functions can be: maximize the metabolite production rate or cells' growth rate; minimize the ATP production rate or substrate uptake rate; maximize growth rate for a given metabolite formation rate.

Another source of additional constraints is usually the introduction of certain types of supplementary measurements. The most useful tool of this type is the **application of isotopic tracer methods**. In isotopic tracer techniques there is a substrate in the cells labeled with an easily detectable isotope of a specific atom, normally ^{14}C, but especially ^{13}C, stable and non radioactive isotopes, to be detected by Nuclear Magnetic Resonance (NMR).

The isotope distribution among the metabolites from a network for a certain labeled substrate and known biochemistry is a function of the *in vivo* metabolic fluxes. This distribution can be obtained by studying the NMR spectra or by measuring the mass isotopomer (the molecules of the same metabolite, but with different labeling characteristics) distribution by Gas / Liquid Chromatography coupled with Mass Spectrometry (GC / LC-MS).

The general model for the determination of metabolic flux distribution is presented in the Fig. 2. The implementation of such flux quantification methods seems simple, but due to the high integrated networks complexity is rather an intensive computer application. There are now important studies of metabolic modeling used to improve the metabolite production in aerobic bioprocesses [17-22].

4. Bioprocess control

The bioprocess control has different goals and objectives, function of bioprocess characteristics and imposed performances. In spite of high non-linearity linear control theory and basic controllers (on/off, PID) are still applied in most industrial applications.

More sophisticated control should rely on models able to correctly represent the biosystems behavior. Due to the complexity of the biological systems, basic models, which are nice to use and help to simplify the underlying mathematics, are not able to reflect the real situations. The large sets of parameters from the complex models need to be experimentally identified, and consequently the e, and consequently the experiments should be carefully designed to provide this valuable information. Taking into account the time-to-market, which must be as short as possible the accepted control solution could be suboptimal based on classical robust control.

Bioprocess reproducibility and living cell systems variability reduction from run to run is to be carefully studied. The media composition optimization and the successful application of PAT (process analytical technologies combining the techniques for in-process monitoring, data-based modeling process control) will contribute to the quality of production improvement.

In bioindustry, bioprocesses are subject to a number of local and / or supervisory control structures. Local controllers are used to get the set-point control of different physical / chemical parameters (e.g. temperature, pH and dissolved oxygen concentration), while supervisory control is necessary for optimizing the feed in a fed batch process or the dilution rate in a continuous one [23].

Various simple feed-control strategies have been applied in the past [12, 24, 25]: (a) *Simple indirect feedback methods:* nutrients (indirect variable) are fed to the bioreactor by an on-off controller when a direct (on-line measured) variable deviates from its set point, e.g., feeding of ammonium by monitoring the pH (pH-stat), or nutrient feeding to keep the dissolved oxygen concentration constant (DO-stat). (b) *Predetermined feeding strategies;* this is a feed-forward strategy based on prior process knowledge, e.g., exponential feeding to grow at a constant biomass-specific growth rate. (c) *Direct feedback;* a substrate concentration can be directly controlled by nutrient feeding when it is measured on-line by sensors inside or outside the bioreactor. (d) *Feed control by state estimation;* the estimation of key-process parameters from on-line measurements can be applied and the control is based on the evolution of the growth rate or the substrate concentration.

Other advanced feed-control strategies may be applied when additional process information is available: *feed-forward model-based control; feedback model-based control* (an extended Kalman filter simultaneously estimates a state variable and adapt the controller); *fuzzy control; neural-network control* (for predictive control); *expert systems* (for supervisory control).

A. Bioprocess control with a priori model (model based process control)

The bioprocess control based on *a priori* model (BCAPM) can be seen as the on-line application of optimal control, where control actions are regularly re-calculated based on a global process model and process information. The global model is used to calculate optimal control actions by a prediction of future outputs over a limited time horizon.

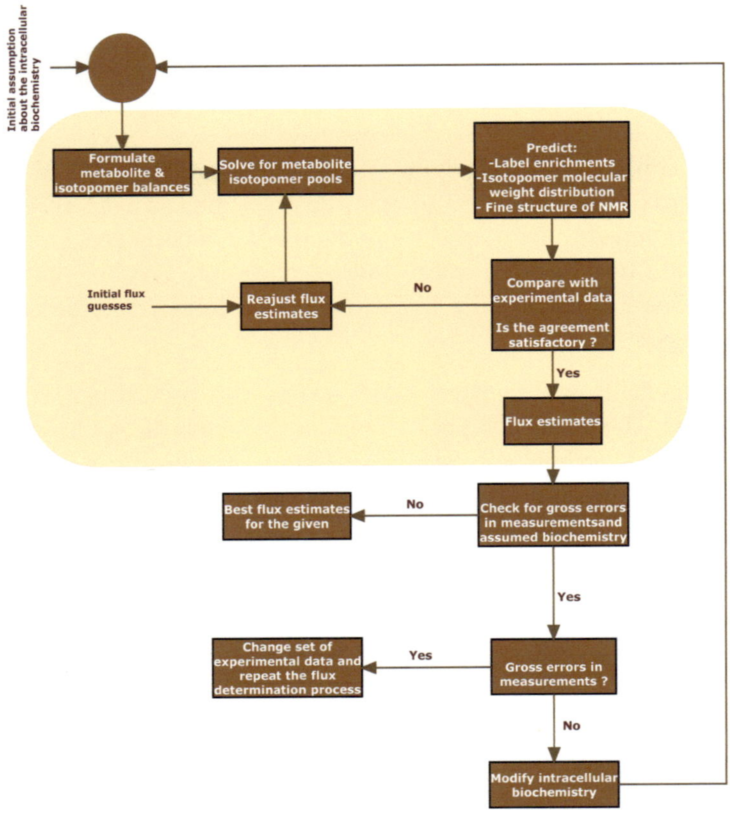

Figure 2. Determination of metabolic flux distribution [14]

For the time being, the unstructured deterministic models (the cells are considered as black-box units) are very used in the bioprocess control [26]. In the future an increase of the structured models role is expected, as a consequence of modern analysis methods development, as well as of the capacity to more adequately describe the phenomena.

The basic concepts of BCAPM consider two main ideas [27, 28]: (1) the explicit use of an *a priori* model to predict the process output(s); (2) the calculation of the future control actions by minimizing a global objective function.

The problem can be solved in different ways: (a) for a linear, time-invariant model, and in the absence of constraints, an explicit analytic solution of the above optimization problem can be obtained; (b) with linear constraints, the above optimization problem is a Quadratic-Programming problem, which can be numerically solved; (c) in the presence of a nonlinear model or nonlinear constraints, a non-convex optimization problem must be solved at each sampling period. So iterative optimization algorithms, (e.g. the Nelder-Mead method) can be used in order to converge to local minima.

There are two major problems which limit the application of BCAPM to bioprocesses [29]: (1) the model must predict the process variables evolution with sufficient precision; (2) given a nonlinear process model, the nonlinear optimization problem is solved for each (sampling) period; hence, the bioprocess model must be linear during these time periods.

The first item obstructs the application of BCAPM to complex or partially known systems, without defined global models. The second item blocks the application to performable systems; otherwise the control techniques are not properly used, due to the short sampling time periods (the second issue can be avoided by reason of large time constants characteristic to bioprocesses).

Recent developments in on-line measurement techniques, parameter and state estimation, in addition to the search of improved quality control, motivated the development of BCAPM. Now the technique was upgraded with better results. For instance [30] the applied BCAPM for feed control in the production of monoclonal antibodies allows to improve the yield with 43%.

B. Bioprocess adaptive control

When the process characteristics change during time, the operation conditions must also be changed: controller parameters and set point values. Moreover, optimal bioprocess evolution is commonly determined off-line, the process conditions are not perfectly known, and the process model is not well defined. Furthermore, it can be a lot of changes in process conditions in conjunction with different microorganisms' life cycles (when the cell concentration increase in time in a batch bioprocess, the oxygen set point must be increased). Hence, there is a need for some feedback mechanisms based on on-line measurements. On-line adaptation is possible when the state variables can be measured online[10] (directly using *hardware sensors* or indirectly by *soft sensors* [31, 32]).

The adaptive control structures are based on the design of different estimation algorithms which are able to determine the off-line parameter values. Many control algorithms were developed based on *minimal* knowledge about bioprocess kinetics (the *minimal modeling* concept) [33-36].

A typical adaptive control system is presented below:

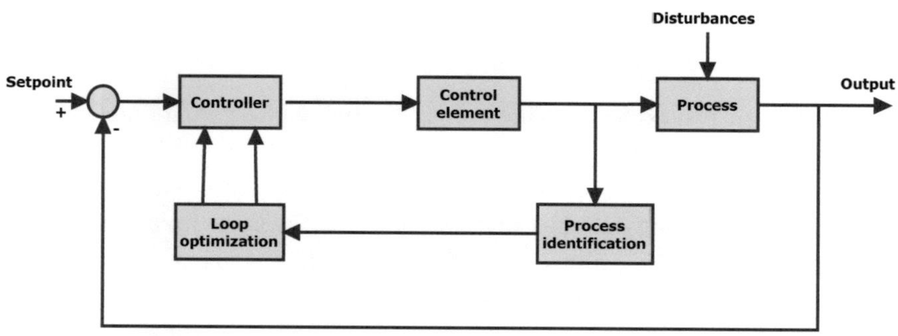

Figure 3. Adaptive control structure

There are two classes of adaptive control (where the adaptation is attained on the basis of on-line parameter observers) [37]: (1) the process changes can be measured – therefore it is possible to systematically adjust the controller settings, based on the measured / anticipated bioprocess changes; (2) the process changes cannot be measured / predicted – hence the controller settings are automatically adjusted by a loop optimizer.

C. Bioprocess control using Artificial Intelligence (AI)

The limitations of the bioprocess control systems do not concern only the measurements or models, but at the same time much valuable human knowledge is only available in a qualitative heuristic form.

Hence, it has been found that the knowledge-based control structures using the human decisional factor (i.e. a subjectively element) offer sometimes better results. Moreover, the computer performances are developed in the detriment of the general knowledge concerning life phenomena and do not promote advanced comprehension upon the metabolic routes of bioprocesses. Consequently, the intelligent techniques (i.e. neural nets, fuzzy structures, genetic algorithms or expert systems) are capable of simulating human expert-like reasoning and decision making, dealing with uncertainties and imprecise information [24].

As the human perception about the bioprocess is commonly altered by the psychological factors, the intelligent control systems founded (only) on the human subjective knowledge is less valuable than the control systems who utilize the objective information fitted by a conceptual model. Hence, the literature recommends the intelligent control techniques utilization only if the control structure based on quantitative models fails.

Frequently, different process parameters are controlled in order to follow predefined transitory trajectories. Such control strategies can be designed by a *trial-and-error* approach in combination with operator's experience and statistical analysis of historic data.

a) One method for automatic bioprocesses control using AI is based on *expert systems (ES)* that reproduce the human operator' rules of action. The literature presents several examples how to transfer the knowledge from operators into knowledge-rule bases [38]. An ES conceptual architecture is presented in Figure 4.

The most used ES systems in bioprocess control are applied for supervisory control, or process monitoring and diagnosis.

Moreover, the ES logic is used to translate human language into a mathematical description. The parameters tuning is then regulated by phase detection based on *if...then* rules, conditional statements representing heuristic reasoning in which *if* expresses the condition to be applied and *then*-the action to be done. Of course, at the same time, it is not possible to calculate optimal parameter' values with this method. For example [39] an ES was developed in order to supervise a conventional control system applied to fed-batch baker's yeast cultivation and to surmount its limitation. Expert system BIOGENES can execute standard process control tasks, but also advanced control tasks: process data classification;

qualitative process state identification (metabolic state, process phase, substrate feeding); supervisory control through corrective actions.

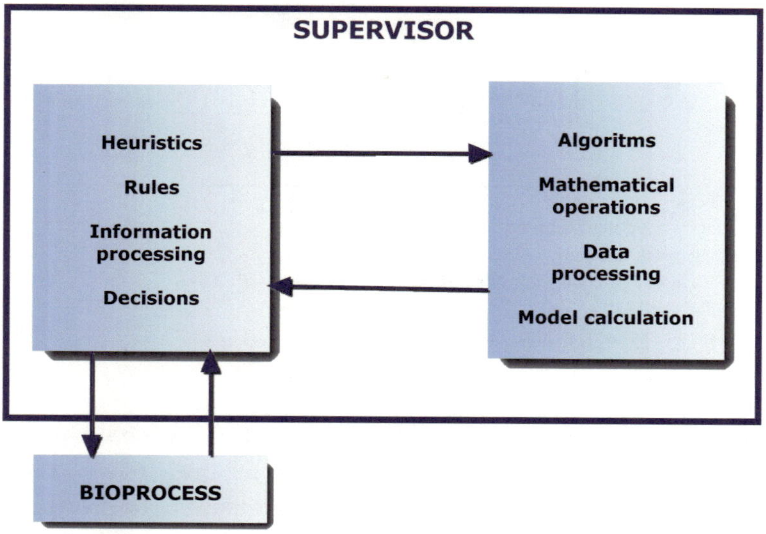

Figure 4. The expert system conceptual architecture

One of the main limits in developing ES is the knowledge acquisition process due to: (1) the linguistic rules formulation by human experts, i.e. the analytical description of their actions during the bioprocess (manual) control; (2) the loss of information during the transfer between the bioprocess expert and the IT specialist; (3) the subjectivity of human expert regarding own decision / rules.

b) Another AI technique, the *fuzzy approach* is based on fuzzy sets and fuzzy reasoning. In actual AI systems, fuzzy rules are often applied together with different types of models / parameter / state estimators. These fuzzy rules can be regarded as problem specific basis function system [40]. Any variable can be a fuzzy variable, particularly recommended when it is not possible to define its value in a given situation. One define fuzzy sets in the form of membership functions (between 0 and 1) in order to express what is likely to be considered as degree / level for a certain characteristic (high, medium, low). Relationships between fuzzy variables can be formulated with fuzzy logic operators (and, or, not) and processed by fuzzy logic. Fuzzy rules reflect the rules of thumb used in everyday practice and can be processed as if…then expressions. With a set of fuzzy rules, considered as universal process approximates, the behavior of a system can be described quite accurately. There are many applications: (a) hierarchical fuzzy models within the framework of orthonormal basis functions41 (Laguerre and Kautz functions); (b) several important use of fuzzy control in the Japanese bioindustry by the companies Ajinomoto, Sankyo or Nippon Roche [42]; (c) the control of the α-amylase fed batch bioprocess with the recombinant E. coli to maintain glucose and ethanol at low concentrations with 2 fuzzy controllers for feed rate control: feed forward and feedback [43] (see Figure 5 below):

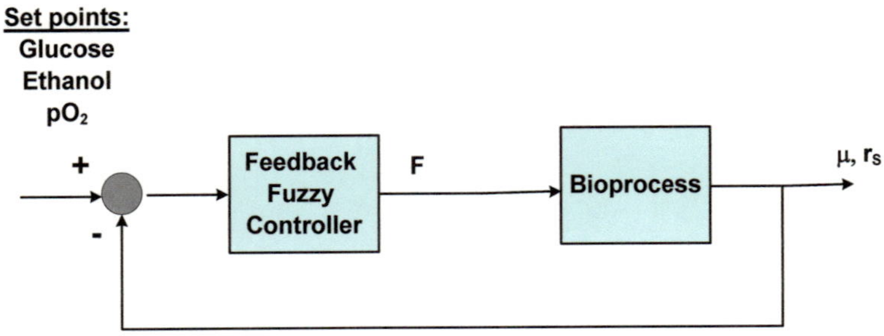

where: F = real glucose feed rate [L/h]
μ = specific growth rate [h^{-1}]
r_s = specific glucose consumption rate [g/L/s]
$Y_{X/S}$ = cellular yield [g/g]
pO_2 = dissolved oxygen concentration [mg/L]

Figure 5. Schematic presentation of the fuzzy control structure

c) One can use *artificial neural networks* (ANN) to get predictions about biosystem behavior. The traditionally used format of ANN is the feed forward. Given a set of process measurements, the output of ANN can be estimated parameters or process variables. The weights applied to the process measurements as inputs are determined through the "training process" of the ANN [44]. To train the ANN it is to get complete process information, corresponding to the NN inputs and outputs, from the data gathered in a set of fermentation runs. This set defines "an experimental space" and the ANN will predict outputs accurately only within this range and not beyond it.

Various applications were studied: (a) biomass and recombinant protein concentration estimation via feed forward NN for a fed batch bioprocess with a recombinant *E. coli* [45, 46]; (b) two types of NN (input/output and continuous externally recurrent) can control the batch and fed batch piruvat production from glucose and acetate with a recombinant strain of *E. coli* [47]; (c) two NN also to control the submerged bioprocess of *Monascus anka* fungus cultivation (the temperature and the dissolved oxygen are the inputs and the controlled outputs are the glucoamylase activity and the concentration of the red pigment [48]; (d) NN based soft sensor for online biomass estimation in fed bioprocess for polyhydroxibutirate production [49]; (e) media formulation optimization with genetic algorithm evaluated by ANN [50].

d) Because all types of information must be used in order to improve the bioprocess control: mathematical / deterministic models, heuristic knowledge, rule-based reasoning, a new control structure is developed in the last years – i.e. *hybrid control system* (HCS). HCS acts on both parts of bioprocess control: conventional control systems (i.e. based on a priori model) merge with AI techniques, in a complementary way: if a priori (mathematical) model exists, it will be preferred; else the linguistic rules (i.e. expert systems / fuzzy techniques) will be used.

Generally it is necessary to design a control system, which can to choose the (intelligent) control strategies, based on analytical models, in order to improve the control performances.

This is an Intelligent Control Structure (ICS) based on Hybrid Control Techniques (HCT). The most widely used hybrid structure combines balance equations with ANN: (a) balance equations for substrate and cell concentrations coupled with ANN for growth rate model in case of bakers 'yeast fed batch cultivation [51]; (b) ANN is responsible for modeling the unknown kinetics in applications with the yeast *Saccharomyces cerevisiae* or activated sludge urban wastewaters bio treatment [52]; (c) batch bioprocess of animal cells [53].

5. Case study [54]

The research objective of this case study was to develop an appropriate control method for a bioprocess and to implement it on a laboratory plant, namely the control of the fed batch cultivation of *Hansenula polymorpha* yeast for alcoholoxydase-containing biomass. At first, the process is described and a mathematical model is proposed and then the control strategy is defined and the intelligent control structure is designed. Finally, the control performances are tested through real data.

A discontinuous fed-batch bioprocess for alcoholoxydase-containing biomass with the methylotrophic yeast *Hansenula polymorpha* CBS - 4732 was operated in an airlift lab - bioreactor The intracellular enzyme, to be separated further on, is used for obtaining a high-specialized kit for methanol/ethanol determination. The yeast was cultivated on a complex medium with $(NH_4)_2SO_4$, KH_2PO_4, Na_2HPO_4, $MgSO_4*7H_2O$, $CaCl_2$, yeast extract or autolysed residual beer yeast as organic N source and microelements (Fe, B, Cu, I, Mn, Zn, Mo).

$$\begin{aligned}
\frac{dV}{dt} &= -\frac{E_S}{\rho_S} - \frac{E_M}{\rho_M} \\
\frac{dX}{dt} &= \frac{\mu_{\max} S}{K_S + S} X + \frac{X}{V}\left(\frac{E_S}{\rho_S} + \frac{E_M}{\rho_M}\right) \\
\frac{dS}{dt} &= -\frac{\mu_{\max} S}{K_S + S}\frac{X}{Y_{X/S}} - \frac{E_S \rho_S}{V} + \frac{S}{V}\left(\frac{E_S}{\rho_S} + \frac{E_M}{\rho_M}\right)
\end{aligned} \tag{31}$$

where: E_S and E_M are the substrate and medium loss by evaporation [g/h]; ϱ_S and ϱ_M are the substrate and medium densities [g/L]; $Y_{X/S}$ is the substrate conversion yield referred to the biomass [g dry matter/ g substrate]; μ is the specific growth rate [1/h]; V is the volume of the cultivation medium in the bioreactor [L]; X and S are the biomass and substrate concentrations [g/L] and *t* is the time [h], μ_{max} represents the maximum specific growth rate [1/h] and K_S is the saturation constant [g/g]. The main process parameters were: continuous temperature control 37°C; a minimal level of pO_2 - 10% from the saturation concentration was maintained during the exponential growth; continuous pH control between 4.5 - 5.0 by addition of NH_4OH (12.5%); no foam control, if the main parameters are optimally controlled. The unique C source, the methanol was introduced function of the yeast growth rate in connection with the substrate consumption rate for avoiding the growth inhibition by substrate concentration. The developed model (1) is based on the mass-balance principle and on the hypothesis of a non-inhibitive substrate effect (i.e. the specific growth rate is defined by the Monod equation). In line with the operation mode (fed-batch with discontinuous substrate feeding), there are discontinuous variations of the main variables due to: substrate feeding, medium feeding (to overcome the loss by evaporation or sample collection) or samples withdraws. That is why the following mass-balance equations are to be added to express each discontinuous modification for volume, and substrate or biomass concentrations:

$$\begin{aligned} &V_k + A_{Sk} + A_{Mk} = P_{Mk} + V_{k+1} \\ &S_k \rho_M V_k + A_{Sk} \rho_S = P_{Mk} \rho_M S_k + S_{k+1} \rho_M V_{k+1} \\ &X_k V_k = P_{Mk} X_k + X_{k+1} V_{k+1} \end{aligned} \tag{32}$$

where: V_k, V_{k+1}=volume before / after modification [L]; A_{Sk}, A_{Mk}=substrate volume and respectively medium volume adding [L]; P_{Mk}=sample withdraw [L]. The same notations are used for S_k, S_{k+1} and X_k, X_{k+1}. We use: ρ_S = 800[g/L]., respectively ρ_M = 1000[g/L]. The identification of the model parameters was carried out based on measured values in order to minimize the modeling error. The identification procedure (i.e. Nelder-Mead algorithm) determines the optimum values for the following process parameters: E_S, E_M, μ_{max}, K_S and $Y_{X/S}$.

For this bioprocess, the overall control objective is to obtain large biomass quantities, based on the assumption that high biomass concentration will assure the obtaining of important alcoholoxydase-active biomass. In this paper a control system based on fuzzy logic is proposed. It is well known that Fuzzy Control Systems (FCS) can manipulate incomplete and uncertain information about the process assuring high control performances [6-8]. The proposed FCS receives information about the state of the bioprocess expressed by the biomass and substrate concentrations. Based on this information, FCS computes the quantity of substrate to be added into the reactor. According to these observations the inputs of FCS are the biomass (X) and substrate (S) concentrations, and the output is the correction to be applied on the substrate addition. The rules of FCS are presented in Table 1.

Rules evaluation by the inference engine is made according to the min-max inference rule and the output defuzzyfication is made based on the centroid defuzzyfication method.

Sk \ Xk	S	M	L
S	Z	PZ	P
M	NZ	Z	PZ
L	N	NZ	Z

Table 6. The rule base

6. Results & discussions

The control loop was implemented in MATLAB, version 7.5. For control loop simulation the proposed mathematical model was used and the simulation results were compared with the experimental data.

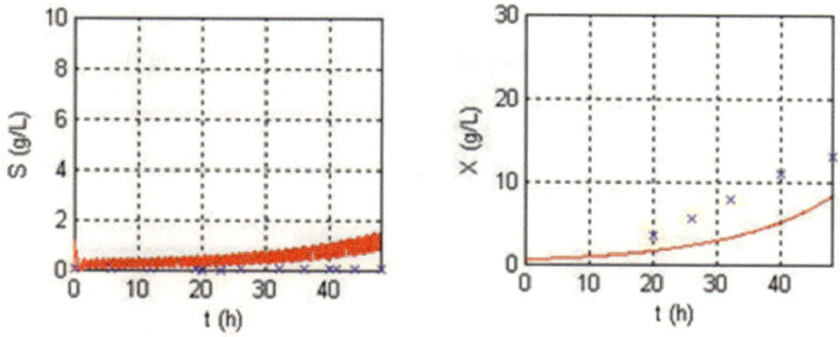

a)

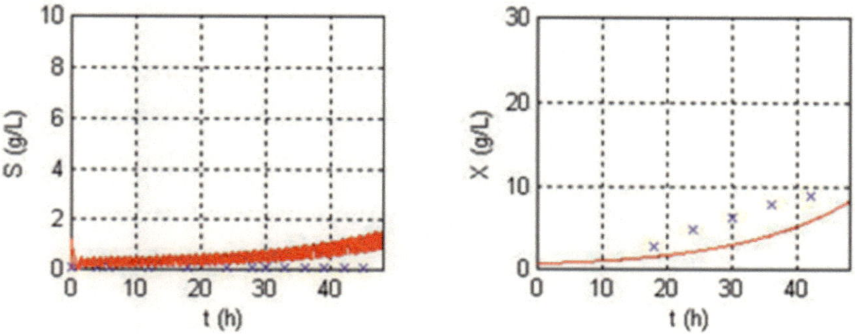

b)

Figure 6. Simulation results of the control loop: a) first experiment; b) second experiment; ('-' – simulation results; 'x' – experimental data)

The simulation results show that the proposed fuzzy control system is capable of computing the substrate feedings needed for cell growth according to the biomass concentration increase. The evolution of the substrate concentration marks the substrate consumption and additions, as well as the increase of the additions along with cell growth. The biomass concentration obtained by simulation follow closely the experimental data. As a conclusion of this case-study, it can be accepted that the success of such a control implementation is critically dependent upon the technical operating conditions of the process.

7. Conclusions

The overview on the current status of bioprocess modeling and control focuses on three main topics: (i) unstructured versus structured and metabolic modeling; (ii) control based on common technique (model based control and adaptive control); (iii) control based on artificial intelligence.

It is finally to underline that the framework of bioprocess modeling & control still offers interesting perspectives to obtain robust control solutions for the aerobic bioprocess. Moreover the future of bioprocesses' optimal control will rely on applying the same concept:

the use of different modeling methods in conjunction with intelligent control techniques. If a simplified representation of the bioprocess exists (i.e. an *a priori* model), this optimal profile can serve as an initial trajectory for intelligent control algorithms when the complexity of the process representation is described in a subjective mode (by human expert).

Author details

Mihai Caramihai and Irina Severin
University POLITEHNICA Bucharest,
Control Engineering & Computer Science Faculty / Engineering & Management of Technological Systems Faculty,
Romania

8. References

[1] A. A. Chirvase, I. Dumitrache, and M. D. Caramihai, *Revue Roumaine de Chimie,* 2007, *52(7),* 621-631
[2] K.H. Bellgardt, "Bioreaction Engineering", Ed. K. Schugerl, and K. H. Bellgardt, Springer Verlag, Berlin, 2000, p. 1-18; 56-72; 391-412
[3] K. Schugerl, "Biotechnology", vol. 4, Ed. H.-J. Rehm, and G. Reed, VCH, Weinheim, 1991, p. 6-28
[4] K. Desai, Y. Badhe, S.S. Tambe, and B. Kulkarin, *Biochem. Eng. J.,* 2006, *27,* 225-239
[5] F. Clementschitsch, J. Kern, F. Potschacher, and K. Bayer, *J.of Biotechn.,* 2005, *120,* 183-196
[6] D. Beohl, D. Solle, B. Hitzmann, and T. Scheper, *J.of Biotechn,* 2003, *105,* 179-188
[7] K. Yagi, *Appl. Microbiol. Biotechn.,* 2007, *73,* 1251-1258
[8] K. Schugerl, *J.of Biotechn,* 2001, *85,* 149-173
[9] A. Moser, *Acta Biotechnologica,* 2004, *15,* 173-195
[10] G. Bastin, and D. Dochain, "On-line estimation and adaptive control of bioreactors", Elsevier, Amsterdam, 1990, p. 20-230
[11] K. H. Bellgardt, "Biotechnology", vol.4, Ed. H.-J. Rehm, and G. Reed, VCH, Weinheim, 1991, p. 267-298
[12] M.A. Boudreau, and G. K. McMillan, "New Directions in Bioprocess Modeling and Control", ISA, New York, 2006, p. 112-181
[13] E. Heinzle, A. P. Biwer, and C. L. Cooney, "Development of Sustainable Bioprocesses: Modeling and Assessment", John Wiley, Chichester, 2006, p. 61-80
[14] M. J. Klappa, and G. Stephanopoulos, "Bioreaction Engineering", Ed. K. Schugerl, and K. H. Bellgardt, Springer Verlag, Berlin, 2000, p. 106-124
[15] L. Wang, and W. Hatzimanikates, *Metabolic Eng.,* 2006, *8,* 133-141
[16] K. G. Gadkar, R. Mahadevan, and F. J. Doyle III, *Automatica,* 2006, *42,* 1723-1733
[17] J. Zhao, and K. Shimizu, *J.of Biotechn,* 2003, *101,* 101-117
[18] C. Zhang, Z. Shi, P. Gao, Z. Duan, and Z. Mao, *Biochem. Eng. J.,* 2005, *27,* 225-239

[19] A. A. de Graaf, L. Eggeling, and H. Sahm, "Advances in Biochem. Eng. / Biotech." vol. 73, Ed. T. Scheper, Springer Verlag, Berlin, 2001, p. 10-30
[20] R. Goodacre, *Metabolomics,* 2005, *1,* 1-2
[21] M.R. Mashego, K. Rumbold, M. De Mey, E. Vandamme, W. Soetaert, and J. J. Heijnen, *Biotechn. Lett.,* 2007, *29,* 1-16
[22] M. J. van der Werf, *Trends in Biotech,* 2005, *23,* 11-16
[23] K. Shimizu, *Computer Chem. Eng.,* 1996, *20,* 915-941
[24] A. Luebbert, and S. Jorgensen, *J.of Biotechn,* 2001, *85,* 187-212
[25] S. Mishra, *WO/2007/18868,* May 4, 2007
[26] C. Brosilow, and B. Joseph, "Techniques of Model-based Control", Prentice Hall PTR, New Jersey, 2002, p. 39-85
[27] V. G. Ryckaert, and J. F. van Impe, "Proceedings of the 1998 IEEE Intern. Conf.", vol. 1, IEEE Print. House, 1998, p. 56-60
[28] A. V. Wouwer, P. Bogaerts, *Bioprocess Biosystems Eng.,* 2005, *27,* 281-282
[29] A. P. Teixeira, C. Alves, P. M. Alves, M. J. T. Carrondo, and R. Oliveira, *BMC Bioinformatics,* 2007, *8,* 30-41
[30] V. Galvanauskas, R. Simutis, and A. Luebbert, *Bioprocess Biosystems Eng.,* 2004, *26,* 393-400
[31] P. Bogaerts, and A. van de Wouwer, *Chem. Eng. Sci.,* 2004, *59,* 2465-2476
[32] A. Cheruy, *J.of Biotechn,* 1997, *52,* 193-199
[33] J.J. Dunn, E. Heinzle, J. Ingham, and J. E. Prenosil, "Automatic Bioprocess Control Fundamentals", Wiley-VCH, Weinheim, 2003, p. 314-367
[34] F. Jadot, G. Bastin, J. F. van Impe, *J.of Biotechn,* 1998, *65,* 61-68
[35] P. A. Lant, M. T. Tham, and G. Montague, *Biotech. Bioeng,* 2004, *42,* 1311-1321
[36] M. Kay, *BioProcess International,* 2006, *4,* 70-76
[37] M.N. Ignatova, V. N. Lyubenova, M. R. Garcia, C. Vilas, and A. A. Alonso, *J. Process Control,* 2008, *18,* 27-35
[38] T. Suzuki, Y. Sakino, M. Nakajima, H. Asama, T. Fujii, K. Sato, H. Kaetsu, and J. Endo, *J.of Biotechn,* 1997, *52,* 277-282
[39] E. Hrncirik, J. Nahlik, and J. Vovsik, *Expert Systems Appl.,* 2002, *23,* 145-153
[40] C.-F. Mandenius, *Bioprocess Biosystems Eng.,* 2004, *26,* 347-351
[41] R. J. G. B. Campello, F. J. von Zuben, W. C. Amaral, L. A. C. Meleiro, and R. M. Filho, *Chem. Eng. Sci.,* 2003, *58,* 4259-4270
[42] J. Horinchi, *J. Bioscience Bioeng,* 2002, *94,* 574-578
[43] H. Honda, and T. Kobayashi, *J. Bioscience Bioeng,* 2000, *89,* 401-408
[44] C. Komives, and R. S. Parker, *Current Opinion Biotech,* 2003, *14,* 468-474
[45] M. R. Warnes, J. Glassey, G. A. Montague, and B. Kara, *Neurocomputing,* 1998, *20,* 67-82
[46] D. Liu, H. Zhang, and S. Hu, *Neurocomputing,* 2008, *71,* 471-473
[47] B. Zelic, N. Bolf, and D. Vasic-Racki, *Bioprocess Biosystems Eng.,* 2006, *29,* 39-47
[48] J.-S. Chang, J.-T. Lee, and A.-C. Chang, *Biochem., Eng. J.,* 2006, *32,* 119-126
[49] S. James, R. Legge, and H. Budman, *J. Process Control,* 2002, *12,* 113-121
[50] E. F. Lara, H. Link, and D. Wenster-Botz, *Process Biochem,* 2006, *41,* 2200-2206

[51] L.F. M. Zorzetto, and J. A. Wilson, *Computer Chem. Eng*, 1996, *20*, S689-S694

[52] J. Peres, R. Oliveira, and S. F. de Azevedo, *Biochem. Eng. J.*, 2007

[53] A. Hanomolo, P. Bogaerts, J. Graefe, M. Cherlet, J. Werenne, and R. Hanns, *Math. Computers Simulation*, 2000, *51*, 375-385

[54] M. Caramihai, A. A. Chirvase, C. Fonteix, I. Marc, F. Fournier, R. Misleanu, and C. Ungureanu, *ESCAPE 17 Computer-Aided Chem. Eng.*, 2007, *24*, 787-792

A Comparative Study on Energy Use and Cost Analysis of Rice Varieties Under Traditional and Semi-Mechanized Farming Systems in North of Iran

Ebrahim Azarpour and Maral Moraditochaee

Additional information is available at the end of the chapter

http://dx.doi.org/10.5772/51165

1. Introduction

Rice is an important food crop for a large proportion of the world's population. It is staple food in the diet of the population of Asia, Latin America, and Africa. Rice provides 35-60% of the dietary calories consumed by more than 3 billion people [12]. Globally, it is also the second most cultivated cereal after wheat. Unlike wheat, 95% of the world's rice is grown in less developed nations, primarily in Asia, Africa, and Latin America. China and India are the largest rice producing and consuming countries in the world. By the year 2025, it is estimated that it will be necessary to produce about 60% more rice than what is currently produced to meet the food needs of a growing world population. In addition, the land available for crop production is decreasing steadily due to urban growth and land degradation. Hence, increases in rice production will have to come from the same or an even less amount of land. This means appropriate rice production practices should be adopted to improve rice yield per unit area [13]. Guilan province has allocated more 35 and 42 percent of paddy production and cultivation land area cultivation area of Iran, respectively. In this province more than 181 exploiters on productive and talented areas with more than 230000 hectares, are busy rice farming [26]. Indeed, rice cultivation is considered the most important agricultural activity in this province and the economy of the province is also based on agriculture, with rice cultivation in top. Most of the under cultivation area of local varieties in Guilan are including Hashemi and Alikazemi. Most of the under cultivation area of breed varieties in Guilan are including Khazar, Hybrid and Gohar.

The system of agricultural productions in the world has been deeply changed because of using mechanization, chemical fertilizers and poisons and reformed seeds and as a result considerable changes in the direction of consumed energy in agricultural section have been created and caused higher relationship to the energy of fossil fuel. This change in the pattern of energy consumption has created problems include warming environment results from green house gases and water and soil pollutions and etc. Nowadays, agricultural sector for providing more food needed the population increase like other sectors has depended to energy sources like electricity and fossil fuels [14]. Energy has been a key input of agriculture since the age of subsistence agriculture. It is an established fact worldwide that agricultural production is positively correlated with energy input [28]. Agriculture is both a producer and consumer of energy. It uses large quantities of locally available non-commercial energy, such as seed, manure and animate energy, as well as commercial energies, directly and indirectly, in the form of diesel, electricity, fertilizer, plant protection, chemical, irrigation water, machinery etc. Efficient use of these energies helps to achieve increased production and productivity and contributes to the profitability and competitiveness of agriculture sustainability in rural living [28]. Energy use in agriculture has been increasing in response to increasing population, limited supply of arable land and a desire for higher standards of living [18]. However, more intensive energy use has brought some important human health and environment problems so efficient use of inputs has become important in terms of sustainable agricultural production [31]. Recently, environmental problems resulting from energy production, conversion and utilization increased public awareness in all sectors of the public, industry and government in both developed and developing countries It is predicted that fossil fuels will be the primary source of energy for the next several decades [8, 9]. The level of fossil fuel dependence differs significantly between developed and developing countries. Although total primary fossil energy input into farm production is comparable between developed countries and developing countries, as illustrated in "Figure 1", developed countries use more than four times the energy per capita (8.0 gigajoules/capita/year) than developing countries (1.7 GJ/capita/year). Moreover, Figure 5 further reveals very different distribution of energy use across agricultural inputs. For developing countries, nitrogen fertilizer accounts for more than half the energy inputs, with fuel and irrigation forming the next largest inputs. By contrast, in developed countries, fuel and machinery account for more than half the inputs, with nitrogen accounting for about one quarter. Efficient use of resources is one of the major assets of eco-efficient and sustainable production, in agriculture [10]. Energy use is one of the key indicators for developing more sustainable agricultural practices [29] and efficient use of energy is one of the principal requirements of sustainable agriculture [18]. It is important, therefore, to analyses cropping systems in energy terms and to evaluate alternative solutions, especially for arable crops, which account for more than half of the primary sector energy consumption [27].

Agricultural systems are complex, and understanding this complexity requires systematic research, but resources for agricultural research are limited. The field experiments investigate a number of variables under a few site-specific conditions. Crop simulation models consider the complex interactions of weather, soil properties, and management factors, which influence crop performance. Mechanistic models are very helpful in deciding the best management options for optimizing crop growth and the yield. In the middle of 1990s, Rice Research Institute of Iran (IRRI), Wageningen University, and the Research Centre developed the ORYZA model series to simulate the growth and development of tropical lowland rice. In 2001, a new version of the ORYZA model was released that improved and incorporated all previous versions into one model called ORYZA2000 [7]. The model ORYZA2000, simulates the growth and development of rice under conditions of potential production, water and nitrogen limitations.

The aims of the study were to survey input energy in local and breed varieties rice production under two farming systems condition (traditional and semi-mechanized), to investigate the energy consumption and to make an economic analysis of rice in Guilan province of Iran.

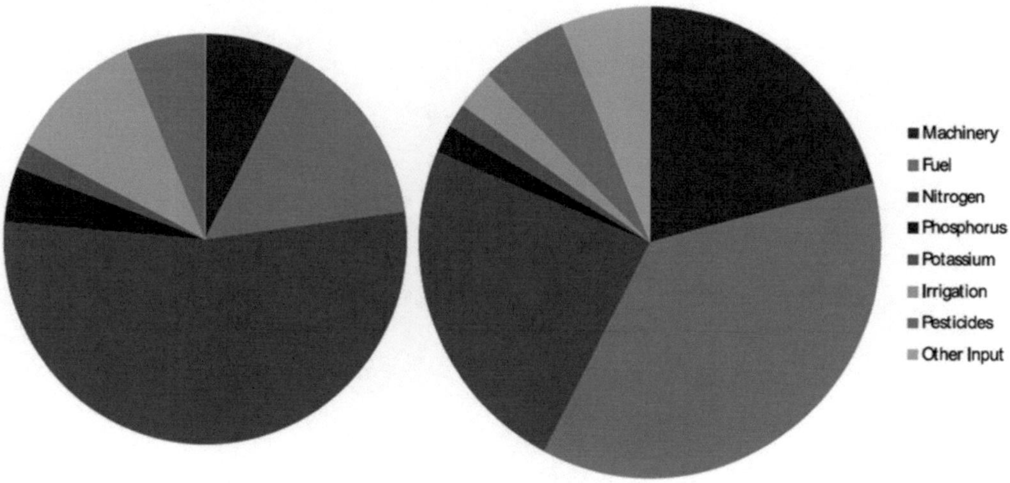

Figure 1. Distribution of farm energy inputs in developing countries (left) and in developed countries (right)

2. Materials and methods

In order to gather the required data in this study, information related to 72 farms in Guilan province during the agricultural year 2010 was studied. The Location of studied region in north of Iran was presented in "Figure 2". The random sampling of production agro ecosystems was done within whole population and the size of each sample was determined by using bottom equation [18].

$$n = \frac{N \times s^2 \times t^2}{(N-1)\, d^2 + s^2 + t^2}$$

In the formula, n is the required sample size, s is the standard deviation, t is the t value at 95% confidence limit (1.96), N is the number of holding in target population and d is the acceptable error.

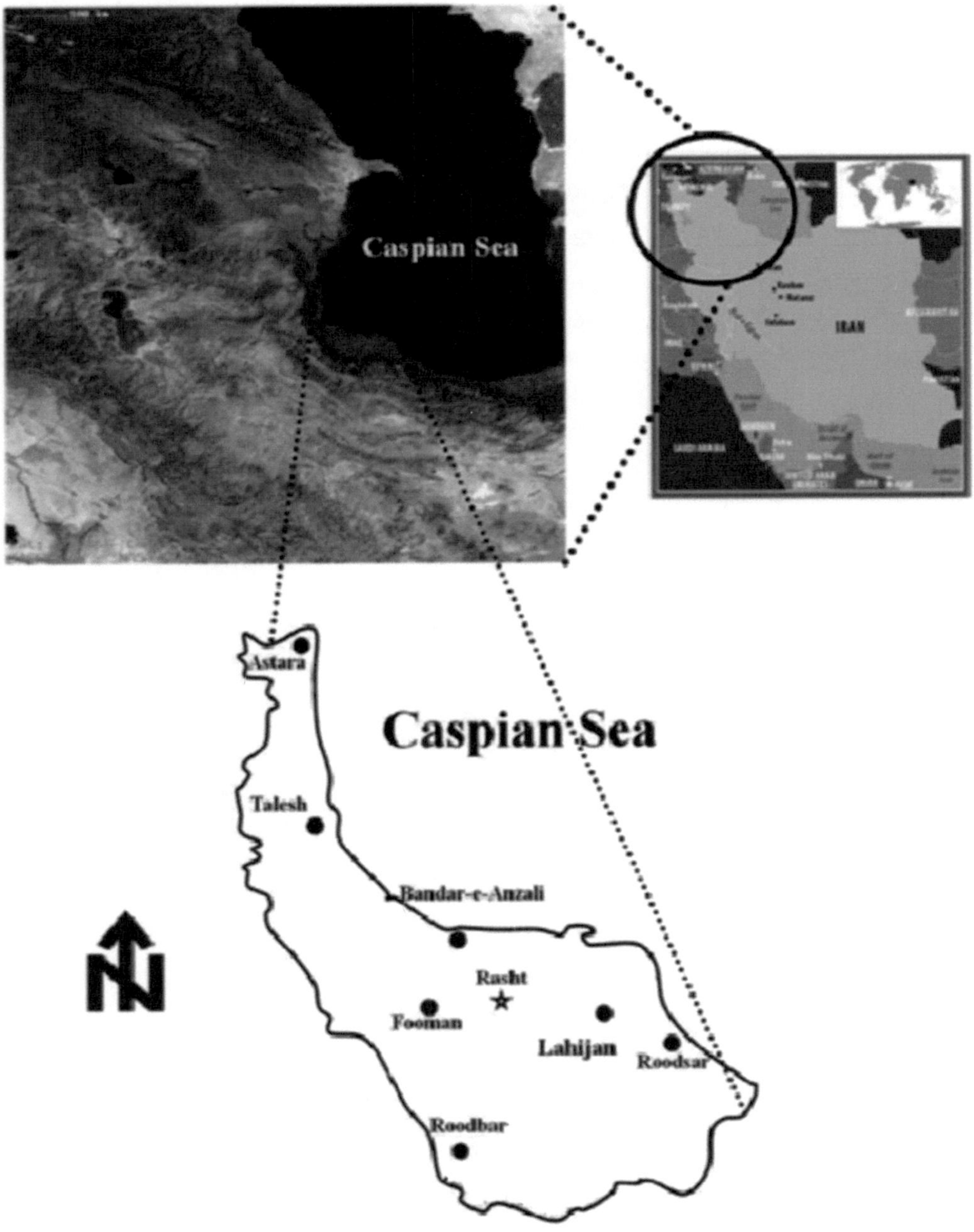

Figure 2. Location of the study area

Cultivated varieties in these farms include local varieties (Hashemi and Alikazemi) and breed varieties (Khazar, Hybrid (GRH1) and Gohar (SA13)). Farming methods in these farms include traditional system and semi-mechanized system. In semi-mechanized system in addition to tiller and thrasher, transforming machine and reaping machine are used for plant out and reaping respectively.

Efficient use of the energy resources is vital in terms of increasing production, productivity, competitiveness of agriculture as well as sustainability of rural living. Energy auditing is one of the most common approaches to examining energy efficiency and environmental impact of the production system. It enables researchers to calculate output-input ratio, relevant indicators, and energy use patterns in an agricultural activity. Moreover, the energy audit provides sufficient data to establish functional forms to investigate the relationship between energy inputs and outputs. The amount of inputs used in agricultural production practices (human labor, machinery, diesel fuel, chemical fertilizers, poison fertilizers, water and seeds) were calculated per hectare and then, these data were converted to forms of energy to evaluate the output-input analysis. In order to calculate output and input energy, these input data and amount of output yield were multiplied with the coefficient of energy equivalent. Energy equivalents of inputs and output were converted into energy on area unit. The previous researches "Table 1" were used to determine the energy equivalents' coefficients [15, 19, 20, 21, 22, 23, 24, 25, 30, 31]. Firstly, the amounts of inputs used in the production of rice were specified in order to calculate the energy equivalences in the study. Energy input include human labor, machinery, diesel fuel, chemical fertilizer, chemical poison, water and seed amounts and output yield include paddy of rice.

In this research, energy indices (energy use efficiency, energy ratio, energy productivity, energy intensity, net energy gain and water and energy productivity) based on the energy equivalents of the inputs and output "Table 2" were calculated according to bottom equations [15, 19, 20, 21, 22, 23, 24, 25, 30, 31].

$$\text{Energy use efficiency} = \frac{\text{Output energy (Mj/ha)}}{\text{Input energy (Mj/ha)}}$$

$$\text{Energy production} = \frac{\text{Grain yield (Kg/ha)}}{\text{Input energy (Mj/ha)}}$$

$$\text{Energy specific} = \frac{\text{Input energy (Mj/ha)}}{\text{Grain yield (Kg/ha)}}$$

$$\text{Water and energy productivity} = \frac{\text{Yield output (Kg/ha)}}{\text{Water applied (M3/ha)} \times \text{Input energy (Mj/ha)}}$$

$$\text{Net energy gain} = \text{Output energy (Mj/ha)} - \text{Input energy (Mj/ha)}$$

The input energy is also classified into direct and indirect and renewable and non-renewable forms energy equivalents for different inputs and outputs in agricultural production. Indirect energy consists of seeds, chemical fertilizer, chemical poison, and machinery energy while direct energy covered human labor, water and diesel fuel used in the rice production. Non-renewable energy includes diesel fuel, chemical fertilizer, chemical poison and machinery and renewable energy consists of human labor, water and seed [2, 4, 5, 6].

Parameter	Hashemi	Alikazemi	Khazar	Hybrid	Gohar	Energy equivalent
			Traditional system			
			Input			
Human labor (h/ha)	94.3	94.3	94.3	94.3	94.3	1.96
Machinery (h/ha)	37.2	37.2	37.2	37.2	37.2	62.7
Diesel fuel (l/ha)	127.2	127.2	127.2	127.2	127.2	56.31
Nitrogen (kg/ha)	125	125	180	230	230	69.5
Phosphorus(kg/ha)	60	60	80	100	100	12.44
Potassium (kg/ha)	110	110	150	200	200	11.15
Herbicide (l/ha)	3	3	3	3	3	85
Fungicide (l/ha)	2	2	2	2	2	160
Insecticide (l/ha)	2	2	1	1	1	99
Water (m^3/ha)	10000	10000	10000	10000	10000	1.02
Seed (kg/ha)	90	90	70	30	30	17
			Output			
Paddy (kg/ha)	3520	4180	4840	6600	8360	14.7
Straw (kg/ha)	4437	5706	6607	9010	11413	12.5
Husk (kg/ha)	813	1045	1210	1650	2090	13.8
Biomass (kg/ha)	8770	10931	12657	17260	21863	13.67
			Semi-mechanized system			
			Input			
Human labor (h/ha)	73.7	73.7	73.7	73.7	73.7	1.96
Machinery (h/ha)	47.3	47.3	47.3	47.3	47.3	62.7
Diesel fuel (l/ha)	142.1	142.1	142.1	142.1	142.1	56.31
Nitrogen (kg/ha)	125	125	180	230	230	69.5
Phosphorus(kg/ha)	60	60	80	100	100	12.44
Potassium (kg/ha)	110	110	150	200	200	11.15
Herbicide (l/ha)	3	3	3	3	3	85
Fungicide (l/ha)	2	2	2	2	2	160
Insecticide (l/ha)	2	2	1	1	1	99
Water (m^3/ha)	10000	10000	10000	10000	10000	1.02
Seed (kg/ha)	70	70	50	20	20	17
			Output			
Paddy (kg/ha)	4000	4750	5500	7500	9500	14.7
Straw (kg/ha)	5461	6485	7508	10239	12969	12.5
Husk (kg/ha)	1000	1188	1375	1875	2375	13.8
Biomass (kg/ha)	10461	12423	14383	19614	24844	13.67

Table 1. Amounts of input-output used and energy equivalent in varieties rice production under traditional system and semi-mechanized system condition

Parameter	Hashemi	Alikazemi	Khazar	Hybrid	Gohar
Traditional system					
Input					
Human labor (h/ha)	184.83	184.83	184.83	184.83	184.83
Machinery (h/ha)	2332.44	2332.44	2332.44	2332.44	2332.44
Diesel fuel (l/ha)	7162.63	7162.63	7162.63	7162.63	7162.63
Nitrogen (kg/ha)	8687.5	8687.5	12510	15985	15985
Phosphorus(kg/ha)	746.4	746.4	995.2	1244	1244
Potassium (kg/ha)	1226.5	1226.5	1672.5	2230	2230
Herbicide (l/ha)	255	255	255	255	255
Fungicide (l/ha)	320	320	320	320	320
Insecticide (l/ha)	198	198	99	99	99
Water (m^3/ha)	10200	10200	10200	10200	10200
Seed (kg/ha)	1530	1530	1190	510	510
Output					
Paddy (kg/ha)	51744	61446	71148	97020	122892
Straw (kg/ha)	55463	71325	82588	112625	142663
Husk (kg/ha)	11219	14421	16698	22770	28842
Biomass (kg/ha)	119857	149390	172979	235887	298794
Semi-mechanized system					
Input					
Human labor (h/ha)	144.45	144.45	144.45	144.45	144.45
Machinery (h/ha)	2965.71	2965.71	2965.71	2965.71	2965.71
Diesel fuel (l/ha)	8001.65	8001.65	8001.65	8001.65	8001.65
Nitrogen (kg/ha)	8687.5	8687.5	12510	15985	15985
Phosphorus(kg/ha)	746.4	746.4	995.2	1244	1244
Potassium (kg/ha)	1226.5	1226.5	1672.5	2230	2230
Herbicide (l/ha)	255	255	255	255	255
Fungicide (l/ha)	320	320	320	320	320
Insecticide (l/ha)	198	198	99	99	99
Water (m^3/ha)	10200	10200	10200	10200	10200
Seed (kg/ha)	1190	1190	850	340	340
Output					
Paddy (kg/ha)	58800	69825	80850	110250	139650
Straw (kg/ha)	68263	81063	93850	127988	162113
Husk (kg/ha)	13800	16394	18975	25875	32775
Biomass (kg/ha)	142967	169781	196568	268058	339535

Table 2. Input-output energy for varieties rice under traditional system and semi-mechanized system condition

In order to calculate energy balance indices, these input data and amount of output yield were multiplied with the coefficient of energy balance equivalent. Energy balance equivalents of inputs and output were converted into energy on area unit. The previous researches "Table 3" were used to determine the energy balance equivalents' coefficients [2, 4, 5, 6] By using of consumed data as inputs and total production as output, and their concern equivalent energy, indicators of energy balance were calculated "Table 4".

Parameter	Hashemi	Alikazemi	Khazar	Hybrid	Gohar	Energy balance equivalent
Traditional system						
Input						
Human labor (h/ha)	848.7	848.7	848.7	848.7	848.7	500
Machinery (h/ha)	37.2	37.2	37.2	37.2	37.2	90000
Diesel fuel (l/ha)	127.2	127.2	127.2	127.2	127.2	9237
Nitrogen (kg/ha)	57.5	57.5	82.8	105.8	105.8	17600
Phosphorus(kg/ha)	12.6	12.6	16.8	21	21	3190
Potassium (kg/ha)	45.1	45.1	61.5	82	82	1600
Chemical Poison (l/ha)	5	5	5	5	5	27170
Water (m^3/ha)	10000	10000	10000	10000	10000	272.2
Seed (kg/ha)	90	90	70	30	30	6513
Depreciation for per diesel fuel (L)	106.85	106.85	106.85	106.85	106.85	9583
Semi-mechanized system						
Input						
Human labor (h/ha)	663.3	663.3	663.3	663.3	663.3	500
Machinery (h/ha)	47.3	47.3	47.3	47.3	47.3	90000
Diesel fuel (l/ha)	142.1	142.1	142.1	142.1	142.1	9237
Nitrogen (kg/ha)	57.5	57.5	82.8	105.8	105.8	17600
Phosphorus(kg/ha)	12.6	12.6	16.8	21	21	3190
Potassium (kg/ha)	45.1	45.1	61.5	82	82	1600
Chemical Poison (l/ha)	5	5	5	5	5	27170
Water (m^3/ha)	10000	10000	10000	10000	10000	272.2
Seed (kg/ha)	70	70	50	20	20	6513
Depreciation for per diesel fuel (L)	119.36	119.36	119.36	119.36	119.36	9583

Table 3. Amounts of input used and energy balance equivalent in varieties rice production under traditional system and semi-mechanized system condition

Parameter	Hashemi	Alikazemi	Khazar	Hybrid	Gohar
		Traditional system			
		Input			
Parameter	Hashemi	Alikazemi	Khazar	Hybrid	Gohar
Human labor (h/ha)	424350	424350	424350	424350	424350
Machinery (h/ha)	3348000	3348000	3348000	3348000	3348000
Diesel fuel (l/ha)	1174946	1174946	1174946	1174946	1174946
Nitrogen (kg/ha)	1012000	1012000	1457280	1862080	1862080
Phosphorus(kg/ha)	40194	40194	53592	66990	66990
Potassium (kg/ha)	72160	72160	98400	131200	131200
Chemical Poison (l/ha)	135850	135850	135850	135850	135850
Water (m^3/ha)	2722000	2722000	2722000	2722000	2722000
Seed (kg/ha)	586170	586170	455910	195390	195390
Depreciation for per diesel fuel (L)	1023924	1023924	1023924	1023924	1023924
		Semi-mechanized system			
		Input			
Human labor (h/ha)	331650	331650	331650	331650	331650
Machinery (h/ha)	4257000	4257000	4257000	4257000	4257000
Diesel fuel (l/ha)	1312578	1312578	1312578	1312578	1312578
Nitrogen (kg/ha)	1012000	1012000	1457280	1862080	1862080
Phosphorus(kg/ha)	40194	40194	53592	66990	66990
Potassium (kg/ha)	72160	72160	98400	131200	131200
Chemical Poison (l/ha)	135850	135850	135850	135850	135850
Water (m^3/ha)	2722000	2722000	2722000	2722000	2722000
Seed (kg/ha)	455910	455910	325650	130260	130260
Depreciation for per diesel fuel (L)	1143865	1143865	1143865	1143865	1143865

Table 4. Input energy in varieties rice production under traditional and semi-mechanized system condition from calculated indicators of energy balance energy

Cluster analysis and correlation analysis of energy indices and balance energy indices for rice production were obtained by SPSS software. Yield function of paddy yield, straw yield, husk yield and biomass yield for rice production was obtained by STATISCA software. Simulation growth indices of rice cultivars were obtained by model ORYZA2000 "Figure 3" [7].

In the last part of the study, the economic analysis of varieties rice production under traditional and semi-mechanized system condition was investigated. Net profit, gross profit and benefit to cost ratio was calculated. The gross value of production, net return and benefit to cost ratio were calculated using the following equations (Mohammadi et al., 2008):

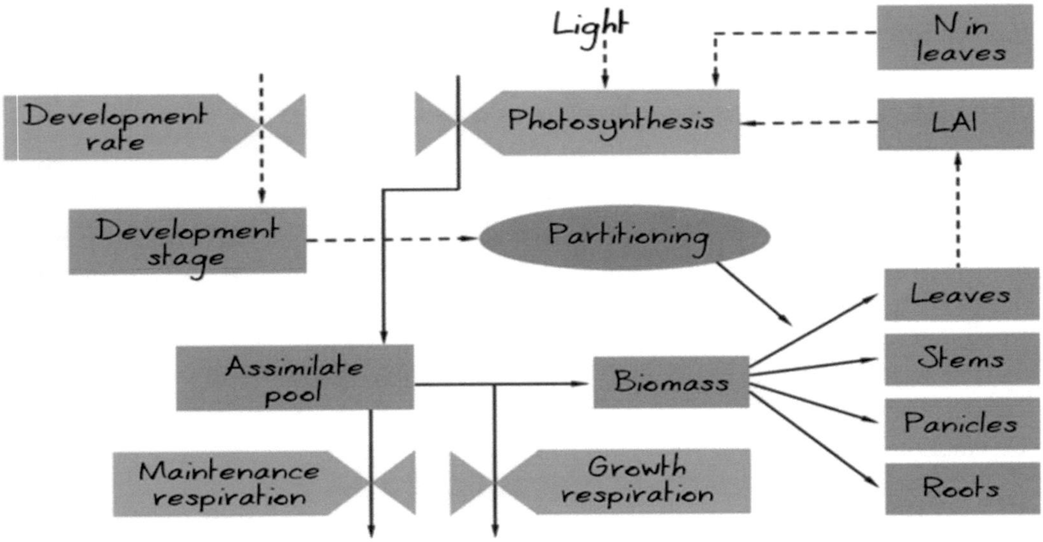

Figure 3. The model ORYZA2000 structure

$$\text{Gross value of production (\$/ha)} = \text{Yield (kg/ha)} \times \text{Sale price (\$/kg)}$$

$$\text{Net return (\$/ha)} = \text{Gross value of production (\$/ha)} - \text{Total cost of production (\$/ha)}$$

$$\text{Productivity (kg/\$)} = \frac{\text{Yield (kg/ha)}}{\text{Total cost of production (\$/ha)}}$$

$$\text{Benefit to cost ratio} = \frac{\text{Gross value of production (\$/ha)}}{\text{Total cost of production (\$/ha)}}$$

3. Results and discussions

3.1. Analysis of energy indices in varieties rice production under traditional and semi-mechanized system condition

In "Figure 4" (traditional system) and "Figure 5" (semi-mechanized system), seven groups of reserves of production of studied figures according to percentage of total energy of reserve is observed. Results showed that highest energy consumption in all varieties was related to chemical fertilizer. The amount of further use of fertilizer and also raising of equivalent amounts of energy in this reserve showed this subject. The energy of water reserve, fuel, poison, machines, seed and human labor are in next grades.

Rice plants require fertilizer during vegetative stage to promote growth and tillering, which in turn, determines potential number of panicles. Fertilizer contributes to spikelet production during early panicle formation stage, and contributes to sink size during the late panicle formation stage. Fertilizer also plays a role in grain filling, improving the photosynthetic capacity, and promoting carbohydrate accumulation in culms and leaf sheaths [1].

Results of "Tables 5 and 6" showed that breed varieties (Khazar, Hybrid and Gohar) because of suitable genetic specifications have higher operation in compared with local varieties (Hashemi and Alikazemi), highest paddy yield (9500 kg/ha), straw yield (12969 kg/ha), husk yield (2375 kg/ha) and biomass yield (24844 kg/ha) of semi-mechanized system and paddy yield (8360 Kg/ha), straw yield (11413 kg/ha), husk yield (2090 kg/ha) and biomass yield (21863 kg/ha) of traditional system observed in Gohar rice.

Breed varieties because of accepting higher fertilizer have further input energy than local varieties under two farming systems condition "Tables 5 and 6". Traditional system because of consumption higher fertilizer and seed has further input energy than semi-mechanized system "Tables 3 and 4".

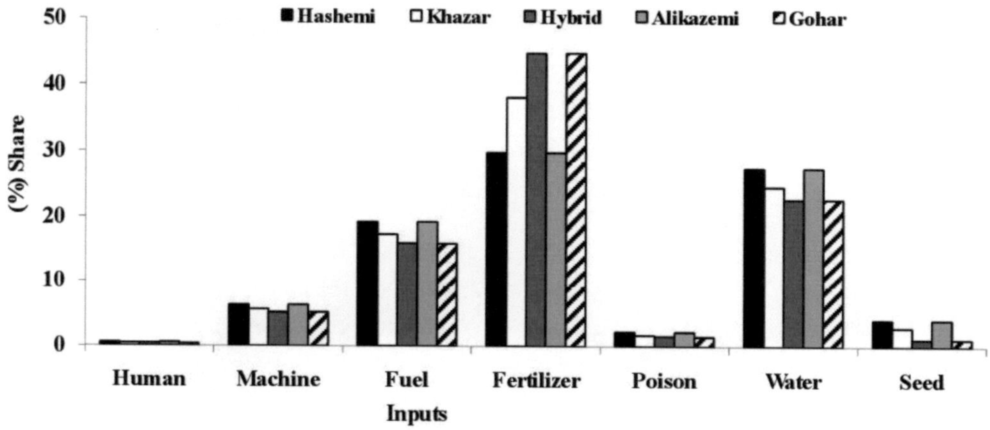

Figure 4. The share (%) production inputs for varieties rice under traditional system condition

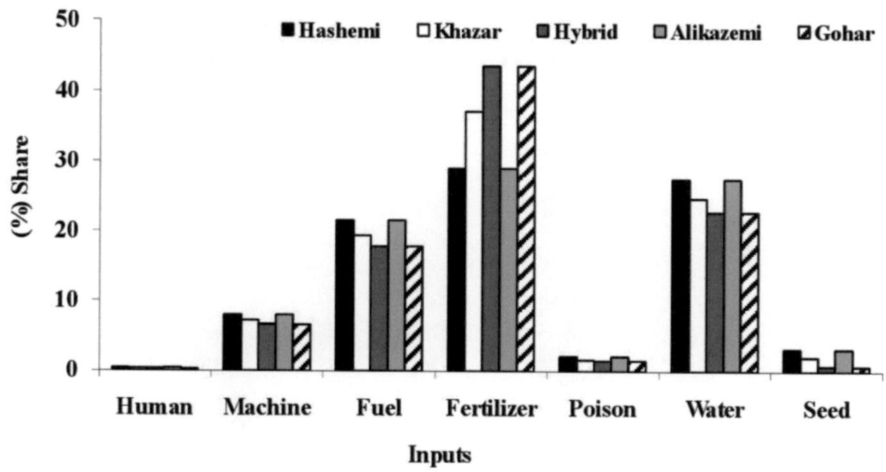

Figure 5. The share (%) production inputs for varieties rice under semi-mechanized system condition

Semi-mechanized system because of producing higher paddy yield, straw yield, husk yield and biomass yield than traditional system of has higher output energy "Tables 5 and 6". Breed varieties (Khazar, Hybrid and Gohar) because of suitable genetic specifications have

Item	Unit	Hashemi	Alikazemi	Khazar	Hybrid	Gohar
Paddy						
Yield	kg/ha	3520	4180	4840	6600	8360
Input energy	MJ/ha	32843	32843	36922	40523	40523
Output energy	MJ/ha	51744	61446	71148	97020	122892
Energy ratio	-	1.58	1.87	1.93	2.39	3.03
Energy intensity	MJ/kg	9.33	7.86	7.63	6.14	4.85
Energy productivity	kg/MJ	0.11	0.13	0.13	0.16	0.21
Net energy gain	MJ/ha	18901	28603	34226	56497	82369
Water and energy productivity	g/m³.MJ	0.011	0.012	0.013	0.016	0.020
Straw						
Yield	kg/ha	4437	5706	6607	9010	11413
Input energy	MJ/ha	32843	32843	36922	40523	40523
Output energy	MJ/ha	55463	71325	82588	112625	142663
Energy ratio	-	1.69	2.17	2.24	2.78	3.52
Energy intensity	MJ/kg	7.40	5.76	5.59	4.50	3.55
Energy productivity	kg/MJ	0.14	0.17	0.18	0.22	0.28
Net energy gain	MJ/ha	22620	38482	45666	72102	102140
Water and energy productivity	g/m³.MJ	0.013	0.017	0.018	0.022	0.028
Husk						
Yield	kg/ha	813	1045	1210	1650	2090
Input energy	MJ/ha	32843	32843	36922	40523	40523
Output energy	MJ/ha	11219	14421	16698	22770	28842
Energy ratio	-	0.34	0.44	0.45	0.56	0.71
Energy intensity	MJ/kg	40.40	31.43	30.51	24.56	19.39
Energy productivity	kg/MJ	0.02	0.03	0.03	0.04	0.05
Net energy gain	MJ/ha	-21624	-18422	-20224	-17753	-11681
Water and energy productivity	g/m³.MJ	0.002	0.003	0.003	0.004	0.005
Biomass						
Yield	kg/ha	8770	10931	12657	17260	21863
Input energy	MJ/ha	32843	32843	36922	40523	40523
Output energy	MJ/ha	119857	149390	172979	235887	298794
Energy ratio	-	3.65	4.55	4.69	5.82	7.37
Energy intensity	MJ/kg	3.74	3.00	2.92	2.35	1.85
Energy productivity	kg/MJ	0.27	0.33	0.34	0.43	0.54
Net energy gain	MJ/ha	87013	116547	136057	195364	258271
Water and energy productivity	g/m³.MJ	0.027	0.033	0.034	0.043	0.054

Table 5. Energy indices for varieties rice under traditional system condition

higher output energy in compared with local varieties (Hashemi and Alikazemi). Highest output energy with averages 139650, 162113, 32775 and 339535 MJ/ha of semi-mechanized system and with averages 122892, 142663, 28842 and 298794 MJ/ha of traditional system observed in Gohar rice "Tables 5 and 6".

Energy ratio in two farming systems and five varieties showed that positive output of energy production and being further of energy output of semi-mechanized system than traditional system and breed varieties than local varieties (tables 5 and 6).

Results of energy intensity under two farming systems condition "Tables 5 and 6" showed that local varieties require of further input from production of paddy yield, straw yield, husk yield and biomass yield than breed varieties.

Results of energy productivity under two farming systems condition "Tables 5 and 6" were showed that in breed varieties lieu of imported energy consumption have higher energy productions than local varieties.

Net energy gain in two farming systems and five varieties showed that highest net energy gain of semi-mechanized system than traditional system and breed varieties than local varieties. Highest net energy gain with averages 97865, 120328, -9010 and 297750 MJ/ha of semi-mechanized system and with averages 82369, 102140, -11681 and 258271 MJ/ha of traditional system observed in Gohar rice "Tables 5 and 6"

Direct, indirect energy, renewable, non-renewable, % direct, % indirect energy, % renewable and % non-renewable in two farming systems and five varieties were showed "Tables 7". In two farming systems and five varieties were showed that direct energy and % direct energy as compared with indirect energy and % indirect energy; renewable energy and % renewable energy as compared with nonrenewable energy and % nonrenewable energy have lower amount "Tables 7". The amount of higher consumption of machinery and diesel fuel in semi-mechanized system lead to increasing indirect energy in this system in compared with traditional system. The amount of higher consumption of chemical fertilizer in breed varieties lead to increasing indirect energy in these varieties in compared with local varieties. Results showed that, lower amount of consumption of seed and human labor in semi-mechanized system in compared with traditional system leads to being lower of renewable energy in semi-mechanized system than traditional system "Tables 7". Lower amount of consumption of seed in breed varieties in compared with local varieties leads to being lower of renewable energy in breed varieties than local varieties. The amount of higher consumption of chemical fertilizer in breed varieties in compared with local varieties leads to increasing nonrenewable energy in these breed varieties than local varieties. The share of direct and indirect energy from total reserve of energy and share of renewable and nonrenewable energies from total reserve of energy "Tables 7" in studied farming systems and varieties were that the percentage of direct energy is lowest than percentage of indirect energy and percentage of renewable energy in producing rice is lowest than nonrenewable energies that this required to consider saving in energy consumption.

Item	Unit	Hashemi	Alikazemi	Khazar	Hybrid	Gohar
Paddy						
Yield	kg/ha	4000	4750	5500	7500	9500
Input energy	MJ/ha	33935	33935	38014	41785	41785
Output energy	MJ/ha	58800	69825	80850	110250	139650
Energy ratio	-	1.73	2.06	2.13	2.64	3.34
Energy intensity	MJ/kg	8.48	7.14	6.91	5.57	4.40
Energy productivity	kg/MJ	0.12	0.14	0.14	0.18	0.23
Net energy gain	MJ/ha	24865	35890	42836	68465	97865
Water and energy productivity	g/m³.MJ	0.012	0.014	0.014	0.018	0.022
Straw						
Yield	kg/ha	5461	6485	7508	10239	12969
Input energy	MJ/ha	33935	33935	38014	41785	41785
Output energy	MJ/ha	68263	81063	93850	127988	162113
Energy ratio	-	2.01	2.39	2.47	3.06	3.88
Energy intensity	MJ/kg	6.21	5.23	5.06	4.08	3.22
Energy productivity	kg/MJ	0.16	0.19	0.20	0.25	0.31
Net energy gain	MJ/ha	34327	47127	55836	86203	120328
Water and energy productivity	g/m³.MJ	0.016	0.019	0.019	0.024	0.030
Husk						
Yield	kg/ha	1000	1188	1375	1875	2375
Input energy	MJ/ha	33935	33935	38014	41785	41785
Output energy	MJ/ha	13800	16394	18975	25875	32775
Energy ratio	-	0.41	0.48	0.50	0.62	0.78
Energy intensity	MJ/kg	33.94	28.56	27.65	22.29	17.59
Energy productivity	kg/MJ	0.03	0.04	0.04	0.04	0.06
Net energy gain	MJ/ha	-20135	-17541	-19039	-15910	-9010
Water and energy productivity	g/m³.MJ	0.003	0.003	0.004	0.004	0.006
Biomass						
Yield	kg/ha	10461	12423	14383	19614	24844
Input energy	MJ/ha	33935	33935	38014	41785	41785
Output energy	MJ/ha	142967	169781	196568	268058	339535
Energy ratio	-	4.21	5.00	5.17	6.42	8.13
Energy intensity	MJ/kg	3.24	2.73	2.64	2.13	1.68
Energy productivity	kg/MJ	0.31	0.37	0.38	0.47	0.59
Net energy gain	MJ/ha	109032	135846	158554	226273	297750
Water and energy productivity	g/m³.MJ	0.031	0.037	0.038	0.047	0.059

Table 6. Energy indices for varieties rice under semi-mechanized system condition

Item	Hashemi	Alikazemi	Khazar	Hybrid	Gohar
Traditional system					
Direct energy (MJ/ha)	17547	17547	17547	17547	17547
Direct energy (%)	53.43	53.43	47.53	43.30	43.30
Indirect energy (MJ/ha)	15296	15296	19375	22976	22976
Indirect energy (%)	46.57	46.57	52.47	56.70	56.70
Renewable energy (MJ/ha)	11915	11915	11575	10895	10895
Renewable energy (%)	36.28	36.28	31.35	26.89	26.89
Nonrenewable energy (MJ/ha)	20928	20928	25347	29628	29628
Nonrenewable energy (%)	63.72	63.72	68.65	73.11	73.11
Semi-mechanized system					
Direct energy (MJ/ha)	18346	18346	18346	18346	18346
Direct energy (%)	54.06	54.06	48.26	43.91	43.91
Indirect energy (MJ/ha)	15589	15589	19667	23439	23439
Indirect energy (%)	45.94	45.94	51.74	56.09	56.09
Renewable energy (MJ/ha)	11534	11534	11194	10684	10684
Renewable energy (%)	33.99	33.99	29.45	25.57	25.57
Nonrenewable energy (MJ/ha)	22401	22401	26819	31100	31100
Nonrenewable energy (%)	66.01	66.01	70.55	74.43	74.43

Table 7. Division of the energy for varieties rice under traditional and semi-mechanized system condition

Moradi and Azarpour [23] with study of energy indices for native and breed rice varieties production in Iran were recorded the highest grain yield, input energy, output energy, energy ratio, energy productivity and Net energy gain obtained from breed varieties as compared with local varieties. Eskandari Cherati et al. [11] with study energy survey of mechanized and traditional rice production system in Mazandaran province of Iran showed that the total energy used for semi-mechanized and traditional rice production system was 67217.95 and 67356.28 MJ/ha, respectively. Based on the results, irrigation and fertilizer in both systems with 50232 and 7610.32 MJ/ha was the most input energy. Total energy output of the traditional method was 127.5 GJ/ha and that of the semi-mechanized was 132.26 GJ/ha. Parallel to the mechanization level of operations that increased, consumption of fuel and machinery energy increased similarly, but the human labor and seed energy consumption dropped. The renewable energy in the traditional and semi-mechanized systems was 3168.3 (4.70% total energy) and 2312.1 MJ/ha (3.44%), respectively. Energy ratio and energy productivity in traditional and semi-mechanized systems was 3 and 3.08, and 0.111 and 0.116 kg/MJ 116.0, respectively. Nonetheless, net energy gain and specific energy showed that energy efficiency of semi-mechanized systems was more than the traditional system. Khan et al. [16] with energy requirement and economic analysis of rice production in western part of Pakistan Energy requirement and economic analysis of rice production in western part of Pakistan revealed that energy consumption and rice yield were 5,756 kWh and 3.23 tons per hectare on Bullock Operated Farms (BOF) and 11,162 kWh and 4.12 tons per hectare on Tractor Operated Farms (TOF). Consumption of animate energy on BOF was more than TOF due to heavy use of animate energy in land preparation operation. Result also showed that energy efficiency i.e. output-input ratio on BOF (6.32) was higher than TOF (4.16). Cost of production remained lower on BOF than TOF, however, the yield and consequently crop values and net return were higher on TOF than BOF.

Khan et al. [17] with study energy requirements and economic analysis of wheat, rice and barley production in Australia revealed that chemical fertilizer consumed 47, 43 and 29 % of the total energy inputs on wheat, rice and barley growing farms, respectively. Wheat consumed 3028, rice 6699 and barley consumed 2175 $kWhha^{-1}$. Similarly, wheat utilized 2852, rice 17754 and barley 856 m^3ha^{-1}. Average energy output of wheat was 27874, rice 44885, and barley obtained 17865 $kWhha^{-1}$. Wheat was most energy efficient crop compared to rice and barley, whereas barley achieved the highest water productivity.

3.2. Analysis of energy indices and balance energy indices in varieties rice production under traditional and semi-mechanized system condition

The inputs used in varieties rice production under two farming system and their energy equivalents and output energy equivalent were illustrated in "Tables 3 and 4". About 848.7 h human labor, 37.2 h machinery power, 1000 m3 water, 5 L chemical poison and 127.2 L diesel fuel for total operations were used in varieties rice production under traditional on a hectare basis; Also 106.85 L depreciation power in this system was used. The highest use of nitrogen fertilizer (105.8 kg/ha), phosphorus (21 kg/ha) and potassium (82 kg/ha) were observed in Gohar rice. The lowest use seed in varieties rice production under traditional was observed in Gohar rice (30 kg/ha). About 663.3 h human labor, 47.3 h machinery power, 1000 m3 water, 5 L chemical poison and 142.1 L diesel fuel for total operations were used in varieties rice production under traditional on a hectare basis; Also 119.36 L depreciation power in this system was used. The highest use of nitrogen fertilizer (105.8 kg/ha), phosphorus (21 kg/ha) and potassium (82 kg/ha) were observed in Gohar rice. The lowest use seed in varieties rice production under traditional was observed in Gohar rice (20 kg/ha).

In "Figure 6" (traditional system) and "Figure 7" (semi-mechanized system), eight groups of reserves of production of studied figures according to percentage of total energy of reserve were observed. Results showed that highest shares of this amount were reported for machinery, water, diesel fuel, chemical fertilizer and depreciation for per diesel fuel in all varieties rice production respectively. The energy inputs of seed, human labor and chemical poison were found to be quite low compared to the other inputs used in all varieties rice production respectively.

The highest percent of compositions, amounts, production energy, and production energy to consumption energy ratio in rice paddy were obtained from starch as compared with protein and fat; the lowest consumption energy to production energy ratio in rice paddy was obtained from starch as compared with protein and fat "Table 8". Results of "Table 8" showed that breed varieties (Khazar, Hybrid and Gohar) because of suitable genetic specifications have higher operation in compared with local varieties (Hashemi and Alikazemi); the highest amounts (protein: 551.76, fat: 183.92 and starch: 6688), production energy (protein: 2207040 kg/ha, fat: 1655280 kg/ha and starch: 26752000 kg/ha), and production energy to consumption energy ratio (protein: 0.20, fat: 0.15 and starch: 2.41) in rice paddy of traditional system and highest amounts (protein: 627, fat: 209 and starch: 7600), production energy (protein: 2508000 kg/ha, fat: 1881000 kg/ha and starch: 30400000 kg/ha), and production energy to consumption energy ratio (protein: 0.21, fat: 0.16 and starch: 2.51) in rice paddy of semi-mechanized observed in Gohar rice.

The highest percent of compositions, amounts, production energy, and production energy to consumption energy ratio in rice husk were obtained from starch as compared with fat and protein; the lowest consumption energy to production energy ratio in rice husk was obtained from starch as compared with fat and protein "Table 9". Results of "Table 9" showed that breed varieties (Khazar, Hybrid and Gohar) because of suitable genetic specifications have higher operation in compared with local varieties (Hashemi and Alikazemi); the highest amounts (protein: 107.22, fat: 107.64 and starch:1045), production energy (protein: 428868 kg/ha, fat: 968715 kg/ha and starch: 4180000 kg/ha), and production energy to consumption energy ratio (protein: 0.04, fat: 0.09 and starch: 0.38) in rice husk of traditional system and highest amounts (protein: 121.84, fat: 122.31 and starch: 1187.50), production energy (protein: 487350 kg/ha, fat: 1100813 kg/ha and starch: 4750000 kg/ha), and production energy to consumption energy ratio (protein: 0.04, fat: 0.09 and starch: 0.39) in rice husk of semi-mechanized observed in Gohar rice.

The highest percent of compositions, amounts, production energy, and production energy to consumption energy ratio in rice straw were obtained from starch as compared with protein and fat; the lowest consumption energy to production energy ratio in rice straw was obtained from starch as compared with protein and fat "Table 10". Results of "Table 10" showed that breed varieties (Khazar, Hybrid and Gohar) because of suitable genetic specifications have higher operation in compared with local varieties (Hashemi and Alikazemi); the highest amounts (protein: 490.76, fat: 148.37 and starch:4941.83), production energy (protein: 1963036 kg/ha, fat: 1335321 kg/ha and starch: 19767316 kg/ha), and production energy to consumption energy ratio (protein: 0.18, fat: 0.12 and starch: 1.87) in rice straw of traditional system and highest amounts (protein:557.67, fat: 168.60 and starch: 6515.68), production energy (protein: 2230668 kg/ha, fat: 1517373 kg/ha and starch: 22462308 kg/ha), and production energy to consumption energy ratio (protein: 0.18, fat: 0.13 and starch: 1.86) in rice straw of semi-mechanized observed in Gohar rice.

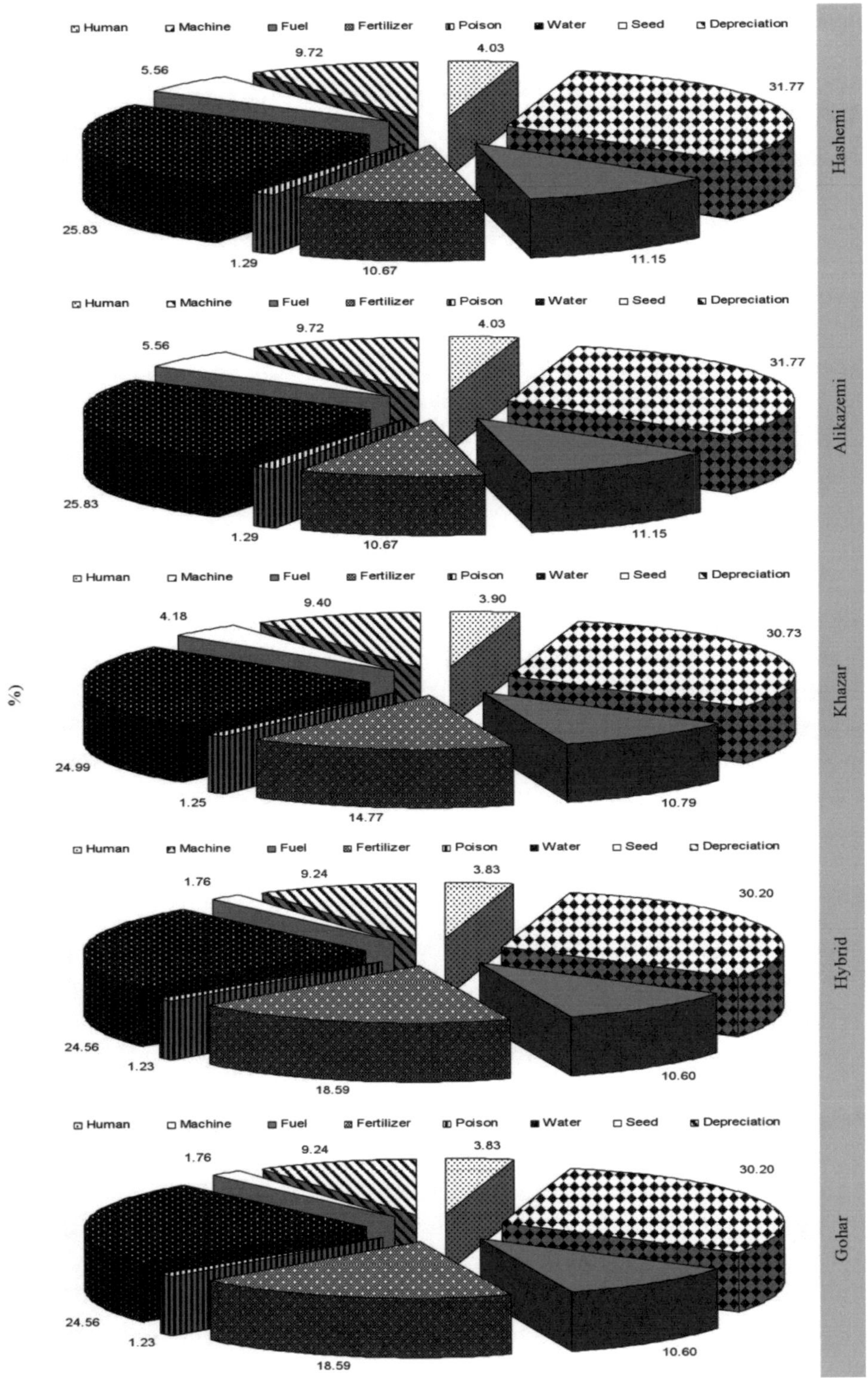

Figure 6. The share (%) production inputs for varieties rice under traditional system condition

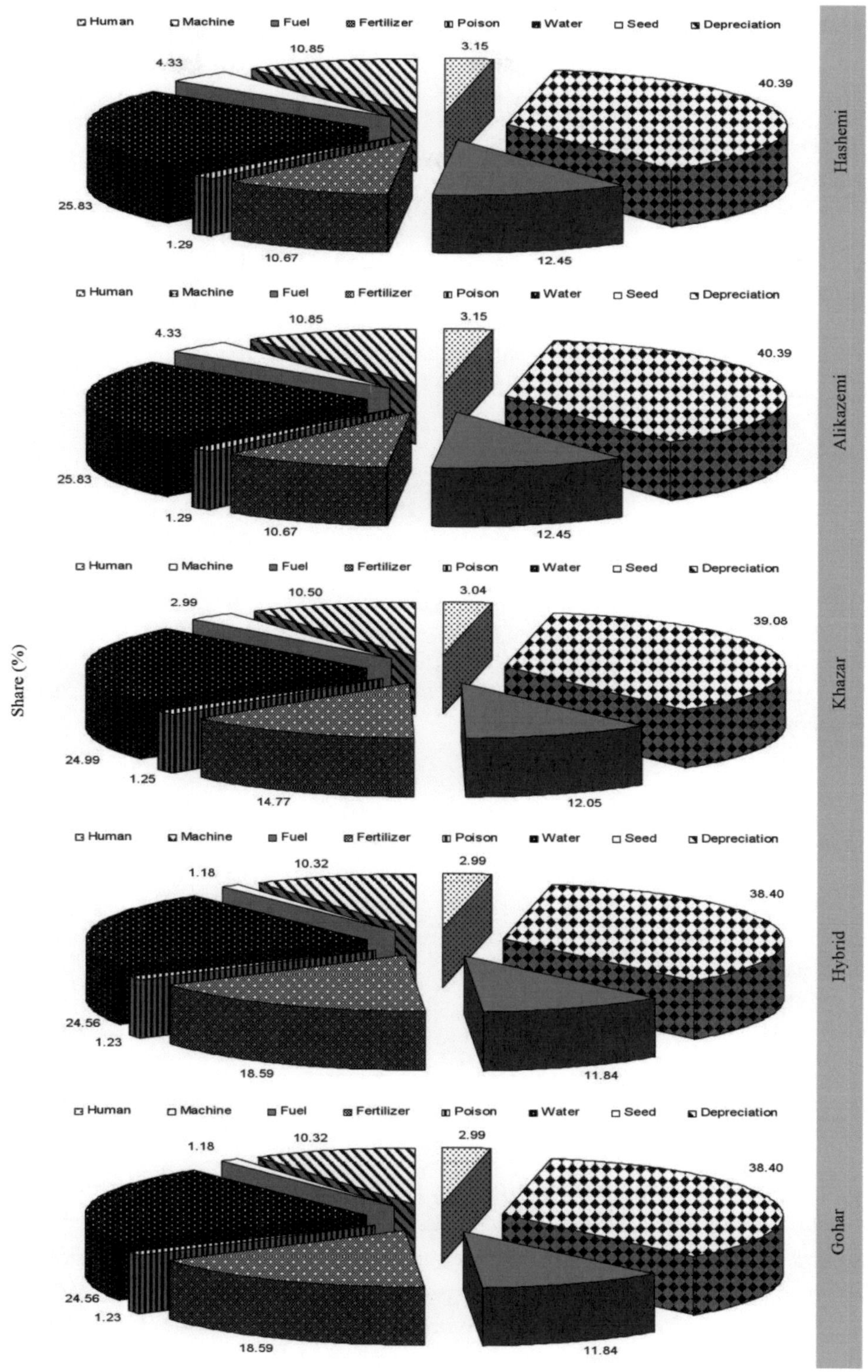

Figure 7. The share (%) production inputs for varieties rice under semi-mechanized system condition

Varieties rice	Item	Percent of compositions	Energy per gram (kcal)	Amounts (kg/ha)	Production energy (kcal/ha)	Production energy / Consumption energy	Consumption energy / Production energy
			Traditional system				
Hashemi	Protein	6.6	4	232.32	929280	0.09	11.34
	Fat	2.2	9	77.44	696960	0.07	15.12
	Starch	80	4	2816	11264000	1.07	0.94
Alikazemi	Protein	6.6	4	275.88	1103520	0.10	9.55
	Fat	2.2	9	91.96	827640	0.08	12.73
	Starch	80	4	3344	13376000	1.27	0.79
Khazar	Protein	6.6	4	319.44	1277760	0.12	8.53
	Fat	2.2	9	106.48	958320	0.09	11.37
	Starch	80	4	3872	15488000	1.42	0.70
Hybrid	Protein	6.6	4	435.6	1742400	0.16	6.36
	Fat	2.2	9	145.2	1306800	0.12	8.48
	Starch	80	4	5280	21120000	1.91	0.52
Gohar	Protein	6.6	4	551.76	2207040	0.20	5.02
	Fat	2.2	9	183.92	1655280	0.15	6.70
	Starch	80	4	6688	26752000	2.41	0.41
			Semi-mechanized system				
Hashemi	Protein	6.6	4	264	1056000	0.09	10.87
	Fat	2.2	9	88	792000	0.07	14.50
	Starch	80	4	3200	12800000	1.11	0.90
Alikazemi	Protein	6.6	4	313.5	1254000	0.11	9.16
	Fat	2.2	9	104.5	940500	0.08	12.21
	Starch	80	4	3800	15200000	1.32	0.76
Khazar	Protein	6.6	4	363	1452000	0.12	8.15
	Fat	2.2	9	121	1089000	0.09	10.87
	Starch	80	4	4400	17600000	1.49	0.67
Hybrid	Protein	6.6	4	495	1980000	0.16	6.11
	Fat	2.2	9	165	1485000	0.12	8.14
	Starch	80	4	6000	24000000	1.98	0.50
Gohar	Protein	6.6	4	627	2508000	0.21	4.82
	Fat	2.2	9	209	1881000	0.16	6.43
	Starch	80	4	7600	30400000	2.51	0.40

Table 8. Items of energy balance indices in rice paddy production under traditional and semi-mechanized system condition

The highest percent of compositions, amounts, production energy, and production energy to consumption energy ratio in rice biomass were obtained from starch as compared with protein and fat; the lowest consumption energy to production energy ratio in rice biomass was obtained from starch as compared with protein and fat "Table 11". Results of "Table 11" showed that breed varieties (Khazar, Hybrid and Gohar) because of suitable genetic specifications have higher operation in compared with local varieties (Hashemi and Alikazemi); the highest amounts (protein: 1087.52, fat: 355.91 and starch:12259.26), production energy (protein: 4350060 kg/ha, fat: 32032260 kg/ha and starch: 49037040 kg/ha), and production energy to consumption energy ratio (protein: 0.41, fat: 0.30 and starch: 4.65) in rice biomass of traditional system and highest amounts (protein:1235.80, fat: 404.44 and starch: 13930.87), production energy (protein: 4943180 kg/ha, fat: 3639978 kg/ha and starch: 55723120 kg/ha), and production energy to consumption energy ratio (protein: 0.47, fat: 0.35 and starch: 5.29) in rice biomass of semi-mechanized observed in Gohar rice.

Varieties rice	Item	Percent of compositions	Energy per gram (kcal)	Amounts (kg/ha)	Production energy (kcal/ha)	Production energy / Consumption energy	Consumption energy / Production energy
			Traditional system				
	Protein	5.13	4	41.71	166828	0.02	63.18
Hashemi	Fat	5.15	9	41.87	376826	0.04	27.97
	Starch	50	4	406.50	1626000	0.15	6.48
	Protein	5.13	4	53.61	214434	0.02	49.15
Alikazemi	Fat	5.15	9	53.82	484358	0.05	21.76
	Starch	50	4	522.50	2090000	0.20	5.04
	Protein	5.13	4	62.07	248292	0.02	43.88
Khazar	Fat	5.15	9	62.32	560835	0.05	19.43
	Starch	50	4	605.00	2420000	0.22	4.50
	Protein	5.13	4	84.65	338580	0.03	32.74
Hybrid	Fat	5.15	9	84.98	764775	0.07	14.49
	Starch	50	4	825.00	3300000	0.30	3.36
	Protein	5.13	4	107.22	428868	0.04	25.85
Gohar	Fat	5.15	9	107.64	968715	0.09	11.44
	Starch	50	4	1045.00	4180000	0.38	2.65
			Semi-mechanized system				
	Protein	5.13	4	51.30	205200	0.02	55.96
Hashemi	Fat	5.15	9	51.50	463500	0.04	24.77
	Starch	50	4	500.00	2000000	0.17	5.74
	Protein	5.13	4	60.94	243778	0.02	47.11
Alikazemi	Fat	5.15	9	61.18	550638	0.05	20.85
	Starch	50	4	594.00	2376000	0.21	4.83
	Protein	5.13	4	70.54	282150	0.02	41.96
Khazar	Fat	5.15	9	70.81	637313	0.05	18.57
	Starch	50	4	687.50	2750000	0.23	4.30
	Protein	5.13	4	96.19	384750	0.03	31.43
Hybrid	Fat	5.15	9	96.56	869063	0.07	13.92
	Starch	50	4	937.50	3750000	0.31	3.22
	Protein	5.13	4	121.84	487350	0.04	24.81
Gohar	Fat	5.15	9	122.31	1100813	0.09	10.99
	Starch	50	4	1187.50	4750000	0.39	2.55

Table 9. Items of energy balance indices in rice husk production under traditional and semi-mechanized system condition

Varieties rice	Item	Percent of compositions	Energy per gram (kcal)	Amounts (kg/ha)	Production energy (kcal/ha)	Production energy / Consumption energy	Consumption energy / Production energy
Traditional system							
	Protein	4.3	4	190.79	763164	0.07	13.81
Hashemi	Fat	1.3	9	57.68	519129	0.05	20.30
	Starch	43	4	1921.22	7684884	0.73	1.37
	Protein	4.3	4	245.36	981432	0.09	10.74
Alikazemi	Fat	1.3	9	74.18	667602	0.06	15.79
	Starch	43	4	2470.70	9882792	0.94	1.07
	Protein	4.3	4	284.10	1136404	0.10	9.59
Khazar	Fat	1.3	9	85.89	773019	0.07	14.09
	Starch	43	4	2860.83	11443324	1.05	0.95
	Protein	4.3	4	387.43	1549720	0.14	7.15
Hybrid	Fat	1.3	9	117.13	1054170	0.10	10.52
	Starch	43	4	3901.33	15605320	1.41	0.71
	Protein	4.3	4	490.76	1963036	0.18	5.65
Gohar	Fat	1.3	9	148.37	1335321	0.12	8.30
	Starch	43	4	4941.83	19767316	1.78	0.56
Semi-mechanized system							
	Protein	4.3	4	234.82	939292	0.08	12.23
Hashemi	Fat	1.3	9	70.99	638937	0.06	17.97
	Starch	43	4	2364.61	9458452	0.82	1.21
	Protein	4.3	4	278.86	1115420	0.10	10.29
Alikazemi	Fat	1.3	9	84.31	758745	0.07	15.13
	Starch	43	4	2808.01	11232020	0.98	1.02
	Protein	4.3	4	322.84	1291376	0.11	9.17
Khazar	Fat	1.3	9	97.60	878436	0.07	13.48
	Starch	43	4	3250.96	13003856	1.10	0.91
	Protein	4.3	4	440.28	1761108	0.15	6.87
Hybrid	Fat	1.3	9	133.11	1197963	0.10	10.10
	Starch	43	4	4433.49	17733948	1.47	0.68
	Protein	4.3	4	557.67	2230668	0.18	5.42
Gohar	Fat	1.3	9	168.60	1517373	0.13	7.97
	Starch	43	4	5615.58	22462308	1.86	0.54

Table 10. Items of energy balance indices in rice straw production under traditional and semi-mechanized system condition

Results of "Table 12" showed that breed varieties (Khazar, Hybrid and Gohar) because of suitable genetic specifications have higher operation in compared with local varieties (Hashemi and Alikazemi); the highest paddy yield (8360 kg/ha), consumption energy (11084731 kcal/ha), production energy (30614320 kcal/ha) and production energy to consumption energy ratio (2.76) in rice paddy of traditional system and highest paddy yield (9500 kg/ha), consumption energy (12093473 kcal/ha), production energy (34789000 kcal/ha) and production energy to consumption energy ratio (2.88) in rice paddy of semi-mechanized observed in Gohar rice. Energy per unit for rice varieties under to farming system was equaled. Highest Consumption energy to production energy ratio for rice varieties under to farming system was observed in Hashemi rice. Energy balance efficiency (production energy to consumption energy ratio) in this study was calculated 2.76 and 2.88; showing the affective use of energy in the agro ecosystems rice paddy production.

Varieties rice	Item	Percent of compositions	Energy per gram (kcal)	Amounts (kg/ha)	Production energy (kcal/ha)	Production energy / Consumption energy	Consumption energy / Production energy
Traditional system							
	Protein	5.5	4	437.64	1750540	0.17	6.02
Hashemi	Fat	1.8	9	143.23	1289034	0.12	8.18
	Starch	62	4	4933.34	19733360	1.87	0.53
	Protein	5.5	4	543.73	2174920	0.21	4.85
Alikazemi	Fat	1.8	9	177.95	1601532	0.15	6.58
	Starch	62	4	6129.32	24517280	2.33	0.43
	Protein	5.5	4	629.59	2518340	0.24	4.33
Khazar	Fat	1.8	9	206.05	1854414	0.18	5.87
	Starch	62	4	7097.14	28388560	2.69	0.38
	Protein	5.5	4	858.55	3434200	0.33	3.23
Hybrid	Fat	1.8	9	280.98	2528820	0.24	4.38
	Starch	62	4	9678.20	38712800	3.67	0.29
	Protein	5.5	4	1087.52	4350060	0.41	2.55
Gohar	Fat	1.8	9	355.91	3203226	0.30	3.46
	Starch	62	4	12259.26	49037040	4.65	0.23
Semi-mechanized system							
	Protein	5.5	4	520.36	2081420	0.20	5.52
Hashemi	Fat	1.8	9	170.30	1532682	0.15	7.49
	Starch	62	4	5865.82	23463280	2.23	0.49
	Protein	5.5	4	617.93	2471700	0.23	4.65
Alikazemi	Fat	1.8	9	202.23	1820070	0.17	6.31
	Starch	62	4	6965.70	27862800	2.64	0.41
	Protein	5.5	4	715.44	2861760	0.27	4.14
Khazar	Fat	1.8	9	234.14	2107296	0.20	5.62
	Starch	62	4	8064.96	32259840	3.06	0.37
	Protein	5.5	4	975.65	3902580	0.37	3.10
Hybrid	Fat	1.8	9	319.30	2873718	0.27	4.21
	Starch	62	4	10998.18	43992720	4.17	0.27
	Protein	5.5	4	1235.80	4943180	0.47	2.45
Gohar	Fat	1.8	9	404.44	3639978	0.35	3.32
	Starch	62	4	13930.78	55723120	5.29	0.22

Table 11. Items of energy balance indices in rice biomass production under traditional and semi-mechanized system condition

Energy balance indices	Hashemi	Alikazemi	Khazar	Hybrid	Gohar
Traditional system					
Grain yield (kg/ha)	3520	4180	4840	6600	8360
Consumption energy (kcal/ha)	10539595	10539595	10894253	11084731	11084731
Production energy (kcal/ha)	12890240	15307160	17724080	24169200	30614320
Energy per unit (kcal)	3662	3662	3662	3662	3662
Production energy/ Consumption energy	1.22	1.45	1.63	2.18	2.76
Consumption energy/ Production energy	27.40	23.07	20.60	15.37	12.13
Semi-mechanized system					
Grain yield (kg/ha)	4000	4750	5500	7500	9500
Consumption energy (kcal/ha)	11483207	11483207	11837865	12093473	12093473
Production energy (kcal/ha)	14648000	17394500	20141000	27465000	34789000
Energy per unit (kcal)	3662	3662	3662	3662	3662
Production energy/ Consumption energy	1.28	1.51	1.70	2.27	2.88
Consumption energy/ Production energy	26.27	22.12	19.70	14.76	11.65

Table 12. Analysis of energy balance indices in rice paddy production under traditional and semi-mechanized system condition

Results of "Table 13" showed that breed varieties (Khazar, Hybrid and Gohar) because of suitable genetic specifications have higher operation in compared with local varieties (Hashemi and Alikazemi); the highest husk yield (2090 kg/ha), consumption energy (11084731 kcal/ha), production energy (5577583 kcal/ha) and production energy to consumption energy ratio (0.50) in rice husk of traditional system and highest husk yield (2357 kg/ha), consumption energy (12093473 kcal/ha), production energy (6338163 kcal/ha) and production energy to consumption energy ratio (0.52) in rice husk of semi-mechanized observed in Gohar rice. Energy per unit for rice varieties under to farming system was equaled. Highest Consumption energy to production energy ratio for rice varieties under to farming system was observed in Hashemi rice. Energy balance efficiency (production energy to consumption energy ratio) in this study was calculated 0.50 and 0.52; showing the affective use of energy in the agro ecosystems rice husk production.

Results of "Table 14" showed that breed varieties (Khazar, Hybrid and Gohar) because of suitable genetic specifications have higher operation in compared with local varieties (Hashemi and Alikazemi); the highest straw yield (11413 kg/ha), consumption energy (11084731 kcal/ha), production energy (23065673 kcal/ha) and production energy to consumption energy ratio (2.08) in rice husk of traditional system and highest paddy yield (12969 kg/ha), consumption energy (12093473 kcal/ha), production energy (26210349 kcal/ha) and production energy to consumption energy ratio (2.17) in rice husk of semi-mechanized observed in Gohar rice. Energy per unit for rice varieties under to farming system was equaled. Highest Consumption energy to production energy ratio for rice varieties under to farming system was observed in Hashemi rice. Energy balance efficiency (production energy to consumption energy ratio) in this study was calculated 2.08 and 2.17; showing the affective use of energy in the agro ecosystems rice straw production.

Energy balance indices	Hashemi	Alikazemi	Khazar	Hybrid	Gohar
Traditional system					
Grain yield (kg/ha)	813	1045	1210	1650	2090
Consumption energy (kcal/ha)	10539595	10539595	10894253	11084731	11084731
Production energy (kcal/ha)	2169653.1	2788792	3229127	4403355	5577583
Energy per unit (kcal)	2669	2669	2669	2669	2669
Production energy/ Consumption energy	0.21	0.26	0.30	0.40	0.50
Consumption energy/ Production energy	97.63	75.95	67.80	50.59	39.94
Semi-mechanized system					
Grain yield (kg/ha)	1000	1188	1375	1875	2375
Consumption energy (kcal/ha)	11483207	11483207	11837865	12093473	12093473
Production energy (kcal/ha)	2668700	3170416	3669463	5003813	6338163
Energy per unit (kcal)	2669	2669	2669	2669	2669
Production energy/ Consumption energy	0.23	0.28	0.31	0.41	0.52
Consumption energy/ Production energy	86.48	72.79	64.84	48.57	38.35

Table 13. Analysis of energy balance indices in rice husk production under traditional and semi-mechanized system condition

Energy balance indices	Hashemi	Alikazemi	Khazar	Hybrid	Gohar
Traditional system					
Grain yield (kg/ha)	4437	5706	6607	9010	11413
Consumption energy (kcal/ha)	10539595	10539595	10894253	11084731	11084731
Production energy (kcal/ha)	8967177	11531826	13352747	18209210	23065673
Energy per unit (kcal)	2021	2021	2021	2021	2021
Production energy/ Consumption energy	0.85	1.09	1.23	1.64	2.08
Consumption energy/ Production energy	35.48	27.59	24.63	18.38	14.51
Semi-mechanized system					
Grain yield (kg/ha)	5461	6485	7508	10239	12969
Consumption energy (kcal/ha)	11483207	11483207	11837865	12093473	12093473
Production energy (kcal/ha)	11036681	13106185	15173668	20693019	26210349
Energy per unit (kcal)	2021	2021	2021	2021	2021
Production energy/ Consumption energy	0.96	1.14	1.28	1.71	2.17
Consumption energy/ Production energy	31.41	26.45	23.55	17.64	13.93

Table 14. Analysis of energy balance indices in rice straw production under traditional and semi-mechanized system condition

Results of "Table 15" showed that breed varieties (Khazar, Hybrid and Gohar) because of suitable genetic specifications have higher operation in compared with local varieties (Hashemi and Alikazemi); the highest biomass yield (19773 kg/ha), consumption energy (11084731 kcal/ha), production energy (56590326 kcal/ha) and production energy to consumption energy ratio (5.37) in rice biomass of traditional system and highest biomass yield (22469 kg/ha), consumption energy (12093473 kcal/ha), production energy (6430278 kcal/ha) and production energy to consumption energy ratio (6.10) in rice biomass of semi-mechanized observed in Gohar rice. Energy per unit for rice varieties under to farming system was equaled. Highest consumption energy to production energy ratio for rice varieties under to farming system was observed in Hashemi rice. Energy balance efficiency (production energy to consumption energy ratio) in this study was calculated 5.37 and 6.10; showing the affective use of energy in the agro ecosystems rice biomass production.

Energy balance indices	Hashemi	Alikazemi	Khazar	Hybrid	Gohar
Traditional system					
Grain yield (kg/ha)	7957	9886	11447	15610	19773
Consumption energy (kcal/ha)	10539595	10539595	10894253	11084731	11084731
Production energy (kcal/ha)	22772934	28293732	32761314	44675820	56590326
Energy per unit (kcal)	2862	2862	2862	2862	2862
Production energy/ Consumption energy	2.16	2.68	3.11	4.24	5.37
Consumption energy/ Production energy	14.73	11.86	10.58	7.90	6.23
Semi-mechanized system					
Grain yield (kg/ha)	9461	11235	13008	17739	22469
Consumption energy (kcal/ha)	11483207	11483207	11837865	12093473	12093473
Production energy (kcal/ha)	27077382	32154570	37228896	50769018	64306278
Energy per unit (kcal)	2862	2862	2862	2862	2862
Production energy/ Consumption energy	2.57	3.05	3.53	4.82	6.10
Consumption energy/ Production energy	13.50	11.37	10.12	7.58	5.99

Table 15. Analysis of energy balance indices in rice biomass production under traditional and semi-mechanized system condition

3.3. Correlation analysis of energy indices and balance energy indices for rice production

Result of "Table 16" (balance energy indices) showed that between paddy yield, straw yield, husk yield and biomass yield with production energy and production energy to consumption energy ratio have a positive and very significant correlation, also between paddy yield, straw yield, husk yield and biomass yield with consumption energy to production energy ratio energy intensity a negative and significant correlation in probability level of 1% were recorded.

Item	Yield	Consumption Energy	Production energy	Production energy / Consumption energy	Consumption energy / Production energy
Paddy yield	1				
Consumption energy	0.58	1			
Production energy	0.99**	0.58	1		
Production energy/ Consumption energy	0.99**	0.48	0.99**	1	
Consumption energy/ Production energy	-0.96**	-0.50**	-0.96**	-0.97**	1
Straw yield	1				
Consumption energy	0.59	1			
Production energy	0.99**	0.59	1		
Production energy/ Consumption energy	0.99**	0.49	0.99**	1	
Consumption energy/ Production energy	-0.96**	-0.52**	-0.96**	-0.96**	1
Husk yield	1				
Consumption energy	0.59	1			
Production energy	0.99**	0.59	1		
Production energy/ Consumption energy	0.99**	0.48	0.99**	1	
Consumption energy/ Production energy	-0.96**	-0.52**	-0.96**	-0.96**	1
Biomass yield	1				
Consumption energy	0.59	1			
Production energy	0.99**	0.59	1		
Production energy/ Consumption energy	0.99**	0.59	0.99**	1	
Consumption energy/ Production energy	-0.96**	-0.51**	-0.96**	-0.96**	1

**and*respectively significant in 1% and 5% area

Table 16. Correlation of energy balance indices for rice production

Result of "Table 17" (energy indices) showed that between paddy yield, straw yield, husk yield and biomass yield with input energy, output energy, energy ratio, energy productivity, net energy gain and water and energy productivity have a positive and very significant correlation, also between paddy yield, straw yield, husk yield and biomass yield with energy intensity a negative and significant correlation in probability level of 1% were recorded.

3.4. Growth analysis of rice varieties

Most climate change studies benefit from crop models. Crop simulation models could provide an alternative, less time-consuming and inexpensive means of determining the optimum crop N requirements under management nitrogen conditions. The model ORYZA2000, which simulates the growth and development of rice under conditions of potential production, water and nitrogen limitations, Results of growth indices analysis of rice varieties "Figure 8" showed that breed varieties (Khazar, Hybrid and Gohar) higher growth indices rather than Hashemi local varieties (Hashemi and Alikazemi). Azarpour et al. [3] with study Evaluation of the ORYZA2000 model of rice cultivars in Guilan climate condition showed that the model ORYZA2000 can satisfactorily in Simulates processes of growth and development and grain yield of rice cultivars under weather conditions of Guilan. Therefore validated ORYZA2000 model can apply to research purposes for rice cultivars under weather conditions of Guilan.

Item	Yield	Input energy	Output energy	Energy Ratio	Energy intensity	Energy productivity	Net energy gain	Water and energy productivity
Paddy yield	1							
Input energy	0.91**	1						
Output energy	0.99**	0.91**	1					
Energy ratio	0.99**	0.86**	0.99**	1				
Energy intensity	-0.97**	-0.90**	-0.97**	-0.97**	1			
Energy productivity	0.98**	0.84**	0.98**	0.99**	-0.96**	1		
Net energy gain	0.99**	0.89**	0.99**	0.99**	-0.97**	0.99**	1	
Water and energy productivity	0.99**	0.87**	0.99**	0.99**	-0.97**	0.99**	0.99**	1
Straw yield	1							
Input energy	0.92**	1						
Output energy	0.99**	0.92**	1					
Energy ratio	0.99**	0.87**	0.99**	1				
Energy intensity	-0.96**	-0.83**	-0.96**	-0.96**	1			
Energy productivity	0.99**	0.88**	0.99**	0.99**	-0.96**	1		
Net energy gain	0.99**	0.90**	0.99**	0.99**	-0.96**	0.99**	1	
Water and energy productivity	0.99**	0.87**	0.99**	0.99**	-0.97**	0.99**	0.99**	1
Husk yield	1							
Input energy	0.92**	1						
Output energy	0.99**	0.92**	1					
Energy ratio	0.99**	0.87**	0.99**	1				
Energy intensity	-0.96**	-0.88**	-0.96**	-0.96**	1			
Energy productivity	0.92**	0.77**	0.92**	0.95**	-0.94**	1		
Net energy gain	0.93**	0.71**	0.93**	0.96**	-0.89**	0.93**	1	
Water and energy productivity	0.95**	0.84**	0.95**	0.96**	-0.93**	0.96**	0.92**	1
Biomass yield	1							
Input energy	0.92**	1						
Output energy	0.99**	0.92**	1					
Energy ratio	0.99**	0.87**	0.99**	1				
Energy intensity	-0.96**	-0.89**	-0.96**	-0.97**	1			
Energy productivity	0.99**	0.97**	0.99**	0.99**	-0.97**	1		
Net energy gain	0.99**	0.91**	0.99**	0.99**	-0.96**	0.99**	1	
Water and energy productivity	0.99**	0.87**	0.99**	0.99**	-0.97**	0.99**	0.99**	1

**and*respectively significant in 1% and 5% area

Table 17. Correlation of energy indices for rice production

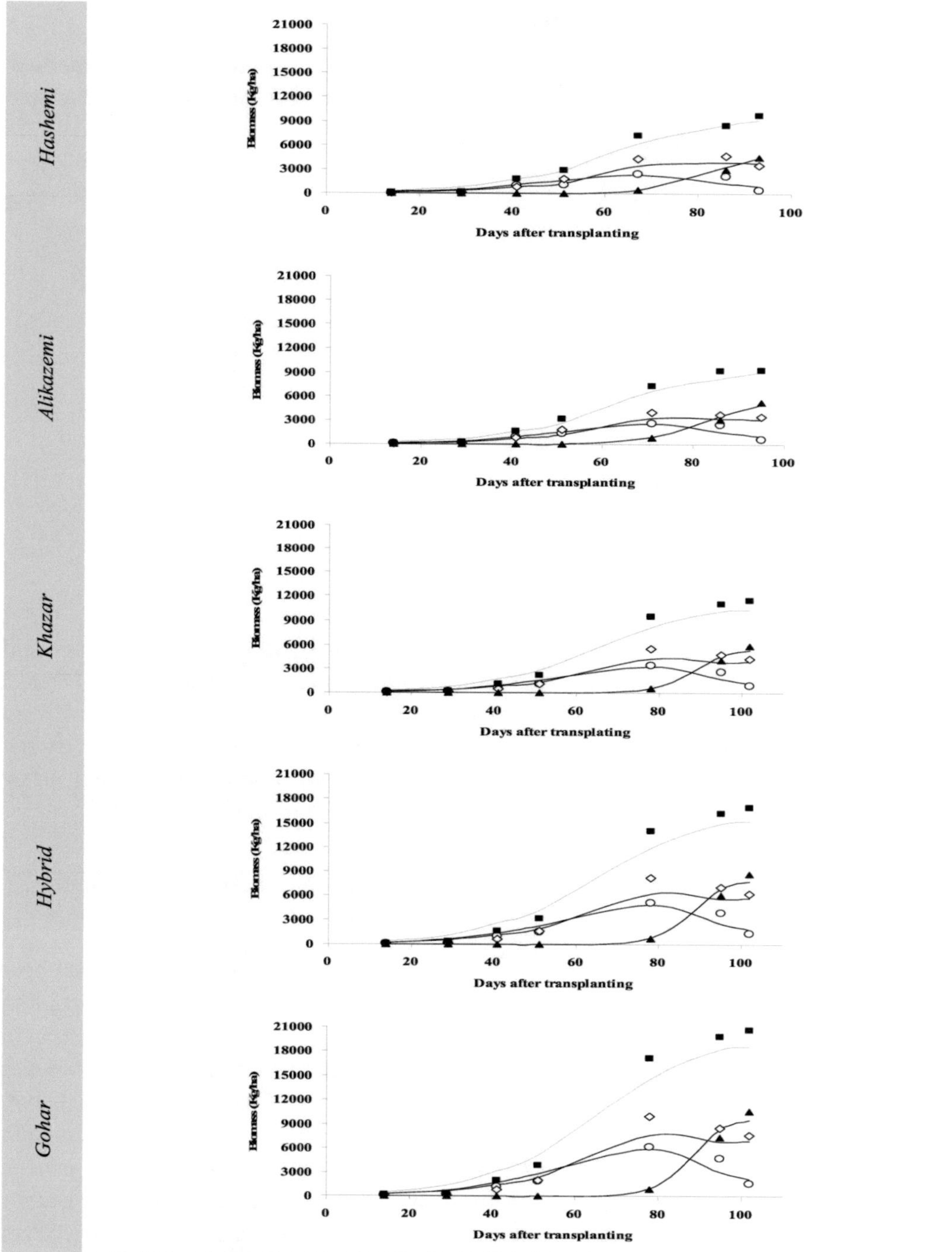

Figure 8. Simulation and measured of biomass of leaves (○), stem (◊), panicles (▲), and total aboveground biomass (■)

3.5. Cluster analysis of energy indices and balance energy indices for rice production

In cluster analysis genotypes were classified into four groups based on Ward's method. Cluster analysis showed that Hybrid and Gohar varieties and Alikazemi, Khazar and Hashemi varieties in group similarities "Figure 9".

```
              Rescaled Distance Cluster Combine

  C A S E     0     5    10    15    20    25

 Label    Num +---------+---------+---------+---------+---------+

 Alikazemi  2   -+-+
 Khazar     3   -+ +---------------------------------------------+
 Hashemi    1   ---+                                             |
 Hybrid     4   ----------+--------------------------------------+
 Gohar    5   -----------+
```

Figure 9. Dendrogram of rice genotypes based on different ward method

3.6. Yield function

Relation between amounts of energy efficiency (energy output to input energy ratio) and energy balance efficiency (production energy to consumption energy ratio) and their effect on paddy yield, straw yield, husk yield and biomass yield were showed in figure 10. Paddy yield, straw yield, husk yield and biomass yield were increased with of use energy efficiency and energy balance efficiency "Figure 10". Yield function of paddy yield, straw yield, husk yield and biomass yield obtained by following relationship "Figure 10".

3.7. Economic analysis of varieties rice production under traditional and semi-mechanized system condition

Crop profitability is the indicator for a farmer to decide what to grow and what and how much should be the energy inputs for growing that specific crop. Total cost of production in two farming systems and five varieties were showed that highest total cost of production in traditional system than semi-mechanized system and local varieties than breed varieties "Figure 11". The amount of higher consumption of human labor, chemical fertilizer, chemical poison and seed in traditional system lead to increasing total cost of production in this system in compared with semi-mechanized system. Also, because of suitable genetic specifications have higher operation in compared with local varieties. The suitable genetic specifications in breed varieties lead to reducing total cost of production in these varieties in compared with local varieties.

Gross value of production in two farming systems and five varieties were showed that highest gross value of production of semi-mechanized system than traditional system and breed varieties than local varieties "Figure 12". Highest gross value of production with average of 11717 $/ha (semi-mechanized system) and 10311 $/ha (traditional system) observed in Gohar rice.

Net return in two farming systems and five varieties were showed that highest net return of semi-mechanized system than traditional system and breed varieties than local varieties "Figure 13". Highest net return with average of 9391 $/ha (semi-mechanized system) and 11239 $/ha (traditional system) observed in Gohar rice.

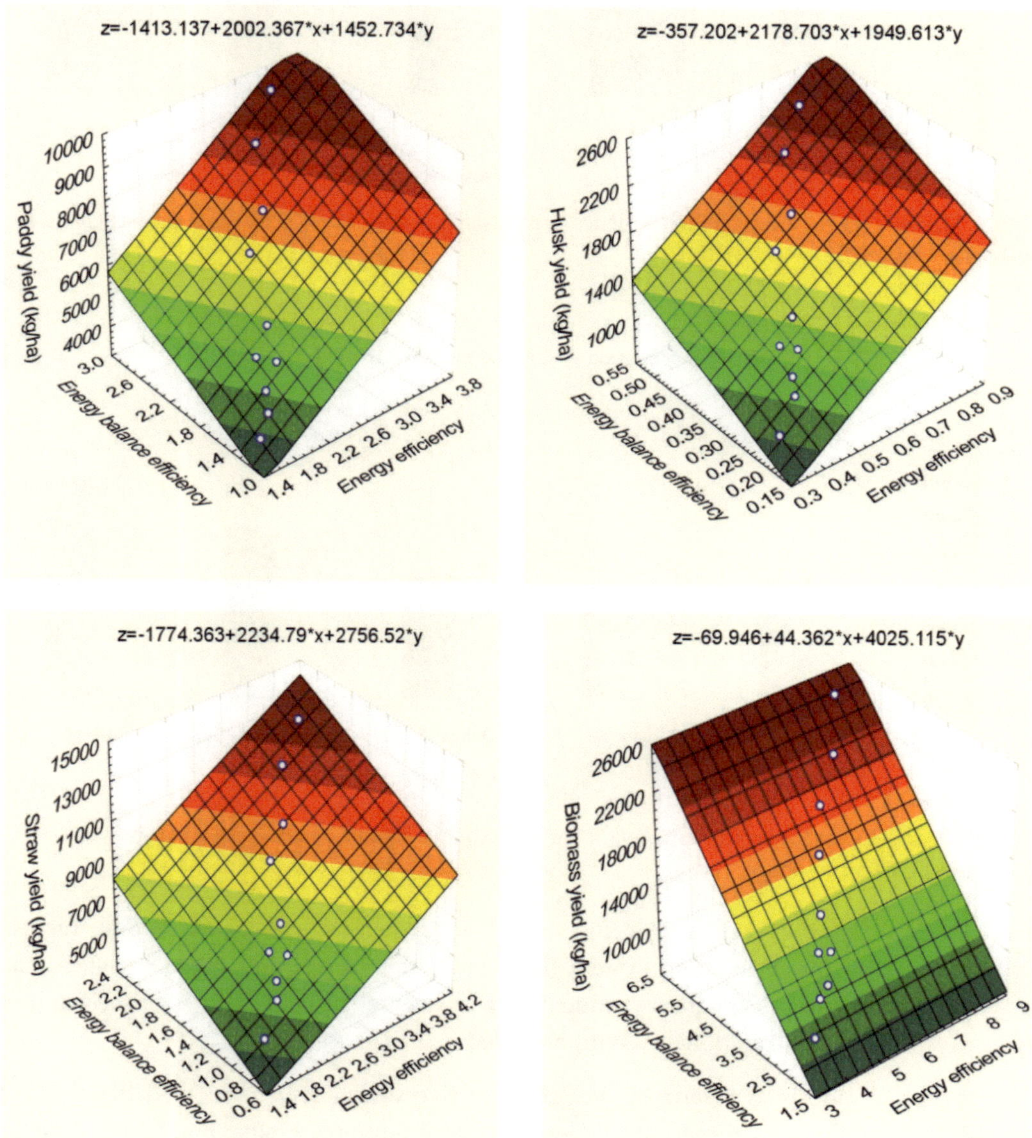

Figure 10. The effect of energy efficiency and energy balance on paddy yield, straw yield, husk yield and biomass yield

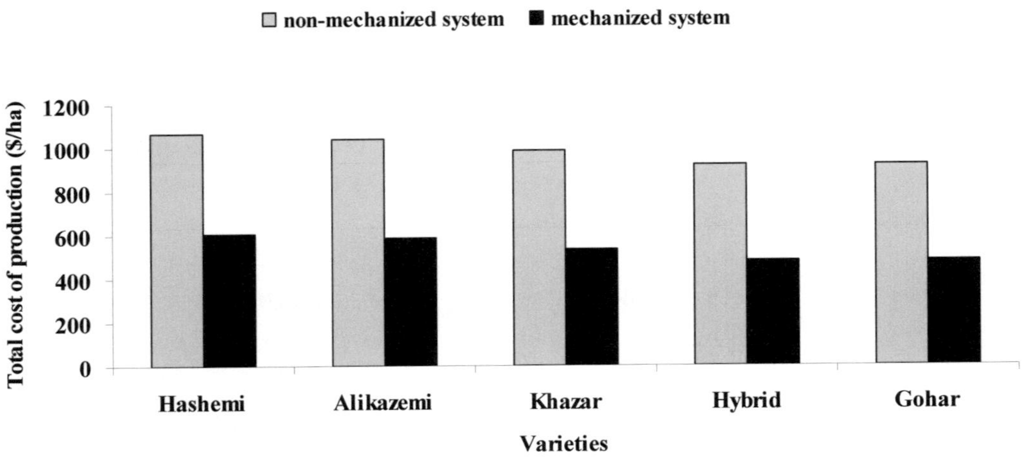

Figure 11. Total cost of production in varieties rice production under traditional and semi-mechanized system condition

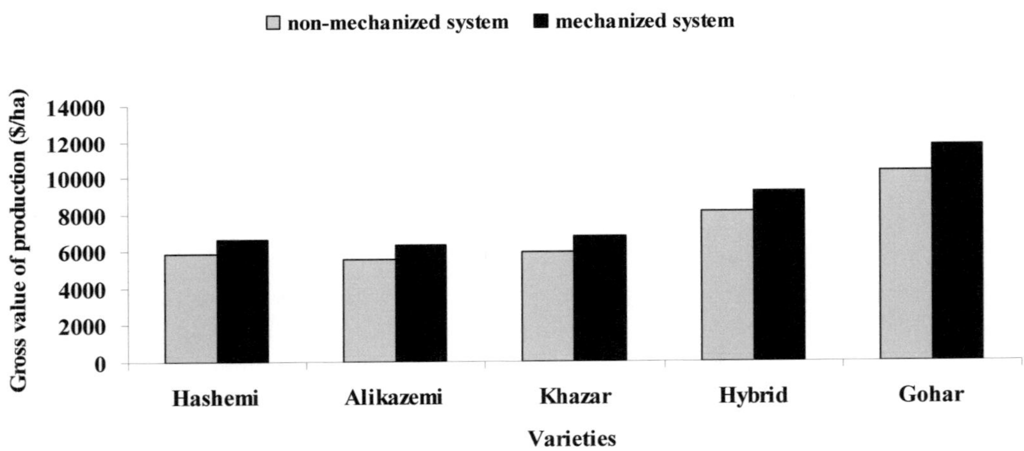

Figure 12. Gross value of production in varieties rice production under traditional and semi-mechanized system condition

Productivity in two farming systems and five varieties were showed that highest productivity of semi-mechanized system than traditional system and breed varieties than local varieties "Figure 14". Highest productivity with average of 19.87 kg/$ (semi-mechanized system) and 9.09 kg/$ (traditional system) observed in Gohar rice.

Benefit to cost ratio in two farming systems and five varieties were showed that highest benefit to cost ratio of semi-mechanized system than traditional system and breed varieties than local varieties "Figure 15". Highest benefit to cost ratio with average of 11.21 (semi-mechanized system) and 24.51 (traditional system) observed in Gohar rice.

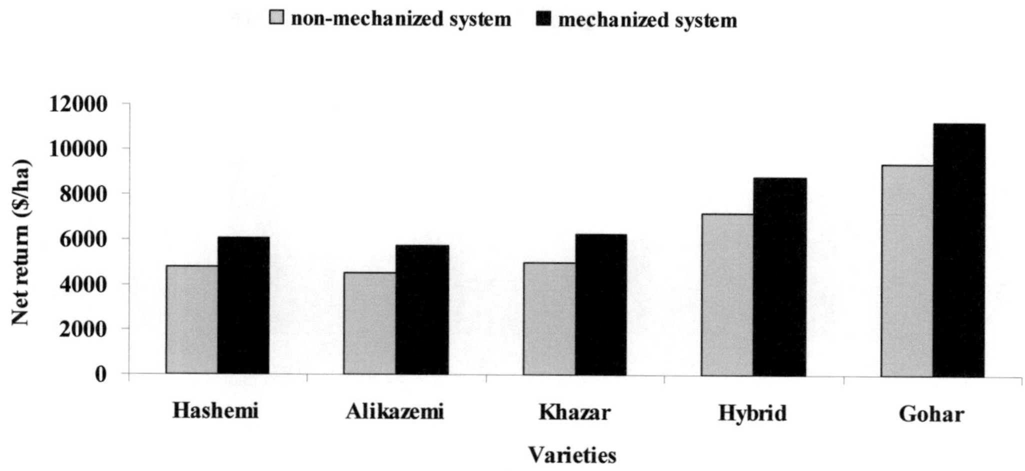

Figure 13. Net return in varieties rice production under traditional and semi-mechanized system condition

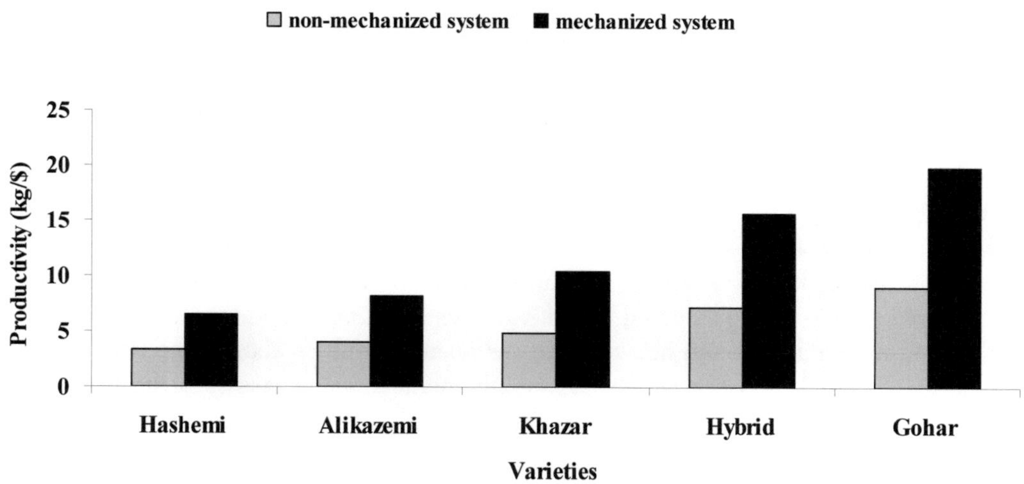

Figure 14. Productivity in varieties rice production under traditional and semi-mechanized system condition

Khan et al. [17] with study energy requirements and economic analysis of wheat, rice and barley production in Australia showed that Cost of production on wheat crop was 323, rice 896 and barley was A$ 246 ha^{-1}. Rice grower obtained the highest return of A$ 2088, as compared to wheat and barley growers, who obtained A$ 589 and 370 ha^{-1}. Therefore, the benefit-cost ratio was the highest on rice farms (3.33) as compared to wheat (2.82) and Barley (2.50). It was concluded that increase in energy consumption at farm level increased yield of rice, hence the farmers with higher cost of production could get better return of their crop [16].

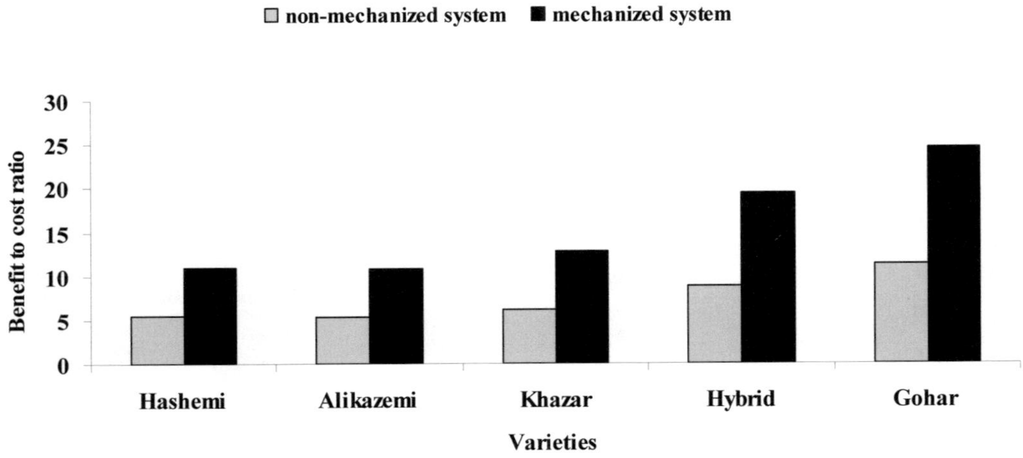

Figure 15. Benefit to cost ratio in varieties rice production under traditional and semi-mechanized system condition

4. Conclusion

Consider that breed varieties rice and semi-mechanized farming system are suitable case for increasing production of rice according to the limitation of rice fields of Guilan province (Iran). Identifying the way of developing and exploitation, energy indicators in agricultural section of Iran either in the light of having weak economical fundamentals or in the light of strict competition in global scene for obtaining better economical condition, helps that we lead our resources and facilities of our production in a direction that can obtain our suitable place in international occasions faster. According to the results of this research and studying the energy and economic analysis, we can say that the condition of the management of energy consumption in producing breed varieties (Khazar, Hybrid (GRH1) and Gohar (SA13)) are more suitable and according to the need of country about producing rice and limitation of energy sources which are mainly nonrenewable energy, producing breed varieties is a step towards sustainable agriculture.

Author details

Ebrahim Azarpour* and Maral Moraditochaee
Young Researchers and Elite Club, Lahijan Branch, Islamic Azad University, Lahijan, Iran

5. References

[1] Artacho P, Bonomelli C, Meza F (2009) Nitrogen Application in Irrigated Rice Grown in Mediterranean Conditions: Effects on Grain Yield, Dry Matter Production, Nitrogen Uptake, and Nitrogen Use Efficiency, Journal of Plant Nutrition. 32: 9, 1574-1593.

* Corresponding Author

[2] Azarpour E (2012). Evaluation energy balance and energy indices of barley production under watered farming in north of Iran. ARPN Journal of Agricultural and Biological Science. 7 (3): 163-168.

[3] Azarpour E, Amiri E, Kashani A, Khodabandeh N, Moradi M (2012) Evaluation of the ORYZA2000 model of rice cultivars in Gilan climat condition. Agronomy and berrding Iran. Islamic Azad University, Lahijan, Iran. 8 (1). In press.

[4] Azarpour E, Moraditochaee M, Bozorgi HR (2012) Evaluation energy balance and energy indices of peanut production in north of Iran. African Journal of Agricultural Research. 7 (16): 2569-2574.

[5] Azarpour E, Moraditochaee M, Bozorgi HR (2012) Evaluation energy balance and energy indices of pumpkin production under dry land farming in north of Iran. Journal of Food, Agriculture & Environment. In press.

[6] Azarpour E, Moraditochaee M, Bozorgi HR (2012) Evaluation energy balance and energy indices of wheat production under rain fed farming in north of Iran. African Journal of Agricultural Research. In press.

[7] Bouman BAM, Kropff MJ, Tuong TP, Wopereis MCS, Ten Berge HFM Van Laar HH (2001) ORYZA2000: Modeling lowland rice. Phillipines: International Rice Research Institute.

[8] Demirbas A (2003) Energy and environmental issues relating to greenhouse gas emissions in Turkey, Energy Conversion and Management. 44: 203-213.

[9] Dincer L (2001) Environmental issues: I-energy utilization. Energy Source. 23: 69-81.

[10] De Jonge AM (2004) Eco-efficiency improvement of a crop protection product: the perspective of the crop protection industry. Crop Protect. 23(12): 1177-1186.

[11] Eskandari Cherati F, Bahrami H, Asakereh A (2011). Energy survey of mechanized and traditional rice production system in Mazandaran Province of Iran. African Journal of Agricultural Research. 6 (11): 2565-2570.

[12] Fageria NK, Slaton NA, Baligar VC (2003) Nutrient management for improving lowland rice productivity and sustainability. Advances in Agronomy. 80: 63-152.

[13] Fageria NK (2007) Yield Physiology of Rice. Journal of Plant Nutrition. 30 (6): 843-879.

[14] Hatirli SA, Ozkan B, Fert K (2005) An econometric analysis of energy input output in Turkish agriculture. Renewable and sustainable energy reviews. 9: 608-623.

[15] Hulsbergen K, Feil J, Diepenbrock W (2002) Rates of nitrogen application required to achieve maximum energy efficiency for various crops: Result of a long- term experiment, Field Crops Research. 77: 113-135.

[16] Khan MA, Awan IU, Zafar J (2009). Energy requirement and economic analysis of rice production in western part of Pakistan. *Soil & Environ. 28(1): 60-67.*

[17] Khan S, Khan MA, Latif N (2010) Energy requirements and economic analysis of wheat, rice and barley production in Australia. Soil & Environ. 29(1): 61-68.

[18] Kizilaslan H (2009) Energy use and input-output energy analysis for apple production in Turkey. J. Food, Agric. Environ. 7(2): 419-423.

[19] Ma H, Oxley L, Gibson J, Kim B (2008) China's energy economy: Technical change, factor demand and interfactor/interfuel substitution. Energy Economics. 30: 2167-2183.

[20] Mandel KG, Saha KP, Ghosh PK, Hati KM, Bandyopadhyay KK (2002) Bioenergy and economic analysis of soybean based crop production systems in central India. Biomass Bioenergy. 23: 337-345.

[21] Mohammadi A, Omid M (2010) Economical analysis and relation between energy inputs and yield of greenhouse cucumber production in Iran. Applied Energy. 87: 191-196.

[22] Mohammadi A, Tabatabaeefar A, Shahin Sh, Rafiee Sh, Keyhani A (2008) Energy use and economical analysis of potato production in Iran a case study: Ardabil province. Energy Conversion and Management. 49: 3566-357

[23] Moradi, M, Azarpour E (2011) Study of energy indices for native and breed rice varieties production in Iran. World applied sciences journal. 13(1): 137-141.

[24] Ozkan B, Akcaoz H, Fert C (2004) Energy input output analysis in Turkish agriculture. Renewable Energy. 29: 39-51.

[25] Ozkan B, Akcaoz H, Karadcniz F (2003) Energy requirement and economic analysis of citrus production in Turkey. Energy Conversion and Management. 44: 46-56.

[26] Peykani GR, Kavoosi Kelashemi M, Sadat Barikani SH, Sasouli MR (2008) Comparison of Production Productivity of 3 Rice Varieties Including Long Grain Good Quality, Long Grain High Yielding and Hybrid Rice in Iran (Case Study: Gilan Province) American-Eurasian J. Agric. & Environ. Sci. 4 (5): 625-632.

[27] Sartori L, Basso B, Bertocco M, Oliviero G (2005) Energy use and economic evaluation of a three year crop rotation for conservation and organic farming in NE Italy. Biosystems Engineering. 9(2): 245-250.

[28] Singh H, Mishra D, Nahar NM (2002). Energy use pattern in production agriculture of a typical village in Arid Zone India-Part. Energy Convers Manage. 43: 2275-2286.

[29] Streimikiene D, Klevas V, Bubeliene J (2007) Use of EU structural funds for sustainable energy development in new EU member states. Renew Sustain Energy Rev. 116: 1167-87.

[30] Taheri Garavand A, Asakereh A, Haghani K (2010) Investigation energy and economic analysis of Soya Bean production in North of Iran. American-Eurasian J. Agric. & Environ. Sci. 7: 648-651.

[31] Yilrnaz L, Akcaoz H, Ozkan B (2005) An analysis of energy use and input costs for cotton production in Turkey, Renewable Energy. 30: 145-155.

Section 3

Biomass for Energy

Chapter 8

Continuous Agave Juice Fermentation for Producing Bioethanol

Dulce María Díaz-Montaño

Additional information is available at the end of the chapter

http://dx.doi.org/10.5772/55923

1. Introduction

The production and utilization of fossil fuels introduce several negative environmental impacts. Bioenergy and biobased products are not a panacea for these problems. However, the environmental burden from use of biorenewable resources is generally much less than from the use of fossil resources. Biofuels include fuels derived from biomass conversion, as well as solid biomass, liquid fuels and various biogases. Forest biomass, agricultural residues and energy crops constitute the three major sources of biomass for energy, with the latter developing into probably the most important source in the 21st century. Land use and the changes thereof is a key issue in sustainable bioenergy production as land availability is ultimately a limiting factor [1]. Biodiesel and bioethanol are the main biofuel. Biodiesel can be made from vegetable oils, microalgae, and animal fats; on the other hand, bioethanol is an alcohol made by fermentation, mostly from carbohydrates produced by sugar or starch crops such as corn or sugarcane, as well as from non-food sources such as agricultural residues. Nevertheless, these processes require as an additional step, prior to saccharification, making production a difficult and expensive. Using agave plants as raw material could be a viable alternative to bioethanol production.

2. Microorganisms involved in the bioethanol production

There is an ever-growing demand for new and improved bioethanol production microorganism strains. Desirable characteristics of bioethanol production microorganisms are listed in Table 1.

Ethanol production microorganisms, mainly *Zymomonas mobilis* and *Saccharomyces cerevisiae,* are potential candidates for bioethanol productions because they showed many of the characteristics presented in the table 1. However, *Zymomonas mobilis* strains have attracted much attention because their growth rate is higher than that of *Saccharomyces cerevisiae,* conventionally used microorganisms for commercial bioethanol production. *Zymomonas mobilis* has been used in tropical areas for making alcoholic beverages from plant sap [2], but its narrow spectrum of fermentable carbohydrates has hampered its industrial exploitation [3]. Several researchers have taken on the challenger on developing recombinant organisms, including: *S. cerevisiae, Z. mobilis, Escherichia coli, Klebsiella oxytoca* and *Erwinia herbicola* [4-5], but the bioethanol production from biomass materials by genetically engineered strains has not yet reached a sufficient level for commercial application [6]. *Zymomonas* cells are gram-negative rods; a minority of the strains are motile, with 1 to 4 polar flagella. These organisms need glucose, fructose, or (for some strains) sucrose in the growth medium. They are very unusual microorganisms since they ferment these sugars anaerobically by way of the Entner-Doudoroff mechanism, followed by pyruvate decarboxylation. The oxidation-reduction balance between G6P dehydrogenase and triosephosphate dehydrogenase on one hand and ethanol dehydrogenase on the other, is mediated through NAD^+. Sugar fermentation is accompanied by formation of a small amount of lactic acid, with traces of acetaldehyde and acetoin [2].

Fermentation Properties	Technological Properties
• Rapid initiation of fermentation	• High genetic stability
• High fermentation efficiency	• Low foam formation
• High ethanol tolerance	• Flocculation properties
• High osmotolerance	• Compacts sediment
• Low temperature optimum	• Low nitrogen demand
• Moderate biomass production	

Table 1. Desirable characteristics of bioethanol production microorganisms

The simplified fermentation process is:

$$\begin{aligned} C_6H_{12}O_6 + (\text{carbon source}) \rightarrow 1.8\ CH_3CH_2OH\ +\ 1.8\ CO_2\ + \\ +\ 0.2CH_3CH(OH)COOH +\ 0.22\ CH_2O\ +\ ATP + 32.7\ \text{kcal} \end{aligned} \quad (1)$$

The molar growth yield indicates that *Zymomonas* is only about 50% efficient in converting its carbon and energy sources. Growth is partially uncoupled. About 2% of the glucose substrate is the source of about half of the cellular carbon. Several amino acids also serve as carbon sources. Some strains grow only anaerobically; others display various degrees of microaerophily. Apparently, the main effect of oxygen is the oxidation of part of the ethanol which converts into acetic acid. Most strains are alcohol tolerant (10%) and grow in up to 40% glucose. The wide pH for growth range from 3.5 to 7.5, and acid tolerance are quite typical. This bacterium has been isolated from fermenting agave sap in Mexico, from fermenting palm saps in Zaire, Nigeria, and Indonesia, from fermenting sugarcane juice in Northeastern Brazil. Undoubtedly, they are important contributors to the fermentation of plant saps in many tropical areas of the America, the Africa, and Asia.

Saccharomyces cerevisiae is a eukaryotic microorganism classified in the fungi kingdom. This yeast is a unicellular microorganism and is defined as basidiomycetes or ascomycetes. *S.*

cerevisiae cells measure 3-7 microns wide and 5-12 microns long. It has elliptic, round and oval shapes and reproduces is by a division process known as budding [7]. It is believed that *S. cerevisiae* was originally isolated from the skin of grapes [8]. Its optimum temperature growth range is 30° C [9]. *S. cerevisiae* is tolerant of a wide pH range (2.4-8.2), being the optimum pH for growth between values of 3.5 to 3.8 [10]. In addition, *S. cerevisiae* is high growth rate (0.5 h-1) in the yeast group. With respect to *S. cerevisiae* nutritional requirements, all strains can grow aerobically on glucose, fructose, sucrose, and maltose and fail to grow on lactose and cellobiose. Also, all strains of *S. cerevisiae* can use ammonia and urea as the sole nitrogen source, but cannot use nitrate since they lack the ability to reduce them to ammonium ions. They can also use most amino acids, small peptides and nitrogen bases as a nitrogen sources [11]. *S. cerevisiae* have a phosphorus requirement, assimilated as a dihydrogen phosphate ion, and sulfur, which can be assimilated as a sulfate ion or as organic sulfur compounds, such as the amino acids: methionine and cysteine. Some metals, such as magnesium, iron, calcium and zinc are also required for good growth of this yeast.

Alcoholic fermentation by yeast consists of three main stages: (1) transporting sugars within the cell, (2) transforming sugars into pyruvate through glycolysis pathway and finally (3) converting acetaldehyde to ethanol.

The simplified fermentation process is:

$$C_6H_{12}O_6 + (\text{carbon source}) \rightarrow 2CH_3CH_2OH + 2\,CO_2 + 2\,ATP + 25.5\,\text{kcal} \tag{2}$$

3. Modes of fermentation process

There are basically three modes of fermentation process: (1) Batch fermentation process. (2) Fed batch fermentation process and (3) Continuous fermentation process (Figure 1).

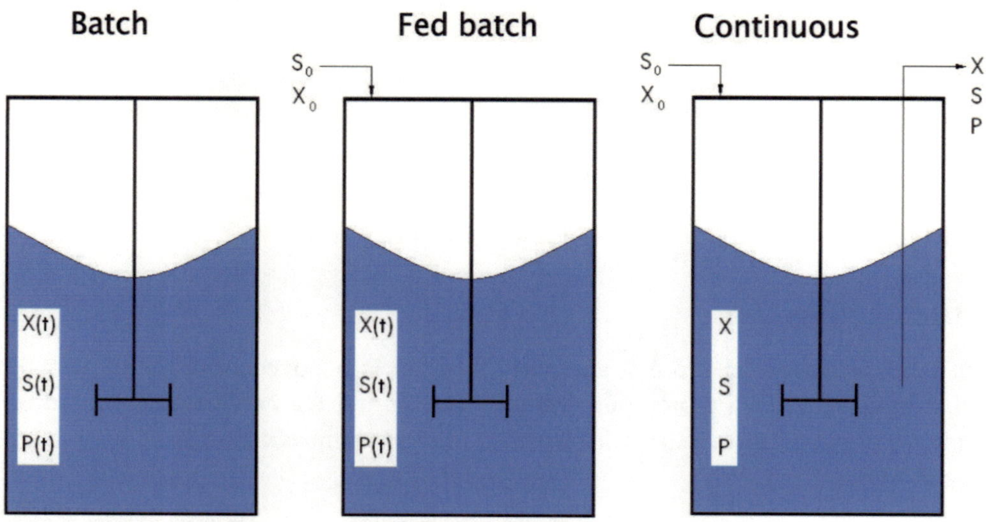

Figure 1. Fermentation process; x: biomass, s: sustrate, p: product, t: time

The mode of operation is dictated by the type of product being produced.

The fermentation process may be divided into six phases:

a. The formulation of media to be used in culturing the process organism during the development of the inoculum and in the production fermenter.
b. The sterilization of the medium, fermenters and ancillary equipment.
c. The production of an active, pure culture in sufficient quantity for inoculating the production vessel.
d. The growth of the microorganism in the production fermenter under optimum conditions for product formation.
e. The extraction of the product and its purification.
f. The disposal of effluents produced by the process.

The interrelationships between the six phases are illustrated in Figure 2.

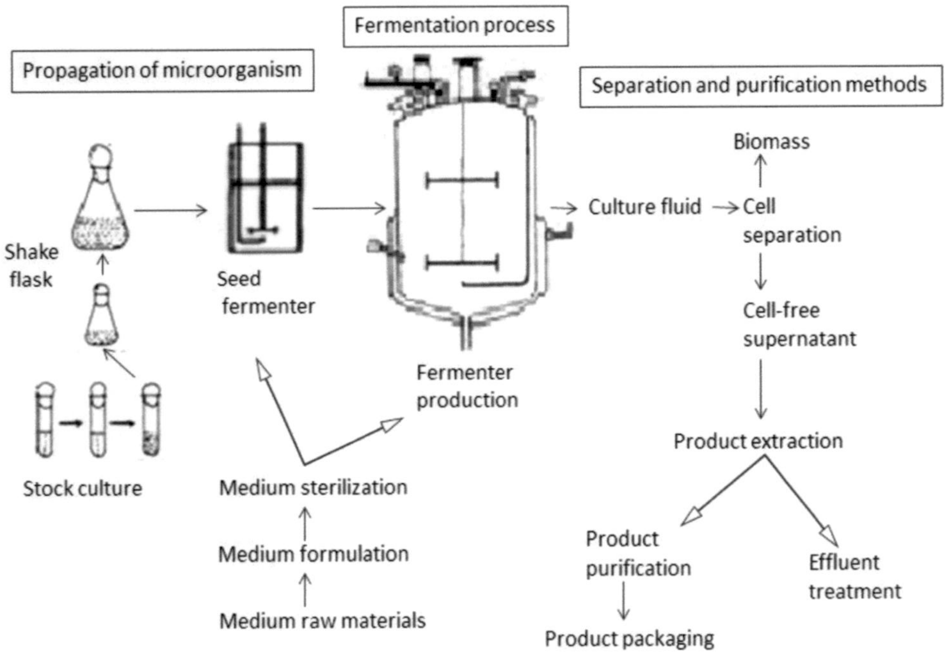

Figure 2. A schematic representation of a typical fermentation process

3.1. Batch fermentation

In the batch fermentation process, the entire medium is removed from the fermentation vessel. The vessel is then thoroughly washed, cleaned and the new batch is started only thereafter. The bioreactor is initially loaded with fresh medium and inoculated with selected microorganism.

During the growth period, no medium is added or removed. The Biomass, nutrients and products concentrations change continuously in time [12].

During the batch fermentation process, various physiological states of the microorganism are observed (Figure 3):

a. Lag phase - Period where microorganisms adapt to the new environment.
b. Positive acceleration phase - Period of slow increase in the population
c. Logarithmic or exponential phase - Period of rapid rise in population due to availability of nutrients. The exponential phase may be described by the following equation:

$$dx/dt = \mu x$$

Where x is the concentration of microbial biomass
t is time, in hours
and μ is the specific growth rate in $hours^{-1}$

d. Negative acceleration phase - Period in which there is a slow rise in population as the environmental resistance increases.
e. Stationary phase - Finally, growth rate becomes stable because mortality and natality rates become equal. During the stationary phase, the organism is still maintaining a certain metabolic activity, while some secondary metabolites are formed (products not associated with microbial growth).
f. Death phase - Finally, environmental stress causes a decrease in metabolic activity of yeast and autolysis.

3.2. Fed batch fermentation

Fed-batch fermentation is described as the type of system where nutrients are added when their concentration falls. In the absence of outlet flow, the volume in the bioreactor will increase linearly. The nutrients are added in several doses to ensure that there are not surplus nutrients in the fermenter at any time. Surplus nutrients may inhibit microorganism growth. By adding nutrients little by little, the reaction can proceed at a high production rate without getting overloaded. The best way to control the addition of the feed is monitoring the concentration of the nutrient itself in the fermenter or reactor vessel.

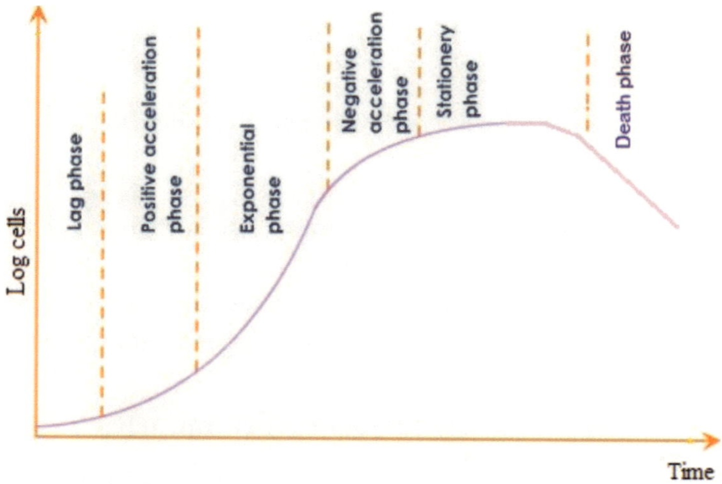

Figure 3. Growth curve of microorganism

The main advantages of the fed batch fermenter are:

a. The extension of the exponential growth phase and production of metabolites of interest.
b. The production of high biomass and product concentrations.
c. The reduced inhibition by the substrate.

However, accumulations of toxic products to the microorganism in the medium and downtime due to charging and discharging (which also occur in batch fermentations) are the main disadvantages of Fed batch fermentation [12].

3.3. Continuous fermentation

Exponential growth in batch fermentation may be prolonged by adding of fresh medium to the vessel. In the continuous fermentation process, the added medium displaced an equal volume of culture from the vessel. Thus, the process of continuous fermentation non-stop and the exponential growth will proceed until the substrate is exhausted. By using proper technique, the desired products are obtained from the removed medium [13].

If medium is fed continuously to such a culture at a suitable rate, a steady state is eventually achieved i. e., the formation of new biomass by the culture is balanced by the loss of cells from the vessel. The flow medium into the vessel is related to the volume of the vessel by the term dilution rate, *D*, defined as:

$$D = {}^{F}/_{V}$$

Where F is the flow rate (volume units/time) and V is the volume (volume units).

The net change in cell concentration over a time period may be expressed as:

$$ {}^{dx}/_{dt} = growth \; - output$$

$$ {}^{dx}/_{dt} = \mu x - Dx$$

Under steady state conditions the cell concentration remains constant, thus ${}^{dx}/_{dt} = 0$ and:

$$\mu = D$$

Thus, under steady state conditions, the specific growth rate is controlled by the dilution rate, which is an experimental variable. It is recalled that under batch culture conditions, an organism will grow at its maximum specific growth rate and, therefore, continuous culture may be operated only at dilution rates below the maximum specific growth rate.

4. Agaves species in the Americas, characteristics and uses

Chiefly Mexican, agaves are also native to the southern and western United States and central and tropical South America. They are succulents with a large rosette of thick, fleshy leaves, each ending generally in a sharp point and with a spiny margin; the stout stem is usually short, the leaves apparently springing from the root. Agave taxa give particulars for all 197 taxa in the two subgenera, Littaea and Agave. The first of a slender form with high in saponin concentration is intended as ornament mainly, except *Dasylirion spp*. Species, which is the raw material to produce Sotol (a Mexican distilled alcoholic beverage). Also the Littaea is used as raw material producing medicinal steroids, since contains smilagenin. In the other hand, the species in the subgenus Agave have been exploited since the ancient pre-Columbian civilization mainly for producing: fiber, fodder, food and alcoholic beverage (Table 2) [14].

5. Alcoholic fermentation process of agave juice

Agave juice bioethanol production from involves multiple steps: at harvest, fermentable sugars are obtained from heads of the agave plant by steaming, milling and pressing. During the steaming process, the polysaccharides (fructans) are hydrolyzed into a mixture of sugars consisting of fructose mainly. After fermentation, the alcohol from the must is purified by distillation and dehydration for obtaining anhydrous ethanol.

Agave species	Main State of Production	Uses	Characteristic
Agave *tequilana* Weber	Jalisco, regions of the states of Nayarit, Michoacán, Tamaulipas, Guanajuato.	Tequila industry	High sugar content
Agave *angustifolia* Haw. Agave *rhodacantha* Trel. Agave *shrevei* Gentry Agave *wocomahi* Gentry Agave *durangensis* Agave *palmeri* Engelm. Agave *zebra* Gentry Agave *asperrima* Jacobi Agave *potatorum* Zucc. Agave *weberi* Cels Agave *tequilana* Weber	Oaxaca, San Luis Potosí, Durango, Jalisco,	Mezcal industry	High sugar content
Agave *angustifolia* Haw.	Sonora	Bacanora Industry	High sugar content
Agave *atrovírens* Kawr Agave *lehmannii* Agave *cochlearís* Agave *lattísíma* Jacobí Agave *mapisaga* Agave *salmiana*	Distrito Federal, Tlaxcala, Hidalgo, Querétaro, Puebla, Morelos, San Luis Potosí	Pulque industry	High sugar content

Agave species	Main State of Production	Uses	Characteristic
Agave *angustifolia* Agave *inaequidens* Agave *maximiliana*	Jalisco	Raicilla industry	
Agave *lechuguilla* Agave *striata* Agave *sisalana*	Yucatan	Fiber industry	Obtained from leaf
Agave *lechuguilla*	Jalisco	Cleaning cloth product	Obtained from agave pulp
Agave *salmiana*	San Luis Potosí	Food and fodder	Obtained from leaf
Agave *sisalana* Agave *fourcroydes*	Yucatan	Paper source	Obtained from leaf
Agave *salmiana* Agave *fourcroydes* Agave *agustifolia* Agave *deweyana*	San Luis Potosi, Jalisco, Yucatan, Sonora	Medicinal uses: steroid drugs	Obtained from leaf High sapogenins concentration

Table 2. Main species of agave with economic importance in México

Alcoholic Fermentation is one of the most important stages in the bioethanol process, as sugars (mainly fructose) are transformed into ethanol and CO_2. Agave juice can be fermented by inoculation (with selected microorganisms) or spontaneously (without inoculums). Significant differences were observed between fermentation conducted with controlled microorganism or inoculated media and spontaneous or no inoculated media. The introduction of selected strains allows fermentation to be regulated and accelerated. Inoculation of culture media with starter cultures allows a high population of selected strain, thereby assuring it dominance. The results are quicker ethanol synthesis, shorter fermentation time, and higher productivity.

Knowledge of physiological behavior of indigenous tequila yeast used in the agave juice alcoholic fermentation process for obtaining bioethanol is still limited. The raw material and physiochemical and biological conditions have significant impact on the productivity fermentation process. For these reasons, a better knowledge of the physiological and metabolic features of these yeasts in agave juice fermentation is required. A study of bioethanol production from *Agave tequilana* Weber var. azul juice fermentations is presented below. For this, the alcoholic fermentation of *Agave tequilana* Weber var. azul juice was carried out in batch and continuous modes of fermentation process.

a. *Agave tequilana* Weber var. azul juice characterization
The *Agave tequilana* Weber juice used in the experimentation was supplied by a distillery. The sugar concentration of the agave juice was 20 °Bx and pH was 4.0. In the distillery, the agave plants are cooked in an autoclave at 95 to 100°C for 4 hours.

The analysis of agave juice amino acids of and of its hydrolyzate was performed and compared to grape juice (Table 3). These results show that agave juice is naturally amino acid poor, even when hydrolyzed [15].

Amino acid (mg/L)	Grape juice[1]	Agave juice[2]	Hydrolyzate Agave juice[2]
L- alanine	58.5*	0.72±0.005	20.98±0.153
L-arginine	255.9±182.3	5.76±0.030	38.68±0.676
L-aspartate	46.4± 22.9	0.41±0.018	25.51±0.322
L-glutamate	91.2± 37.7	0.12±0.001	42.12±0.117
L-glutamine	122.9± 93.9	nq	nq
L-glycine	4.1± 3.1	0.44±0.016	21.75±0.526
L-histidine	103.9± 85.9	0.19±0.008	10.09±0.301
L-isoleucine	13.4*	0.06±0.003	11.70±0.196
L-leucine	13.4*	0.14±0.003	21.28±0.524
L-lysine	7.6± 6.67	0.06±0.002	6.59±0.150
L-metionine	24.2± 13.9	nd	4.10±0.126
L-phenylalanine	16.9± 11.3	0.06±0.003	12.44±0.100
L-serine	53.1± 23.4	1.34±0.024	32.52±0.306
L-threonine	51.6± 25.1	0.32±0.014	18.54±0.270
L-tyrosine	13.3*	0.22±0.010	13.97±0.109
L-valine	17.7*	0.14±0.004	21.49±1.058

[1]: amino acid concentration of 11 grape varieties must [16]; [2]: Each value represents the average ± standard deviation of duplicate determinations, the method limited detection is 1 pmols/mL; *: amino acid concentration constant in the 11 varieties of grape [16]; nd: not detected; nq: not quantified.

Amino acid analyses were determined by HPLC [17]. The acid hydrolysis of agave juice was performed as reported by Umagath et al. [18].

Table 3. Amino acid composition of grape and agave juices.

b. Batch fermentation process

The bioethanol production from agave juice batch fermentation process is shown. For this work, three yeast strains isolated from agave juice were studied for their fermentative capacity. The strains (S1, S2 and S3) were identified by biochemical and molecular tests [15]. The experiments were performed using agave juice supplemented with sufficient ammonium sulphate, for maintaining a good performance of the yeast strains. For fermentation medium, sugar concentration of the agave juice was adjusted to 12 ºBrix (95±5 g/L reducing sugar) and then supplemented with 1g/L of ammonium sulphate. Culture media were sterilized at 121 °C for 15 min. The pH of the unadjusted juice was 4.2. This fermentation medium was similar to the must typically used in industrial distilleries for obtain alcoholic beverage. The fermentations were carried out under anaerobic conditions at 35 °C and 250 rpm in a 3 L bioreactor (Applikon, Netherlands). The inoculation level was 20 million cells/mL. Two fermentations were performed with each yeast.

Each must was fermented for 72 h, and sampling was performed every 2 h during the first 12 h of fermentation, then every 4 h during the following 48 h, until the last sampling event at 72 h. Biomass concentration was obtained by dry weight measurement. Reducing sugar concentration was determined by the DNS method modified and glucose, fructose and glycerol concentration was determined by HPLC [15]. Samples were micro-distilled and ethanol concentration was determined in distillates by using the potassium dichromate method [19].

Fermentation Kinetic Analysis - The evolution of biomass, sugar consumption and ethanol production versus time were plotted in Fig. 1 and Table 1, showing the kinetic parameters of each strain. All *Saccharomyces* strains grew faster reaching a biomass concentration level of 4-5.3 g/L by approximately 12 h and sugar was completely depleted by 18-24 h of the fermentation (Figure 4). The S1 and S2 strains showed a higher ethanol concentration and sugar consumption than S3 (Figure 4 and Table 4).

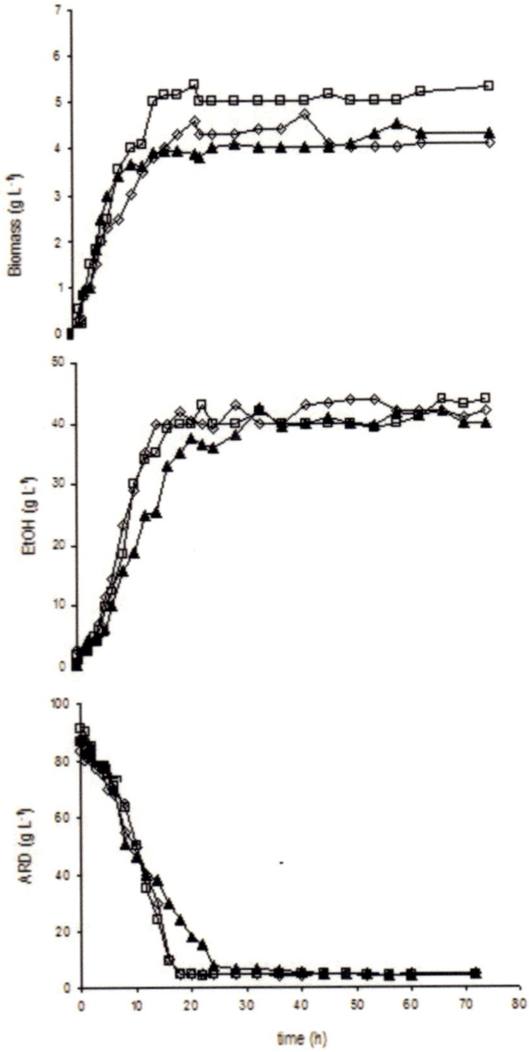

Figure 4. Kinetic profiles of the fermentation of S1(◊), S2(□) and S3(▯) strains in a *Agave tequilana* Weber blue variety juice medium at 12 ºBx, supplemented with ammonium sulfate (1g/L). Biomass: biomass concentration profile; ARD: reduction sugar concentration profile; ETOH: ethanol concentration profile.

Growth and ethanol yields were different: 0.046-0.059 g/g and 0.47-0.49 g/g, respectively (Table 4). Statistical analysis (95% LSD) showed significant differences between yeast strains in all kinetic parameters (Table 4). *S. cerevisiae* S1 strain presented a higher value of maximum specific growth and sugar consumption than S2 and S3 strains. Likewise, S1 and S3 strains showed a high maximum specific ethanol rate (Table 4).

	Kinetics parameters							
Strain	μ_{max} (h^{-1})	q_{smax} (g/gh^{-1})	q_{pmax} ($g/g\ h^{-1}$)	$Y_{x/s}$ (g/g)	$Y_{p/s}$ (g/g)	X_f (g/L)	S_c (g/L)	$Etoh_f$ (g/L)
S1	0.43±.016	4.28±.27	1.56±.12	0.050±.004	0.49±.027	4.34±.26	86.7±2.0	42.6±1.0
S2	0.33±.030	2.85±.15	1.34±.06	0.055±.004	0.49±.001	4.86±.44	87.4±1.2	43.5±.55
S3	0.35±.020	3.74±.27	1.52±.06	0.052±.001	0.47±.015	4.35±.10	83.9±.30	39.9±1.4

μ_{max}: maximum specific growth rate; q_{smax}: maximum specific sugar consumption rate; q_{pmax}: maximum specific ethanol production rate; $Y_{x/s}$ and $Y_{p/s}$: yields of biomass and ethanol; S_c: consumed substrate concentration; X_f: final biomass concentration; $Etoh_f$: final ethanol concentration. Each value represents the average ± standard deviation of duplicate determinations of two fermentations.

Table 4. Comparison of kinetic parameters and final concentration of biomass, consumed substrate and ethanol for the different strains.

c. Continuous fermentation process

Bioethanol production from agave juice continuous fermentation process is shown below. In continuous fermentation process, the effects of dilution rate, nitrogen and phosphorus source addition and micro-aeration on growth, and synthesis of ethanol of two native *Saccharomyces cerevisiae* S1 and S2 strains were studied.

Continuous cultures were carried out in a 3 L bioreactor (Applikon, The Netherlands) with a 2 L working volume. Cultures were started in a batch mode, by inoculating fermentation medium with 3.5 x 10^6 cells/mL (97±2 %. initial viability) and incubating at 30 °C and 250 rpm for 12 h. Afterwards, the culture was fed with fermentation medium (12 °Brix = 95 ± 5 g/L reducing sugar and 1 g/L of ammonium sulfate). Culture media were sterilized at 121 °C for 15 min.

To reach the steady state in each studied condition, the culture was maintained during five residence times and samples were taken every 6 h. A steady state was reached, when the variation in the concentrations of biomass, residual sugars and ethanol were less than 5%. Data presented on tables and figures are the mean ± standard deviation of three assays at the steady state.

Effect of the dilution rate on S. cerevisiae strains fermentative capability in continuous cultures

Both yeast strains (S1 and S2) were used and fermentation medium was fed at different D (0.04, 0.08, 0.12 and 0.16 h^{-1}) for studying the effect of dilution rate (D) on the kinetic parameters and concentrations of biomass, residual reducing sugar and ethanol at a steady state of agave juice continuous fermentation process (Table 5 and Figure 5).

Concentrations of biomass and ethanol decreased as D increased for both strains cultures while residual reducing sugars increased parallel with the increase of D (Figure 5).

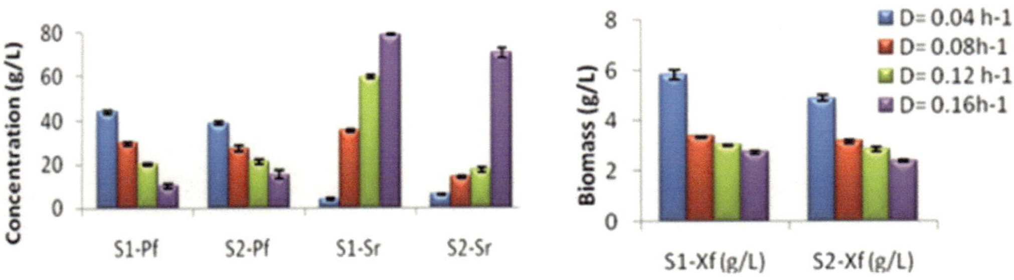

Figure 5. Concentration of Residual reducing sugar (Sr), Ethanol (Pf) and Biomass (Xf) at the steady state of continuous culture of two strains of S. cerevisiae (S1 and S2) fed with agave juice at different dilution rate (D). Data are presented as mean ± standard deviation of four assays at the steady state.

Although, *S. cerevisiae* S2 consumed more reducing sugars than S1 for each D, ethanol yields reached by S1 were higher than those obtained by S2, which were near the theoretical value (0.51) with no significant differences among the different D tested ($p>0.05$) (Table 5).

At D = 0.04 h^{-1}, S1 and S2 strains reached the highest ethanol productions (43.92 and 38.71 g/L, respectively) and sugar consumptions (96.06 and 94.07 g/L, respectively) which were similar to those obtained using batch fermentations (see *Batch fermentation process* section). The low fermentative capacities displayed by both strains at higher D than 0.04 h^{-1} could be due to a low content of nutrients and/or toxic compounds in agave juice cooked [15].

Both strain cultures reached maximal ethanol production rates at 0.12 h^{-1} (2.37 and 2.53 g/L·h, respectively for S1 and S2), maximal growth rates were achieved at 0.16 h^{-1} (0.44 and 0.38 g/L·h, respectively for S1 and S2) and maximal sugar consumption rates were obtained at 0.08 h^{-1} (5.08 g/L·h) for S1 and at 0.12 h^{-1} (9.96 g/L·h) for S2 (Table 5 and Figure 6).

Effect of the pH value on the fermentative capacity of S1 and S2 strains - The effect of pH was observed, switching from a controlled pH (at 4) to an uncontrolled pH (naturally set at 2.5±0.3). Figure 7 shows biomass and ethanol productions for strain S1, in non-aerated or aerated (0.01 vvm) systems fed with sterilized medium. Results did not show significant differences on the biomass or ethanol productions ($P > 0.05$) between the fermentations with control (4) and with no control (2.5) of pH. Conversely, biomass and ethanol productions increased on aerated culture compared to that non aerated, for both pH levels studied. These results agreed with those reported by Díaz-Montaño et al. [20]. These results are important, since the operation of a continuous culture naturally adjusted to a low pH would limit the growth of other yeasts [21, 22] or bacteria [23, 24], indicating the feasibility of working with non-sterilized media on an industrial scale. Another advantage of not controlling the pH is that instrumentation for this operation is not required, thus removing it from the initial investment [25].

Parameter	Strain	D (h^{-1})			
		0.04	0.08	0.12	0.16
Biomass (g/L)	S1	5.83 ± 0.21	3.38 ± 0.03	3.04 ± 0.04	2.75 ± 0.07
	S2	4.89 ± 0.12	3.18 ± 0.08	2.86 ± 0.08	2.39 ± 0.06
Ethanol (g/L)	S1	43.92 ± 0.81	29.63 ± 0.79	19.76 ± 0.32	9.95 ± 0.39
	S2	38.71 ± 0.74	27.33 ± 1.60	21.10 ± 0.48	15.20 ± 0.51
RS (g/L)	S1	3.94 ± 0.53	35.34 ± 0.94	59.75 ± 0.81	79.08 ± 1.08
	S2	5.93 ± 1.16	13.69 ± 1.70	16.96 ± 0.43	70.70 ± 2.17
Glucose (g/L)	S1	nd	1.41 ± 0.06	2.32 ± 0.06	3.07 ± 0.16
	S2	nd	0.43 ± 0.03	0.65 ± 0.04	3.46 ± 0.48
Fructose (g/L)	S1	2.79 ± 0.57	32.12 ± 0.85	51.48 ± 0.28	65.94 ± 1.39
	S2	2.14 ± 0.05	10.54 ± 0.37	15.74 ± 0.50	63.10 ± 2.82
Glycerol (g/L)	S1	2.44 ± 0.28	1.94 ± 0.04	1.70 ± 0.03	1.86 ± 0.26
	S2	2.09 ± 0.09	2.34 ± 0.07	2.54 ± 0.08	1.32 ± 0.05
$Y_{X/S}$ (g/g)	S1	0.06 ± 0.00	0.05 ± 0.00	0.08 ± 0.00	0.17 ± 0.01
	S2	0.05 ± 0.00	0.03 ± 0.00	0.03 ± 0.00	0.07 ± 0.01
$Y_{P/S}$ (g/g)	S1	0.46 ± 0.01	0.47 ± 0.02	0.49 ± 0.01	0.47 ± 0.01
	S2	0.39 ± 0.01	0.30 ± 0.02	0.24 ± 0.00	0.44 ± 0.04
r_X (g/Lh)	S1	0.23 ± 0.01	0.27 ± 0.00	0.36 ± 0.01	0.44 ± 0.01
	S2	0.19 ± 0.01	0.25 ± 0.01	0.34 ± 0.01	0.38 ± 0.01
r_S (g/Lh)	S1	3.80 ± 0.02	5.08 ± 0.08	4.69 ± 0.10	2.52 ± 0.17
	S2	3.96 ± 0.05	6.91 ± 0.14	9.96 ± 0.05	4.69 ± 0.35
r_P (g/Lh)	S1	1.76 ± 0.03	2.37 ± 0.06	2.37 ± 0.04	1.59 ± 0.06
	S2	1.55 ± 0.03	2.19 ± 0.13	2.53 ± 0.06	2.43 ± 0.08

RS: Residual reducing sugar concentration, $Y_{X/S}$: yield of biomass, $Y_{P/S}$: yield of ethanol, r_X: growth rate, r_S: reducing sugars consumption rate, r_P: ethanol production rate, nd: not detected at the assayed conditions. Data are presented as mean ± standard deviation of four assays at the steady state.

Table 5. Kinetic parameters at the steady state of continuous cultures of two strains of *S. cerevisiae* (S1 and S2) fed with agave juice at different dilution rates (*D*).

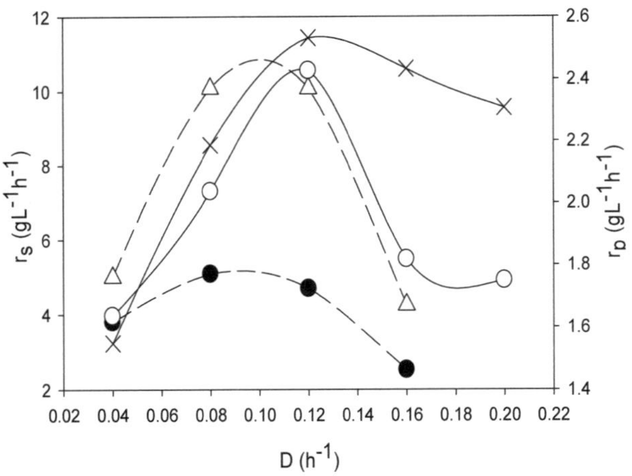

Figure 6. Ethanol production and reducing sugars consumption rates at different dilution rates for *S. cerevisiae* S1 (r_P -Δ- and r_s -•-) and S2 (r_P -x- and r_s -o-).

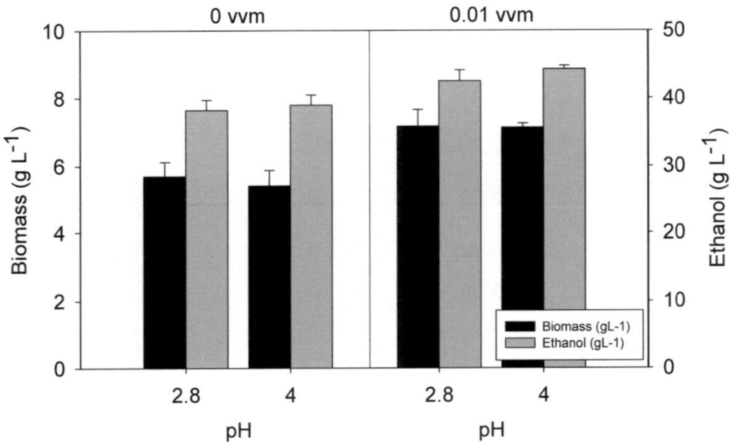

Figure 7. Effect of controlling (at 4) or not controlling (2.5 ± 0.3) pH, in the production of biomass and ethanol at aeration rates of 0 or 0.01 vvm during the culture of S1 strain.

Effect of the nitrogen and phosphorus supplementation on S. cerevisiae S1 sugar consumption

Since both *S. cerevisiae* strains were unable to consume sugars efficiently in cultures fed at D higher than 0.04 h^{-1}, a nutritional limitation and/or some inhibitory substances formed in the agave cooking step (Maillard compounds), which can act on *S. cerevisiae* strain activity. In fact, *Agave tequilana* juice is deficient in nitrogen sources (Table 3). Amino acids are the most important nitrogen source in agave juice; however, their natural concentrations (0.02 mg N/L) are not enough to support balanced yeast growth and the complete fermentation of sugars [26]. Therefore, agave juice supplemented with ammonium sulfate at 1 g/L could be insufficient. Several authors point out the importance of nitrogen sources (type and concentration) for achieving a complete fermentation, since they improve cell viability, yeast growth rate, sugar consumption and ethanol production (11; 20). It is worth noting that ammonium phosphate (AP) was chosen as a nitrogen source, since the two macronutrientes frequently implied in the causes of stuck fermentation when present in small quantities are nitrogen and phosphate (see the reviews by Bisson [11]).

Therefore, the effect of the ammonium phosphate (AP) addition on *S. cerevisiae* S1 sugar consumption was studied in a continuous culture (Figure 8). To study the effect of nitrogen and phosphorus source addition on the agave juice fermentation by *S. cerevisiae,* S1 strain was used and fermentation medium was fed at D of 0.08 h^{-1}, while after the steady state was reached, the ammonium phosphate (AP) concentration was gradually increased, as follows: 1g/L (first addition), 2 g/L (second addition), 3 g/L (third addition) and 4 g/L (fourth addition).

The fermentation was started in batch mode using the fermentation medium. After 12 h, the culture was fed using medium supplemented with 1 g/L of AP (first addition). At the steady state, residual concentrations of sugars and ammonium nitrogen were 29.42 and 0.08 g/L, respectively. These results were not significantly different ($p>0.05$) from the condition previously tested for the same strain (at D = 0.08 h^{-1}), feeding an unsupplemented fermentation medium (Figure 5).

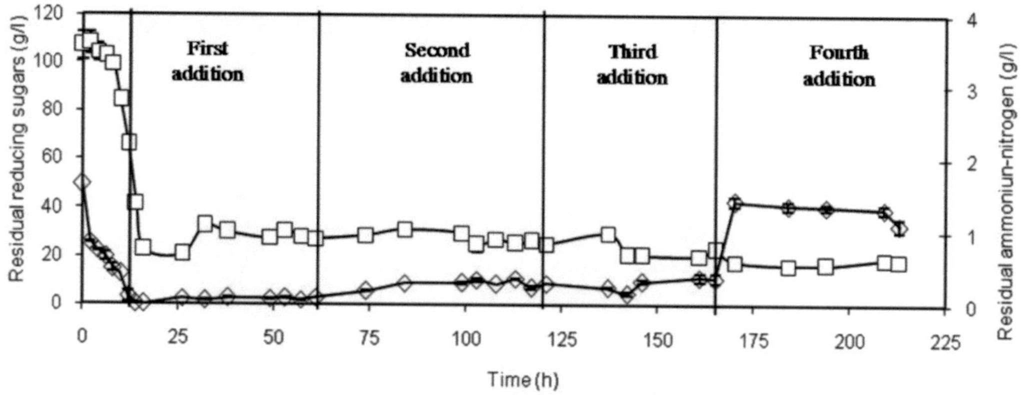

Figure 8. Effect of the addition of ammonium phosphate to the agave juice fed to *S. cerevisiae* S1 chemostat culture (at D=0.08 h^{-1}), on the consumptions of reducing sugars (□) and ammonium-nitrogen (◊). First addition: 1 g/L; Second addition: 2 g/L; Third addition: 3 g/L; Fourth addition: 4 g/L.

Those residual concentrations of reducing sugars (high) and ammonium nitrogen (low) indicate the necessity of adding more AP. At the steady states of the second (2 g/L), third (3 g/L) and fourth (4 g/L) additions of AP, the residual sugars concentrations were 25.96, 21.25 and 17.60 g/L, respectively. This indicates that the residual ammonium nitrogen concentrations were 0.31, 0.36 and 1.29 g/L, respectively; indicating that the AP addition improved *S. cerevisiae* S1 fermentative capability, but other nutritional deficiencies still existed [27].

Effect of the micro-aeration rate on S. cerevisiae S1 fermentative capability - Lack of oxygen has proved to be a main limiting factor to fermentation [11], since yeasts require low amounts of oxygen for synthesizing some essential lipids to assure cell membrane integrity [28]. Because *S. cerevisiae* is Crabtree-positive, alcoholic fermentation is privileged in culture media containing high sugars concentrations, even in the presence of oxygen [29]. The effect of the micro-aeration rate (0, 0.01 and 0.02 vvm) on the fermentative capacity of *S. cerevisiae* S1 (at D = 0.08 h^{-1}) was studied for investigating the yeast oxygen requirement during the continuous fermentation, using the last fermentation medium supplemented with 4 g/L of AP for feeding at D of 0.08 h^{-1}. Biomass and ethanol concentrations increased as air flow increased, reaching at the steady state, 5.66, 7.18 and 8.04 g/L, and 40.08, 44.00 and 45.91 g/L, respectively for 0, 0.01 and 0.02 vvm (Figure 9).

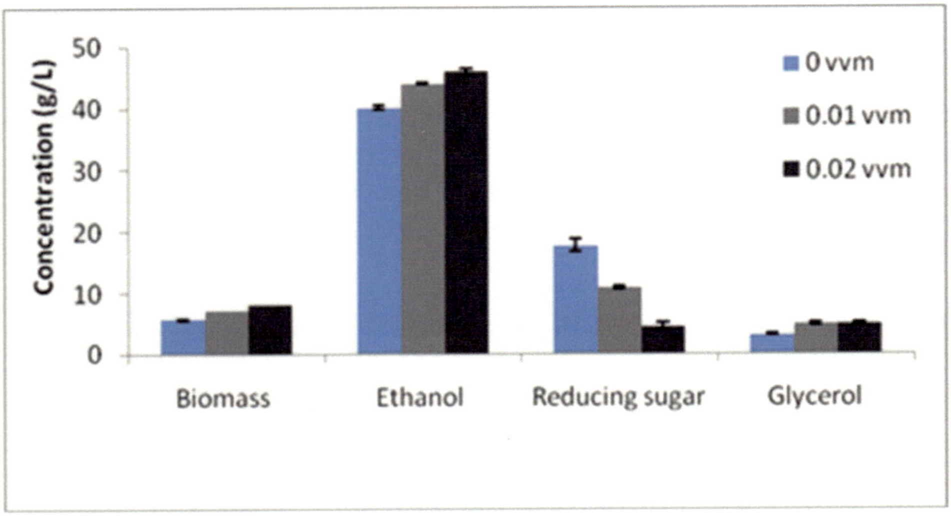

Figure 9. Concentration of Residual reducing sugar, Ethanol and Biomass at the steady state of continuous culture of two strains of S. cerevisiae S1 fed with agave juice (*D* = 0.08 h-1) at different micro-aeration rates. Data are presented as mean ± standard deviation of four assays at the steady state.

Meanwhile, residual sugars decreased as micro-aeration increased, reaching 17.67, 10.71 and 4.48 g/L, respectively for 0, 0.01 and 0.02 vvm; showing an improvement in the fermentation process due the dissolved oxygen in the must. However, statistical differences were not found in biomass and ethanol yields at the different tested aeration rates ($p>.05$) (Table 6). In addition, sugars consumption rates and ethanol and biomass productions increased as micro-aeration increased, achieving a faster fermentation (Table 6). These results were in accordance to those reported by Díaz-Montaño [20]. Viability of the S1 strain was 100% in aeration experiments.

Glycerol is a metabolite providing yeast metabolic activity information. In fact, yeasts produce glycerol mainly for reoxidating the NADH generated by glycolysis. Since the citric acid cycle and the respiratory chain are slightly activated by micro-aeration, NAD might be partially regenerated, and consequently, glycerol concentration decreases [30]. However, in this work, glycerol concentration increased as aeration increased (Table 6). Given that biomass concentration and fermentation efficiency also increase as aeration increases, glycerol production could contribute to faster NAD regeneration.

Parameter	Micro-aeration rates (vvm)		
	0.00	0.01	0.02
$Y_{X/S}$ (g/g)	0.06 ± 0.00	0.07 ± 0.00	0.08 ± 0.00
$Y_{P/S}$ (g/g)	0.48 ± 0.01	0.49 ± 0.00	0.48 ± 0.01
r_X (g/Lh)	0.45 ± 0.01	0.57 ± 0.00	0.64 ± 0.01
r_S (g/Lh)	6.55 ± 0.08	7.14± 0.02	7.64 ± 0.05
r_P (g/Lh)	3.21 ± 0.03	3.52 ± 0.02	3.67 ± 0.04

$Y_{X/S}$: yield of biomass, $Y_{P/S}$: yield of ethanol, r_X: growth rate, r_S: reducing sugars consumption rate, r_P: ethanol production rate. Data are presented as the mean ± standard deviation of four assays at each steady state.

Table 6. Kinetic parameters of *S. cerevisiae* S1 continuous cultures at steady state fed with agave juice ($D = 0.08$ h^{-1}) at different micro-aeration rates.

Effect of feeding non-sterilized medium on the fermentative capability of S. cerevisiae strains

Non-sterilized medium (NSM) was fed to S1 and S2 continuous cultures and the aeration rate was gradually increased from 0 to 0.02 vvm. For these experiments, pH was controlled at 4 for S2 strain and not controlled for S1 strain. Ethanol production increased significantly ($P < 0.05$) as the aeration rate increased during S1 fermentations fed with SM or NSM. In contrast, aeration did not have any effect on ethanol or biomass production during the S2 fermentation fed with NSM (Figure 10-B). For S1 continuous fermentation, medium type (SM or NSM) did not show a significant difference in the production of ethanol ($P > 0.05$), but it had a significant difference in the production of biomass ($P < 0.05$). Multiple range tests divided S1 fermentations in aerated (0.01 and 0.02 vvm) and non-aerated systems, indicating higher biomass and ethanol productions in aerated cultures. Nevertheless, no significant difference was found in the productions of biomass or ethanol ($P > 0.05$) between experiments aerated at 0.01 and those aerated at 0.02 vvm. These results could be attributed to the lower pH (2.3) observed at 0.02 vvm, which could have reduced cell viability. Interestingly, S1 strain flocculation was not observed for 0.02 vvm and biomass retention time was lowered, decreasing the cell population (Figure 10-B).

For all the fermentation conditions, the consumption of reducing sugars was significantly augmented ($P < 0.05$) as aeration rate increased, reaching 4 ± 2 g L^{-1} of residual reducing sugars at 0.02 vvm for both medium types. It has been reported that more than 12% of total sugars contained in agave juice are non-fermentable, since fructans hydrolysis is not complete during the cooking step. In this study, oligosaccharides might be taken into account as residual reducing sugars, because they are difficult to degrade by *S. cerevisiae*.

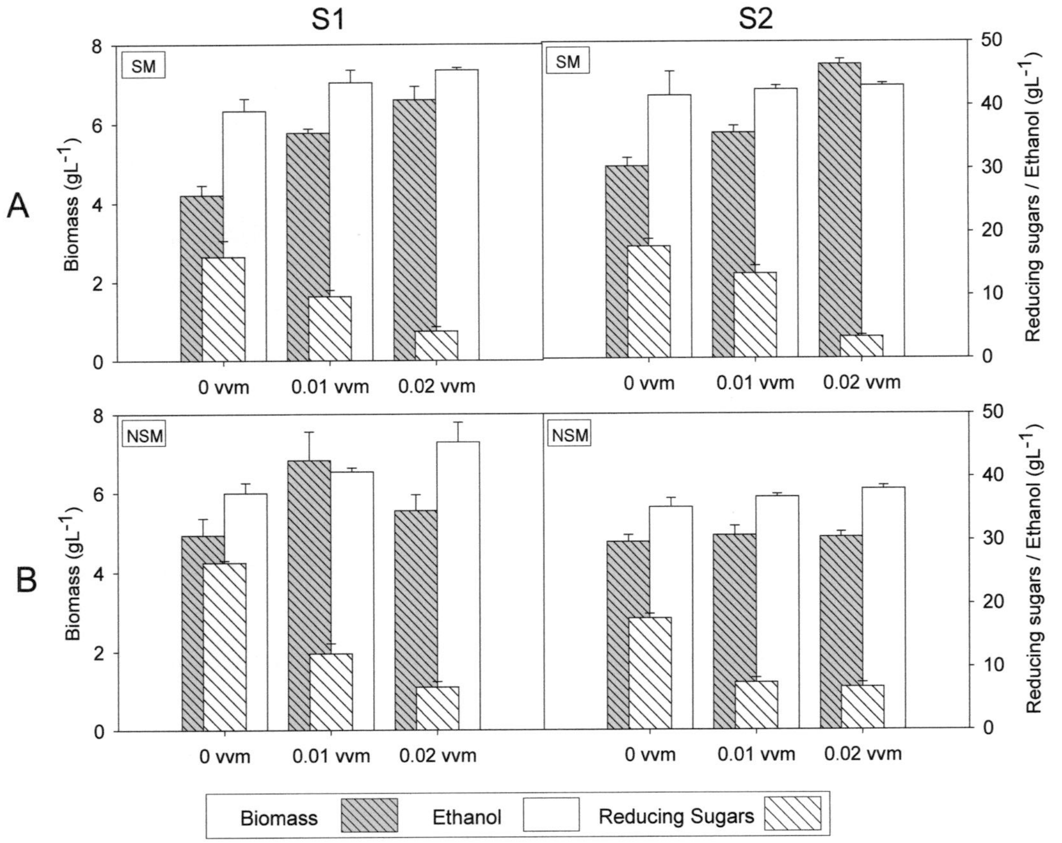

Figure 10. Effect of the aeration on the productions of biomass and ethanol of two *S. cerevisiae* strains (S1 and S2) using the continuous addition of A) sterilized (SM) and B) non-sterilized (NSM) media, pH was 4 and 2.5 ± 0.3 for S1 and S2 strain cultures.

S2 continuous fermentations were divided by the multiple range test, according to the aeration rates (0, 0.01 and 0.02 vvm), showing an increase in the fermentative capability of the S2 strain as aeration increased. The type of medium led to a significant difference ($P < 0.05$) in ethanol and biomass production. Nevertheless, no significant differences ($P > 0.05$) were found in the consumption of reducing sugars between both types of medium. Higher biomass and ethanol production was observed during SM fermentations. Differences between cultures with different types of medium (NSM and SM) could not be attributed to changes in medium composition during sterilization (121 .C, 15 min), since the cooking of agave heads is a more aggressive treatment (100 .C, 36 h). Furthermore, Maillard reactions during the heating are not favored since agave juice nitrogen source content is low (Table 3). Work is ongoing to answer this phenomenon; however, those changes could be attributed to a possible contamination of wild yeast carried by the non-sterilized agave juice. Nevertheless, microscopy did not show any bacterial contamination for fermentation of either strain. Moreover, the pH during S2 continuous fermentation was controlled at 4 for all the experimental conditions in comparison to S1 fermentation, which was not controlled and reached lowered pH values, which could have limited the microbial contamination. In addition, compared to S2, the capacity of S1 to flocculate could be an advantage for this strain to be retained longer inside the bioreactor. Several studies have proved the capability of inoculated *S. cerevisiae* strains in continuous fermentations to resist contamination by wild yeast. Cocolin *et al.* showed by molecular methods that the starters strain was able to drive the fermentation until the end of the process (12 days). On the other hand, de Souza Liberal *et al.* identified *Dekkera bruxellensis* as the major contaminant yeast, even though its growth rate is lower than that of *S. cerevisiae* in batch fermentations. They indicated the possibility that *D. bruxellensis* grows faster than *S. cerevisiae* in a continuous culture under certain conditions.

6. Conclusion

Agave plants could be a viable alternative as an accessible raw material for bioethanol production, since high concentration of fermentable sugar is released when agave plant fructans is cooked and/or hydrolyzed. This mixture of sugars, mainly fructose, could be converted into ethanol by microorganism action.

The present study examined the use of batch and continuous fermentation processes for investigating bioethanol production from *Agave tequilana* Weber var. azul. juice.

The fermentable sugars of agave juice fermentation in batch culture were depleted between 18-24 hours by indigenous tequila *S. cerevisiae* strains. The ethanol productivity obtained in batch fermentation was 2.36, 2.42 and 1.66 g/Lh for S1, S2 and S3 yeast strains respectively. Agave juice continuous fermentation was examined for increasing ethanol productivity in the fermentation process. For this, a chemostat system was used for investigating the impact of the dilution rate, pH value, nitrogen and phosphorus source addition, micro-aeration and non-sterilized medium on growth, sugar consumption and ethanol production of two *S. cerevisiae* strains. The dilution rate and nutrient addition have a significant impact on the physiology of the *S. cerevisiae* yeast strains. When S1 and S2 yeast strains are used in continuous cultures, they show low sugar consumption at $D \geq 0.08 h^{-1}$. The study revealed a nutritional limitation on the agave juice, which was corrected by adding of nitrogen sources and oxygen, achieving *S. cerevisiae* S1 strain complete sugar consumption with high ethanol conversion at $0.08 h^{-1}$. The pH did not have a significant effect on the fermentative capability of *S. cerevisiae* S1 strain at the levels studied. Uncontrolled pH fermentations naturally reached acid values (pH ◎2.5 ± 0.3), which is advisable, since bacteria or yeasts contamination could be limited. The type of agave juice tested (SM and NSM) did not have a significant effect on ethanol production in S1 cultures, but did have an effect on ethanol production in S2 cultures. These results could be attributed to the higher pH fermentation during S2 continuous cultures, which could have favored the proliferation of contaminant wild yeasts. The ethanol productivity obtained in S1 strain agave juice continuous fermentation process was 3.6 g/Lh. Thus, the ethanol productivity in continuous fermentation is higher, 34.4% more than in S1 strain batch fermentation.

These results showed the possibility of performing agave juice fermentations in continuous culture feeding non-sterilized medium and taking advantage of the possible improvements that continuous fermentations and agave plant could offer to the bioethanol industry, such as high productivity with full sugar consumption.

Author details

Dulce María Díaz-Montaño
Universidad Autónoma de Guadalajara, Guadalajara, México

7. References

[1] Scott N, Felby C (2012) Biomass for energy in the European Union - a review of bioenergy resource assessments. Biotechnology for Biofuels. doi:10.1186/1754-6834-5-25.
[2] Swings J, De Ley J (1977) The biology of Zymomonas. acteriol Rev. 41(1): 1–46
[3] Skotnicki ML, Warr RG, Goodman AE, Lee KJ, Rogers PL (1983). High-productivity alcohol fermentations using Zymomonas mobilis. Biochem Soc Symp.48:53-86.
[4] Dien BS, · Cotta MA, Jeffries TW (2003) Bacteria engineered for fuel ethanol production: current status. Appl Microbiol Biotechnol.63: 258 -266.
[5] Fujita Y, Ito J, Ueda M, Fukuda H, Kondo A (2004) Synergistic Saccharification, and Direct Fermentation to Ethanol, of Amorphous Cellulose by Use of an Engineered Yeast Strain Codisplaying Three Types of Cellulolytic Enzyme. Appl. Environ. Microbiol. 70: 1207-1212.
[6] Yanase H, Nozaki K, Okamoto K (2005) Ethanol production from cellulosic materials by genetically engineered Zymomonas mobilis. Biotechnol Lett. 27(4):259-63.
[7] Pelczar M, Reid RD, Chan ECS (1990). Microbiología. McGraw Hill, México.
[8] Pretorius IS 2000 Tailoring wine yeast for the new millennium: novel approaches to the ancient art of winemaking. Yeast (Chichester, England), 16(8), 675-729.
[9] Meyer JP (1979) Les levures adaptees aux basses temperatures et leur selection. Revue Fran.aise d fOEnologie, No. 76 :45-49
[10] Gray WD (1984) Studies on the alcoholic tolerance of yeasts. Journal of Bacteriology, 42:561-574.
[11] Bisson L (1999) Stuck and sluggish fermentations. Am J Enol Vitic. 50: 107-118.
[12] Dunn, I.J., et al., *Biological Reaction Engineering*. 2nd Edition ed. 2003, New York: Wiley-VCH.
[13] Stanbury PF, Whitaker A, Hall SJ (1999) Principles Fermentation Technology. Butterworth-Heinemann second edition.
[14] Scott H (1982). Agaves of Continental North America. The University of Arizona Press. Tucson, Arizona. ISBN 0-8165-0775-9

[15] Díaz-Montaño D Délia ML, Estarrón M, Strehaiano P (2008) Fermentative capability and aroma compound production by yeast strain isolated from *Agave tequilana* Weber juice. Enzyme and Microbial Technology. 42:608-616.

[16] Hernández-Orte P, Cacho JF, Ferreira V. (2002) Relationship between varietals amino acid profile of grapes and wine aromatic compounds. Experiments with model solutions and chemometric study. J Agric Food Chem. 50: 2892-2899.

[17] Vázquez-Ortiz FA, Caire G, Higuera-Ciapara I, Hernández G (1995) High Performance Liquid Chromatographic Determination of Free Amino Acids in Shrimp. J Liq Chrom. 18: 2059-2068.

[18] Umagath H, Kucera P, Wen L (1982) Total Amino Acid Analysis Using Pre-Column Fluorescence Derivatization. J Chromatogr. 239: 463-474.

[19] Bohringer P, Jacob A (1964) The determination of alcohol using chromic acid. Zeitschr Flussiges Abst. 31: 233-236.

[20] Díaz-Montaño DM (2004). Estudio fisiológico y cinético de dos cepas de levadura involucradas en la etapa fermentativa de la elaboración de tequila. No. Registro: 2172. Doctorat de L′ I. N. P. T. et de L′Université de Guadalajara. Difusión electrónica del SCD-INP de Toulouse Francia.

[21] Arroyo-López FN, Orlic S, Querol A, Barrio E (2009) Effects of temperature, pH and sugar concentration on the growth parameters of Saccharomyces cerevisiae, S. kudriavzevii and their interspecific hybrid. Int J FoodMicrobiol 131:120–127.

[22] Betts GD, Linton P, Betteridge RJ (1999) Food spoilage yeasts: effects of pH, NaCl and temperature on growth. Food Control 10:27–33.

[23] Russell JB, Diez-Gonzalez F, Poole RK (1997) The Effects of Fermentation Acids on Bacterial Growth. *AdvMicrob Physiol* 39:205–234.

[24] Adamberg K, Kask S, Laht T-M and Paalme T (2003) The effect of temperature and pH on the growth of lactic acid bacteria: a pH-auxostat study. *Int J FoodMicrobiol* 85:171–183.

[25] Hernandez Cortes G, Cordova Lopez J, Herrera Lopez EJ, Díaz-Montaño D.M. (2010) Effect of aeration, pH and feeding non-sterilized agave juice in continuous tequila fermentation. J of Science of food and Agriculture. 90: 1423-1428.

[26] Valle-Rodríguez JO, Córdova-López JA, Hernández-Cortés G, Estarron-Espinosa M, Díaz-Montaño, DM (2011) Effect of the amino-acids supplementation on the Agave *tequilana* juice fermentation by *Kloeckera africana* in batch and continuous cultures. Antonie van Leeuwenhoek Journal of Microbiology. Antonie van Leeuwenhoek DOI 10.1007/s10482-011-9622-x. ISSN 0003-6072.

[27] Moran-Marroquí GA, Córdova-López J, Valle-Rodriguez JO, Estarrón-Espinoza M, Díaz-Montaño DM (2011). Effect of dilution rate and nutrients addition on the fermentative capability and synthesis of aromatic compounds of two indigenous strains of *Saccharomyces cerevisiae* in continuous cultures fed with *Agave tequilana* juice. International Journal of Food Microbiology.151: 87–92.

[28] Andreasen AA, Stier TJ, (1953) Anaerobic nutrition of Saccharomyces cerevisiae. I. Ergosterol requirement for growth in a defined medium. Journal of Cellular and Comparative Physiology. 41(1): p. 23-36.

[29] van Urk H, Voll WSL, Scheffers WA, van Dijken J (1990) Transient-state analysis of metabolic fluxes in Crabtree-positive and Crabtree-negative yeasts. Applied and Environmental Microbiology 56, 281–287.

[30] Kuriyama H, Kobayashi H (1993) Effects of oxygen supply on yeast growth and metabolism in continuous fermentation. Journal of Fermentation and Bioengineering 75, 364–367.

Chapter 9

Sequential Anaerobic-Aerobic Phase Strategy Using Microbial Granular Sludge for Textile Wastewater Treatment

Khalida Muda, Azmi Aris, Mohd Razman Salim and Zaharah Ibrahim

Additional information is available at the end of the chapter

http://dx.doi.org/10.5772/54458

1. Introduction

The textile industry involves a long chain of complex activities, from processing raw materials up to finishing the fabrics. These industries have created job opportunities to millions of people and have become one of the major incomes to many countries in the world. Unfortunately, the industry is also one of the major contributors to water pollution. The textile wastewater contains not only the colorant, which is one of the main pollutants, but also other chemicals that are added throughout the textile processing. The dye compounds present in textile wastewater are able to impose a major impact to a receiving water body even in small quantities.

Due to the non-biodegradable nature of textile wastewater, a conventional aerobic biological process is incapable of treating the wastewater. For a complete degradation of textile wastewater, a combination of anaerobic and aerobic reaction phases is necessary.

This chapter briefly reviews the characteristics of textile wastewater and available technologies. This is followed by an in-depth discussion on biogranulation technology and the application of a hybrid biogranular system in treating the textile wastewater.

2. The textile industry

The fabrics, either in the form of natural or chemical fibres, have reached millions of tonnes of production and have provided huge advantages to world economic values (Aizenshtein, 2004). In social terms it has provided benefits to more than 2.2 million workers through 114,000 textile-related companies. In 2001, the European textile and clothing industries contributed to about 3.4% of the EU manufacturing industrial revenue and granted 6.9% of the work opportunities to the citizens (IPPC, 2003). According to recent statistics, the global textile market is worth more than US$400 billions (Directory of Textile Manufacturers and Suppliers - http://www.teonline.com/industry-overview.html). It is predicted that the global textile production will grow up to 50% by 2014 as compared to the fabrication in 2005.

Globally, Malaysia is also known for its high quality textile and apparels. Since the early 1970s, when the country started to embark on being an export-oriented country, the growth of Malaysian's textile and apparel industry has increased tremendously and now provides an export value of 3.5 billion USD. This has listed the textile industry as the ninth largest contributor to total earnings of the manufactured exports in 2007. The industry has provided more than 67,000 work opportunities through 637 licensed textile production companies with investments of 2.6 billion USD (MIDA, 2007).

In Malaysia and many other developing countries, most of the textile mills are of small and medium scale. For these mills, the full installation of a wastewater treatment plant is quite difficult due to economic reasons. Hence, the mills have been discharging significant quantities of pollutants into the streams with fiber manufacturing and dyeing sectors being the predominant ones (Haroun and Azni, 2009).

2.1. Characteristics of textile wastewater

The textile industry consumes the largest portion of the colorant available in the world market. Due to the high customer demand, more than 100,000 commercial dyes exist in the market causing more than 700,000 tonnes of dyes to be produced annually (McMullan et al., 2001; Pearce et al., 2003). The result is a very high production of colored wastewater. The characteristics of textile wastewater (either quantitatively or qualitatively) vary greatly depending on the type of raw materials, chemicals, techniques or specific process operations at the mill, the equipment used and the production design of the textile processes (Bisschops and Spanjers, 2003; Dos Santos et al., 2006).

The textile industry consumes huge amounts of water in its wet processes. The average wastewater generation from a dyeing facility is estimated between 3800 and 7600 million m^3 per day. Desizing, scouring, bleaching, mercerizing and dyeing are the common wet textile process operations. Among these, the mercerizing and dyeing processes consume the biggest specific volumes of water with a water usage of 230-310 L/kg and 8-300 L/kg of textile processed, respectively (Dos Santos et al. 2007).

Due to inefficiency of the textile processing activities, only 10% of the chemicals in the pre-treatment and dyes in dyeing operations remain on the fabric. In other words, about 90% of chemical substances will be discharged as textile effluent (IPPC, 2003). Others have reported that between 50 and 95% of the dyes are fixed on the fiber while the remainder is discarded in the subsequent textile-washing operations (EPA, 1997; Trovaslet et al., 2007). The amount of dye lost into the wastewater depends upon the type of dyestuff used, as well as the methods and application routes of the textile processing operation. Additionally, it depends on the intended color intensity that is required for each particular design (Willmott et al., 1998).

Textile wastewater is characterized with high chemical and biochemical oxygen demand, suspended solids, high values of conductivity and turbidity and intense color. This is caused by the presence of dye residues or intermediates and auxiliary chemicals added in the many stages in textile processing (Mohan et al., 2007a; Miranda et al., 2009). Textile processes with natural fibers generate higher pollution load as compared to synthetic fibers mainly due to the use of pesticides for preservation of the natural fibers (Correia et al., 1994).

Textile dyeing wastewater is also characterized by high salt content, which also imposes potential environmental problems. Typical cotton batch dyeing operations use quantities of salt that range from 20 to 80% of the weight of goods dyed, with common concentrations between 2,000 mg/L and 3,000 mg/L. Sodium chloride and sodium sulfate constitute the majority of the total salts used. Magnesium chloride and potassium chloride are used as raw materials in lower concentrations (EPA, 1997).

Common characteristics of textile wastewater from cotton textile wet processing for different processing categories are shown in Table 1. The highest concentration of organic pollutants (in terms of COD) is generated from bleaching while the highest concentration of total solids comes from the desizing process. The highest concentration of color, ranging from 1450-4750 ADMI, is generated from the dyeing process (Bisschops and Spanjers, 2003; Dos Santos et al., 2007). Metals such as copper, cadmium, chromium, nickel and zinc are also found in textile effluents, as they are the functional groups that form the integral part of the dye molecule (IPPC, 2003).

2.2. Treatment technology

At present, treatment of textile wastewater mainly involves physical and/or chemical processes. These include coagulation and flocculation (Harrelkas et al., 2009), precipitation (Solmaz et al., 2007), adsorption (Sayed and Ashtoukhy, 2009), membrane filtration and nanofiltration (Miranda et al., 2009), ion exchange (Wu et al., 2008), ultrasonic mineralization (Maezawa et al., 2007) and electrolysis (De Jonge et al., 1996). While these methods are often costly, they remove the pollutants by transferring them from one phase to another. Some of them generate highly concentrated sludge, hence creating disposal problems (Pearce et al., 2003) that may lead to soil contaminations. Excessive use of chemicals in dye treatment creates secondary pollution problems to the environment.

Process	COD (g/L)	BOD (g/L)	TS (g/L)	TDS (g/L)	pH	Color (ADMI)
Desizing	4.6-5.9	1.7-5.2	16.0-32.0	-	-	-
Scouring	8	0.1-2.9	7.6-17.4	-	10--13	694
Bleaching	6.7-13.5	0.1-1.7	2.3-14.4	4.8-19.5	8.5-9.6	153
Mercerising	1.6	0.05-0.10	0.6-1.9	4.3-4.6	5.5-9.5	-
Dyeing	1.1-4.6	0.01-1.80	0.5-14.1	0.05	5-10	1450-4750
Bleaching and Dyeing*	0.2-5.5	2.0-3.0	0.1-5.0	-	2-10	280-2000

*Characterization of textile wastewater in Malaysia (Ahmed et al., 2005; Ibrahim et al., 2009)

Table 1. Characteristics of textile wastewater (Bisschops and Spanjers, 2003; Dos Santos et al., 2006)

Treatment using ozonation, Fenton's reagent, electrochemical destruction and photocatalysis are some of the emerging techniques reported to have potential use for decolorization (Faouzi et al., 2006; Ay et al., 2009). However, such technologies usually involve complicated procedures and are economically unattainable (Chang and Lin, 2000).

Among the available techniques, the one that can offer effective pollutant removal at a lower cost is the desirable alternative. Of these, biological treatment is the obvious choice due to the relatively low operating cost.

While a conventional aerobic biological process is incapable of treating textile wastewater, studies have shown that the integration of anaerobic and aerobic processes are able to provide complete mineralization of colored substances (Knackmuss, 1996; Melgoza et al., 2004; van der Zee and Villaverde, 2005). It can be done by using either two separate anaerobic and aerobic reactors (Khelifi et al., 2008) or using integrated anaerobic/aerobic treatment in a single reactor (Frijters et al., 2006; Cinar et al., 2008). The wastewater is initially treated under an anaerobic condition followed by an aerobic condition. Under the anaerobic condition, the N=N bond of the azo dyes are cleavaged, leading to the production of amines, the colorless byproducts. This is followed by complete mineralization under the aerobic condition. Different forms of biomass (i.e. suspension, film and granules) have been used in different types of reactor in the studies.

2.3. The water quality issues

The textile industry, in particular the wet industry, has been considered as one of the major water environment polluters. This is mainly due to the enormous amount of water and the complexity of the chemicals used in the manufacturing processes that end up in the wastewater. The poorly treated wastewater is still highly colored comprising of significant amounts of nonbiodegradable chemicals that are hazardous to the environment. Under anaerobic condition, some of the organics i.e. the azo dyes are transformed into more toxic chemicals (i.e. amines) that worsen the condition. The color will make a river inhabitable to a majority of aquatic plants and animals.

While there are many technologies available in treating the wastewater, a majority of them are relatively expensive to be applied by the small and mid-size industries. Furthermore, many of the physico-chemical technologies only transform the pollutants from one form or one phase to another and therefore do not provide any ultimate solution to the problem.

A conventional aerobic bioprocess fails to treat the wastewater due to the non-biodegradable nature of the wastewater. However, recent research and advancement in biological processes show that there is a huge potential of these new findings in providing low cost yet efficient technology to solve the textile wastewater problem.

3. Biogranulation treatment technology

Microbial granules form a self-immobilization community that is formed with or without support material. They are defined as discrete macroscopic aggregates containing dense microbial consortia packed with different bacterial species. Each biogranule consists of millions of microorganisms per gram of biomass (Weber et al., 2007), formed via biological, physical and chemical forces. According to Calleja (1984), microbial granulation is a multicellular association in a physiological state that is causing the mixture of cells into a fairly stable and contiguous structure.

The main advantages of biogranules systems are mainly due to the biogranules good settling property and the fact that biogranules are formed without the need of any biomass carrier. The relatively large size and high-density biogranules give them a rapid settling rate, which enhances the separation of the treated effluent from the biomass and results in high solid retention time (SRT) (Ahn and Richard, 2003; Liu and Tay, 2004). Due to a better settling rate, the system also shows low suspended solid content discharged in the effluent (Wirtz and Dague, 1996).

Within the biogranules, the microorganisms are closely lumped together, hence generating syntrophic associations between the cells. This relationship occurs due to optimum distances between the cells at appropriate substrate levels and such condition enables high and stable performance of metabolism activities (Batstone et al., 2004).

The granulation system is first recognized in an up-flow anaerobic sludge blanket (UASB) system characterized by anaerobic biogranules. Much research has been carried out using innovative upflow sludge bed (USB) type reactors (Bachman et al., 1985; Lettinga et al., 1997). The applications of anaerobic granulation systems have been successfully demonstrated particularly in removing biodegradable organic matter from industrial wastewaters (Lettinga et al., 1980; Schmidt and Ahring, 1996). Later the attention has also been diverted to the development and applications of aerobic biogranules. The reason has been several drawbacks that have been observed in the anaerobic biogranules system, including long start-up periods, relatively high temperature requirements and ineffectiveness in dealing with nutrient and low organic strength wastewater (Liu and Tay, 2004).

Aerobic granulation systems have been used for organics, nitrogen, phosphorus and toxic substances removal, especially high strength wastewater (Yi et al., 2008; Kishida et al., 2009). In most cases, the system is in the form of a sequencing batch reactor (SBR) (Beun et al., 1999; Kim et al., 2008). The reaction phase of the system has been carried out either in anaerobic, aerobic or anoxic conditions, with or without mixing, depending on the purpose of the treatment.

3.1. Development of biogranules

Bacteria normally do not aggregate naturally to each other due to repulsive electrostatic forces via the presence of negatively charged protein compounds of the cell wall (Voet and Voet, 2004). However, under selective environmental conditions, microorganisms are capable to attach to one another and thus form aggregates.

Development of biogranules involves integration of physical, chemical and biological processes occuring in multiple stages (Calleja, 1984; Liu and Tay, 2002; Linlin et al., 2005;

Weber et al., 2007). The first stage of a biogranulation process is initiated by several forces, which include diffusion of mass transfer, hydrodynamic and gravitational forces, thermodynamic effects, as well as the tendency of cells to move towards one another. These forces result in cell-to-cell or cell-to-solid surface interactions. The second stage involves several physical forces (e.g. Van der Waals forces, surface tension, hydrophobicity, opposite charge attractions, thermodynamic of surface free energy, bridges by filamentous bacteria), and chemical and biochemical forces (e.g. cell surface dehydration, cell membrane fusions and signals among microbial communities). At this stage, the multicell connections are stabilized. The third stage is the maturing stage, which involves the production of substances that facilitate more cell-to-cell interactions; at this stage, highly organized microbial structures are formed. Several mechanisms of metabolite production will also change, such as higher production of extracellular polymer, growth of cellular cluster, metabolite change and environmental-induced genetic effects. The final stage involves shaping of the three dimensional granules by hydrodynamic shear forces.

Beun et al. (1999) have also described the path of aerobic granules formation in a reactor as illustrated in Figure 1. Immediately after inoculation, bacteria and fungi will be dominating the reactor system. At this early stage, mycelial pellets manage to retain in the reactor due to their good settling ability. Bacteria, which do not hold this characteristic, are discarded with the effluent. Due to the shear force imposed by air bubbles during the aeration phase, the filaments will be detached from the surface of pellets. The pellets then grow bigger until they reach a diameter of up to 5-6 mm. When the sizes of the pellets have grown even larger, self-defragmentation will take place due to the limitation of oxygen transfer in the inner parts of the grown pellets. The fragmented mycelial pellet will act as a matrix for bacteria to grow and form new colonies. The bacterial colonies grow larger and will form granules. As the granules are formed, the whole system will be governed by bacterial growth.

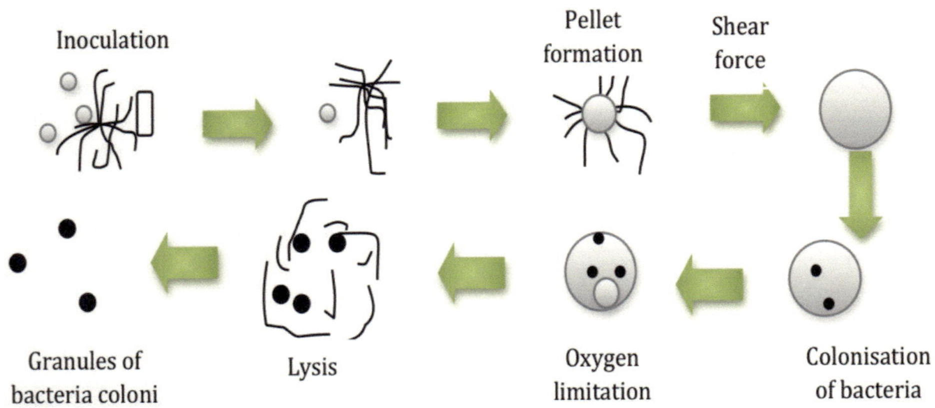

Figure 1. Schematic diagram of aerobic granulation developed without any carrier material (Beun et al., 1999)

Weber et al. (2007) have illustrated the involvement of several eukaryotic organisms in three consecutive phases. Microscopic analysis has revealed that eukaryotic organisms play a key role in aerobic granule formation. Stalked ciliates of the subclass *Peritrichia* and occasionally, the fungi, are found to be involved in the biogranulation process development.

Development of biogranules seeded with anaerobic granular sludge in an SBR system has been demonstrated by Linlin et al. (2005). At the initial stage, the anaerobic granular seeds disintegrate into smaller flocs and debris due to the hydrodynamic shear force created by the air bubbles during the aerobic phase. Lighter and small sized flocs or debris will be washed out in the effluent during the decanting stage. The remaining heavier anaerobic granules remain and act as precursors that initiate the growth of new aerobic granules. The optimal combination of the shear force and the growth of the microorganisms within the aggregates govern the stable structur of the biogranules (Chen et al., 2008). The morphology of these aerobic granules is slightly different as compared to the aerobic granules as described by Beun et al. (1999).

3.2. Characteristics of biogranules

Biogranules are known for their outstanding features of excellent stability and high removal efficiency making biogranulation an innovative modern technology for wastewater treatment. The size of the biogranules is an important aspect that may influence the stability and performance of the reactor system. Biogranules with bigger sizes can easily be defragmented under high shear force resulting in high biomass washout. Meanwhile, if the size is too small, the biogranules cannot develop good settling properties, resulting in higher suspended substances in the effluent. Bigger biogranules with loose structure will be developed in an SBR system supplied with low superficial air velocity. Smaller biogranules but with high strength structures are observed being formed in systems aerated at higher superficial air velocity (Chen et al., 2007). Granular sizes range from 0.3 mm to 8.8 mm in diameter possessing different granular characteristics (Dangcong et al., 1999; Zheng et al., 2005).

The hydrodynamic shear force imposed through the aeration rate of the reactor system will control the development of biogranules (Chisti, 1999). The size of biogranules is the net result of the balance between the growth and the hydrodynamic shear force imposed by superficial air velocity (Yang et al., 2004). For the optimal performance and economic purposes, the operational diameter range for effective aerobic SBR granular sludge should be in the range of 1.0-3.0 mm (Toh et al., 2003)

The usual structure of an aerobic granule is normally spherical in shape with smooth surface areas, which can be influenced by the concentration and type of substrate used in the media compositions (Zhu and Wilderer, 2003; Adav and Lee, 2008). Based on electron microscope (SEM) observations, glucose-fed granules appear with fluffy outer surface due to the predominance growth of filamentous bacteria. On the other hand, the acetate-fed granules show a more compact microstructure with smooth surface. The non-filamentous and rodlike bacteria were observed dominating the acetate-fed granules that are tightly linked together (Tay et al., 2001).

Settleability of a biogranular sludge shows the capacity of the biogranules to settle within a specified period of time. Such properties will allow fast and clear separation between sludge biomass and effluent. The settling velocity of aerobic granules is in the range of 30 to 70 m/h depending on the size and structure of the biogranules, which is comparable to the anaerobic granules. Settling velocity of activated sludge flocs is in the range of 8 to 10 m/h that is three times lower than to those of aerobic granules. Good settleability of sludge biomass is desirable in wastewater treatment plants to facilitate high percentage of sludge retention in a reactor system. Superior characteristics of settleability assist to maintain the stable performance, high removal efficiency and can handle high hydraulic loading of wastewater (Tay et al., 2001). Good settling property of biogranules is also shown by a low value of the SVI. The SVI of biogranules is lower than 100 mL/g (Peng et al., 1999 and Qin et al., 2004), much lower compared to the SVI of flocs (above 150 mL/g). The observed density of microbial aggregates is the consequence of balance interaction between cells (Liu and Tay, 2004). The density of the aerobic granule is reported to be in the range of 32 to 110 g VSS/L (Beun et al., 2002; Arrojo et al., 2006) and the specific gravity is in the range of 1.004 to 1.065 (Etterer and Wilderer, 2001 and Yang et al., 2004).

When biogranules grow bigger, the compactness of the granules decreases. This can be detected via a less solid and loose architectural assembly (Toh et al., 2003). Biogranules with high physical strength can withstand high abrasion and shear force. The physical strength of the biogranules is expressed as an integrity coefficient. This coefficient is an indirect quantitative measurement of the ability of the biogranules to withstand the hydrodynamic shear force (Ghangrekar et al., 2005). A good granular strength is indicated by an integrity coefficient of less than 20.

Biogranules are also characterized by high cell hydrophobicity and high EPS content. The former aspect is postulated to be the main triggering force in the initial stage of the biogranulation process and is a measure of the cell-to-cell interaction (Liu et al., 2003). The latter characteristic is postulated to be responsible for the aggregation between cells (Liu et al., 2004).

The presence of the EPS will enhance the polymeric interaction, which is one of the attractive forces that can promote the adhesion of bacterial cells. The networking between cell and EPS will assist the formation of biogranules (Zhang et al., 2007).

4. A hybrid biogranular system for textile wastewater treatment

The application of hybrid biogranular system in treating textile wastewater is reported in this section. In this study, the development of biogranules during the treatment of textile wastewater is investigated. The changes on the physical characteristics of the biogranules as well as the system performance in the removal of organic compound and color intensity of the textile wastewater are further discussed.

4.1. The system

The schematic representation of the reactor design is given in Figure 2. The design of the reactor is based on Wang et al. (2004) and Zheng et al. (2005) with several modifications. The column of the reactor has a working volume of 4 L with internal diameter of 8 cm and a total height of 100 cm. The reactor is designed with a water-jacketed column for the purpose of temperature control. This can be achieved by allowing the circulation of hot water from a water heating circulation system to the water jacketed column of the system. The temperature of the heating system was set at 30^{0}C. Air was supplied into the reactor by a fine air bubble diffuser located at the bottom of the reactor column. The reactor system was equipped with dissolved oxygen and pH sensors for the continuous monitoring throughout the experiment. The wastewater was fed into the reactor from the bottom of the reactor. The decanting of the wastewater took place via an outlet sampling port located at 40 cm above the bottom of the reactor. The reactor system has been designed with volumetric exchange rate (VER) of 50%. This means that only particles with settling velocity larger than 4.8 m/h remained in the column. Particles having smaller settling velocity will be washed out in the effluent. All operations of peristaltic pumps, circulation of influent, air diffuser and decanting process were controlled by means of a timer.

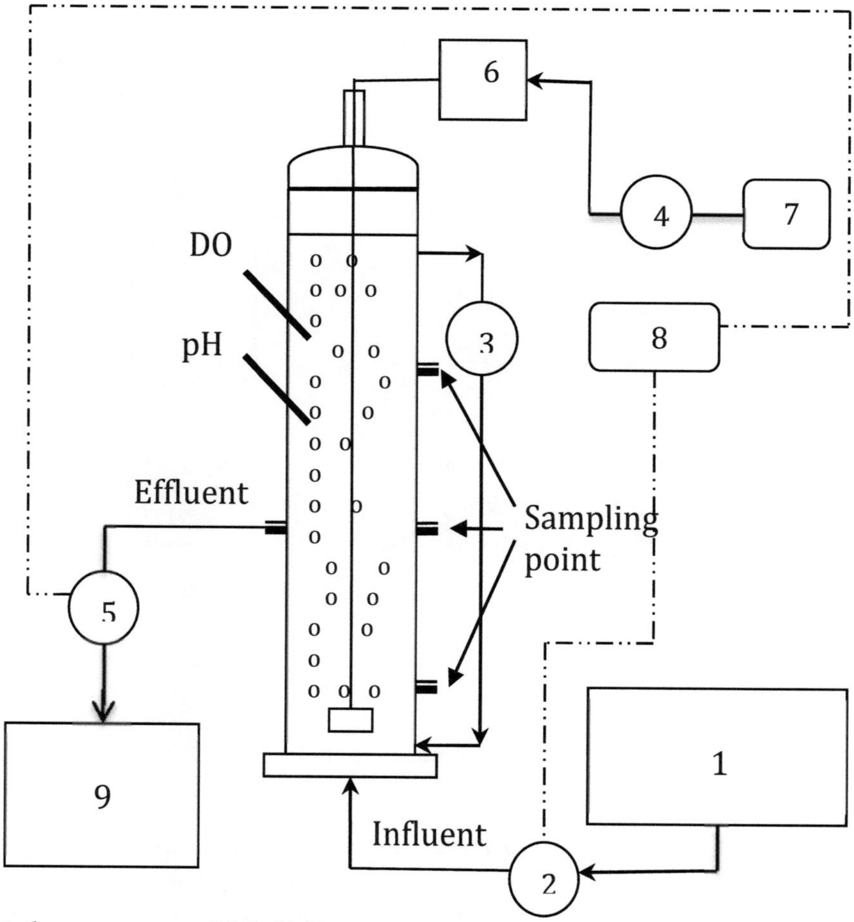

1. Influent tank
6. Mass-flow controller
8. Timer controller

2-5. Peristaltic pumps
7. Air pump
9. Effluent tank

Figure 2. Schematic layout of the hybrid biogranular system

4.2. The operation and analysis

During the start-up period, 2 L of mixed sludge and 2 L of synthetic textile wastewater were added into the reactor system giving the working volume of 4 L with 5.5 g/L of sludge concentration after inoculation. The system was supplied with external carbon sources consisting of glucose, sodium acetate and ethanol with substrate loading rate of 2.4 kg COD/m³·d. The operation of the system started with 5 min filling of wastewater entering from the bottom of the reactor. The operation then continued with the react phase followed by 5 min settling, 5 min decanting and 5 min of idle time. The react time varies depending on the hydraulic retention time set for the system. Figure 3 shows the steps involved in one complete cycle of the hybrid biogranular system. During the biogranules development, the HRT of the reactor was set for 6 hours for one complete cycle. This will give a react time of 340 minutes. The react phase is divided into equal anaerobic and aerobic react periods. Table 2 shows the successive phase for one complete cycle of the reactor system.

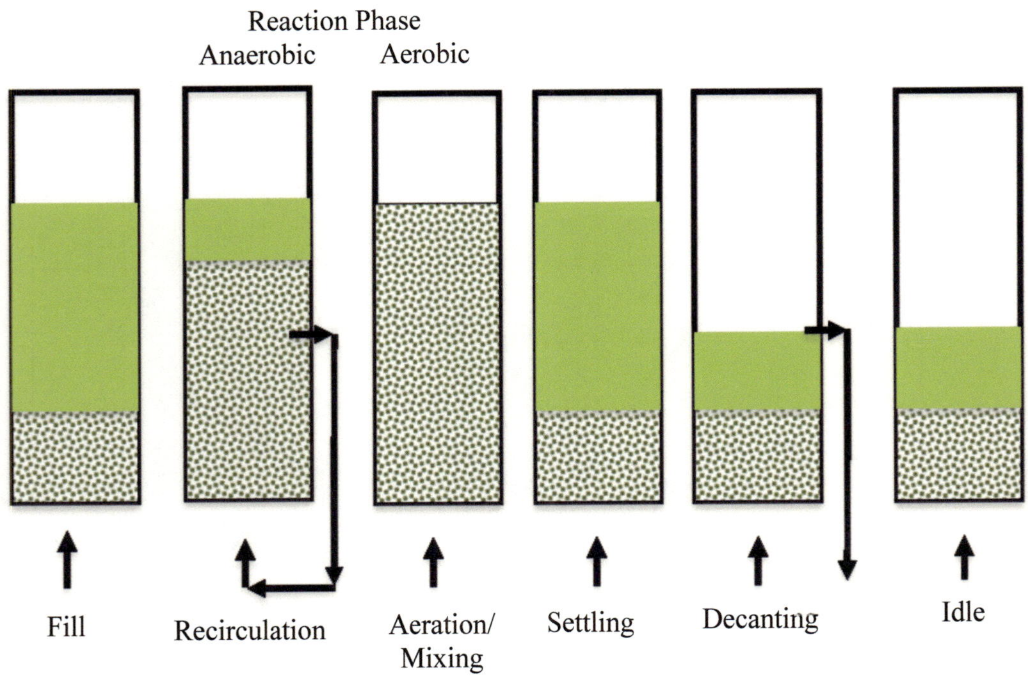

Figure 3. One complete cycle of the Hybrid Biogranular System

The operation of the reactor system was designed with intermittent anaerobic and aerobic react phases. The reaction phase started with an anaerobic phase followed by an aerobic phase. The reaction phase was repeated twice. During the anaerobic react phase, the wastewater was allowed to circulate from the upper level of the reactor and returned back through a valve located at the bottom of the system. The circulation process was carried out using a peristaltic pump at a rate of 18 L/h. The circulation system was stopped at the end of the anaerobic phase. The circulation process is required to achieve a homogeneous distribution of substrate as well as a uniform distribution of the granular biomass and restricts the concentration gradient. The DO concentrations remained low during the anaerobic condition (0.2 mg/L) and reached saturation during the aerobic phase. The superficial air velocity during the aerobic phase was 1.6 cm/s. The system was operated without pH control causing variation in the range of 6.0 to 7.8 during the react phase.

Successive Phases	**One complete cycle (6 hours)**		
Filling	5 min		
React		Anaerobic	Aerobic
	1st phase	40	130
	2nd phase	40	130
Settling	5 min		
Decant	5 min		
Idle	5 min		
Total cycle length	360 min		

Table 2. One complete cycle of the hybrid biogranule system

In order to observe the changes on the characteristics of the biogranules due to the variations of HRT during textile wastewater treatment, the development of biogranules with sizes in the range of 0.3-2.5 mm was inoculated into the bioreactor at a ratio of 1:4 of the working volume of the reactor system. 1 L of acclimated mixed sludge was also added into the reactor system. The MLSS and MLVSS concentrations during the start-up of the experiment were 23.2 g/L and 18.4 g/L respectively. The operation steps of one complete cycle of the reactor system are shown in Table 3.

Sequence phase	**Phase period**	**Air supply**	**Recirculation**
Fill	15 min	Off	Off
Reaction			
Anaerobic	Varies*	Off	On
Aerobic	Varies*	On	Off
Settle	5 min	Off	Off
Decant	5 min	Off	Off
Idle	5 min	Off	Off

Table 3. Operation steps during single cycle operation

The morphological and structural observations of the granules were carried out using a stereo microscope equipped with digital image management and analyzer (PAX-ITv6, ARC PAX-CAM). The microbial compositions of the biogranules were observed qualitatively with a scanning electronic microscope (FESEM-Zeiss Supra 35 VPFESEM). The biogranules were left dried at room temperature prior to gold sputter coating (Bio Rad Polaron Division SEM Coating System) with coating current of 20 mM for 45 s. The microbial activity of the biogranules was determined by measuring the oxygen utilization rate (OUR) following Standard Methods (APHA, 2005). The physical characteristics of the biogranules including settling velocity, sludge volume index, granular strength were measured throughout the experiment.

An initial value of 15 mL influent sample was taken from the influent tank before a new cycle operation started, while another 15 mL of the effluent sample was taken from the effluent tank after the effluent was released during the decanting phase as the final values. Samples were centrifuged for 5 min at 4000 rpm at 4^{0}C in order to pellet down all of the suspended particles from the samples. The supernatant was used to measure the removal performance of the COD, color and ammonia removal. All of the measurements for COD, color and ammonia were performed according to Standard Method (APHA, 2005).

10 mL of sample was taken from the top portion of the reactor about 10 minutes after the filling stage ended and 10 mL of sample was taken from the effluent after the decanting stage for the measurement of the suspended particles in the influent and effluent. Another 10 mL of sample volume was taken during the aeration phase for the analysis of MLSS and MLVSS, which were measured according to Standard Methods (APHA, 2005).

The bed height of the biomass in the reactor was measured twice a week in order to estimate the SVI. The bed height was determined immediately after the settling time ended and before the wastewater was drained out during the decanting time. The SVI value can be calculated by measuring the bed volume of the sludge biomass in the reactor divided with the dry weight of the biomass in the reactor. The bed volume is the bed height of the sludge biomass that settled in the reactor 5 minutes after the aeration phase stopped. The bed volume is obtained by multiplying the bed height with the surface area of the bed column. The measurement of the SVI and the sludge retention time were calculated according to Beun et al. (1999). The settling velocity was determined by recording the average time taken for an individual granule to settle at a certain height in a glass column filled with tap water (Linlin et al., 2005).

Determination of the biogranules' strength was based on Ghangrekar et al. (1996). Shear force on the biogranules was introduced through agitation using an orbital shaker at 200 rpm for 5 minutes. At a certain degree of the shear force, parts of the biogranules that are not strongly attached within the biogranules will detach. The ruptured biogranules were separated by allowing the fractions to settle for 1 minute in a 150 ml measuring cylinder. The dry weight of the settled biogranules and the residual biogranules in the supernatant were measured. The ratio of the solids in the supernatant to the total weight of the biogranular sludge used for biogranular strength measurement was expressed as the integrity coefficient (IC) in percent. This percentage indirectly represents the strength of the biogranules. The higher the IC values the lesser the strength of the biogranules and vice versa.

4.3. Morphology and cellular characterization of biogranules

Development of biogranules was obtained within 66 days of operation period with 6 hours HRT. Morphology of the biogranules was investigated via visual and microscopic observations. At the initial development stage, the biomass was composed more of loosely clumped sludge, which can easily break up into pieces under vigorous shaking. Within a week, the anaerobic seed granules underwent morphological changes from spherical in shape and black in color with average diameter of 1 mm into smaller grey granules due to exposure to the shear force during the aerobic react phase. On day 30, two different types of granules were clearly observed in the reactor as shown in Figure 4.

Figure 4a shows mainly irregular-shaped with yellow colored biogranules (Type A) that are solely developed from the activated sludge. In Figure 4b, the anaerobic granules that have fragmented into smaller pieces have formed different sizes of biogranules (Type B) containing pieces of anaerobic granules. The outer layer of the latter were yellow in color indicating the domination of aerobic or facultative microorganisms while the darker spots within the granules indicate the presence of anaerobic fragments originated from the anaerobic granules. The formation of Type A biogranules can be elucidated by the mechanisms explained by Beun et al. (1999). The development was initiated from the mycelial pellets that were retained in the reactor due to high settling velocity. These mycelial pellets eventually become the support matrix for the bacterial growth. Bacteria that were able to attach to this matrix were retained and suppressed the growth of filamentous microorganisms and became the dominant species in the reactor.

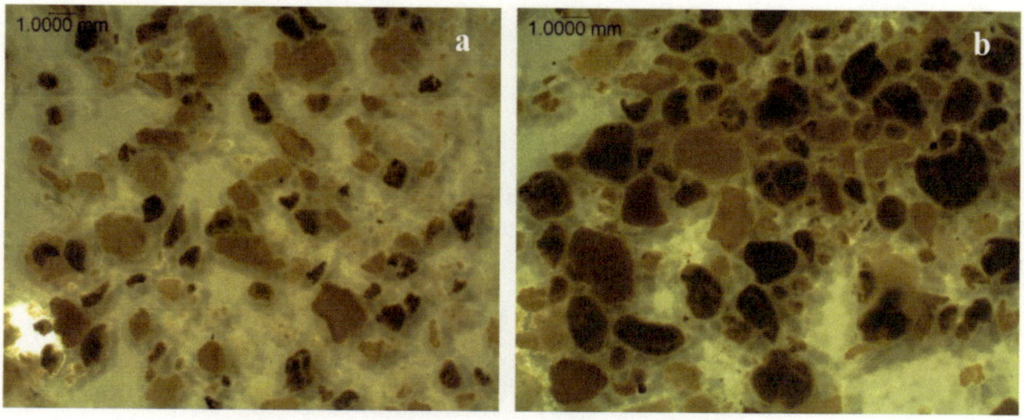

Figure 4. The morphological development of biogranules (scale bar at steady-state equals to 1mm). Pictures were taken using a stereo microscope with magnification of 6.3X. (a) Biogranules developed from the activated sludge. (b) Biogranules developed from anaerobic granules patches.

The formation of Type B granules has been discussed by Linlin et al. (2005). These biogranules were formed through a series of physical and morphological changes. The anaerobic granules initially disintegrated into smaller size flocs and debris when exposed to aeration forces in the reactor column. Some of the granules and debris that were too small were washed out with the effluent while the heavier ones were retained in the column and acted as nuclei for the formation of new granules. Having these combinations of aerobic and anaerobic portions within the biogranules will increase the possibility of complete degradation through the anaerobic/aerobic degradation process. Figure 5 shows the obvious morphological differences between sludge particles with average sludge particles of 0.02 ± 0.01 mm (Figure 5a) during the initial stage of the experiment and matured biogranules (Figure 5b) at the final stage with average diameter of 2.3 ± 1.0 mm.

The microstructure of the biogranules was examined using SEM (Figure 6). The SEM observation of the mature biogranules shows the domination of non-filamentous coccoid bacteria. The bacteria are tightly linked and embedded to one another and form a rounded shape on the surface of the biogranule and covered with extracellular polysaccharides substances (EPS) (Figure 6a). Figure 6b shows the presence of cavities between the clumped bacteria. These cavities are anticipated to be responsible to allow a smooth mass transfer of substrates or metabolite products into and out of the granules (Tay et al., 2003 and Toh et al., 2003).

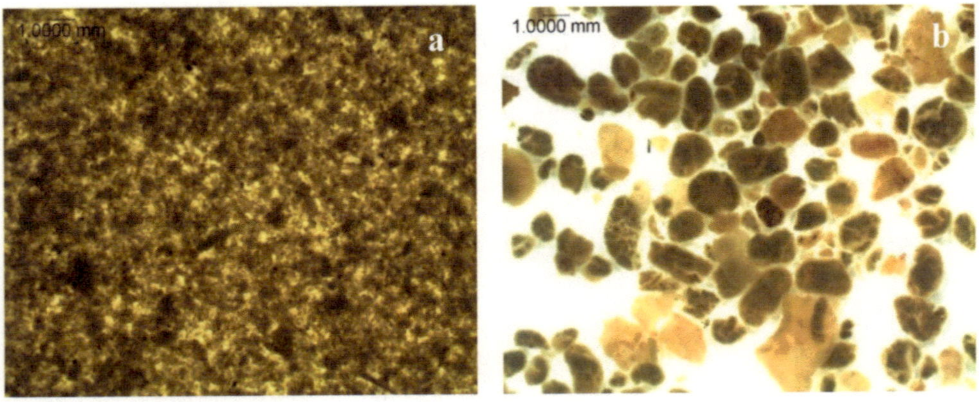

Figure 5. Pictures of sludge particles during the initial stage of the experiment (a) and matured biogranules at the 66 days of the experiment (b). Pictures were taken using a stereo microscope with magnification of 6.3X (scale bar equals to 1 mm)

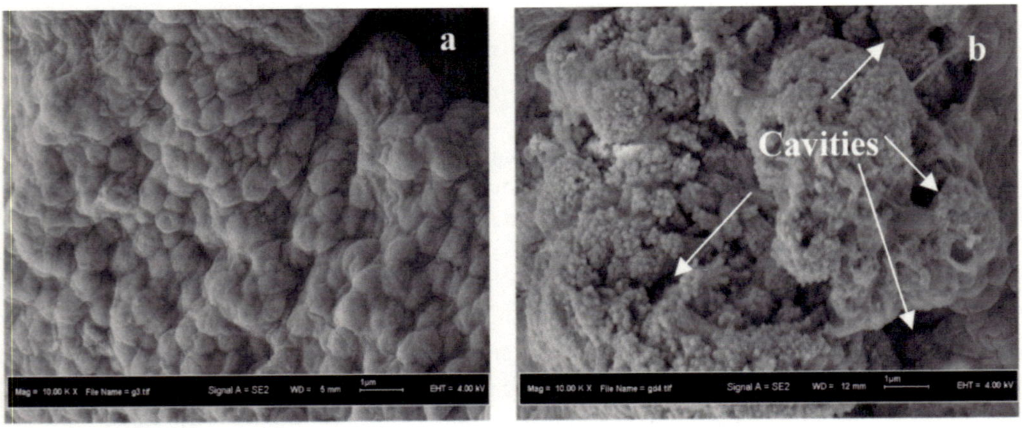

Figure 6. FESEM microstructure observations on mature biogranules under the magnification of 10,000K. (a) Coccoid bacteria tightly linked to one another. (b) Cavities that appear between bacteria clumped inside the biogranules

4.4. Physical characteristics of biogranules

The shear force imposed in the development of granules in this experiment, in terms of superficial upflow air velocity (i.e. 1.6 cm/s), resulted in the development of biogranules with an average diameter of 2.25 mm. The strong shearing force produced by aeration during the aerobic phase prevents the development of bigger aerobic granules. However, reduction in famine period may also lead to the formation of bigger aerobic granular sizes (Liu and Tay, 2006).

The average settling velocity of the sludge and anaerobic granular sludge used as the seeding were 9.9 ± 0.7 m/h and 42 ± 8 m/h respectively. The settling velocity of the biogranules increased from 17.8 ± 2.6 m/h to 83.6 ± 2.6 m/h at the end of experiment. The average settling velocity of the mature biogranules reached almost 80 ± 7.6 m/h, which was nearly three times greater than the settling velocity of the aerobic granules reported by Zheng et al. (2005).

The increase in settling velocity has given significant impact on the biomass concentration in the reactor. The relationship between the concentration of the MLSS and settling velocity of the granules is shown in Figure 7. Despite the short settling time (5 min), the high settling velocity possessed by the developed biogranules enabled the biogranules to escape from being flushed out during the decanting phase. Such conditions have caused more biogranules to retain in the reactor and resulted in the increase of biomass concentration.

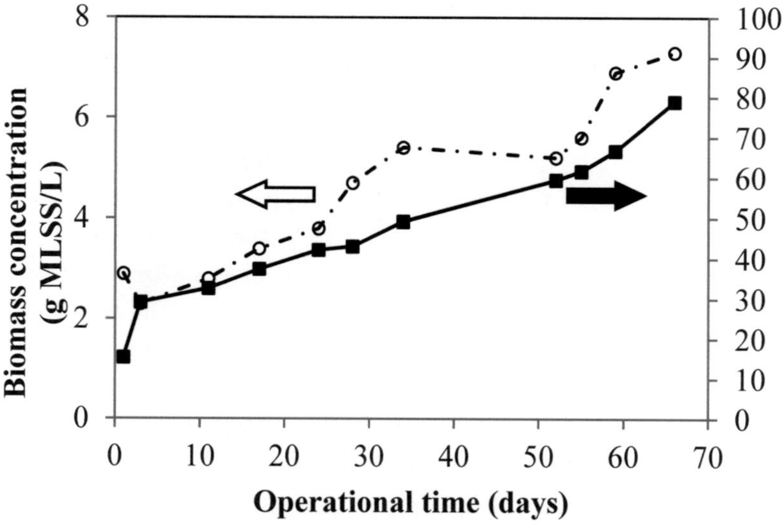

Figure 7. The relationship between the biomass concentrations retained in the reactor with the settling velocity of the biogranules (□) Settling velocity; (□) Biomass concentration

The SVI value has also improved from 277 mL/g at the initial stage to 69 mL/g at the mature development of biogranules. This indicates good settling properties of the biogranules, which is favorable in wastewater treatment plant operation. Figure 8 demonstrates the SVI profile along with the settling velocity. As the SVI value improved, the granular settling properties increased from 50 m/h to about 80 m/h. The SVI of biogranules seems to vary depending on the settling time of the reactor system. McSwain et al. (2004) reported the SVI of biogranules improved from 115 ± 36 ml/g to 47 ± 6 ml/g when the settling time decreased from 2 to 10 min. Biogranules developed with anaerobic seeding, showed higher settling velocity and improved SVI.

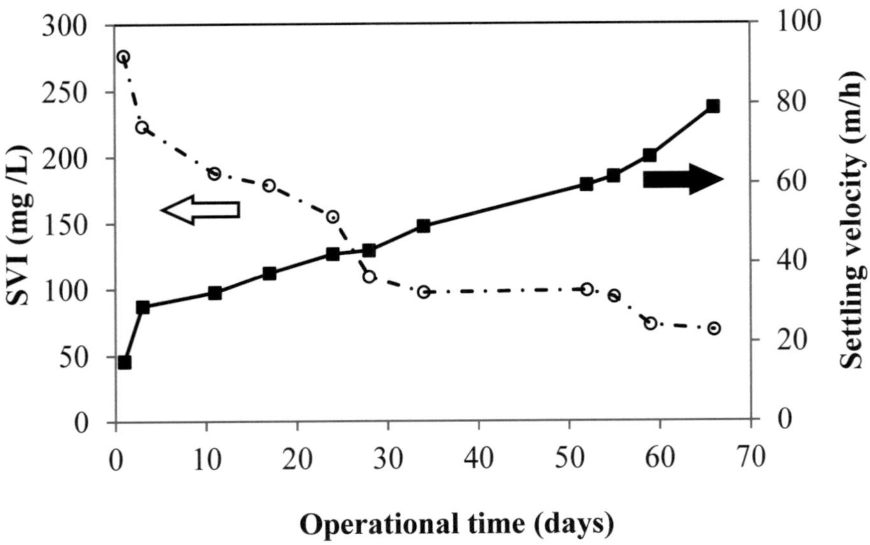

Figure 8. The relationship between the SVI values and settling velocity of the biogranules (▯) SVI, (▯) Settling velocity

The granular strength of the biogranules was measured based on the integrity coefficient (IC) defined earlier. The smaller the value of IC, the higher the strength and ability of the biogranules to clump together and being prevented to break due to shear force of the aeration. Figure 9 shows the profile of IC of the developed biogranules as a function of time. The IC reduced as the biogranules developed. The initial value of IC was 30. Then the IC was reduced to about 9 as it reached a mature stage. According to Ghangrekar et al. (2005), biogranules with integrity coefficient of less than 20 were considered high strength granules. The reduction in IC value indicates the increase in the strength of the bond that holds the microorganisms together within the developed biogranules.

During the initial development, the microbes within the biogranules were loosely bounded to each other. At this stage, the biogranules may consist of more cavities causing the biogranules become less dense, as manifested by low settling velocity. As more microbes are linked together, the biogranules increase in size. Under certain selective pressures (i.e. short settling time, hydrodynamic shear force, starvation of the microbial cell), microbes may produce more extrapolysaccarides (EPS) (Lin et al., 2003; Qin et al., 2004). As reported by Zhang et al. (2007) and Adav and Lee, (2008), the EPS contribute greatly to the strength and the stability of aerobic granules. When microbial cells produce more EPS, they form a cross-linked network and further strengthen the structural integrity of the granules. The cavities within the biogranules will be filled with EPS as it is a major component of the biogranules matrix material. This caused the biogranules to become denser and stronger as shown by their high settling velocity and lower IC value. The physical characteristics of the seed sludge and the matured biogranules are summarized in Table 4. The developed biogranules possess desirable biomass characteristics in the biological wastewater treatment system.

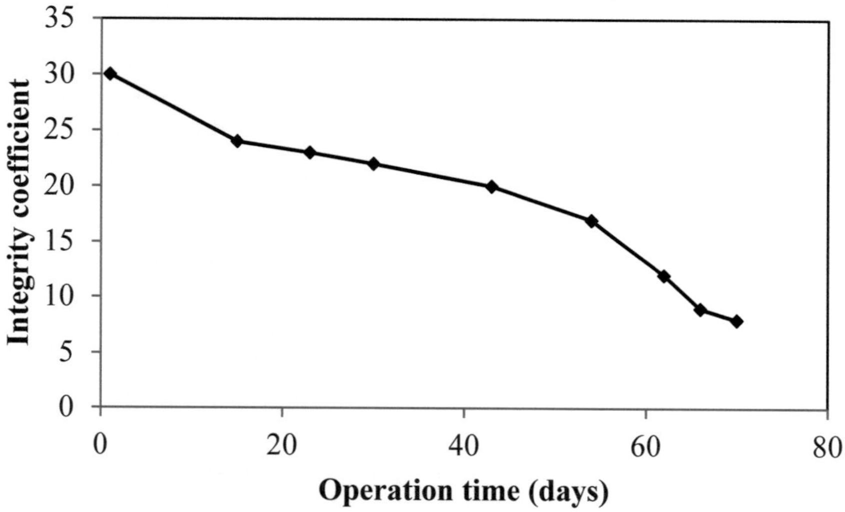

Figure 9. The profile of integrity coefficient representing the granular strength of the biogranules

Characteristics	Seed Sludge	Biogranules
SVI (mL/g)	277	69
Average diameter (mm)	0.02 ± 0.01	2.3 ± 1.0
Average settling velocity (m/h)	9.9 ± 0.7	80 ± 8
IC	92 ± 6	9.4 ± 0.5
MLSS (g/L)	2.9 ± 0.8	7.3 ± 0.9
MLVSS (g/L)	1.9 ± 0.5	5.6 ± 0.8

Table 4. Characteristics of seed sludge and biogranules

The profile of the biomass concentration (i.e. MLSS) after seeding with the anaerobic granules is shown in Figure 10. During the first few days, almost half of the sludge was washed out from the reactor causing a rapid decrease in the biomass concentration. The MLSS reduced from initial concentrations of 5.5 g/L to 2.9 g/L mainly due to the short settling time used in the cycle (i.e 5 min). During this initial stage, the anaerobic granules were also observed to disintegrate into smaller fragmented biogranules and debris resulted from shear force caused by aeration. These small fragments have poor settling ability and were washed out from the reactor causing an increase of suspended solids concentration in the effluent.

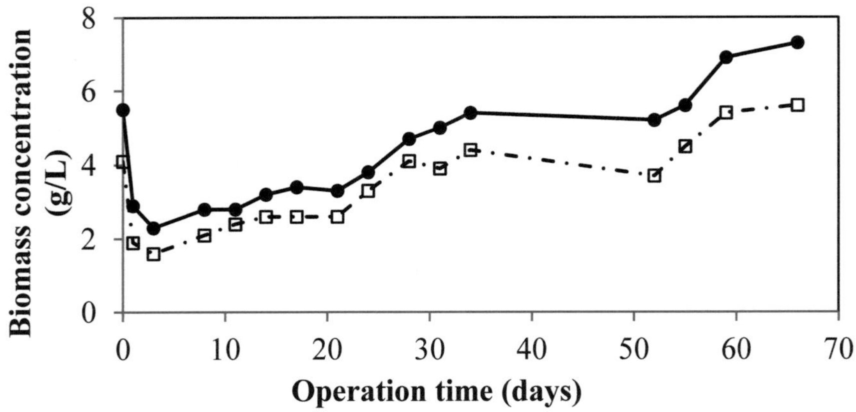

Figure 10. The profile of biomass concentration in the SBR. (•) MLSS, (□) MLVSS

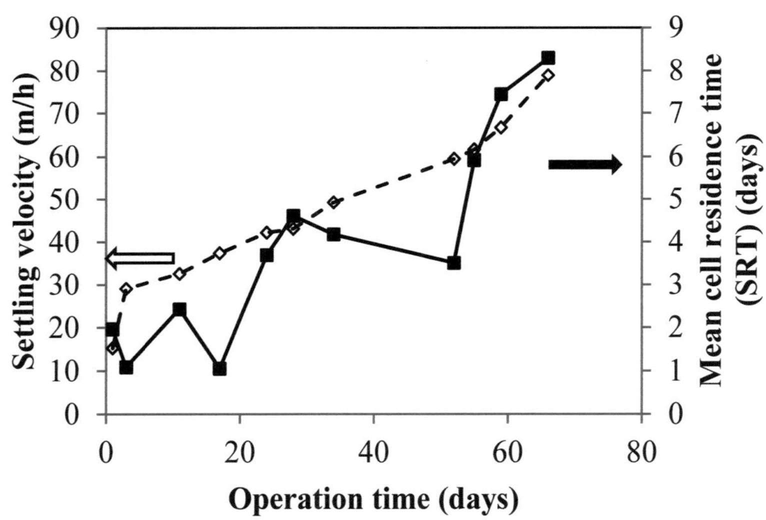

Figure 11. The settling velocity profile in relation to mean cell residence time (SRT). (▯) SVI, (▯) SRT

As the experiment continued, the concentration of the biomass increased and reached 7.3 g MLSS/L on the 66th day. The profile of MLVSS follows the same trend of MLSS, ranging from 1.9 g/L to 5.6 g/L. The mean cell residence time (SRT) also increased from 1.4 days at the initial stage to 8.3 days on the 66th day, indicating the accumulation of the biomass in the reactor. As less biomass was washed out during the decanting period, the increase in SRT is another manifestation of good settling characteristics resulting from the high settling velocity. Nonetheless, the benefit of high SRT will depend on the goal of the treatment process (Tchobanoglous et al., 2004). The SRT is affected by the settling velocity. The profiles of the settling velocity and the SRT as functions of time are given in Figure 11.

4.5. System profile

Changes in the HRT of the reactor system caused variation of the anaerobic and aerobic react times. It also may affect the loading rate imposed to the system if the substrate concentration is maintained. These conditions will affect the microbial activity within the biogranules and may influence the performance of the reactor system. The details of the experimental conditions of the reactor system are shown in Table 5.

The microbial activity was measured based on the OUR of a complete one cycle operation. The OUR was measured several times before each of the stages ended and showed that most of the external substrate was consumed more or less within the first 30 minutes of each aerobic reaction phase. Figures 12 and 13 show the profiles of the OUR throughout the experiment from Stage I to Stage VI.

Stage	Days covered	Phase (hours)				HRT (hrs)	OLR (kg COD/ m^3·day)
		1st		2nd			
		Anaerobic	Aerobic	Anaerobic	Aerobic		
I	49	1.42	1.42	1.42	1.42	6	2.5
II	43	2.92	2.92	2.92	2.92	12	1.3
III	51	5.92	5.92	5.92	5.92	24	0.6
IV	43	5.92	5.92	5.92	5.92	24	0.8
V	46	8.92	2.92	8.92	2.92	24	0.8
VI	46	2.92	8.92	2.92	8.92	24	0.8

$OLR=\frac{X}{V_{total}}\times\frac{V_{add}}{T}$, where X = COD concentration of the influent (mg/L); V_{add}= Volume of influent added in each cycle operation (mL); V_{total} = Total working volume of the experiment (mL); T = Hydraulic retention time (hour).

Table 5. Details of experimental conditions of the reactor system

The OUR profile (Figure 12) shows that the initial measurement of the OUR was reduced as the HRT increased (Stage I to Stage III). This is due to the reduction in the OLR as the HRT increased. Less oxygen is required as the organic load concentration is reduced. After a sharp increase of OUR at the beginning of each cycle in all stages, the OUR measurement was consistently low until the end of the cycle. The low value of the OUR indicates that most of the external substrates have been consumed. It also means that the microorganisms in the reactor system are under starvation phase. At this phase, no further degradation was observed even though the HRT was extended. During the starvation phase, endogenous respiration will take place, except at the beginning of the second phase of aerobic reaction where there was a short increase in the OUR. This increase is caused by the mineralization of amines, the byproduct of dye degradation during the second anaerobic reaction phase. As the duration of anaerobic reaction phase increased, the short pulse increased as shown in Figure 13 (a and b) of Stage IV and V, respectively. Stage IV and Stage V were operated with the same HRT and organic loading but were different in the anaerobic and aerobic reaction phase ratio.

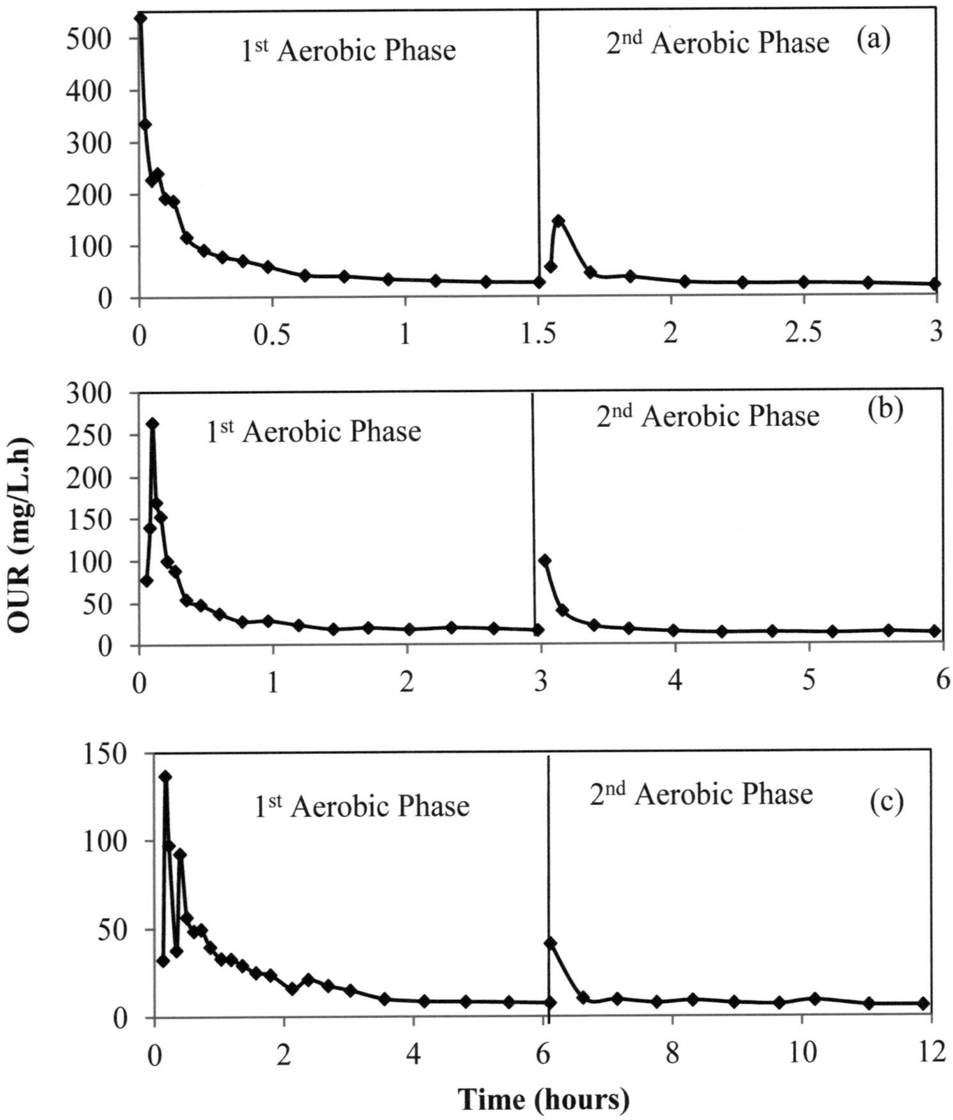

Figure 12. OUR profile of (a) Stage I (Aerobic phase 2.84 hours), (b) Stage II (Aerobic phase 5.84 hours) and (c) Stage III (Aerobic phase 11.84 hours)

The changes in the HRT will also affect the biomass accumulated within the reactor system. The HRT was increased from 6 hours in Stage I to 24 hours in Stage III, without the addition of any substrate. This resulted in the reduction of OLR supplemented into the reactor system from 2.5 to 0.6 kg COD/m³ day. The HRT for Stage III to VI was kept constant i.e. 24 hours, but the duration of anaerobic and aerobic react phases was varied. From Stage III onwards, the OLR was increased to 0.8 kg COD/m³ day by increasing the concentration of the carbon sources in the synthetic textile dyeing wastewater.

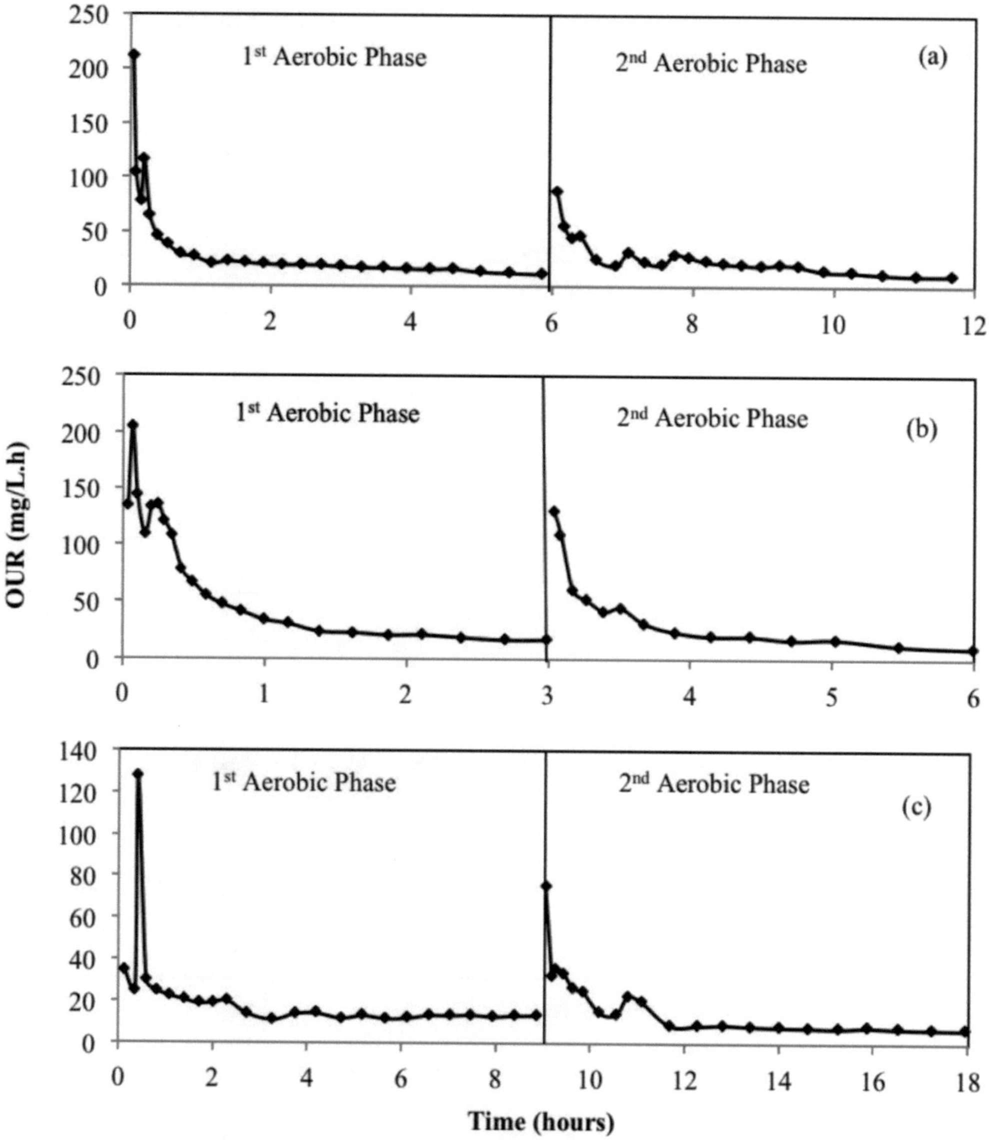

Figure 13. OUR profile of (a) Stage IV (Aerobic phase 11.84 hours), (b) Stage V (Aerobic phase 5.84 hours), (c) Stage VI (Aerobic phase 17.84 hours)

Table 6 shows the oxidation-reduction potential (ORP) values measured during the second phase of the anaerobic and aerobic reactions during the experiments. The ORP profile of all the stage corresponded very well with the dissolved oxygen. As the anaerobic react phase increased, more of negative values of the ORP were recorded. During the aerobic phase the ORP varies between +98 to +177 mV.

The biomass profile at steady state with stepwise increment of HRT (Stage I to III) and variation of react phases (Stage IV to VI) are shown in Table 8. As shown in Table 7, it is apparent that the biomass concentration (MLSS) in the reactor decreased and the VSS in the effluent were also reduced with the increase in the HRT (Stage I to III). The reduction of the biomass concentration in the reactor may be due to the lower value of OLR applied in the reactor system as the HRT increased.

Stage	Anaerobic React Phase	Aerobic React Phase
I	-124 ± 27	125 ± 19
II	-219 ± 33	129 ± 24
III	-358 ± 29	174 ± 34
IV	-355 ± 51	151 ± 17
V	-407 ± 21	112 ± 21
VI	-225 ± 28	177 ± 15

Table 6. Oxidation Reduction Potential

React Phase	Stage					
	I	II	III	IV	V	VI
Anaerobic (hours)	2.8	5.8	11.8	11.8	17.8	5.8
Aerobic (hours)	2.8	5.8	11.8	11.8	5.8	17.8
MLSS (g/L)	35.3 ± 1.6	28.7 ± 0.6	25.2 ± 1.8	30.5 ± 3.4	31.6 ± 3.7	23.3 ±0.8
MLVSS (g/L)	31.9 ± 1.8	24.5 ± 2.2	18.5 ± 2.2	26.0 ± 3.4	22.4 ± 2.0	20.2 ± 0.8
VSS/SS	0.90	0.85	0.73	0.85	0.71	0.87
Effluent (VSS g/L)	0.34 ± 0.16	0.31 ± 0.11	0.26 ± 0.19	0.34 ± 0.11	0.33 ± 0.10	0.55 ± 0.22
SRT (day)	27.6 ± 13.4	42.4 ± 10.2	78.9 ± 23.9	70.1 ± 23.9	72.5 ± 23.3	41.6 ± 18.4

Table 7. Biomass concentrations at different stages of the experiment

When the OLR was increased to 0.8 kg COD/m^3·day, there was an improvement in the biomass concentration where the biomass concentration have increased to 30.5 ± 3.4 g/L and 31.6 ± 3.7 g/L in Stage IV and Stage V as compared to 25.2 ± 1.8 g/L of biomass concentration in Stage III which run at the same HRT (24 hours) but with OLR 0.6 kg COD/m^3·day. The increase in OLR has caused an increment in the biomass concentration in the reactor. A slight increase in the biomass concentration was also observed along with the longer period of the anaerobic phase (Stage V), i.e. 18 hours.

The ratio of the volatile biomass (MLVSS) to total biomass (MLSS) reduced from Stage I to Stage III mainly due to decrease in the OLR as the HRT increased from 6 to 24 hours, whereas the MLVSS/MLSS ratio of the Stage III and Stage IV with 12 hours aerobic reaction phase was observed higher with the ratio of 0.73 and 0.85, respectively. The increment may be due to the increase of the OLR from 0.6 to 0.8 kg COD/m^3 day (Stage III to Stage IV). Increase in the OLR means more carbon sources were supplied to the microorganisms in the reactor. When more food is available, more growth will take place and this is indicated by the increase in the MLVSS/MLSS ratio.

However, when the anaerobic period of the HRT is extended, the MLVSS/MLSS ratio decreased (0.71). Decrease in MLVSS/MLSS ratio may indicate an increase of inorganic accumulation within the granulation biomass. When the duration of aeration phase was increased up to 18 hours, the biomass started to reduce again (Stage VI) and increase of VSS in the effluent was once again observed. This may give an indication that too long of aerobic reaction phase is not suitable for granular biomass system. Prolong of aeration time may result in instability of the reactor performance. The profile of biomass concentration of the reactor system is given in Figure 14.

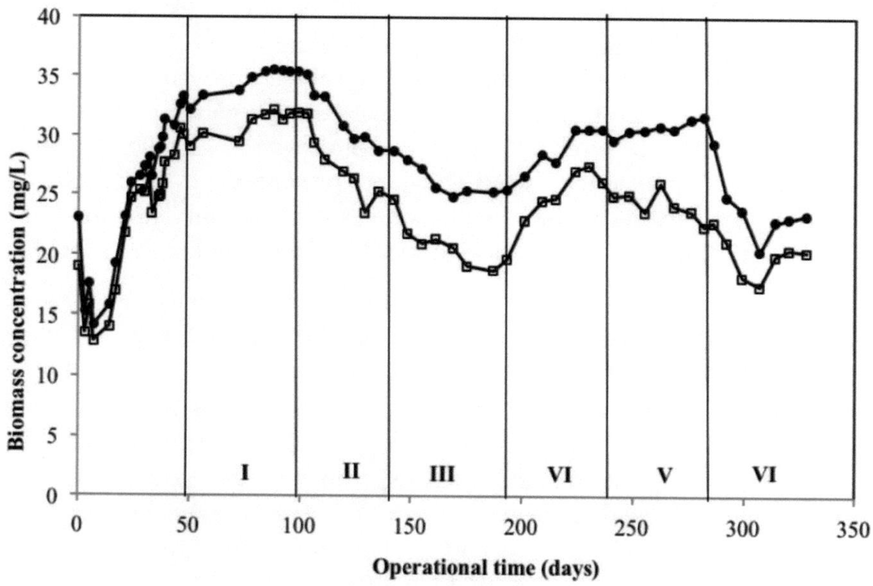

Figure 14. Profile of biomass concentration at different stages of the experiment. (•) MLSS, (□) MLVSS. Stage I: anaerobic (2.8 h): aerobic (2.8 h); Stage II: anaerobic (5.8 h): aerobic (5.8 h); Stage III and Stage IV: anaerobic (11.8 h): aerobic (11.8 h); Stage V: anaerobic (17.8 h): aerobic (5.8 h); Stage V: anaerobic (5.8 h): aerobic (17.8 h)

The SRT of the reactor system increased from 27.6 ± 13.4 to 78.9 ± 30.8 d when the length of the HRT increased from 6 to 24 hours (Stage I to Stage III). With HRT of 24 hours, increase of anaerobic reaction phase up to 18 hours (Stage IV to Stage V) has slightly increased the SRT from 70.1 ± 23.9 to 72.5 ± 23.3 d. The SRT value changes in each stage of the experiment. According to Wijffels and Tramper (1995), the favorable sludge age for high removal efficiency for COD and nitrification process is more than 4 days. Based on the SRT obtained, this biogranular system is capable of the simultaneous degradation of nitrification process and COD removal. Since the treatment goal is to remove recalcitrant dyeing compound, the SRT value of all stages evaluated in this experiment was in the acceptable range from degradation of xenobiotic compounds (Grady et al. 1999).

React Phase	Stage					
	I	II	III	IV	V	VI
Anaerobic (hours)	2.8	5.8	11.8	11.8	17.8	5.8
Aerobic (hours)	2.8	5.8	11.8	11.8	5.8	17.8
SVI (mL/g)	13.1 ± 0.4	18.8 ± 1.5	21.4 ± 1.6	16.8 ± 1.3	15.5 ± 1.3	24.8 ± 0.9
SV (m/h)	41.3 ± 3.1	35.1 ± 0.8	24.5 ± 1.1	28.4 ± 1.3	33.4 ± 2.5	21.3 ± 0.5

Table 8. Physical properties of the biogranules at different stages of the react phase

The SVI value of the biogranules was used to evaluate the biogranules settling ability. It is anticipated that bigger biogranules will have higher settling velocity and hence, reduce the SVI value, indicating good settling ability. The SVI value improved when the anaerobic react phase was prolonged in Stage V indicating such reaction pattern will help to develop granules with better settling profile. According to Panswad *et al.* (2001a), inert biomass increased as the anoxic/anaerobic condition was prolonged. It is possible that the accumulation of inert particles within the biogranules increased and resulted in improved SVI properties. Table 8 showed the physical properties of the biogranules at different stages of the react phase.

Figure 15 shows the profile of SVI of the reactor system. The SVI value in Stage V was reduced from 16.8 ± 1.3 mL/g (in Stage IV) to 15.5 ± 1.3 mL/g. This is expected to be due to the accumulation of more inert solids within the biogranules as shown with low levels of MLVSS/MLSS ratio in Stage V (0.71). Despite changes in HRT that caused decrease in the size of biogranules, the SVI values of the whole experiments were good except for Stage VI. During Stage VI, the prolonged of the aerobic phase (i.e. 17.8 hours), which was operated at high superficial air velocity (2.5 cm/s), cause the biogranules to rupture. At this stage, the size of biogranules becomes smaller causing the settleability of the particles to reduce and was demonstrated with increase in SVI value.

Hydraulic retention time is an important parameter that control the contact time between the biomass and the wastewater in a reactor system. The HRT of a system must be long enough for the degradation process to take place. However, in the application of biogranules in the treatment system, the HRT should not be too long as it may cause the disintegration of the granules. According to Tay et al. (2002) and Wang et al. (2005), a short

HRT is favorable for rapid granulation process, while too long HRTs may lead to granulation system failure due to high biomass lost (Pan et al., 2004). An optimum HRT of biogranulation systems will be able to stabilize the reactor performance with good biomass retention and high removal performance. According to Pan et al. (2004), the optimum HRT for aerobic granulation systems ranging from 2 to 12 hours where stable aerobic granules with good settleability and microbial activities. However, the optimum HRT for treating different types of wastewater may vary depending on the type of wastewater and the targeted degradation compound.

4.6. Removal of color

Color removal was observed to increase from 66.7 ± 1.6 % to 76.5 ± 0.8 % as the HRT increased from Stage I to Stage III. Increase in the HRT allows longer contact time between the biogranules and the wastewater resulting in better color removal. Furthermore, when the OLR was increased from 0.6 kg COD/m^3·day (Stage III) to 0.8 kg COD/m^3·day (Stage IV), a significant improvement in color removal from 76.5 ± 0.8 % to 83.1 ± 1.4 % was observed. This may be due to the increase in the microbial population. Ong et al. (2005) reported that the percentage of color removal efficiency increased by 16% in anaerobic and 50% in aerobic SBR reactor systems when the OLR rate was increased from 2.66 to 5.32 g COD/L·day. An increase of color removal efficiency from 82% to 90% was also observed by Talarposhiti et al. (2001) when the COD loading was increased in a two-phase anaerobic packed bed reactor from 0.25 to 1 kg COD/m^3·day.

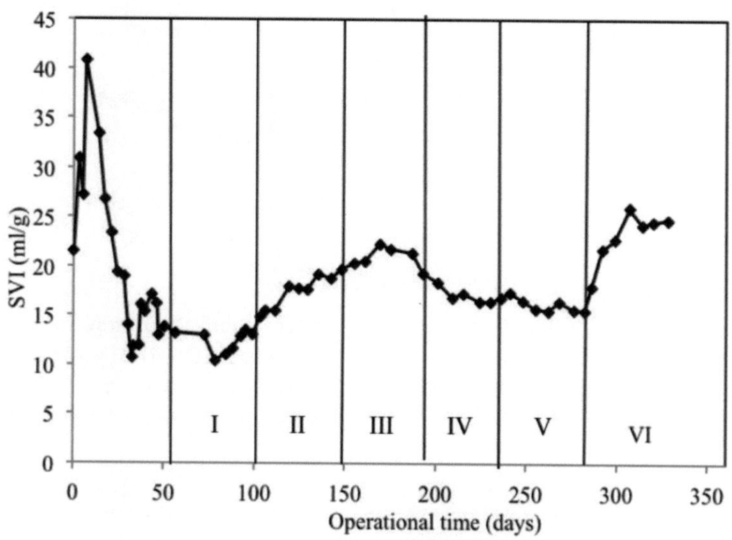

Figure 15. Sludge volume index profile of biogranules. Stage I: anaerobic (2.8 h): aerobic (2.8 h); Stage II: anaerobic (5.8 h): aerobic (5.8 h); Stage III and Stage IV: anaerobic (11.8 h): aerobic (11.8 h); Stage V: anaerobic (17.8 h): aerobic (5.8 h); Stage V: anaerobic (5.8 h): aerobic (17.8 h)

Since more color removal took place in the anaerobic condition (Banat et al., 1996; Dos Santos et al., 2007), the percentage of color removal was once again increased from Stage IV

(83.1 ± 1.4%) to Stage V (86.5 ± 0.5%) when the anaerobic reaction phase was extended from 12 to 18 hours of the 24 hours reaction cycle. Improved decolorization process that occurs during the anaerobic stage enhances the overall wastewater biodegradation since more readily biodegradable substances can be degraded in the following aerobic treatment (Stolz, 2001). Figure 16 shows the profile of the color removal performance.

With respect to the mechanisms that are involved in color degradation, the addition of electron–donating substrate can considerably improve the decolorization reductive rate (Bras et al., 2001, Dos Santos et al., 2005). In anaerobic and aerobic sequential wastewater treatment system, the anaerobic stage was the main step for color degradation while the aerobic phase acted as the polishing step and enhancement in COD removal. Higher initial COD concentration did not improve color removal but caused deterioration in COD removal in the anaerobic-aerobic SBR system (Kapdan and Oztekin, 2006).

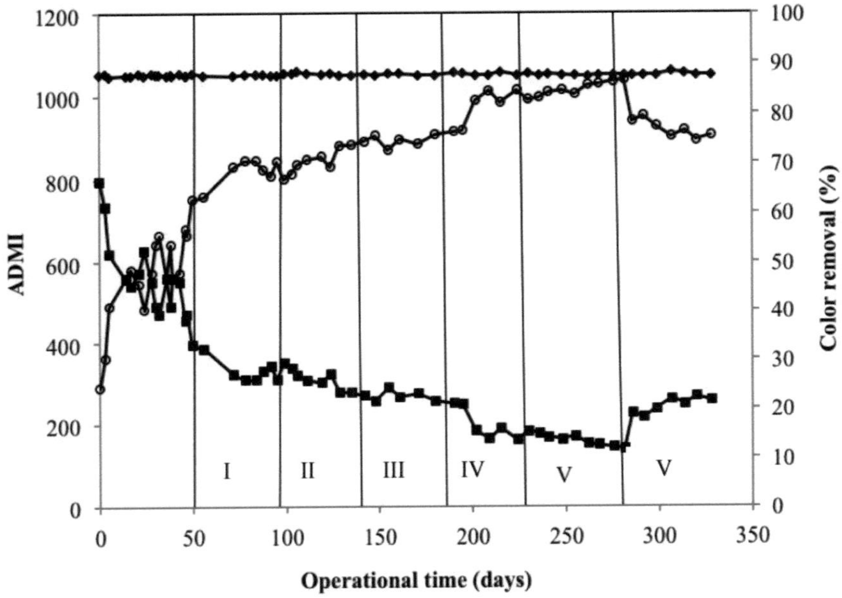

Figure 16. Profile of color removal performance of the reactor system at different stages of the experiment. (□) Influent color, (□) Effluent color, (□) Color removal. (100 ADMI ≈ 1 Pt-Co). Stage I: anaerobic (2.8 h): aerobic (2.8 h); Stage II: anaerobic (5.8 h): aerobic (5.8 h); Stage III and Stage IV: anaerobic (11.8 h): aerobic (11.8 h); Stage V: anaerobic (17.8 h): aerobic (5.8 h); Stage V: anaerobic (5.8 h): aerobic (17.8 h)

Psukphun and Vinitnantharat (2003) reported that the duration of the anaerobic phase should be long enough to obtain better COD and color removal. An increase in the HRT will provide enough time of the COD and inter-metabolites of simulated textile wastewater in anaerobic or/and anaerobic/aerobic systems (Isik and Sponza, 2008). Biodegradation of the azo bonds may require a certain contact time in order to achieve high removal efficiency. Depending only on the filling stage to provide anaerobic condition for the cleavage of azo bond compounds may not be adequate for textile wastewater treatment. However, the time required for the cleavage of the azo bond may be affected by the complexity of the dye molecule structures. The suitable contact time of anaerobic and aerobic reaction phases may provide high removal performance for the cleavage of the N=N bond (anaerobic condition) and mineralization of aromatic amines (aerobic phase). Furthermore, the reduction of COD is more effective during the aerobic stage as compared to the anaerobic reaction condition (Smith et al., 2007). It shows that having longer anaerobic (18 hours) and shorter aerobic (6 hours) react phase resulted in the highest removal for color and a slight improvement in the efficiency of COD removal.

4.7. Removal of chemical oxygen demand

The time history of the COD concentration in the influent and effluent and the removal rate for all six stages is given in Figure 17. The biogranular system showed consistent COD degradation performance with 84.2 ± 0.9% removals after about 50 days of start-up period (acclimatization phase). The overall performance was almost consistent despite the fact that the duration of the experimental process was increased from 6 hours to 24 hours. This phenomenon may be due to the decreasing biomass concentration and reduction in the OLR as mentioned earlier. When the OLR was increased from 0.6 kg COD/m^3·day to 0.8 kg COD/m^3·day on the 194th day of the experiment (Stage III to Stage IV), the COD removal efficiency increased from about 84.4 ± 0.4% at the end of Stage III (day 193) to 90.7 ± 0.2% at the end of Stage IV (day 236).

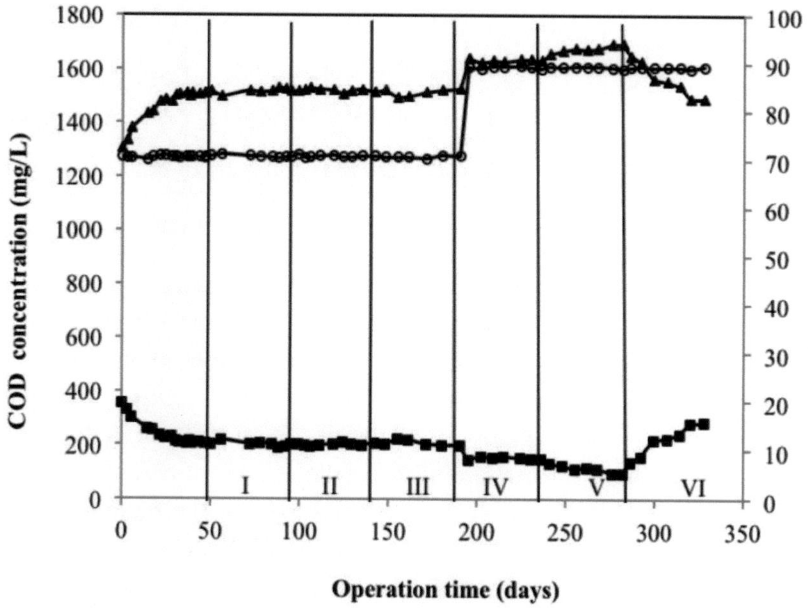

Figure 17. Profile of COD removal performance of the reactor system at different stages of the experiment. (▯) Influent COD; (▯) Effluent COD, (▯) COD removal. Stage I: anaerobic (2.8 h): aerobic (2.8 h); Stage II: anaerobic (5.8 h): aerobic (5.8 h); Stage III and Stage IV: anaerobic (11.8 h): aerobic (11.8 h); Stage V: anaerobic (17.8 h): aerobic (5.8 h); Stage V: anaerobic (5.8 h): aerobic (17.8 h)

At the final stage (Stage VI) of the experiment, a surge drop of COD removal efficiency was observed. As the aeration time was increased from 6 to 18 hours, the COD removal was reduced from 94.1 ± 0.6% to 82.6 ± 0.8%. The drop in the COD removal efficiency was due to the increase in biomass loss into the effluent. The MLSS in Stage VI was 23.3 ± 0.8 g/L as compared to 31.6 ± 3.7 g/L observed in the previous stages. The effect of HRT on the COD and color removal performance by the biogranules at different stages is given in Table 10.

An increase in the percentage of COD removal efficiency was also observed when the period of anaerobic phase was increased from 12 hours to 18 hours. As noted in Table 9 the removal increased from 90.7 % in Stage IV to 94.1% in Stage V. Psukphun and Vinitnantharat (2003) claimed that the increase in the non-aeration phase in the SBR system will cause an alteration in the population of anaerobic microorganisms in the system. The state is expected to produce good COD and color removal for textile wastewater. However, according to Kapdan and Oztekin (2006), when the duration of anaerobic phase is too long, the contribution of aerobic react phase can be decreased. This is possibly due to the toxic effect of aromatic amines produced during dye degradation.

Reaction Phase	Stage					
	I	II	III	IV	V	VI
Anaerobic (hours)	2.8	5.8	11.8	11.8	17.8	5.8
Aerobic (hours)	2.8	5.8	11.8	11.8	5.8	17.8
COD (%)	84.2 ± 0.9	84.6 ± 1.1	84.4 ± 0.4	90.7 ± 0.2	94.1 ± 0.6	82.6 ± 0.8
Color (%)	66.7 ± 1.6	74.3 ± 0.4	76.5 ± 0.8	83.1 ± 1.4	86.5 ± 0.5	75.4 ± 0.3

Table 9. Profile of COD and color removal percentage at different stages of experiment

Owing to the condition in the reactor where different react phases occur in the same column, too long anaerobic reaction periods will cause high accumulation of aromatic amine in the same compartment. High concentrations of aromatic amines may inhibit the activity of aerobic microorganisms during the aerobic phase. In this study, even though the anaerobic react phase was extended up to 18 hours, there was no reduction in COD removal. This shows that there was no inhibition on the activity of aerobic microorganisms by the long accumulation of the byproducts produced from anaerobic degradation of the dye compound. The reason can be that the concentration of dye used was not sufficiently high to produce any toxicity effect on the microorganisms within the biogranules. Furthermore, the biogranules may not be affected by the dyestuff degradation byproducts due to the structural form of the biogranules. The biogranules structure, which consisted of EPS acts as a shield for microorganisms within the granules against any shock loading or toxic compound.

5. Conclusions

Stable biogranules can be cultivated in the SBR system with the application of intermittent anaerobic and aerobic reaction modes during the react phase. The matured biogranules showed the domination of non-filamentous bacteria that were tightly linked and embedded to one another and covered with EPS. The use of seed sludge in the development process affects the morphology of the developed granules. Matured biogranules had an average diameter of 2.3 ± 1.0 mm and a settling velocity of 80 ± 8 m/h and a low IC value of 9.4 ± 0.5. This indicates successful development of excellent settling properties of biogranules. The cultivation of biogranules seeded with anaerobic granules resulted in better granules formation. The OUR/SOUR and SMA analysis proved the presence of anaerobic and aerobic microorganisms activities in the biogranules. They are capable of performing degradation both in anaerobic and aerobic conditions. The size and the SVI of the biogranules were very much affected by the variations of the HRT. An increase in the aeration react time resulted in the disintegration of the biogranules. Too long aerobic reaction times exposed the biogranules under prolonged starvation condition causing instability of the granular structure that lead to disruption of the biogranules.

The percentage of COD removal was not likely affected by the increase in the HRT, but mainly caused by the decrease in the granular biomass and OLR. However, the COD removal was improved with the increase in the anaerobic reaction phase. The percentage of color removal has improved with the increase of the HRT. An HRT with a prolonged anaerobic react time and reduced aerobic reaction time is considered as the best condition for the removal of color and the organic compound, resulting in maximum color and COD removal.

Author details

Khalida Muda, Azmi Aris, Mohd Razman Salim
Department of Environmental Engineering,
Faculty of Civil Engineering,
Universiti Teknologi Malaysia, Skudai, Johor,
Malaysia

Zaharah Ibrahim

Department of Biological Sciences, Faculty of Biosciences and Bioengineering,
Universiti Teknologi Malaysia, Skudai, Johor, Malaysia

Acknowledgement

The authors wish to thank the Ministry of Science, Technology and Innovation (MOSTI), Ministry of High Education (MOHE) and Universiti Teknologi Malaysia for the financial supports of this research (Grants No.: 79137, 78211 and 75221)

6. References

Adav, S.S. & Lee, D.J. (2008). Physiological Characterization and Interactions of Isolates in Phenol Degrading Aerobic Granules. *Appl. Microbiol. Biotechnol.* 78, 899-905.

Ahmed, M., Idris, A. & Adam, A. (2005). Nitrogen Removal of Textile Wastewater by Combined Anaerobic-Aerobic System. *Suranaree J. Sci. Technol.* 12(4), 286-295.

Ahn, Y.H. & Richard, E.S. (2003). Settleability Assessment Protocol for Anaerobic Granular Sludge and Its Application. *Water SA.* 29 (4), 419-426.

Aizenshtein, E.M. (2004). World Production of Textile Raw Materials in 2002. *Fibre. Chem.* 36, 1–6.

American Public Health Association, 2005. *Standard Methods for the Examination of Water and Wastewater.* (21th ed). Washington, DC, USA American Public Health Association.

Arrojo, B., Mosquera-Corral, A., Campos, J.L. & M'endez, R. (2006). Effects of Mechanical Stress on Anammox Granules in a Sequencing Batch Reactor (SBR). *J. Biotechnol.* 123, 453-463.

Ay, F., Catalkaya, E.C. & Kargi, F. (2009). A Statistical Experiment Design Approach for Advanced Oxidation of Direct Red Azo-Dye by Photo-Fenton Treatment. *J. Hazard. Materials.* 162, 230-236.

Bachman, A., Beard, V.L. & McCarty, P.L. (1985). Performance Characteristics of the Anaerobic Baffled Reactor. *Water Res.* 19 (1), 99-106.

Banat, I.M., Nigam, P., Singh, D. & Marchant, R. (1996). Microbial Decolorization of Textile-Dye-Containing Effluents. *Biores. Technol.* 58, 217-227.

Batstone, D. J., Keller, J. & Blackall, L.L. (2004). The Influence of Substrate Kinetics on the Microbial Community Structure in Granular Anaerobic Biomass. *Water Res.* 38, 1390-1404.

Beun, H.J., Hendricks, A., van Loosdrecht, M.C.M., Morgenroth, E., Wilderer, P.A. & Heijnen, J.J. (1999). Aerobic Granulationin a Sequencing Batch Reactor. *Water Res.* 33, 2283-2290.

Beun, J.J., van Loosdrecht, M.C.M. & Heijnen, J.J. (2002). Aerobic Granulation in a Sequencing Batch Airlift Reactor. *Water Res.* 36, 702–712.

Bisschops, I.A.E. & Spanjers, H. (2003). Literature Review on Textile Wastewater Characterisation. *Environ. Technol.* 24, 1399-1411.

Bras, R., Ferra, M.I.A., Pinheiro, H.M. & Goncalves, I.C. (2001). Batch Tests for Assessing Decolorization of Azo Dyes by Methanogenic and Mixed Cultures. *J. Biotechnol.* 89, 155-162.

Calleja, G.B. (1984). *Microbial Aggregation.* Florida: CRC.

Chang, J.S. & Lin, Y.C. (2000). Fed-Batch Bioreactor Strategies for Microbial Decolorization of Azo Dye using a *Pseudomonas luteola* Strain. *Biotechnol. Prog.* 16, 979-85.

Chen, M.Y., Lee, D.J. & Tay, J.H. (2007). Distribution of Extracellular Polymeric Substances in Aerobic Granules. *Appl. Microbiol. Biotechnol.* 73, 1463–1469.

Chen, Y., Jiang, W. & Liang, D.T. (2008). Biodegradation and Kinetics of Aerobic Granules Under High Organic Loading Rates in Sequencing Batch Reactor. *App. Microbiol. Biotechnol.* 79, 301–308.

Chisti, Y. (1999). Shear Sensitivity. In: Flickinger, M.C. and Drew, S.W., (Eds). Encyclopedia of Bioprocess Technology: Fermentation, Biocatalysis and Biosep. *New York: Wiley.* 5, 2379-406.

Cinar, O., Yasara, S., Kertmena, M., Demirozb, K., Yigit, N.O. & Kitisb, M. (2008). Effect of Cycle Time on Biodegradation of Azo Dye in Sequencing Batch Reactor. *Proc. Safe. and Environ. Protection.* 86, 455-460.

Correia, V.M., Stephenson, T. & Judd, S.J. (1994). Characterisation of Textile Wastewaters - A Review. *Environ. Technol.* 15(10), 917-929.

Dangcong, P., Bernet, N., Delgenes, J.P. & Moletta, R. (1999). Aerobic Granular Sludge - A Case Report. *Water Res.* 33(3), 890-893.

Directory of Textile Manufacturers and Suppliers - http://www.teonline.com/industry-overview.html

Dos Santos, A.B., Bisschops, I.A.E. & Cervantes, F.J. (2006). Closing Process Water Cycles and Product Recovery in Textile Industry: Perspective for Biological Treatment. In: Cervantes, F.J., van Haandel, A.C. & Pavlostathis, S.G. (Eds.). Advanced Biological Treatment Processes for Industrial Wastewaters, Vol.1. *Inter. Water Assoc., London.* 298-320.

Dos Santos, A.B., Cervantes, F.J. & van Lier, J.B. (2007). Review Paper on Current Technologies for Decolourisation of Textile Wastewaters: Perspectives for Anaerobic Biotechnology. *Biores. Technol.* 98, 2369-2385.

Dos Santos, A.B., Madrid, M.P., Stams, A.J.M., van Lier, J.B. & Cervantes, F.J. (2005). Azo Dye Reduction by Mesophilic and Thermophilic Anaerobic Consortia. *Biotechnol. Prog.* 21, 1140-1145.

EPA. (1997). *Profile of the Textile Industry.* Washington, USA: Environmental Protection Agency.

Etterer, T. & Wilderer, P.A. (2001). Generation and Properties of Aerobic Granular Sludge. *Water Sci. Technol.* 43, 19-26.

Faouzi, M., Canizares, P., Gadri, A., Lobato, J., Nasr, B., Paza, *R.* Rodrigo, M.A. & Saez, C. (2006). Advanced Oxidation Processes for the Treatment of Wastes Polluted with Azoic Dyes. *Electrochimica Acta.* 52, 325-331.

Frijters, C.T.M.J., Vos, R.H., Scheffer, G. & Mulder, R. (2006). Decolorizing and Detoxifying Textile Wastewater, Containing Both Soluble and Insoluble Dyes, in a Full Scale Combined Anaerobic/Aerobic System. *Water Res.* 40,1249-1257.

Ghangrekar, M.M., Asolekar, S.R., Ranganathan, K.R. & Joshi, S.G. (1996). Experience with UASB reactor start-up under different operating conditions. *Water Sci. Technol.* 34 (5–6), 421–428.

Ghangrekar, M.M., Asolekar, S.R. & Joshi, S.G. (2005). Characteristics of Sludge Developed Under Different Loading Conditions During UASB Reactor Start-Up and Granulation. *Water Res.,* 39, 1123-1133.

Grady, C.P.L., Daigger, G.T. & Lim, H.C. (1999). *Biological Wastewater Treatment.* (2nd ed.). New York: Marcel Dekker.

Haroun, M. & Azni Idris, A. (2009). Treatment of Textile Wastewater with an Anaerobic Fluidized Bed Reactor. *Desalination.* 237, 357-366.

Harrelkas, F., Azizi, A., Yaacoubi, A., Benhammou, A. & Pons, M.N. (2009). Treatment of Textile Dye Effluents using Coagulation- Flocculation Coupled with Membrane Processes or Adsorption on Powdered Activated Carbon. *Desalination.* 235, 330-339.

Ibrahim, Z., Amin, M.F.M., Yahya, A., Aris, A., Umor, A., Muda, K. & Sofian, N.S. (2009). Characterisation of Microbial Flocs Formed from Raw Textile Wastewater in Aerobic Biofilm Reactor (ABR), *Water Sci. Technol.,* 60(3): 683-688.

IPPC. (2003). Integrated Pollution Prevention and Control in Textile Industry

Isik, M. & Sponza, D.T. (2006). Biological Treatment of Acid Dyeing Wastewater using a Sequential Anaerobic/Aerobic Reactor System. *Enzyme and Microbial. Technol.* 38, 887–892.

Isik, M. & Sponza, D.T. (2008). Anaerobic/Aerobic Treatment of a Simulated Textile Wastewater. *Sep. & Purif. Technol.* 60, 64-72.

Kapdan, I.K. & Oztekin, R. (2006). The Effect of Hydraulic Residence Time and Initial COD Concentration on Color and COD Removal Performance of the Anaerobic- Aerobic SBR System. *J. Hazard. Mat.* B136, 896-901.

Khelifi, E., Gannoun, H., Touhami, Y., Bouallagui, H. & Hamdi, M. (2008). Aerobic Decolourization of the Indigo Dye-Containing Textile Wastewater using Continuous Combined Bioreactors. *J. Hazard. Mat.* 152, 683-689.

Kim, I.S., Kim, S.M. & Jang, A. (2008). Characterization of Aerobic Granules by MicrobialDensity at Different COD Loading Rates. *Biores. Technol.* 99, 18-25.

Kishida, N., Tsuneda, S., Kim, J.H. & Sudo, R. (2009). Simultaneous Nitrogen and Phosphorus Removal from High-Strength Industrial Wastewater Using Aerobic Granular Sludge. *J. Environ. Eng.* 135, 153-158.

Lettinga, G., Field, J., van Lier, J., Zeeman, G. & Hulshoff Pol, L.W. (1997). AdvanceAnaerobic Wastewater Treatment in the Near Future. *Water Sci. & Technol.* 35(10), 5-12.

Lettinga, G., Hobma, S.W., Klapwijk, A., van Velsen, A.F.M. & de Zeeuw, W.J. (1980). Use of the Upflow Sludge Blanket (USB) Reactor Concept for Biological Wastewater Treatment. *Biotechnol. Bioeng.* 22, 699-734.

Libra, J.A., Maren Borchert, M., Vigelahn, L. & Storm, T. (2004). Two Stage BiologicalTreatment of a Diazo Reactive Textile Dye and the Fate of the Dye Metabolites. *Chemosphere.* 56, 167–180.

Lin, Y.M., Liu, Y. & Tay, J.H. (2003). Development and Characteristics of PhosphorusAccumulating Microbial Granules in Sequencing Batch Reactors. *Appl. Microbiol. Biotechnol.*62, 430–435.

Linlin, H., Jianlong, W., Xianghua, W. & Yi, Q. (2005). The Formation and Characteristics of Aerobic Granules in Sequencing Batch Reactor (SBR) by Seeding Anaerobic Granules. *Proc. Biochem.* 40, 5–11.

Liu, Q.S., Tay, J.H. & Liu, Y. (2003). Substrate Concentration-Independent AerobicGranulation in Sequential Aerobic Sludge Blanket Reactor. *Environ. Technol.* 24, 1235-43.

Liu, Y. & Tay, J.H. (2002). The Essential Role of Hydrodynamic Shear Force in the Formation of Biofilm and Granular Sludge. *Water Res.* 36, 1653-65.

Liu, Y. & Tay, J.H. (2004). State of the Art Biogranulation Technology for WastewaterTreatment. *Biotechnol. Adv.* 22, 533-563.

Liu, Y.Q. & Tay, J.H. (2006). Variable Aeration in Sequencing Batch Reactor with Aerobic Granular Sludge. *J. Biotechnol.* 124, 338–346.

Liu, Y.Q., Liu, Y. & Tay, J.H. (2004). The Effects of Extracellular Polymeric Substances on the Formation and Stability of Bioganules. *Appl. Microbiol. Biotechnol.* 65, 143-148.

McMullan, G., Meehan, C., Conneely, A., Nirby, N., Robinson, T., Nigam, P.I.M., Marchant, R. & Smyth, W.F. (2001). Mini Review: Microbiol Decolourisation and Degradation of Textile Dyes. *Appl. Microbiol. Technol.* 56, 81-87.

McSwain, B.S., Irvine, R.L. & Wilderer, P.A. (2004). The Influence of Settling Time on the Formation of Aerobic Granules. *Water Sci. & Technol.* 50 (10), 195-202.

Melgoza, R.M., Cruz, A. & Buitro, G. (2004). Anaerobic/Aerobic Treatment of ColorantsPresent in Textile Effluents. *Water Sci. & Technol.* 50 (2), 149-155.

MIDA (2007). Annual Report 2007.

Miranda, M.I.A., Damas, S.B., Pi, A.B., Clar, M.I.I., Clar, A.I. & Roca, J.A.M. (2009). Nanofiltration as a Final Step Towards Textile Wastewater Reclamation. *Desalination.* 240, 290-297.

Mohan, N., Balasubramaniam, N. & Ahmed, B. (2007). Electrochemical Oxidation of Textile Wastewater and its Reuse. *J. Hazard Mat.* 147, 644–651.

Mohanty, S., Dafale, N. & Rao, N.N. (2006). Microbial Decolorization of Reactive Black-5 in a Two-Stage Anaerobic–Aerobic Reactor using Acclimatized Activated Textile Sludge. *Biodegradation.* 17, 403–413.

Ong, S.A., Toorisaka, E., Hirata, M. & Hano, T. (2005). Treatment of Azo Dye Orange II in Aerobic and Anaerobic-SBR Systems. *Proc. Biochem.* 40, 2907–2914.

Pan, S., Tay, J.H., He, Y.X. & Tay, S.T.L. (2004). The Effect of Hydraulic Retention Time on the Stability of Aerobically Grown Microbial Granules. *Lett. Appl. Microbiol.* 38, 158-163.

Panswad, T., Iamsamer, K. & Anotai, J. (2001). Decolorisation of Azo-Reactive Dye by Polyphospate and Glycogen-Accumulating Organisms ina an Anaerobic-Aerobic Sequencing Batch Reactor. *Biores. Technol.* 76, 151-159.

Pearce, C.I., Lloyd, J.R. & Guthrie, J.T. (2003). The Removal of Colour from Textile Wastewater using Whole Bacterial Cells: A Review. *Dyes Pigment.* 58, 179-196.

Peng, D., Bernet, N., Delgenes, J.P. & Moletta, R. (1999). Aerobic Granular Sludge – A Case Report. *Water Res.* 33, 890-893.

Pasukphun, N. & Vinitnantharat, S. (2003). Degradation of Organic Substances and Reactive Dye in an Immobilized-Cell Sequencing Batch Reactor Operation on Simulated Textile Wastewater. *J. Env. Sci. Health, Part A: Toxic/Hazard. Substances and Environ. Eng.*, 1532-4117. 38 (10), 2019-2028.

Qin, L., Tay, J.H. & Liu, Y. (2004). Selection Pressure is a Driving Force of Aerobic Granulation in Sequencing Batch Reactors. *Proc. Biochem.* 39, 579-584.

Savin, I.I. & Butnaru, R. (2008). Wastewater Characteristics in Textile Finishing Mills. *Environ. Eng. & Manag. J.* 7(60), 859-864.

Sayed, Z. & Ashtoukhy, E. (2009). Loofa Egyptiaca as a Novel Adsorbent for Removal of Direct Blue Dye from Aqueous Solution. *J. Environ. Manag.* 90, 2755–2761.

Schmidt, E. & Ahring, B.R. (1996). Granular Sludge Formation in Upflow Anaerobic Sludge Blanket (UASB) Reactors. *Biotechnol. Bioeng.* 49, 229–246.

Smith, B., O'Neal, G., Boyter, H. & Pisczek, J. (2007). Decolorizing Textile Dye Wastewater by Anoxic/Aerobic Treatment. *J. Chem. Technol. & Biotechnol.* 82, 16-24.

Solmaz, S.K.A., Ustun, G.E., Birgul, A. & Tasdemir, Y. (2007). Treatability Studies with Chemical Precipitation and Ion Exchange for an Organized Industrial District (OID) Effluent in Bursa, Turkey. *Desalination.* 217, 301-312.

Stolz A. (2001). Basic and Applied Aspects in the Microbial Degradation of Azo Dyes. *Appl. Microbiol. & Biotechnol.* 56. 69–80.

Talarposhti, A.M., Donnelly, T. & Anderson, G.K. (2001). Colour Removal from a Simulated Dye Wastewater using a Two-Phase Anaerobic Packed Bed Reactor. *Water Res.* 35, 425-432.

Tay, J.H., Liu, Q.S. & Liu, Y. (2001). Microscopic Observation of Aerobic Granulation in Sequential Aerobic Sludge Reactor. *J. Appl. Microbiol.* 91, 168-175.

Tay, S.T.L., Ivanov, V., Yi, S., Zhuang, W.Q. & Tay, J.H. (2002). Presence of Anaerobic Bacteroides in Aerobically Grown Microbial Granules. *Microb. Ecol.* 44, 278-285.

Toh, S.K., Tay, J.H., Moy, B.Y.P., Ivanov,V & Tay, S.T.L. (2003). Size-Effect on the Physical Characteristics of the Aerobic Granule in a SBR. *Appl. Microbiol. Biotechnol*. 60, 6876-95.

Trovaslet, M., Enaud, E., Guiavarch, Y., Corbisier, A.M. & Vanhulle, S. (2007). Potential of a *Pycnoporus sanguineus* laccase in Bioremediation of Wastewater and Kinetic Activation in the Presence of an Anthraquinonic Acid Dye. *Enzyme and Microbial Technol*. 41, 368-376.

van der Zee, F.P. & Villaverde, S. (2005). Combined Anaerobic-Aerobic Treatment of Azo Dyes: A Short Review of Bioreactor Studies. *Water Res*. 39, 1425-1440.

Voet, D. & Voet, J.G. (2004). *Biochemistry*. 1, Hoboken, New Jersey: John Wiley & Sons Inc.

Wang, F., Yang, F., Zhang, X.W., Liu, Y., Zhang, H. & Zhou, J. (2005). Effects of Cycle Time on Properties of Aerobic Granules in Sequencing Batch Airlift Reactors. *World J. Microbiol. Biotechnol*. 21, 1379-84.

Wang, Q., Dua, G. & Chena, J. (2004). Aerobic Granular Sludge Cultivated under the Selective Pressure as a Driving Force. *Proc. Biochem*. 39, 557–563.

Wang, S.G., Liu, X.W., Zhang, H.Y. & Gong, W.X. (2007). Aerobic Granulation for 2,4 Dichlorophenol Biodegradation in a Sequencing Batch Reactor. *Chemosphere*. 69, 769-75.

Weber, S.D., Ludwig, W., Schleifer, K.H., & Fried, J. (2007). Microbial Composition and Structure of Aerobic Granular Sewage Biofilms. *Appl. & Environ. Microbiol*. 6233–6240.

Wijffels, R. H. and Tramper, J. (1995). Nitrification by Immobilized Cells. *Enzyme & Microbial. Technol*. 17, 482–492.

Willmott, N.J., Guthrie, J.T. & Nelson, G. (1998). The Biotechnology Approach to colour Removal from Textile Effluent. *J. Soc. of Dyers & Colour*. 114, 38-41.

Wirtz, R.A. & Dague, R.R. (1996). Enhancement of Granulation and Start-Up in Anaerobic Sequencing Batch Reactor. *Water Environ. Res*. 68, 883-892.

Wu, J.S., Liu, C.H. Chu, K.H. & Suen, S.Y. (2008). Removal of Cationic Dye Methyl Violet 2B from Water by Cation Exchange Membranes. *J. Membr. Sci*. 309, 239–245.

Yang, S.F., Liu, Q.S., Tay, J.H. & Liu, Y. (2004). Growth Kinetics of Aerobic Granules Developed in Sequencing Batch Reactors. *Lett. Appl. Microbiol*. 38, 106–112.

Yi, L., Feng, W., Siqing, X. & Jianfu, Z. (2008). Study of 4-*t*-octylphenol Degradation and Microbial Community in Granular Sludge. *J. Environ. Sci*. 20, 167–171.

Zhang, Z.P., Show, K.Y., Tay, J.H., Liang, D.T., Lee, D.J. & Jiang, W.J. (2007). Rapid Formation of Hydrogen-Producing Granules in an Anaerobic Continuous Stirred Tank Reactor Induced by Acid Incubation. *Biotechnol. & Bioeng*., 96 (6), 1040-1050.

Zheng, Y.M., Yu, H.Q. & Sheng, G.P. (2005). Physical and Chemical Characteristics of Granular Activated Sludge from a Sequencing Batch Airlift Reactor. *Proc. Biochem*. 40, 645-650.

Zhu, J. & Wilderer, P.A. (2003). Effect of Extended Idle Conditions on Structure and Activity of Granular Activated Sludge. *Water Res*. 37, 2013-2018.

Zollinger, H. (1987). *Color Chemistry-Syntheses, Properties and Applications of Organic Dyes and Pigments*. New York: VCH.

Chapter 10

Second Generation Ethanol from Residual Biomass: Research and Perspectives in Ecuador

Enrique Javier Carvajal Barriga, Cristina Guamán-Burneo, Patricia Portero Barahona, Edgar Salas, Carolina Tufiño, Bernardo Bastidas

Additional information is available at the end of the chapter

http://dx.doi.org/10.5772/51951

1. Introduction

Ecuador is located between 1°N and 5°S on the west coast of South America. Although relatively small in size, mainland Ecuador can be subdivided nevertheless into three different and quite distinctive climatic regions: the Pacific coastal plain, the Andean highlands and the Amazon basin. In addition, Ecuador possesses a fourth region, namely the Galapagos Islands.

Climatically, the Pacific coastal plain is hot all year, with a rainy season between December and May. In the Andean highlands, the climate is markedly cooler, varying according to altitude. In contrast, the Amazon basin is hot, humid and wet all year round, while the Galapagos Islands are dry, with an annual average temperature of 25° C (77° F).

These characteristics provide Ecuador with a huge potential to develop second generation ethanol from industrial biomass, to replace a portion of the gasoline needed and, thus, the reduction of CO_2 emissions. The climatic conditions as well as the photoperiods and rainfall along the year make this country an excellent candidate to develop second generation biofuels technology from biomass.

Tropical cultures such as bananas, oil palm, sugar cane, and others that are produced mainly in the coastal region of the country generates each year enough cellulose to produce almost all the ethanol the country needs. The current situation in terms of the use of these lignocellulosic materials is still in its very beginning and much work is to be developed to establish a market for the lignocellulosic residues.

Additionally, microbial biodiversity and its research is becoming one important issue in terms of the development of innovative technologies based on biotechnology, pointing out the search for novel genes and metabolic abilities especially in wild yeasts studied in natural environments all around the country. Local researchers are devoted to the metabolic engineering of yeasts to improve the fermentation yields.

In this chapter we report some results from the or Sustainable Resources for Ethanol (RESETA) project, from the quantification and characterization of the most important cultures in Ecuador, its residues and characteristics, to the development of genetic engineered yeasts and the design and construction of a biorefinery at pilot scale.

The above mentioned project involves one of the most important researches the Ecuadorian Government has founded since 2008. The advances and results of this project can be taken as models for other tropical countries in the world.

Finally, we present the economic viability analysis of second generation ethanol projects in large scale in Ecuador, looking forward the industrial production of ethanol, biogas, biofertilizers and renewable chemicals in biorefineries in Ecuador.

2. Biomass: A qualitative and quantitative approach to the concept

The concept of biomass is largely extended among the bioenergy, agricultural, biotechnology and other specialists. When a farmer talks about biomass, he or she is referring to the foliage, fruits, grains, stems or waste materials produced in crops.

In the animal husbandry activities, biomass is referred to the manure and purines excreted by farm animals. On the other hand, talking about cell's biotechnology, biomass is referred to the cell production in a culture: biomass of yeast produced during fermentative processes.

In other terms, biomass constitutes a broad range of biological matter including the vegetal coverage of the planet, the micro and macro organisms living on the planet, including humankind.

Displaying in this manner the concept of biomass, allows the broadening of its quantitative and qualitative applications and uses. This way of thinking on biomass plays with the absolute and relative values of the concept and may certainly boost the current scheme of exploitation of biomass sources.

3. The cellulosic biomass potential worldwide

Cellulose is the most abundant carbohydrate polymer in the planet. This remarkable molecule is composed by monomers of glucose linked up by glycosidic (β 1—4) boundaries that provides this molecule with the capacity of forming linear, fibrous shapes of straight chains. These simple chains are joined by hydrogen bonds, provoking the formation of resistant and strong fibers which are the main components of plant cell walls.

Cellulose, itself is not only a strong, flexible and resistant fiber, but also insoluble and hard to be decomposed molecule. Nevertheless, fungal and bacterial micro flora which is abundant in nature has developed certain enzymes groups that allow them to decompose the glycosidic links and use the released glucose contained in cellulose. Thus, cellulose is recycled very efficiently in nature, sustaining the carbon equilibrium in biosphere, which relies fundamentally in anabolic and catabolic cycles of cellulose, where microbes play a main role.

Cellulose is formed during photosynthesis, where CO_2 and the energy from the sun are taken by plant cells, where this transformation takes place. Cellulose naturally is a structural molecule synthesized by plants to allow them to grow. Glucose is also a fuel molecule in as much as the bonds are in the α configuration, thus forming fuel reserve molecules such as amylose and amylopectin, both molecules components of starch.

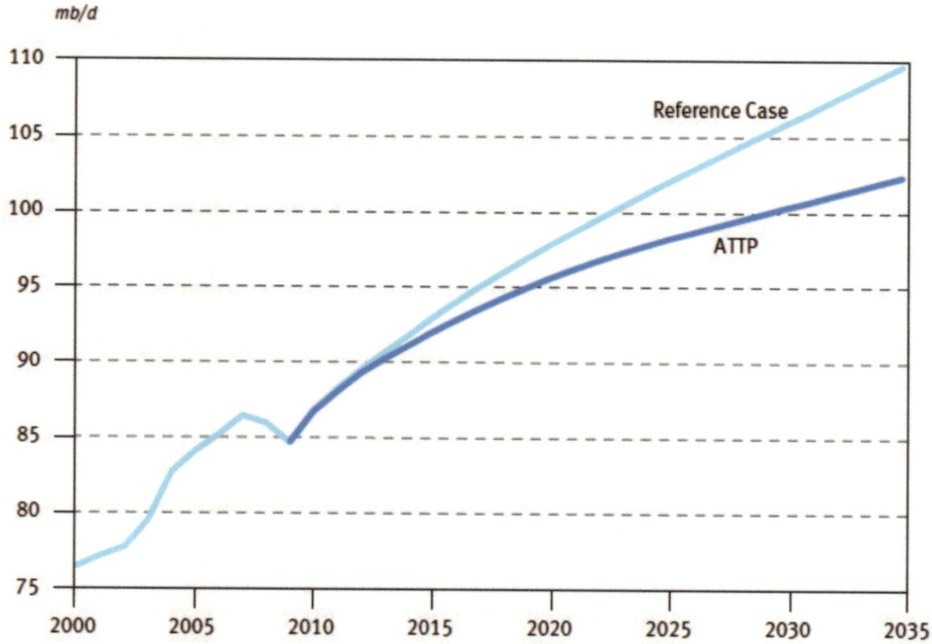

Figure 1. The projection for the crude oil needs of the OPEC (Organization of Petroleum Exporters Countries) in the ATTP scenario (Accelerated Transportation Technology and Policy) for 2035. [2]

Cellulose recycling in nature is in the order of 10^{15} kg per year [1]. This number is so high that we could make enough fuel ethanol to supply 100 times the energy requirements of the entire world in a rampant development scenario projection for 2035 [2]. In other words, we may probably need "only" 1 per cent of the cellulose synthesized by plants in one year, so we can have enough liquid fuel to run our vehicles and industries. In this calculation, it is not taken into account the biogas (to run power plants) and biofertilizers (to return minerals to soil) production if all the ethanol from cellulose were produced within biorefineries. If humankind had the technology, cultural and economic conditions to efficiently pick this cellulose up, we may probably reach the sustainable and clean environmental goals for future. Nevertheless, this utopia is not yet possible in current historical conditions, so we need to focus on much less ambitious objectives, such as the recovery and technological transformation of agricultural feedstocks and industrial leftovers. Figure 1 shows the crude oil we'll need in 2035. The amount is very impressive, that is 110 million barrels crude per day to keep running our industrial and mobilization needs.

Before the worldwide petroleum reserves are depleted, humankind must change the energetic matrix based on oil and look for sustainable and clean sources to produce fuels. An analysis made by Bruce Dale from the Department of Chemical Engineering and Materials Science at Michigan State University (USA), show a clear disadvantage in terms of energy input of gasoline production compared to first and second generation ethanol (Table 1). In this scenario, the issues are not necessarily lying on the economic or technical feasibility of the conversion processes, but in political and ethical issues.

	petroleum	natural gas	coal	other	total	GHG emissions
gasoline	1.10	0.03	0.05	0.01	1.19	94
ethanol today	0.04	0.28	0.41	0.04	0.77	77
cellulosic ethanol	0.08	0.02	0.02 *	0.02	0.10	11

Table 1. Energy Inputs of various energy carriers in MJ per MJ of fuel produced and Greenhouse Gas (GHG). Outputs in kg of carbon dioxide equivalents per MJ fuel produced for various fuels. * Credit for coal not consumed due to process residues being burned to provide heat [3]

The energy input to produce gasoline is 0.19 times higher than the energy harvested; the GHG emission are the highest (94 kg/MJ). The cellulosic ethanol case exhibits a much higher energy output than the energy required to produce it. Moreover, GHG emissions are much lower than gasoline case (11 kg/MJ). In the case of corn ethanol the energy balance is still positive even though the margin remains narrow.

3.1. Is there enough biomass to replace oil?

There is a tendency to believe that there is no enough biomass in the world to replace fossil fuels that nowadays fulfill our energetic necessities. Taking into account that in the planet every year about 10^{15} kg of cellulose are naturally recycled, we can do a simple calculation to better understand the real potential of biomass in a hypothetical scenario where we can transform this cellulose material into second generation ethanol.

That huge amount of cellulose is understood that is produced in forests, seas, rivers and crops all around the planet. In this calculation it is not considered the economic feasibility of getting that cellulose; nevertheless it can illuminate a possible future scenario.

In Table 2 it is shown a calculation of the crude necessities for 2035

Talking in terms of volume may allow us to understand the ethanol potential production from cellulose, but it does not explain the energetic issues concerning to both energy carriers when being compared. Ethanol is a fuel that possesses less calorific energy per kilogram than crude oil. If we want to calculate the amount of cellulosic fuel ethanol needed to replace oil it is necessary to take a look at the equivalences.

CRUDE OIL			CELLULOSIC ETHANOL			
Crude necessities mb/d (data OPEC)	L crude/day 2035 (1 bbl. Crude =158,98 liters)	Liters crude oil/year (2035)	Kg cellulose recycled/ year	Enzymatic cellulose hydrolysis (Yield~70%)	Conversion to ethanol (0,45kg ethanol/kg glucose)	Liters ethanol/year (ethanol density 0,789Kg/l)
110	$1{,}75\times10^{10}$	**$6{,}38\times10^{12}$**	1×10^{15}	7×10^{14}	$3{,}15\times10^{14}$	**$3{,}99\times10^{14}$**

Table 2. The volume of crude oil needs projected to 2035 vs. the potential cellulosic ethanol from cellulose annual production on Earth. Bold numbers show that the cellulosic ethanol potential volume is 100 times more than the oil necessities.

Crude Oil calorific value is about 40,000 MJ/Kg., while fuel ethanol's calorific value is about 28,800 MJ/Kg. Then, we can consider that one Kg. of ethanol is equivalent to 0.72 Kg of crude oil. In other words, crude oil's calorific power is about 1.4 times higher than fuel ethanol.

In Table 2, it is important to notice that the calculation intends to show that in nature there is enough cellulose to fulfill the necessities of energy to replace oil (at least in theory), but not the energy contained in both fuels; the cellulosic ethanol production from such a huge amount of cellulose may overwhelm in two orders of magnitude the figure of the crude oil requirements by 2035.

As a resulting corollary of this analysis, it is possible to infer that the need of cellulose produced naturally on Earth in order to replace crude oil may be only 1% of its total weight. However, it still represents an immense amount of cellulose to be collected and technologically processed in an efficient and financially feasible way (i.e. ~10 billion MT of pure cellulose/year).

In this scenario, the easiest and more efficient way to start producing second generation ethanol is by utilizing the residual lignocellulosic feedstocks such as by-products from agriculture or industrial activities.

In this analysis, we have taken into account only cellulose—a glucose polymer—to be converted into fuel ethanol by alcoholic fermentation. Nonetheless, most of cellulose in nature is associated with hemicellulose and lignin. Hemicellulose is mainly composed by pentoses such as xylose and arabinose, while lignin is composed mainly by aromatic compounds.

4. Availability of residual biomass in Ecuador

Ecuador is a biodiverse country with rich and fertile natural regions. In the coastal zone of Ecuador there is the large scale agriculture of a wide variety of crops which have positioned this country as one of the most important producers of bananas, palmito (palm heart), oil palm and other valuable products in South America. Moreover, Ecuador has unique vegetal species that are being exploited in small scale, presenting novel and potential sources of lignocellulose for the future.

In terms of abundance of lignocellulosic residues, the most conspicuous industries producing leftovers—as a consequence of the harvest or the extraction of valuable commodities—are the bananas farms, and the oil palm and sugar cane mills. There are still other important industries located mainly in the highlands such as flowers and cereals that produce lignocellulosic material potentially usable. Nevertheless, the amounts of these residues are not enough for huge biorefining installations, nor even available in an economical and technical way.

As for the availability of residues, studies carried out by researchers from the Neotropical Center for the Biomass Research at the Pontificia Universidad Católica del Ecuador, reveal that there is a very high potential for lignocellulosic ethanol and biorefineries setting up in Ecuador. Nevertheless, there still exist constraints due to the disperse areas where the agricultural and industrial lignocellulosic materials are disposed; the local roads infrastructure and networks; the lack of development of markets for certain specific residues; the traditional uses and ways of final disposal; the physical and chemical composition of residues; and, the prices per dry ton. There are also social and environmental components to be taken into account when projecting lignocellulosic biorefineries to take the most of the agricultural and industrial byproducts, leftovers or residual material. In our survey we have considered the above-mentioned factors to develop the feasibility study for a biorefinery based on local lignocellulosic residues in the country.

In this survey we have pursued the following general objectives:

1. Evaluate the abundance and the potential of the main crops produced in Ecuador.
2. Determine the utilization, destiny, and availability of the agricultural residues.
3. Estimate the evolution of the agricultural production and residues generation in 5 years (until 2014).

Moreover, we have focused the following specific goals:

1. Determine the main crops in Ecuador, its exact geographical location, and the quantity of biomass residues produced per year.
2. Establish the temporality of crops and harvest.
3. Take current and historical data on volume of waste biomass produced to project future volumes, considering a period of five years. Analyze the succession of crops
4. Determine on the basis of the previous information, the more adequate zones where to install a future biorefinery plant.

In Ecuador there are three crops that worth to be studied with biorefining ends, because of their characteristics in terms of composition, final disposal, abundance and lack of sustainable use. These crops are: bananas, sugar cane and oil palm.

Table 3 shows the complete results of our survey on 13 different crops in Ecuador, the calculation of its dry mass and cellulose average contents as well as the potential for ethanol production. As it is going to be seen, Ecuador potentially could provide at least half of the ethanol needs for replacing gasoline in vehicles if the cellulose contained in agricultural residues were transformed into ethanol.

This suggests that biorefinery plants can be a reasonable and sustainable option for the post oil economy in Ecuador. Moreover, there still exists a huge potential for power generation if biogas from stills and residual lignin are burned in biorefineries.

There still exist other valuable products from biorefining of second generation ethanol such that can be produced from the residual water generated after distillation which, after an anaerobic digestion process, yields biogas, liquid and solid fertilizers (sludge). The non-hydrolyzed fibers as well as yeast biomass obtained after fermentation can be dried and sold as animal feed solid matter. The solid matter that can be recovered from fermenters before distillation is really considerable. Moreover, carbon dioxide from fermentation can be collected and treated to be sold in as much as during the fermentation for ethanol production, almost the same amount of CO2 is released. Theoretically, the production ratio of ethanol to CO2 in fermentation is 92:88. The uses for this gas are very wide including food, drink and chemical industries. CO2 is widely used in soft drinks and beer to carbonation of these beverages. It is also used to fill packs of vegetables and meet to keep it fresh. CO2 can also be used as raw material for the synthesis of methanol, formic acid, and urea. Other applications of CO2 include its use as a medium in supercritical CO2 extraction and in fire extinguishing equipment [4].

POTENTIAL OF SECOND GENERATION ETHANOL PRODUCTION FROM AGRICULTURAL RESIDUES IN ECUADOR					
Residues by Product	**Dry weight (MT/year)**	**Average cellulose content (MT/year)**	**Theoretical potential ethanol (Gal)**	**Potentially supplied vehicles/year**	**Percent of potentially supplied vehicles per year** (Total number of cars: 1.4 MM to 2014)
Soy bean	19,873	7,949	1'510,192	3,020	0,2
Palmito	24,285	9,714	1'845,509	3,691	0,3
Flowers	29,259	11,704	2'223,489	4,447	0,3
Potatoes	66,790	26,716	5'075,609	10,151	1
Rice	90,742	36,297	6'895,808	13,792	1
Plantain	138,787	55,515	10'546,915	21,094	2
Soft corn	288,340	115,336	21'911,914	43,824	3
Sugar cane	327,422	130,969	24'881,855	49,764	4
Cocoa	343,249	137,300	26'084,624	52,169	4
Bananas	351,031	140,412	26'675,973	53,352	4
Dry corn	447,365	178,946	33'996,714	67,993	5
Coffee	568,736	227,494	43'220,137	86,440	6
Oil palm	2'071,995	828,798	157'457,762	314,916	22
TOTALS	**4'767,873**	**1'907,149**	**362'326,502**	**724,653**	**51,8**

Table 3. The hypothetical potential of lignocellulosic biomass in Ecuador to produce cellulosic ethanol

5. Second generation ethanol from residual feed stocks generated in industrial activities

5.1. Xylose fermenting yeast in natural environments of Ecuador

Lignocellulosic biomass is composed by mainly three different fractions of molecules: cellulose, hemicellulose and lignin. We have already talked about some applications for cellulose and lignin, mainly as ethanol and fuel biomass. Nevertheless, hemicellulose which is mainly composed of xylose, a five carbon sugar, is a very important and abundant source that accounts for 23% to 32% of the dry lignocellulosic biomass weight. This sugar can also be used for the production of ethanol as well as other valuable products.

As part of a survey in biodiversity in Ecuador, the CLQCA or Catholic University Yeast Collection has collected some isolates of yeast that exhibit fermentation skills when xylose and a Nitrogen base are mixed up in culture broths.

Xylose fermenting yeasts have been collected from different provinces of Ecuador, including the Galapagos Islands and the Amazonia. Nevertheless, none of these yeast isolates present high ethanol tolerance nor quick fermentation rates, which make these organisms not suitable for industrial processes. It can probably discourage someone to study natural occurring xylose fermenting yeasts, nonetheless, the genes involved in this physiological processes are still useful for metabolic engineering approaches.

ISOLATE CODE	YEST SPECIES	SUSTRATE	XYLOSE ASSIMILATION	XYLOSE FERMENTATION
CLQCA-24SC-002	*Yamadazyma mexicana*	*Inga Vera* (MUCILAGE)	S	W
CLQCA-24SC-016	*Yamadazyma mexicana*	*Bursera graveolens* (EXUDE)	W	W
CLQCA-24SC-312	*Galactomyces geotrichum*	*Scalesia* sp. (ROTTEN WOOD)	+	W
CLQCA-24SC-320	*Scheffersomyces stipitis*	*Scalesia* sp. (ROTTEN WOOD)	+	+
CLQCA-24SC-321	*Scheffersomyces stipitis*	*Scalesia* sp. (ROTTEN WOOD)	+	+

Table 4. Xylose fermenting yeast isolates collected in Ecuador (Galapagos Islands) and deposited at the Catholic University Yeast Collection (CLQCA). S: slow positive, W: weakly positive, +: positive.

As seen on table 4, only two strains of *Schffersomyces (Pichia) stipitis* are positive to ferment xylose. This yeast species has been reported to ferment xylose as will be seen further in this chapter.

In terms of ethanol production, there have been a lot of different approaches; in the last times the metabolic engineering of *Saccharomyces cerevisiae* was regarded as a suitable solution to ferment xylose, arabinose and other non-conventional sugars.

One example of this line of research is to be shown in the following scheme, where there are three different genes in charge of the transport, isomerization and phosphorylation in the process to ferment xylose. This construct has been designed at the Neotropical Center for the Biomass Research as part of the RESETA project. This genetic tool is very versatile and has been thought to be used as a genetic platform where to assay a wide variety of genes. The design of this construct was made by Carvajal et al. in 2011.

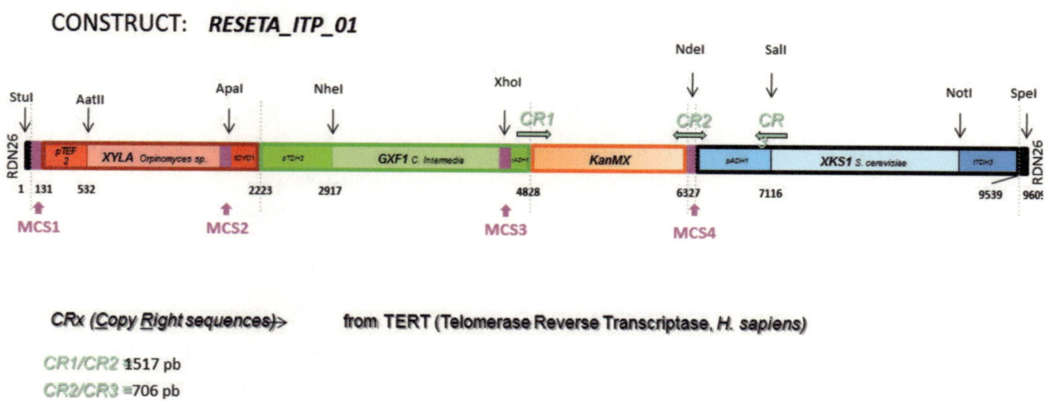

Figure 2. RESETA ITP 01 Construct. Composed by a *XYLA* gene from *Orpinomyces* sp.; *GFXC*, from *Candida intermedia*; and, *XKS1*, from *Saccharomyces cerevisiae*. As a resistant marker it has been used the *KanMX4* gene which provides resistance against Geneticin.

This genetic construction is being tested in laboratory conditions, integrated in industrial *Saccharomyces cerevisiae*. This first step forward to the metabolic engineering is expected to give new perspectives to the residual biomass transformation into valuable products in the context of biorefineries in Ecuador and worldwide.

5.2. Xylose to Xylitol fermentation

Xylose to ethanol conversion in nature is been done by a few organisms such as yeasts like *Scheffersomyces stipitis*, an ascomycetous yeast that has been extensively investigated for the fermentation of xylose to ethanol, L-lactic acid for its further polymerization into PLA (Poliy Lactic Acid) and other products from hemicellulose, the second abundant component of cellulosic biomass [5, 6, 7]. Laboratory strains of *Schef. stipitis*, which are amenable to genetic and physiological manipulation, have been developed by metabolic engineering for xylose utilization [5, 7, 8]. The genome sequence was recently obtained for this yeast and will provide a valuable resource for enhancement of xylose utilization and other industrial attributes by *Schef. stipites* [8]. Genes from Schef. stipitis have also been introduced into *Saccharomyces cerevisiae* to enable fermentation of pentoses [5, 7, 8].

5.3. Agro-industrial residues for xylitol production

Xylitol is a five-carbon sugar alcohol of high value added as a sweetener for high power, anticariogenic properties and insulin metabolism independent that guarantee its application in food and pharmaceutical industries. The power as a sweetener is similar to sucrose, and higher than ordinary polyols in addition to reduce caloric value, can be tolerated by diabetics. It is an anticariogenic and cariostatic compound that is not metabolized by microorganisms of the oral microbiota; thus, it is used in the manufactures of sweets. Can be used clinically for the prevention of otitis media because it inhibits the growth and adhesion of *Pneumococcus* spp and *Haemophilus influenzae* in nasopharynx cells, and it has skin smoothing properties [9, 10, 11, 12, 13, 14]. Owing to all these characteristics, xylitol is a feedstock of particular interest to the food, odontological and pharmaceutical industries.

Xylose is a sugar widely distributed in nature. Plants and fruits contain relatively low concentration, and extraction from natural sources is usually not profitable [10]. Nowadays, xylitol is derived industrially via a chemical process from hydrolyzates of lignocellulosic wastes by either chemical reduction or microbial fermentation [9]. However, due to a requirement for several chemical purification steps, such a process is very expensive. Therefore, this conversion could be alternatively performed by bacteria, filamentous fungi, yeasts or purified enzymes from these microorganisms which are capable of reducing xylose to xylitol as a first step in D-xylose metabolism [15]. Nevertheless, to make this process exploitable and economical at an industrial level, the bioconversion must be rapid, offer high yield, employ an alternative and cheap culture media and allow for results comparable to those of the present technology.

Lignocelluloses are the most abundant organic mass in the biosphere, which accounts for approximately 50% of the biomass. In nature the annual production of biomass is estimated to 10 to 50 × 10^9 tons [16] Their major components, cellulose, hemicellulose and lignin, vary with plant species. The pentose fraction, composed of D-xylose (usually not less than 95%) and L-arabinose is much higher in hardwoods (19 to 33%) than in softwoods (10 to 12%)[17]

Hemicellulose is a branched polymer, which is composed of both linear and branched heteropolymers of D-xylose, L-arabinose, D-mannose, D-glucose, D-galactose and D-glucuronic acid with a high content of xylans, that consist essentially in β-1,4 links with branching variables; due to it heterogeneous structure and low degree of polymerization, it is easily hydrolyzed to xylose [15, 16]. Xylan accounts for 11-35% (dry weight basis) of lignocellulosic materials such as hardwoods and agricultural residues, such as sugarcane bagasse [14, 18], rice straw [19], and soy hulls [20, 21] which are xylan-rich substrates and have been satisfactorily used as alternative media for xylitol production through different treatments [22] and cultivation conditions [23], aimed at increasing process yield and productivity.

D-xilose, also can be converted into a range of substances of industrial interest such as fuels and solvents (ethanol, butanol, 2,3-butanediol, acetone and 2-propanol), alditols (xylitol and glycerol) and organic acids (latic, acetic and butyric acid). It can also be used as a substrate for production of glucose isomerase [14].

For this process of xylitol production, pure xylose is necessary. The process starts with the production of xylose from xylan after acid-catalysed hydrolysis from hard-wood; however, the chemical process requires several purification steps, because only pure xylose can be used for chemical reduction. Therefore, overall xylitol yield is relatively low (50 – 60 %) from the total xylan content of the wood hemicelluloses [24, 11, 15].

Furthermore, the choice of cultivation and/or conversion system is another crucial point for the success of this bioprocess. Different bench-scale cultivation systems were investigated, utilizing batch, fed-batch and continuous processes [15]. Another important factors which affect the xylitol production is the quantity of inocula, substrate, media, temperature, pH and aeration [17, 25]

On the other hand, the biotechnological procedures are based on the utilization of microorganisms and/or enzymes. These procedures are interesting because they do not require a pure xylose solution as is the case when xylitol is produced by the chemical pathway. The bioconversion process would hold more promises of both hexoses and pentose sugars from lignocellulosic materials. The promising yeast species include the generous *Candida, Pichia, Debaryomyces* and *Pachysolen* [9, 26, 14, 16] by NADPH-dependent xylose reductase, enzyme which can ferment hemicelluloses hydrolysate from woody plant materials (Figure 3).

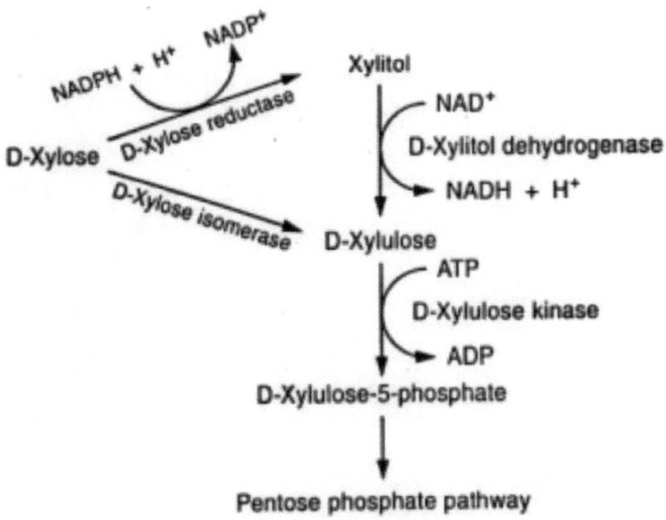

Figure 3. Pathway for microbial xylose utilization [9]

The first step in the metabolism of D-xylose is the transport of the sugar across the cell membrane. Once inside the yeast cell, D-xylose is reduced to xylitol by either NADH- or NADPH-dependent xylose reductase. Xylitol is either secreted from the cell or oxidized to xylulose by NAD- or NADP-dependent xylitol dehydrogenase. The first two reactions are considered to be limiting in D-xylose fermentation. The phosphorylation of xylulose to xylulose 5-phosphate is catalyzed by xylulokinase, which is a prerequisite for its utilization by the central catabolic pathways [23, 17, 16].

In most studies on xylitol production by fermentative processes, xylose of analytical grade is commonly the major substrate. The main problem in the fermentation of these hydrolysates is the presence of toxic compounds released from the lignocellulosic structure during the hydrolytic process, as well as those originated from the sugar degradation, which inhibit the microbial growth and the fermentative activity of the yeasts. In this way, several approaches have been assayed to minimize this effect. According to Silva *et al*, 1998 the maximum xylitol production (54 g/L) occurs when the hydrolysate is first treated with CaO until reaching pH 8.4 and then treated with H_3PO_4 until the pH decreases to 6.0. Thus, pH is an important factor to take into account for the xylitol fermentation. Its effect is related to the acetic acid concentration in the hydrolysate, which concentration, if it is higher than 3.0 g/L, can inhibit the yeasts capability to convert xylose into xylitol [27]. Nonionized acetic acid, which is found in the medium at pH < 7.0, has been found to be the main inhibitor compound in yeast metabolism [28]

The hemmicellulosic hydrolysates from agroforest residues can be efficiently utilized in fermentative processes for xylitol production after an initial treatment designed to remove or reduce the compounds known to be toxic to cell metabolism. This technology is still in its research and development stage, but the results attained points that it may be feasible to

6. The biorefinery concept from the perspective of RESETA project

The RESETA project has focused a basic model of biorefinery for producing up to 45,000 liters ethanol/year; 450.000 liters fertilizer. This concept includes the production of ethanol, biogas, biodiesel, fertilizers and animal food (Figure 4). The whole plant is installed in an area of 1,200 m^2.

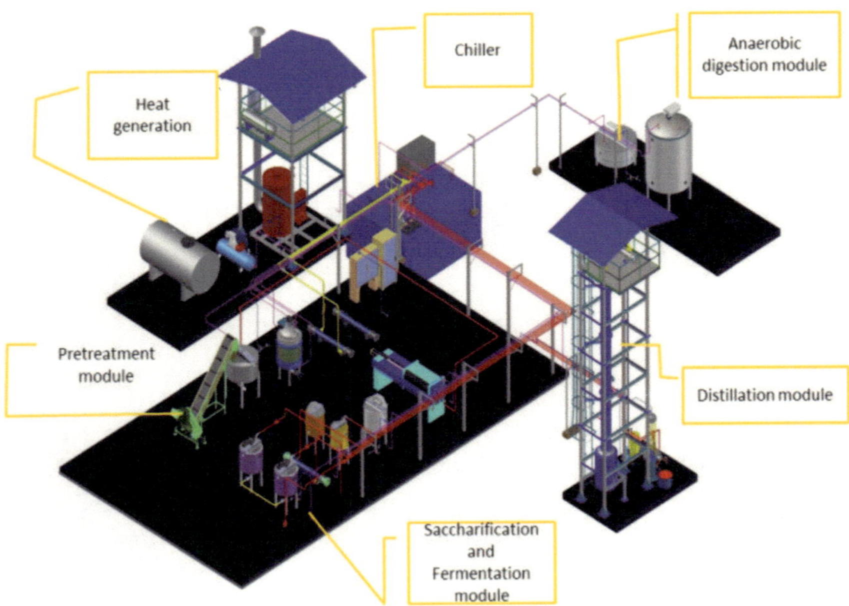

Figure 4. Isometric plan of the biorefinery built for the RESETA project.

This biorefinery is also able to produce first and second generation ethanol from sugar, starch or cellulose containing feedstocks. From that point of view, the biorefinery concept can be applied not only to large scale, but also in mid and small scale production. In Ecuador, there are communities that produce a wide variety of residues that can be utilized. The biorefinery concept is applicable to promote a sustainable economy in vulnerable and underdeveloped zones.

An important achievement of this R&D project was the development of local technology in order to reduce the dependence of foreign technicians. The RESETA project biorefinery is completely automated which permits the operators scoring historical data of the trials performed during experimentation. Researchers and engineers are working together in the optimization of processes and designs. The resulting ideas and philosophy are being of great value looking forward the future technological independence in these strategic issues for Ecuador.

7. Economical and financing analysis on a hypothetical biorefinery in Ecuador

Ecuadorian National Planning and Development Department, has determined a new energetic matrix where it is established the use of 5% ethanol in regular gasoline to be used in vehicles until 2020 in the whole country [29].

To accomplish this endeavor, in 2010 Ecuador's fuel ethanol needs were 32'187,903 gallons; nonetheless, the ethanol production reached was just 1'195,427 gallons [30]. This is, only 3.71% of the national requirements were fulfilled, which demonstrates that there exists a huge unsatisfied demand of fuel ethanol, representing an interesting market for ethanol producers as seen on Figure 5.

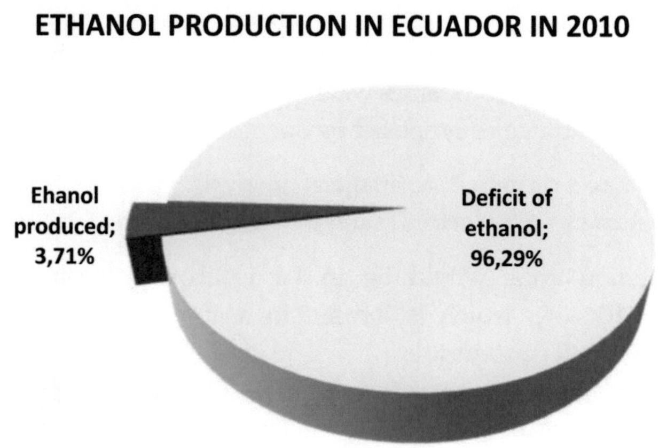

Figure 5. In Ecuador, only 3.71% of the demand of ethanol was satisfied in 2010 [30]

The price of ethanol per liter has been fixed in 0.76 Dollars by the Ecuadorian Government. This is a very competitive price that the sugar mills are interested in producing this renewable fuel using molasses as feedstock; molasses are byproducts rich in sugars that represent a valuable source of energy in animal husbandry.

The RESETA Project (the acronym of Sustainable Sources for Ethanol) studies the second generation ethanol production from lignocellulosic materials as well as leftovers and industrial effluents. One of the objectives of such project is the development of a financial model based on a baseline survey on lignocellulosic materials in Ecuador [31]. By means of the data obtained in such survey, it was possible to determine that there are three candidate raw materials taken into account its amount, localization and availability. The sources are banana stems, sugar cane bagasse and empty fruit bunches form oil palm.

Laboratory analysis demonstrated that the high level of moisture in banana stems, as well as the final disposition of the biomass, which is very dispersed in the field, makes its use an unfeasible task in terms of economy. On the other hand, sugar cane bagasse is already been used by sugar cane mills to generate power, thus, just a few biomass of this kind is really available. Finally, empty fruit bunch from oil palms present good characteristics for being used as feedstock in an industrial scale biorefinery. Empty fruit bunches account for about 45% cellulose dry weigh; which is enough to provide the raw material to run about 22% gasoline cars in Ecuador (Table 1).

One of the key factors influencing the feasibility studies to install a biorefinery is the price of the raw materials. Currently, the lignocellulosic residues represent a big environmental problem in Ecuador as well as an unsolved trouble to farmers and industries. One of the most common fates for lignocellulosics is the composting or decomposition on soil to provide organic matter for cultures and as mulch in a number of crops. The uncontrolled decomposition generates lixiviates, as well as greenhouse gases that finally are disposed in the subsoil and the atmosphere [31]. It is necessary to provoke a change in the current situation that makes a real shift in terms of the use of the lignocellulosics. One strategy proposed by the RESETA project is pushing to the creation of a market for the lignocellulosics in sustainable and fair conditions for the owners of the biomass as well as for the biorefinery project to be successful. The establishment of companies where the owners of the biomass can participate as shareholders is one of the strategies proposed by our study.

In these terms, we have designed a financial analysis based on the data obtained by laboratory and field surveys. The factors taken into account were:

a. The implementation zone would be in La Concordia town (Latitude -0.000017°; Longitude - 79,383563°), which is located in a strategic point for its big oil palm production and oil mill concentration.
b. The raw material in terms of empty fruit bunch is about 640 MT per day.
c. Enzymes, chemicals, yeasts, are calculated in 0.62 Dollars/Gal of ethanol.
d. Other costs are about 0.45 Dollars/Gal, as it is detailed in Table 5. Additionally, it has been established an income per by products commercialization. These by products are represented by liquid and solid fertilizers.
e. Another issue to be taken into account is the energy savings by means of the use of power co-generated using biogas from the biorefinery's processes.

COSTS OF RAW MATERIAL	USD/gal.
Empty fruit bunch oil palm	0.28
Enzymes	0.55
Reactive	0.04
Yeast	0.03
TOTAL COSTS PER RAW MATERIAL	**0,9**
OTHER COSTS	**USD/gal.**
Transport	0.11
Electricity	0.05
Fuel	0.11
Waste management	0.02
Water	0.03
Chemicals	0.02
Research & Development	0.05
Administration costs	0.04
Others	0.02
TOTAL OTHER COSTS	**0.45**
TOTAL COST PER GALLON	**1.35**

Table 5. Detailed costs of raw material and other inputs for ethanol production from empty fruit bunch palm in Ecuador.

A similar European project (i.e. PERESEO project from Spain) establishes that the cost for the construction of a biorefinery plant for 500 MT of biomass per day is about 12 to 15 million Euros. In terms of the RESETA project, we may need to process about 640 MT/day. The criteria for building up a biorefinery are to consider a future growth, so the capacity must be overestimated in about 30% [32]. Then, the investments may overcome in about 1.6 times those of the European PERSEO project. In terms of Dollars, the total investment is among 24.96 to 31.2 million Dollars. We will use the highest number for our calculations.

It has been determined the cost per liter ethanol from lignocellulosics from empty fruit bunch of oil palm in 0,49 Dollars, which is shown in Table 6.

PRODUCTION COSTS USD/L etanol	
DEPRECIATION	$ 0,10
MAINTENANCE	$ 0,01
OTHER FIXED COSTS	$ 0,12
TOTAL FIXED COSTS	**$ 0,23**
RAW MATERIAL	$ 0,24
DIRECT LABOR	$ 0,02
TOTAL VARIABLE COSTS	**$ 0,26**
TOTAL PRODUCTION COSTS	**$ 0,49**

Table 6. Production costs of ethanol from empty fruit bunch palm in Ecuador.

In this calculation it has been taken into account a 1% growth in costs and incomes with the aim of using the whole installed capacity at the end of the project. Following in Table 7, it is shown the cash flow of the project, where the initial investment (31,2 million Dollars), the investment recovery period; the IRR and NPV are shown. Table 8 is a resume of such values.

CASH FLOW									
Year	Investment	Cash Inflows (Income)	Cash Outflows (Expenses)	Gross profit	Utilities employees	Income Tax	Depreciation	Net Income	Accumulated Cash
0	(31.200.000,00)							(31.200.000,00)	(31.200.000,00)
1		12.480.156,38	7.748.346,30	4.731.810,08	709.771,51	1.005.509,64	1.560.000,00	4.576.528,93	(26.623.471,07)
2		12.604.957,94	7.825.829,76	4.779.128,18	716.869,23	1.015.564,74	1.560.000,00	4.606.694,22	(22.016.776,86)
3		12.731.007,52	7.904.088,06	4.826.919,46	724.037,92	1.025.720,39	1.560.000,00	4.637.161,16	(17.379.615,70)
4		12.858.317,60	7.983.128,94	4.875.188,66	731.278,30	1.035.977,59	1.560.000,00	4.667.932,77	(12.711.682,93)
5		12.986.900,77	8.062.960,23	4.923.940,54	738.591,08	1.046.337,37	1.560.000,00	4.699.012,10	(8.012.670,84)
6		13.116.769,78	8.143.589,83	4.973.179,95	745.976,99	1.056.800,74	1.560.000,00	4.730.402,22	(3.282.268,62)
7		13.247.937,48	8.225.025,73	5.022.911,75	753.436,76	1.067.368,75	1.560.000,00	4.762.106,24	1.479.837,62
8		13.380.416,85	8.307.275,99	5.073.140,87	760.971,13	1.078.042,43	1.560.000,00	4.794.127,30	6.273.964,92
9		13.514.221,02	8.390.348,75	5.123.872,27	768.580,84	1.088.822,86	1.560.000,00	4.826.468,58	11.100.433,50
10		13.649.363,23	8.474.252,23	5.175.111,00	776.266,65	1.099.711,09	1.560.000,00	4.859.133,26	15.959.566,76
11		13.785.856,86	8.558.994,76	5.226.862,11	784.029,32	1.110.708,20	1.560.000,00	4.892.124,59	20.851.691,35
12		13.923.715,43	8.644.584,70	5.279.130,73	791.869,61	1.121.815,28	1.560.000,00	4.925.445,84	25.777.137,19
13		14.062.952,59	8.731.030,55	5.331.922,04	799.788,31	1.133.033,43	1.560.000,00	4.959.100,30	30.736.237,49
14		14.203.582,11	8.818.340,86	5.385.241,26	807.786,19	1.144.363,77	1.560.000,00	4.993.091,30	35.729.328,79
15		14.345.617,93	8.906.524,26	5.439.093,67	815.864,05	1.155.807,40	1.560.000,00	5.027.422,21	40.756.751,00
16		14.489.074,11	8.995.589,51	5.493.484,61	824.022,69	1.167.365,48	1.560.000,00	5.062.096,44	45.818.847,44
17		14.633.964,85	9.085.545,40	5.548.419,45	832.262,92	1.179.039,13	1.560.000,00	5.097.117,40	50.915.964,84
18		14.780.304,50	9.176.400,86	5.603.903,65	840.585,55	1.190.829,52	1.560.000,00	5.132.488,57	56.048.453,41
19		14.928.107,55	9.268.164,86	5.659.942,68	848.991,40	1.202.737,82	1.560.000,00	5.168.213,46	61.216.666,87
20		15.077.388,62	9.360.846,51	5.716.542,11	857.481,32	1.214.765,20	1.560.000,00	5.204.295,59	66.420.962,47

Table 7. Cash flow of the project

With such calculations and with a discount rate of 13.76%, calculated in basis to the established formula shown next, we have attained an IRR biggest than the discount rate.

Discount Rate = Rf+ β(Rm – Rf) + Rp,
Rf: Risk of the treasure founds bonus in the USA.
β: Sensitivity or risk of the Project in the market
Rm: Stock market index.
Rp: Country-risk

DISCOUNT RATE	13,96%
IRR	14,15%
NPV	$ 307.655,43
INVESTMENT RECOVERY PERIOD	6 years

Table 8. Resume of the main financial indicators of the project

This Project exhibits a positive NPV, combined with a reasonable IRR value; both values can be interpreted as indicators of a feasible project. This financial study doesn't consider the externalities and the social benefits a biorefinery can give the surrounding populations and environment.

8. Perspectives and conclusions

Ecuador is a very rich country in terms of biomass, produced by agricultural activities as well as biomass from natural forests. Nevertheless, this biomass is now misused and underestimated, because of the lack of mature technologies to take the best from it.

Ecuador has a high potential for the production of lignocellulosic ethanol, if we take a look of the amount of lignocellulosic waste materials that are being produced every year in crops such as bananas, palm oil, sugar cane, etc. Nevertheless it is to be understood that social and economic costumes are deeply rooted so it's going to be a big task changing the current uses of several waste materials. The case of empty fruit bunches is a typical example where it is possible to analyze the behaviors and preferences of the farmers and industry people.

It is necessary to establish rules and laws to regulate the raising biomass markets, moreover if the biomass is going to be transformed in valuable and useful products such as xylitol, ethanol, foods and renewable chemicals.

The technologies are being improved to get better yields and lower production costs. A demonstrative scale biorefinery has recently sat up in Ecuador at the Neotropical Center for the Biomass Research, under the Pontificia Universidad Catolica del Ecuador. This biorefinery was completely designed and constructed in Ecuador and is the very first in its genus in the country as well as one of the few in South America.

Author details

Enrique Javier Carvajal Barriga, Patricia Portero Barahona, Edgar Salas, Carolina Tufiño and Bernardo Bastidas
Pontificia Universidad Católica del Ecuador, Centro Neotropical para la Investigación de la Biomasa, Quito-Ecuador

Cristina Guamán-Burneo
Departamento de Microbiologia, ICB, C.P. 486, Universidade Federal de Minas Gerais, Belo Horizonte, Brazil

Acknowledgement

The authors want to acknowledge the Pontificia Universidad Católica del Ecuador and the SENESCYT (Secretariat of Science and Technology of Ecuador) for financing the RESETA project and scientific activities at the Neotropical Center for the Biomass Research.

9. References

[1] Berg, Jeremy Mark, Tymoczko, Jhon L. and Lubert, Stryer. *Biochemistry.* New York: Freedman and Company, 2007.

[2] OPEC. *World Oil Outlook.* Vienna, Austria: Secretariat OPEC, 2011.

[3] DALE, BRUCE. Biofuels: Thinking Clearly about the Issues. *Journal of Agricultural and Food Chemistry*. 56, 2008, 3885–3891.

[4] Cheng, J.J., et al., et al. *Biomass to Renewable Energy Processes*. New Yersey: Taylor & Francis Group, 2010.

[5] *High activity of xylose reductase and xylitol dehydrogenase improves xylose fermentation by recombinant Saccharomyces cerevisiae*. Karhumaa, K., et al., et al. 2007, Appl. Microbiol. Biotechnol., pp. 73, 1039–1046.

[6] *Efficient production of L-lactic acid from xylose by Pichia stipitis*. Ilmén, M., et al., et al. 2007, Appl. Environ. Microbiol., pp. 73, 117–123.

[7] *Engineering yeasts for xylose metabolism*. Jeffries, T.W. 2006., Curr. Opin. Biotechnol., pp. 17, 320–326.

[8] *Genome sequence of the lignocellulose-bioconverting and xylose-fermenting yeas*. Jeffries, T.W., et al., et al. 2007, Nat. Biotechnol, pp. 25, 319-326.

[9] Affleck, R.P. *Recovery of Xylitol from Fermentation of Model Hemicellulose Hydrolysates Using Membrane Technology*. Virginia, USA: Dissertation for the degree of Master of Science in Biological Systems Engineering at Virginia Polytechnic Institute and State University, 2000.

[10] Sampaio, F. *Seleção de Microrganismos para a Conversão de Xilose em Xilitol*. Brasilia: Dissertação apresentada na Universidade de Brasília para obtenção do título do Mestre em Microbiologia Agrícola., 2001.

[11] Granström, T. *Biotechnological production of xylitol with Candida yeasts*. Espoo, Finland: Dissertation for the degree of Doctor in Chemical Technology at Helsinki University of Technology, 7th of June, 2002.

[12] *Xilitol: Edulcorante com efeitos benéficos para a saúde humana*. Mussatto, S. I. and Roberto, I. C. 2002: s.n., 2002, Revista Brasileira de Ciências Farmacêuticas., pp. 38:401-413.

[13] Lima, L. H. *Estudos moleculares dos genes XYL1 e XYL2 de Candida tropicalis visando a produção de xilitol*. Brasília: Dissertação apresentada a Universidade de Brasília para obtenção do título do Doutor em Ciências Biológicas- Biologia Molecular, 2006.

[14] *Crescimento e consumo de Xilose de Candida guilliermondii na fermentação submersa utilizando-se bagaço de cana-de-açúcar*. Jucoski, M. C., et al., et al. 2007, Evidência, Joaçaba, pp. V. 7, N.2, P. 119-130.

[15] *Bioconversion of D-xylose to xylitol by Debaryomyces hansenii UFV-170: Product formation versus growth*. Sampaio, F. C., et al., et al. 2005, Process Biochemistry., pp. 40, 3600–3606.

[16] *Review, Bioconversion of pentose sugars into ethanol: A review and future directions*. Chandel, A. K., et al., et al. 2010, Biotechnology and Molecular Biology Review, Vol. 6(1), pp. 008-020.

[17] *Microbial conversion of D-Xylose to Xylitol*. Winkelhausen, E. and Kuzmanova, S. 1998, Journal of Fermentation and Bioengineering., pp. 86: 1-14.

[18] *Evaluation of Sugarcane bagasse acid hydrolizate treatments for xylitol production*. Gurgel, P. V., et al., et al. 1998, Brazilian Journal of Chemical Engineering vol.15, n.3, pp. 309-312.

[19] *Improvement of xylitol production by Candida guillermondii FTI 20037 previously adapted to rice straw hemicellulosic hydrolisate.* Silva, C.J.S. and Roberto, I.C. 2001, Letters in Applied Microbiology 2001, pp. 32, 248-252.

[20] Schmirmer, A.C. *Produção biotecnológica de xilitol e etanol a partir de hidrolisado de casca de soja.* Rio Grande do Sul: Dissertação apresentada na Universidade Federal de Rio Grande do Sul para obtenção do mestre em Ciência e Tecnologia de Alimentos., 2007.

[21] Ribeiro, A. *Otimização da hidrólise da casca de soja (Glycine max) e avaliação da capacidade de produção de xilitol e etanol por microrganismos sobre este hidrolisado.* Rio Grande do Sul: Dissertação apresentada na Universidade Federal de Rio Grande do Sul para obtenção do mestre em Ciencia e Tecnologia de Alimentos, 2010.

[22] *Pretreatment of sugarcane bagasse hemicellulose hydrolysate for xylitol production by Candida guilliermondii.* Alves, L. A., et al., et al. 1998, Applied Biochemical Biotechnology, pp. 70–72: 89–97.

[23] *Factors that affect the biosynthesis of Xylitol by Xylose-Fermenting Yeasts.* Silva, S.S., Felipe, M.G.A. and Mancilha, I.M. 1998, Applied Biochemistry and Biotechnology. São Paulo., pp. vol. 70-72.

[24] *Biotechnological production of xylitol. Part 3: Operation in culture media made from lignocellulose hydrolysates.* Parajó, J. C., Domínguez, H. and Domínguez, J. M. 1998b, Bioresource Technology., pp. 66: 25-40.

[25] *Chemostat study of xylitol production by Candida guilliermondii.* Granströn, T., Ojamo, H. and Leisola, M. 2001, Applied Microbiology Biotechnology, pp. 55: 36-42.

[26] *Reclassification of the Candida guilliermondii FTI20037 as Candida tropicalis based on molecular phylogenetic analysis.* Lima, L. H., Felipe, M. G.F. and Gonçalves, F. A. 2003, Brazilian Journal of Microbiology., pp. 34 (Suppl.1) 96-98.

[27] *Effect of acetic acid on xylose fermentation to xylitol by Candida guilliermondii.* Felipe, M. G. A., et al., et al. 1995, Journal of Basic Microbiology, pp. vol 35: 171-177.

[28] *Biotechnological Production of Xylitol from Agroindustrial Residues. Evaluation of Bioprocesses.* Rodrigues, D. C., et al., et al. 1998, Applied Biochemistry and BiotechnologyY., pp. Vol. 70-72.

[29] Erazo, Paúl Mena. http://www.bbc.co.uk/mundo/ciencia_tecnologia/2010/01/100112_0245_ecuador_gasolina_verde_jrg.shtml. *Ecuador arranca venta de biocombustible.* enero 13, 2010, p. digital.

[30] PETROECUADOR. *INFORME CIFRAS PETROLERAS, PERIODO ENERO - DICIEMBRE 2010.* Quito: EP PETROECUADOR, 2010.

[31] SmartResearch. *Estudio desechos agricolas en el Ecuador, Proyecto RESETA: Recursos Sustentables para Etanol.* Quito: PUCE, 2009.

[32] Cedeño, N. M., León, J. E. and Cedeño, J. C. *Modelo de Estimación de Precio para un Combustible de Composición 95% Gasolina Extra y 5% Etanol, que Ayude a Disminuir la Contaminación Ambiental en la ciudad de Guayaquil.* Guayaquil: ESCUELA SUPERIOR POLITECNICA DEL LITORAL, 2010.

[33] *Biotechnological production of xylitol. Part 1: Interest of xylitol and fundaments of its biosynthesis.* Parajó, J. C., Domínguez, H. and Domínguez, J. M. 1998a, Bioresource Technology., pp. 65, 191-2111.

[34] *High capacity xylose transport in Candida intermedia PYCC 4715.* Gárdonyi, M., et al., et al. 2003, FEMS Yeast, pp. Res. 3, 45–52.

Chapter 11

Catalytic Decomposition of Biomass Tars at Low-Temperature

Le Duc Dung, Kayoko Morishita and Takayuki Takarada

Additional information is available at the end of the chapter

http://dx.doi.org/10.5772/55356

1. Introduction

Tar is a viscous black liquid derived from pyrolysis of organic matter as well as a complex mixture of hydrocarbons. The presence of tar in product gas may cause blockage and corrosion of equipment and be responsible for fouling or reducing overall efficiency of processes. By far, tar removal is the most problematic during biomass gasification. Hence, the successful implementation of gasification technology for gas engine, gas turbine or fuel cell based power projects depends much on the effective and efficient removal or conversion of tar from the product gas. Beside that the catalytic steam reforming tar is one of the most promising methods to suppress the problems. Biomass product gas is usually low high heating value; therefore enhancement of product gas quality is other important target. We propose a research topic that use of nickel loaded brown coal char as a new catalyst for decomposing tar from biomass gasification in fluidized bed gasifier. The method is promising to achieve some advantages of low cost by use of low rank coal as catalyst support material, high catalyst activity and enhancement of product gas quality. Yallourn brown coal has been selected for preparing catalyst support. The coal is low rank with high moisture content, low heat value and high oxygen content. It is hard to use for generating energy. However, it has many outstanding features such as less ash and sulfur content, and including abundant of oxygen–containing functional groups such as carboxyl and phenol groups which are available for ion–exchange with metals. In this research work, a nickel loaded brown coal char (Ni/BCC) was prepared by ion–exchange method, dried at 380 K in nitrogen for 24 h, and then pyrolysed at 923 K in nitrogen for 90 min. The works have been carried out is that using nickel loaded brown coal to decompose tar in pyrolysis and steam gasification process. It was carried out in a two–stage fixed–bed reactor and a lab scale fluidized bed gasifier under mild conditions (temperature, steam, space velocity, operation time). Inside of gasifier is constructed by two beds, the primary one is a fluidized bed with sand, and the second one is a catalyst bed. The new catalyst has shown high catalytic activity and stable activity and given the high quality of product gas in presence of steam, approximately 90% of biomass tar was decomposed and useful gas components (CH_4, CO, and H_2) yields were higher than those of Ni/Al_2O_3 catalyst. Ni/BCC catalyst was characterized and exhibited good dispersion of nickel particles, ultra–fine Ni less than 15 nm and having a large surface area about 350 m^2/g. Moreover, at the end of catalyst life span, the catalyst can be disposed of simply by gasifying/burning the coal char, during which the energy value of the char support can be recovered. Also, the agglomerated nickel residues could be used as functional materials of powder metallurgy and battery development. The general results suggest that the Ni/BCC catalyst offers a potential to be used as a tar steam reforming catalyst in biomass gasification.

1.1. Tar and tar removal

There are still many questions related to tar and the problems they may cause. Tar is a viscous black liquid derived from pyrolysis of organic matter as well as a complex mixture of hydrocarbons [1]. Various research groups are defining tar differently. In the EU/IEA/US-DOE meeting on tar measurement protocol held in Brussels in the year 1998, it was agreed by a number of experts to define tar as all organic contaminants with a molecular weight larger than benzene [2]

The presence of tar in product gas may cause blockage and corrosion of equipment and be responsible for fouling or reducing overall efficiency of processes. Tar is formed when biomass is heated the molecular bonds of the biomass break; the smallest molecules gaseous, the larger molecules are called primary tars. These primary tars, which are always fragments of the original material, can react to secondary tars by further reactions at the same temperature and to tertiary tars at high temperature [3, 4, 5, 6] Figure 1 show tar is quite complex and hard to decompose. By far, tar removal is the most problematic during biomass gasification. Hence, the successful implementation of gasification technology for gas engine/turbine based power projects depends much on the effective and efficient removal/conversion of tar from the producer gas. Up to now, a great amount of work concerning tar reduction or reforming has been reported with abundant technologies to remove tar from biomass product gas.

J. Han, and H. Kim had divided tar removal methods into five groups: mechanical methods (using cyclone, filters ceramic), granular beds, Electrostatic precipitators and Scrubbers; self-modification, selecting optimal operation parameters for gasifier or using a low tar gasifier; Catalytic cracking; Thermal cracking and Plasma methods [5]. The review shows that the primary use of mechanism methods is to capture the fly ash or particles from product gas; the effect on tar removal is also very good. However, these methods only remove or capture the tar from product gases, while the energy in tar is lost. The self–modification and other methods can not only reduce the tar but also convert the tar into useful gases. The self–modification methods include: selecting better gasifer, and optimizing operation parameters. Tar reduced by modifying operation parameter is at the expense of reducing the heat value of gases. Catalyst cracking and thermal cracking are generally used to decompose or reduce tar though there are still some disadvantages. Plasma technology cannot only

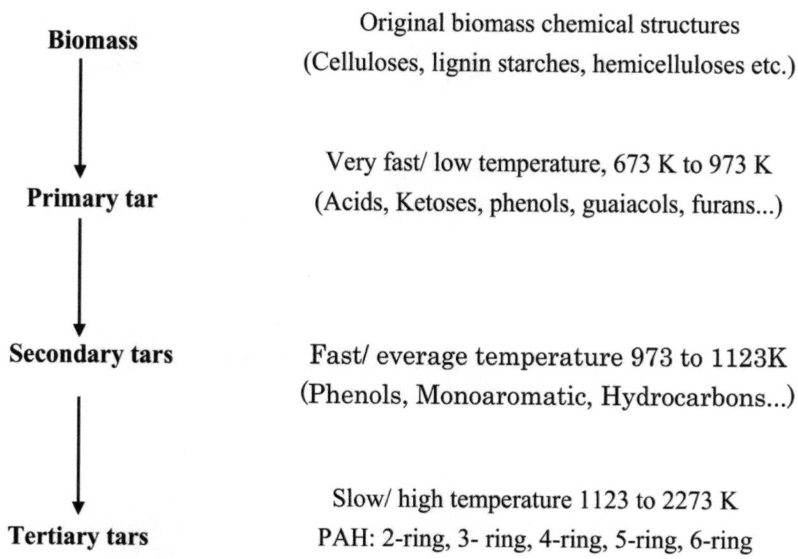

Figure 1. Formation of biomass tars and compounds formed

effectively remove fly ash, NOx and SO_2, but also sharply decrease the formation of tar during biomass gasification. In order to get highly efficient tar decomposition, the temperature of thermal cracking needs to be very high, which results in operating cost increase. Catalyst cracking can modify the composition of product gases at low temperature with high carbon conversion efficiency. Nevertheless, there still exists a short coming such as: the commercial Ni-based and alkali metal catalysts will be inactive by deposited carbon, and H_2S. The newly developed novel catalyst can overcome the disadvantages by use of expensive metals (Co, Pt, Ru, Pd and Rh), and also catalyst supports (Al_2O_3, Al, SiO, TiO_2, ZrO_2, MgO or WO_3) and perform tar removal with high and stable activity even under the presence of high concentration of H_2S in some cases. In order to satisfying both high and stable activity and good price, the development of catalyst meets the need to be continuing.

1.2. Catalysts

Due to the advantages of converting tar into useful gases and adjusting the compositions of product gases, catalyst cracking has been of interest since the middle 1980s. The simplified mechanism for catalyst tar reforming can be described as follows [7–9]. First, methane or other hydrocarbons are dissociatively adsorbed onto a metal site where metal catalyzed dehydrogenation occurs. Water is also dissociatively adsorbed onto the ceramic support, hydroxylating the surface. At the appropriate temperature, the OH radicals migrate to the metal sites, leading to oxidation of the intermediate hydrocarbon fragments and surface carbon to CO + H_2. David [9,10] summarized the criteria for catalyst as follows:

1. the catalysts must be effective in removing tar;
2. if the desired product was syngas, the catalysts must be capable of reforming methane;
3. The catalysts should provide a suitable syngas ratio for the intended process;

4. the catalysts should be resistant to deactivation as a result of carbon fouling and sintering;
5. the catalysts should be easily regenerated.
6. The catalysts should be strong; and
7. the catalysts should be inexpensive.

Moulijn J.A. [11] has classified main causes of the deactivation into five reasons that are poisoning, fouling, thermal degradation (sintering, evaporation) initiated by the often high temperature, mechanical damage and corrosion/leaching by the reaction mixture. The deactivation phenomenon inside a catalyst particle is described on Figure 2 [11]. Among them, thermal degradation reason often occurs during catalyst reforming tar at relative high temperature (Figure 3).

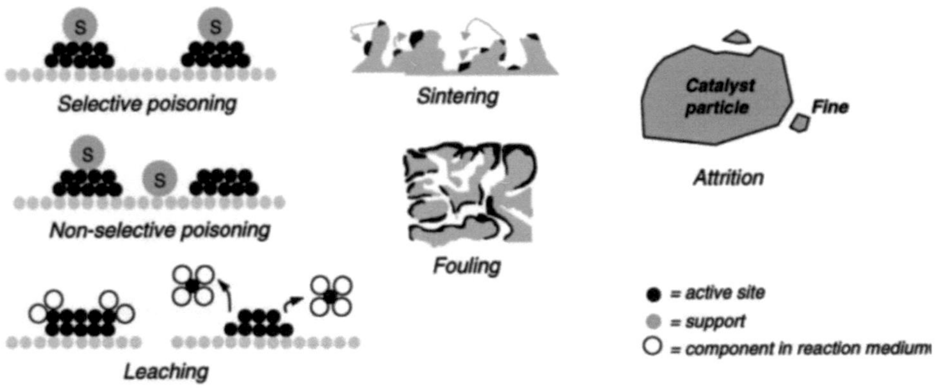

Figure 2. Major types of deactivation in heterogeneous catalysis

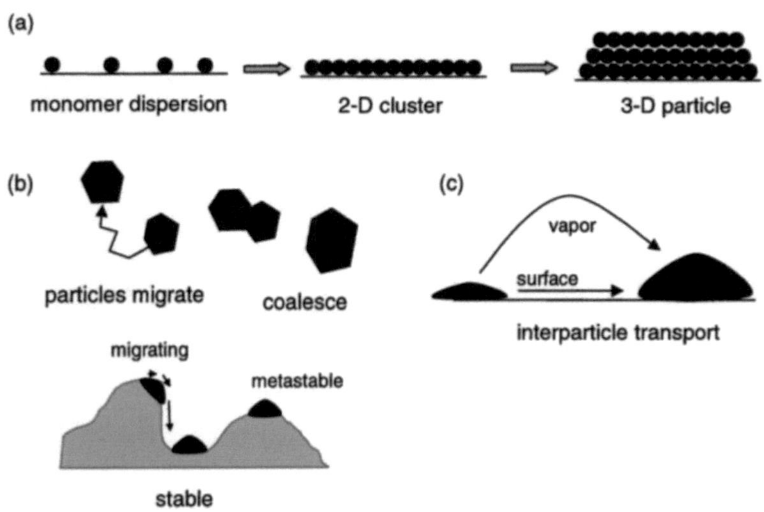

Figure 3. Schematic of the various stages in the formation and growth of particles from monomer dispersion ((a): Clusters of atoms (or small metal particles); two–dimensional clusters, and; three–dimensional particles; (b): Particles might move and coalesce; (c): Atoms move from one particle to another, either by volatilisation or by surface migration.)

Nikel based catalysts

Nickel has been developed with various promoters and carriers for decomposing tar and tar models [7–27].

Zhang [13] investigated tar catalytic destruction in a tar conversion system consisting of a guard bed and catalytic reactor. Three Ni based catalysts (ICI46–1, Z409 and RZ409) were proven to be effective in eliminating heavy tars (99% destruction efficiency). The experimental results demonstrated that space velocity (1500 –6000) had little effect on gas compositions, while increasing temperature boosted hydrogen yield and reduced light hydrocarbons (CH_4 and C_2H_4) formation, which suggested that tar decomposition was controlled by chemical kinetics.

Furusawa *et al.* [14] reported that 10 wt% Ni/MgO (873 K) catalyst showed the best performance. Nickel supported on silica was active for tar catalyst cracking methane at relatively low temperature (823 K) was described by Zhang [9].

Srinakruang *et al.*[15,16] has developed Ni/Dolomite as highly efficient sulphur and coking resistance catalyst, and reported that calcining at 500 °C exhibited the most effective catalyst among Ni/SiO_2 and Ni/Al_2O_3; the poisoning effect was enhanced by increasing the reaction temperature and steam/C ratio; higher activity, durability and coking resistance.

Sato *et al.* [17], has developed Ni–WO_3/MgO–CaO catalyst for naphthalene and toluene reforming. The results exhibited a better resistance to sulfur and coking catalyst; tar reforming to better than 90% and 100 h steady tar reforming operation (in H_2S) at 800–850 °C.

Dou *et al.*[18] compared five catalysts on tar removal from fuel gases in a fixed–bed reactor. The Y–zeolite and NiMo catalysts were found to be the most effective about 100% tar removal can be achieved at 550 °C. It was also observed that process variables like temperature and space velocity had very significant effect on tar removal.

Baker [19] also mentioned the phenomena in their experiments. In order to overcome the shortcoming of the commercial Ni–based catalyst, many Ni–based catalysts were developed.

Miyazawa *et al.* [20] has prepared Ni (Ni/Al_2O_3, Ni/ZrO_2, Ni/TiO_2, Ni/CeO_2 and Ni/MgO) catalyst to reformed tar in the partial oxidation (POT) and steam introduction. Results have been achieved: the order of the performance at 823 K was as follows: Ni/Al_2O_3 > Ni/ZrO_2 > Ni/TiO_2 > Ni/CeO_2 > Ni/MgO > no catalyst; Ni/CeO_2 showed smaller amount of coke than other catalysts; in the POT, much higher tar conversion and lower coke yield were obtained than that in SRT using fixed bed reactor.

1.3. Catalytic brown coal gasification

It is very important to increase the thermal efficiency of coal conversion for not only protecting the limited coal resources but also reducing CO_2 and air pollutant emission. Steam gasification of coal is one of the most promising energy conversion technologies for producing hydrogen.

Brown Coal

Brown coal or lignite is a low rank with high moisture content of around 60 %, low heat value and high oxygen content. Therefore, it is hard to use for converted to useful energy. However, it is concluding many outstanding features such as less ash and sulfur content, and especially, including abundant of oxygen–containing functional groups such as carboxyl and phenol groups which are available for ion–exchange with metals. The structural unit of coal models is shown in Figure 4d.

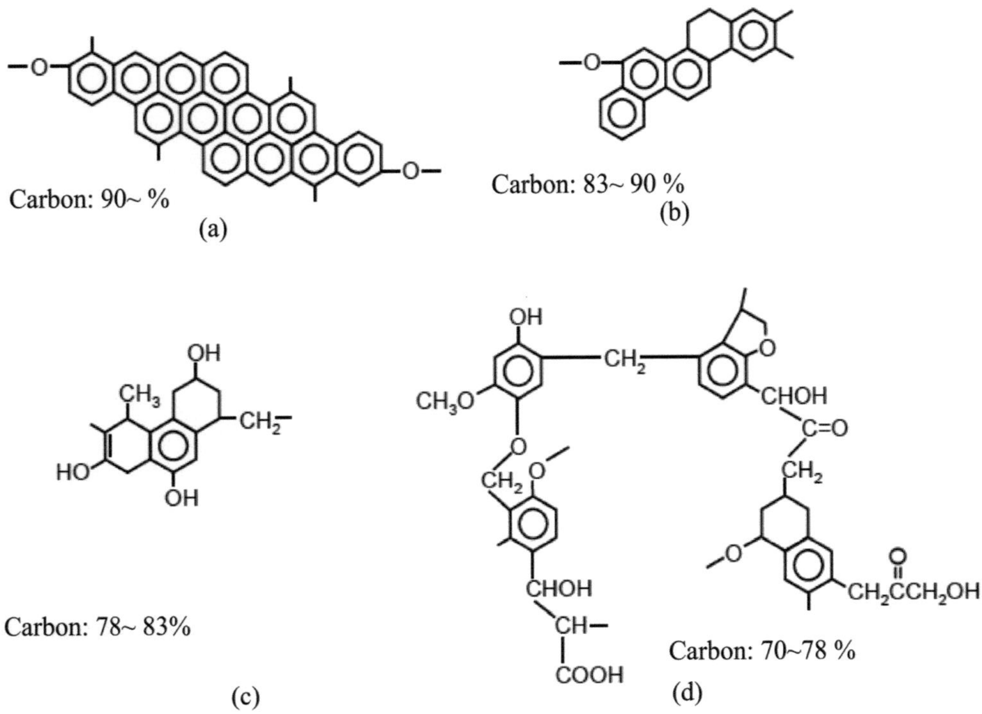

Figure 4. Structural unit of coal models (a: anthracite coal; b: bituminous coal; c: bituminous coal; d: brown coal)

Catalytic coal gasification

Tomita [21, 22] has reported low temperature gasification of Yallourn coal catalysed by nickel. By the different way of prepared catalyst as usual like conventional impregnation methods, coal was mixed with an aqueous solution of hexamine nickel carbonate. This mixture gave a perfect homogeneous, catalyst bearing coal. Analysis product gas he found that the gases in the rapid stage of char gasification at 723 K presented mainly hydrogen and carbon dioxide. The study demonstrates the first time that nickel catalyst can enhance gasification reactivity in a similar manner as observed for activate carbon.

Tomita [23, 24] continued to carry out nickel brown coal gasification with various gasification conditions such as amount weight of nickel loaded on coal, various temperatures. The good results exhibited carbon conversion reached 85%, with 30 min for steam gasification at low temperature as 723 K.

Ohtsuka [25] reported calcium catalysed steam gasification of Yallourn brown coal with the same way of preparation catalyst as Tomita have done. The results also showed that high activity for calcium catalyst steam gasification of brown coal.

Takarada [26] investigated catalyst steam gasification of coal by mixing of K–exchange brown coal. The results gave evident that rate of enhancement of K–exchange Yallourn coal by physical mixing method is independent of the caking property of higher rank coal; Potassium is a highly suitable catalyst for catalytic gasification.

Recently, Miki [27] also carried out pyrolysis and gasification of coal which was loaded Ni by ion– exchanged method, and again proved nickel loaded brown coal has a high activity in coal gasification.

2. Nickel catalyst preparation

2.1. Ni/BCC catalyst

The procedure of ion exchange is shown in Figure 5 and Figure 6. Yallourn (YL) coal char was used as catalyst support. YL coal (Australian brown coal) received in the form of briquettes. The coal was crushed, sieved to a particle size range of 1 – 2 mm in diameter, and then dried at 380 K for 12 h. The nickel addition to the coal matrix was achieved by ion–exchange with a solution of basic hexa ammine nickel carbonate $(NH_3)_6NiCO_3$ (Figure 5, 6). The coal was mixed with the $(NH_3)_6NiCO_3$ solution for 24 h and then recovered by filtration. The recovered solid was washed with distilled water and filtered again. Then, the washed solid was dried under N_2 flow at 380 K for 24 h. Last, raw catalysts were produced.

Catalysts was characterized by powder X–ray diffraction on XRD; M03XHF22, Mac Science Co., Ltd, using CuKα radiation(40 kV, 30 mA) in order to identify the potential evolution of the crystalline phases during catalyst pyrolysis tests. The diffractograms were recorded a step time of 10 sec. SEM analysis was applied to study the surface structure of Ni/BCC catalyst (FE–SEM; JSM–6700F, JEOL Datum Ltd.). An atomic absorption flame emission spectrophotometer (AA–6400F, Shimadzu Corp.) was used to examine the amount of Ni on raw Ni/BCC (Ni 9 ± 1%–dry). After pyrolyzing with nitrogen gas at 923 K for 90, its weigh loss is approximately 54 %, and therefore, nickel loaded in coal char could be estimated 19.6 ± 2%–char base. The intimacy of the contact between coal char and catalyst was so effective that the reactivity was considerably higher than those prepared by a conventional impregnation method [21–24]. Nitrogen adsorption characterization of catalysts was performed on equipment for automatic gas and vapor adsorption measurement (BELSORP–max, BEL Japan Co. Ltd.). Prior to adsorption measurement, the catalysts were degassed at 573 K for 3 h under a dynamic vacuum. The surface areas of fresh catalysts (Ni/BCC), which were obtained after pyrolysis of raw Ni/BCC at 923 K for 90 min is 350 m^2/g [28,29].

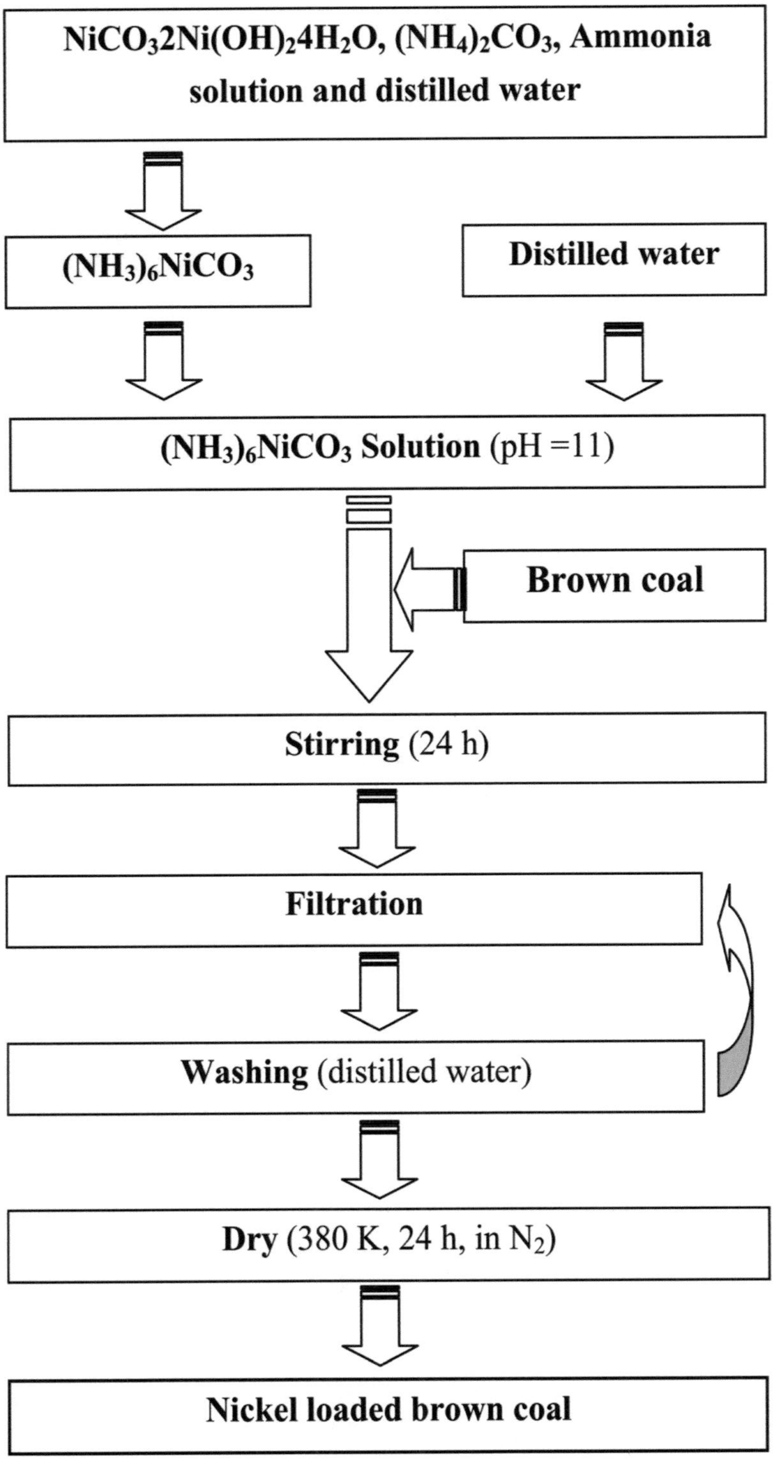

Figure 5. Schematic flow diagram of nickel and brown coal ion–exchange procedure

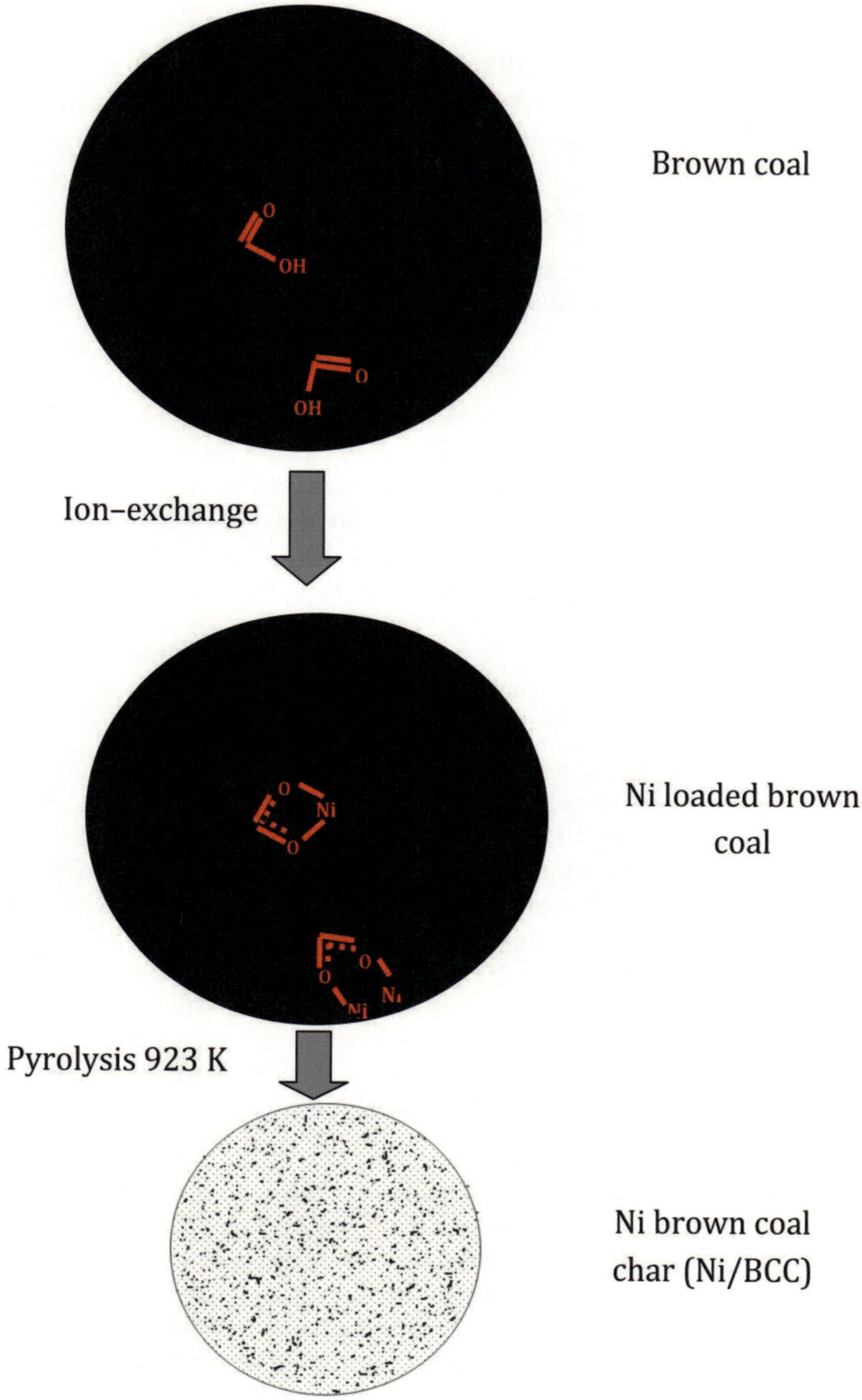

Figure 6. schematic diagram of nickel loaded brown coal char with structure unit of Ni/BCC.

2.2. Ni/Al_2O_3 catalyst

A conventional nickel catalyst (No.C13–4, Ni 20±2 wt % SÜD–CHEMIE CATALYSTS JAPAN, Inc.) that was supported with alumina was also used to compare with Ni/BCC catalyst. It was crushed and sieved to the fraction of 0.5 – 1 mm.

3. Catalytic decomposition of tar from woody biomass pyrolysis

Pyrolysis is an important process in energy recovery from biomass and also as a previous stage to other processes such as gasification. Valuable gases, such as H_2 and CO, can also be generated by pyrolysis. These gases can be useful, among other applications, in chemical synthesis and high efficiency combustion systems such as fuel cells.

The hydrogen–rich product gas from biomass pyrolysis is believed to become a valuable energy source with natural carbon dioxide. However, biomass has low energy density, so the enhancement of the product gas quality from biomass gasification is necessary. Beside a particular problem which has not been completely solved so far is tar formation. Catalytic processes are considered as the most promising method with the highest potential to contribute a solution to this problem. Temperature is an important variable in thermal decomposition processes of biomass, significantly influencing the product distribution. Pyrolysis is an endothermic process, and the use of low temperatures in this process decreases the input energy for a system that is very positive from an energetic point of view and the system operation is easier than high temperature systems.

For both catalytic activity and economic reason, a nickel catalyst is the most suitable choice and the most widely used in the industry among metals like Co, Pt, Ru and Rh, which were investigated by many authors [10,17–23]. Moreover, nickel based catalysts are reported to be quite effective not only for tar reduction, but also for decreasing the amount of nitrogenous compounds such as ammonia [24]. In most reports, conventional nickel catalysts, which had been developed for steam tar reforming, were tested [6,9,13,15,25,26]. However, the investigations are still limited because of coking [19,22] or the use of expensive materials as catalyst supports (CeO_2, Al_2O_3, Al, SiO_2, TiO_2, ZrO_2, MgO or WO_3). Among the above investigations, the most interesting one concerning the current study is brown coal gasification by the addition of a nickel catalyst in fluidized bed gasification at low temperature [21–23]. However, nickel catalyst has only been used for coal gasification itself.

To satisfy both high quality of product gas and use of a cheaper catalyst support (brown coal char), in this study, nickel loaded brown coal char catalyst has been developed for the new purpose of decomposing tarry material from woody biomass pyrolysis. A suitable temperature for catalytic tar decomposition was investigated in a two–stage fixed bed reactor. The effect of temperature on gas yield and carbon conversion have been discussed in detail and compared in the case of non catalyst and Ni/BCC catalyst. The better temperature is a reference result, being used in fluidized bed gasification which is available for continuous tests to assess durability. Catalytic activity was tested, evaluated by woody biomass pyrolysis in a fluidized bed gasifier for both of Ni/BCC and reference catalyst

Ni/Al_2O_3. In a lab scale fluidized bed gasifier (FBG) Experiments, inside of FBG reactor is constructed by two beds, the primary one is a fluidized bed with sand where biomass was fed to produce tar, and the other is a catalyst bed that is used to evaluate and to compare catalytic activity between the new catalyst and a conventional Ni/Al_2O_3 catalyst. The Ni/BCC catalyst is prepared by ion exchange method, dried at 380 K in nitrogen for 24 h, and is then calcined at 923 K in nitrogen for 90 min. Sample for characterization of catalyst was prepared on the fixed–bed reactor under various conditions such as nickel loaded brown coal particle size range of 0.5 to 2 mm, pyrolysis temperature range of 823 to 1023 K that are needed to investigate the effect both of catalyst particle size and pyrolysis temperature on crystallite size of Ni/BCC. The temperature as a function of gas yield and stable activity of catalyst absence of steam are investigated in this chapter.

3.1. Experimental facilities

The two–stage fixed–bed quartz reactor

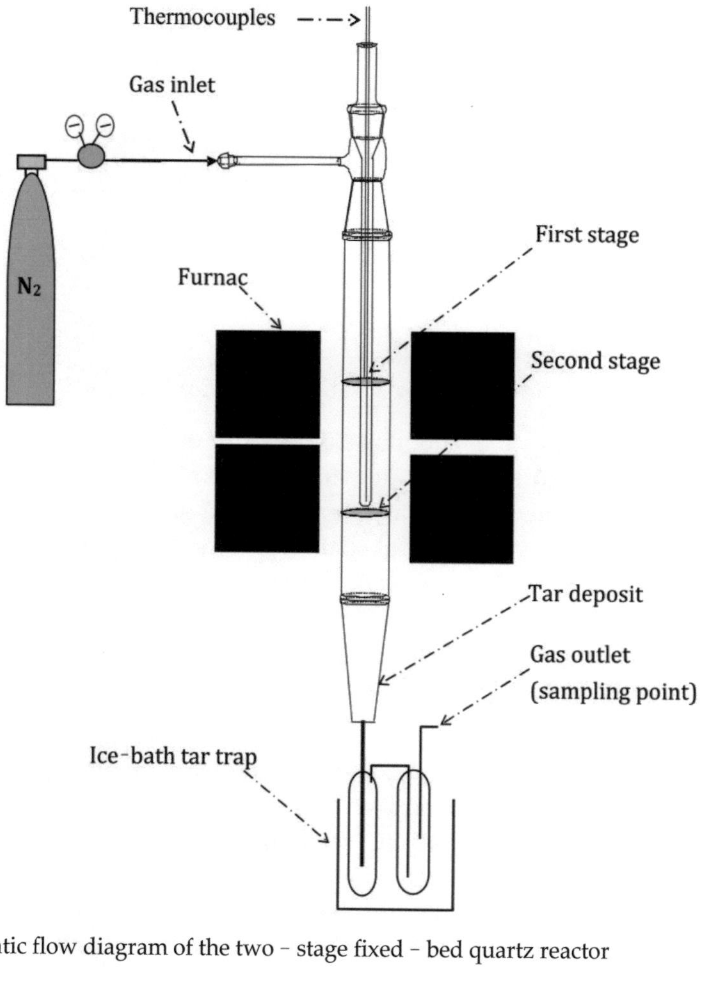

Figure 7. Schematic flow diagram of the two - stage fixed - bed quartz reactor

The fluidized bed gasifier

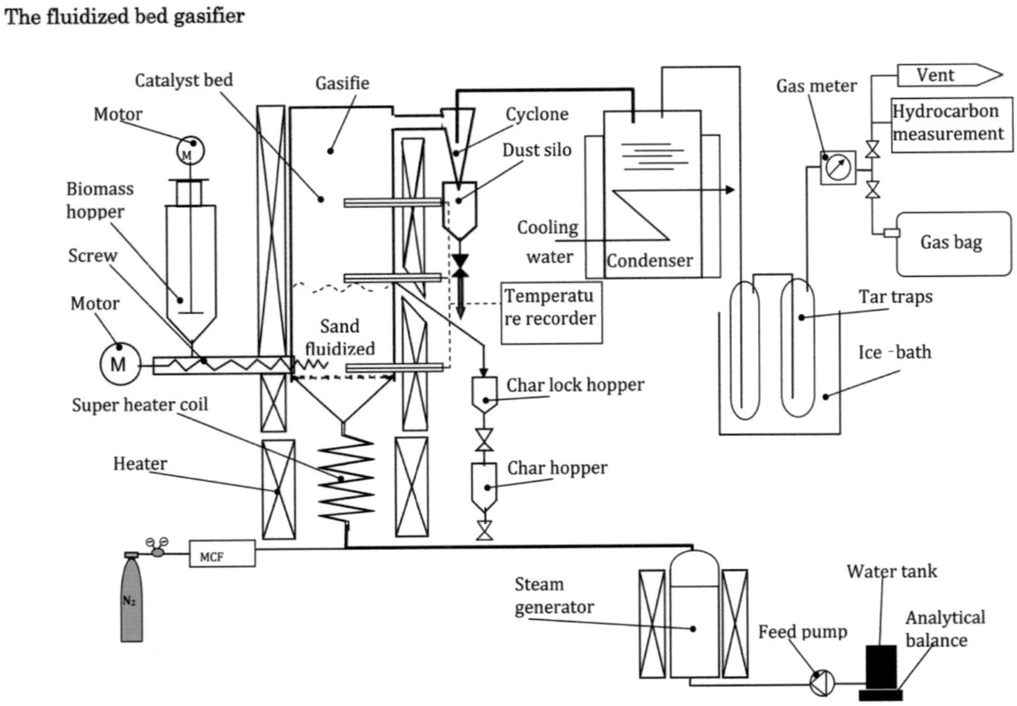

Figure 8. Schematic flow diagram of fluidized bed gasifier

3.2. Results and discussion

3.2.1. Catalyst Evaluation in a Two–stage Fixed–bed Reactor

Figure 9 illustrates the gas yield and the biomass carbon balance of woody red pine pyrolysis in a two fixed–bed quartz reactor. In the case Ni/BCC catalyst, total gas yield increased drastically at a catalyst bed temperature of 923 K, at which the yield of CO and H_2 achieved was 21.2 and 29.5 [mmol/g–sample daf], respectively, approximately three and six times in comparison to sand (Figure 9(a)). It was considered that tarry material was efficiently decomposed by the Ni/BCC catalyst. If we consider the effect of catalytic pyrolysis temperature on gas yield, Figure 9 (a) also shows that the gas yield increased by increasing temperature from 823 to 923 K, thus suggesting tar decomposition can be controlled by chemical kinetics.

Although there was no direct measurement of tar, we have the biomass carbon balance, which is illustrated in Figure 9(b). Among total carbon in biomass, percentages of carbon in product gas (C_ gas) and carbon in char (C_char) could be obtained by analyzing product gas and product char, respectively. Carbon in tar (C_tar) was estimated fairly by a different method: C_tar = 100 – (C_ gas + C_char). In the case of Ni/BCC, we could assume that the total carbon of product gas was released from biomass pyrolysis because the pyrolysis time

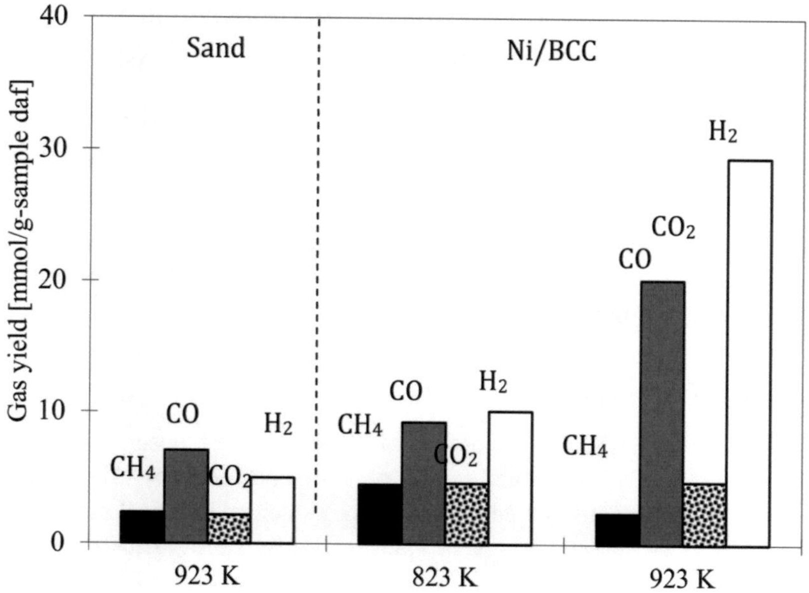

(a)

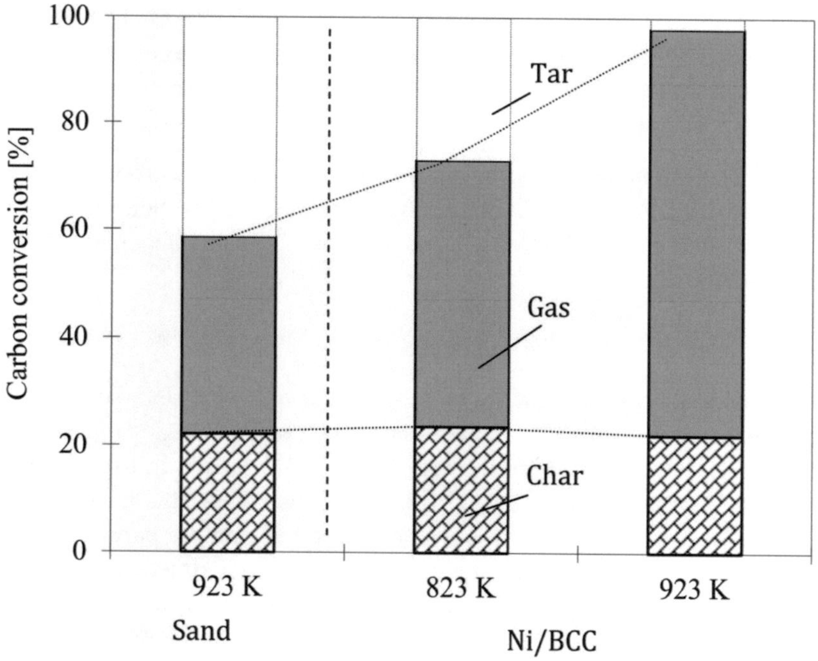

(b)

Figure 9. Effect of temperature on catalyst pyrolysis: (a) gas yields and (b) biomass carbon balance (space velocity 3000) of 90 min was enough to release most releasable carbon in Ni/BCC at 923 K. The amount of C_chars was almost constant in all cases, because the char is accumulated in the first bed without contacting the catalyst particles at the same temperature of 1173 K. In the case of catalytic tar decomposition, the amount of C_gas increased drastically compared to no catalyst at 923 K. That is to say, the tar was decomposed over Ni/BCC catalyst by Equation, Tar $\rightarrow CO + H_2 + CO_2 + CH_4 + C_2H_4 +$ other hydrocarbon.

When using Ni/BCC catalyst, C_tar approaches zero at 923 K. Moreover, we did not observe tar adhered on the reactor. Thus, it suggests almost all of the tar was decomposed at 923 K under the pyrolysis experimental conditions.

3.2.2. *Catalyst Evaluation in a Fluidized Bed Gasifer*

In the catalytic activity tests, the formation of products were observed for 60 min, and significant heavy tar was not observed on the pipeline and tar traps. All experiments were performed at 923 K under nitrogen carry gas, space velocity 11000. All calculated results of gas yield and C_gas were the average of specific results from various specific sampling times, which started at 10 min after feeding biomass and then in 20 min intervals. The effects of the catalyst on gas yields (CH_4, CO, CO_2, H_2 yield) are illustrated in Figure 10 (a). The bars from left to right show the results for non–catalyst, Ni/Al_2O_3 and Ni/BCC catalyst. Using Ni/BCC catalyst, CH_4, CO, CO_2 and H_2 yields were almost the same as those of Ni/Al_2O_3: 2.8, 15.6, 6.3, 23.1 [mmol/g–sample daf], respectively. Especially, both CO and CO_2 yields increased drastically by 2 times and H_2 by approximately 5 times compared to those of non–catalyst. This result indicates that Ni catalysts are quite effective to decompose tar to useful gases such as CO and hydrogen.

Biomass carbon balance is illustrated in Figure 10 (b). A detailed carbon balance could not be carried out because of difficulty in accurately estimating the tar yield. In a similar way as described above section, we defined C_gas, C_char, C_coke (Deposited carbon on the catalyst) and calculated C_tar:

$$C\_tar = 100 - (C\_\ gas + C\_char + C\_coke.$$

In the case of no catalyst, C_coke was not observed at all, because coke is assigned to the carbon deposited on the catalyst surface. For the case Ni/BCC catalyst, C_coke was estimated by the difference of carbon in fresh Ni/BCC and carbon in used Ni/BCC catalyst. The amount of C_char was almost constant in all of the cases. This is because the char is accumulated in the fluidized bed without contacting the catalyst particles. In the case of catalytic tar decomposition, the amount of C_gas increased drastically compared to no catalyst. The blank on the top of each bar in Figure 10(b) can be considered as a percentage of C_tar. For Ni/BCC and Ni/Al_2O_3 catalysts, C_ tar was 12.3 and 8.9% and C_gas was 54.9 and 55.1%, respectively. The results show that the Ni/BCC catalyst could not perform as well as the Ni/Al_2O_3 catalyst to decompose tar under pyrolysis process. This result might be affected by a part of the deposited carbon being on some of the reactive surface of the Ni catalyst, while the raw Ni/BCC catalyst was calcined due to volatile release from the brown coal. However, the results show that both catalysts are quite active to decompose tar.

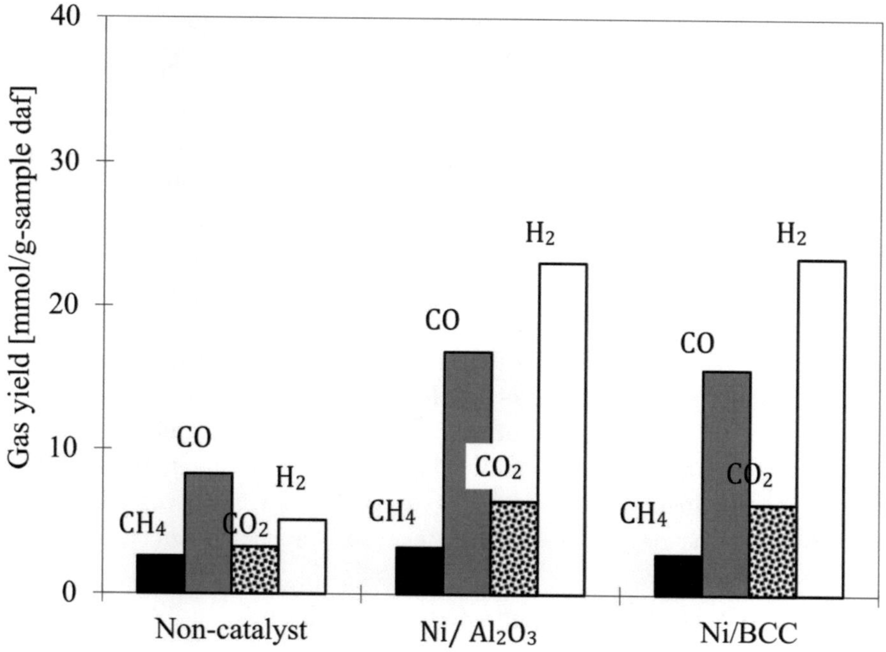

(a)

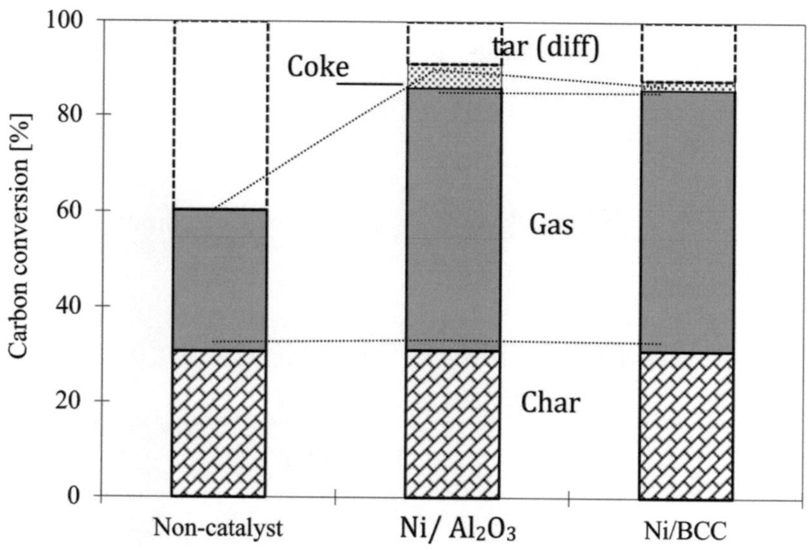

(b)

Figure 10. Comparison of different catalysts and non–catalyst without steam: (a) gas yields and (b) biomass carbon balance at 923 K and no steam (923 K, sv = 11000)

4. Catalytic steam reforming tar from woody biomass gasification

The aim of current study is to increase the coking resistance ability as well as steam gasification of deposited carbon on Ni/BCC catalyst. It was available to regenerate activity of catalyst. By the way, product gases propose to achieve an enhancement of the product gas quality by not only recovering energy from tar reforming but also addition by–product gas from steam gasification of the Ni/BCC char at relative low temperature. To decompose tar of biomass gasification by the use of Ni/BCC catalyst has been investigated under mild conditions in a laboratory scale fluidized bed gasifier with introducing steam as a gasifying agent and nitrogen as the product gas carrier. A conventional Ni/Al_2O_3 catalyst also was selected to compare with the Ni/BCC catalyst in the presence of steam. In this study, the Ni/BCC catalyst was consumed at different steam feed rate so as determine the effect of steam feed rate on the crystallite size of catalyst; catalytic tar reforming temperature, the space velocity as a function of the gas yield and biomass carbon conversions in fluidized bed gasifier were investigated. The product gas components was discussed in detail and compared between both cases of the absent of steam and the presence of steam.

4.1. Discussion of catalytic steam tar reforming pathway

Figure 11 shows simple pathway of the woody biomass gasification process with Ni/BCC catalyst. The woody biomass was first pyrolysis to form gas, tar and char at 923 K. Both useful gas and tar passed through the catalyst particles. Tar was dissociatively adsorbed onto a nickel site where nickel catalyzed dehydrogenation occurs. With steam injection tar would be cracked and reformed follow the mechanism.

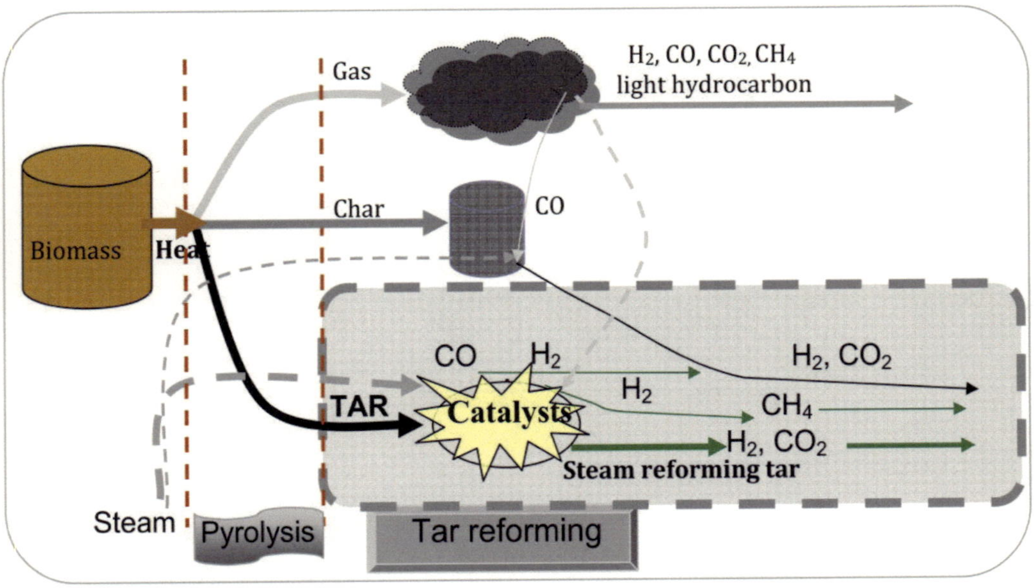

Figure 11. Schematic pathway of biomass pyrolysis and tar reforming using Ni/BCC catalyst

The chemistry of coal (biomass) gasification is usually depicted to involve the following reactions of carbon and steam [29]. The standard enthalpy change (gram molecules) at 298 K is shown for each reaction. The most important reactions are listed in Table 1[29–31]

Process	ΔH^{O}_{298} (kJ/mol)	
Steam reforming		
$CH_4 + H_2O = CO + 3H_2$	+ 206	(7-1)
$C_nH_m + nH_2O = nCO + n + (m/2)H_2$	+ 1175[a]	(7-2)
CO_2 reforming		
$CH_4 + CO_2 = 2CO + 2H_2$	+ 247	(7-3)
Gasification		
$C + H_2O \rightarrow CO + H_2$	+ 131.3	(7-4)
$C + 2H_2O \rightarrow CO_2 + H_2$	+ 90.2	(7-5)
$C + CO_2 \rightarrow 2CO$	+ 172.4	(7-6)
Water–gas shift reaction		
$CO + H_2O \rightarrow CO_2 + H_2$	- 41.1	(7-7)
Methanation		
$2CO + 2H_2 \rightarrow CH_4 + CO_2$	- 247.3	(7-8)
$CO + 3H_2 \rightarrow CH_4 + H_2O$	- 206.1	(7-9)
$CO_2 + 4H_2 \rightarrow CH_4 + H_2O$	- 165.0	(7-10)
$C + 2H_2 \rightarrow CH_4$	- 74.8	(7-11)

a for nC_7H_{16}

Table 1. Synthesis gas reactions

4.2. Catalyst evaluation in steam gasification process in fluidized bed gasification

The conventional Ni/Al_2O_3 catalyst and Ni/BCC catalyst available for steam reforming were used to test tar reforming performance. As mentioned in section 7.1 and discussed in section 7.3, the deposited carbon may cause for deactivating catalysts due to covering activate site of catalysts. In this section, all experiments were performed under steam injection with s/c: 0.6 mol/mol. The added steam was expected to suppress the deposited carbon on activate surface of catalysts. In this section, the effect of steam addition on tar conversion, gas yields, and carbon conversion were investigated. The reactivity both of the Ni/BCC and Ni/Al_2O_3 have been compared and discussed in detail.

In the activity tests, the formation of products were observed for 120 min, all calculated results of the gas yields and C_gas were the average of the specific results from various specific sampling times, which started at 20 min after feeding biomass and then in 20 min intervals.

As illustrated in Figure 12 (a), the gas yields are shown lowest for non–catalyst, while higher gas yields have achieved for the catalysts. The great improvement of product gas for the case of Ni/BCC catalyst should be given more attention. Most main gas components (CH_4, CO, CO_2, H_2) were higher than those of Ni/Al_2O_3 catalyst. Especially, in the case of the Ni/BCC catalyst, CO and H_2 yield were 10.8 and 12.3 [mmol/ g–sample daf] higher than those of the Ni/Al_2O_3 catalyst. These satisfactory results could be explained by a part of the deposited carbon on the Ni/BCC catalyst and Ni/ BCC char had been gasified in the presence of steam according to the reaction pathway as following reaction equations (Eqs. (7-4), (7-5), and (7-6)) in the Table 1.

Steam might also produce a larger active surface of Ni/BCC by steam gasification of deposited carbon on the surface of catalyst, which is also evidenced by BET data of used catalyst. After 1 h operation, total free surface of the Ni/Al_2O_3 decreased from 104 to 32 m^2/g due to reduction of nanopores by blockage of deposited carbon and catalyst particle growth. While, total free surface of Ni/BCC lightly reduced from 350 to 339 m^2/g, this is due to characteristic porosity of brown coal char. The results indicate that steam plays a very important factor to regenerate activity of the new catalyst by steam gasification of deposited carbon on catalyst and to significantly enhance the quality of product gas of woody biomass gasification.

Biomass carbon balance is illustrated in Figure 12 (b). It was carried out in a similar way as described in section 7.3 The blank on the top of each bar can be considered as a percentage of the C_tar which was calculated by equation 3-5 in section 7.3.

It is different from the pyrolysis process, approximately 16.5% carbon in the fresh Ni/BCC catalyst was gasified in the presence of steam. Its percentage was defined by comparing between carbon in the fresh Ni/BCC catalyst and carbon in used Ni/BCC catalyst. In the presence of the Ni/BCC, biomass carbon conversion (C_gas) was calculated by subtraction between carbon of total product gas and conversion carbon of fresh Ni/BCC, which is mentioned on above. Using that method, we found that highest C_gas and lowest C_tar were achieved as 66 and 4.4% for Ni/BCC catalyst test, respectively, while the C_gas and C_tar obtained were only 59.9% and 7.4% for Ni/Al_2O_3 catalyst test, respectively. Biomass tar conversion obtained was approximately 88.9% in Ni/BCC catalyst. The results indicate better catalyst activity for Ni/BCC. The detailed mechanism for this high activity is unclear at the present, however, it can be explained that some of the following characteristics of the Ni/BCC catalyst might be associated with this activity: well distribution of nickel particles due to carbon functional group in brown coal, high porosity of the catalyst, mineral component. In addition, Tomita *et al.* [32] reported that in the presence of steam, tar might be absorbed on catalyst and then be gasified without forming soot. Even if carbon was formed on the catalyst surface, it could be easily gasified. He also found that the carbon deposited over nickel was rapidly gasified with hydrogen at 873 K by reaction 7-11 in Table 1 [33]. This fact that can be observed both of CH_4 and H_2 yields are higher than that of the Ni/A_2O_3 catalyst.

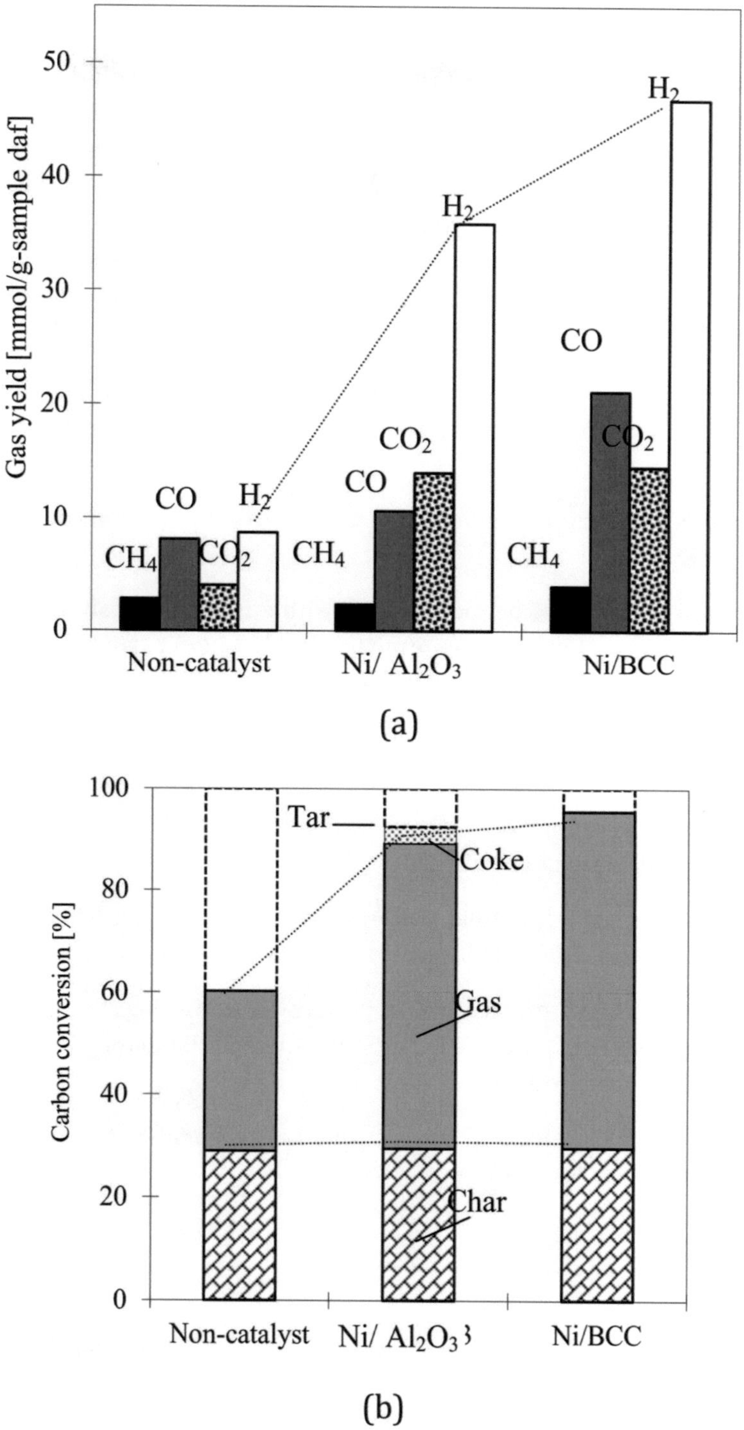

Figure 12. Comparison of different catalysts and non–catalyst in the presence of steam: (a) gas yields and (b) biomass carbon balance (923 K, sv = 10000, s/c =0.6)

5. Summary and conclusions

Nickel loaded brown coal char acts a new catalyst for decomposing tar of woody biomass gasification in a two-stage fixed-bed and fluidized bed gasifier has been investigated. With the advantages of catalytic steam tar reforming is carry out at low temperature. On the other hand, catalytic reforming tar methods have significant possibilities in low temperature gasification processing for high product gas quality. This chapter attempts to a comprehensive knowledge for low temperature pyrolysis and gasification process covering study of operation conditions affecting catalytic activity behaviors of nickel loaded brown coal char catalyst.

For the effect of pyrolysis temperature on the crystalline size of nickel particle size, it is slightly affected by temperature lower than 923 K, but great affected by temperature higher than 973 K.

Two-stage Fixed-bed Reactor has been identified as processing the good activity even at low temperature 923 K.

The Ni/BCC catalyst could not perform as well as the Ni/Al_2O_3 catalyst to decompose tar under pyrolysis process. However, the results show that both catalysts are good active to decompose tar from biomass pyrolysis at 923 K.

The experimental results show a new catalyst having good catalytic activity and stability in the presence of steam at 923 K.

In the new catalyst application with the presence of steam, Ni/BCC catalyst exhibited more activate than conventional catalyst Ni/Al_2O_3.

It was found that, catalyst has a good performance and stability at 923 K. Approximately 89.5 % of biomass tar was reformed to useful gas components (CO, H_2, CH_4).

Steam has already proved to be very important in activating the new catalyst and significantly enhances the quality of product gas of woody biomass gasification with high hydrogen concentration of product gas.

The results suggest that the Ni/BCC catalyst offers a potential to be used as a tar steam reforming catalyst in biomass gasification.

Author details

Le Duc Dung
School of Heat Engineering and Refrigeration, Hanoi University of Science and Technology, Vietnam

Le Duc Dung, Kayoko Morishita and Takayuki Takarada
Department of Chemical and Environmental Engineering, Gunma University Faculty of Engineering, Japan

Acknowledgement

We would like to express my sincere thanks to all who have contributed to this work. We would like to express my gratitude to Ms. Yukiko Ogawa for her help in performing proximate and ultimate analyses of samples. I would like to thank Mrs. Miyoko Kakuage, Mrs. Mayumi Tanaka, Dr. Xianbin Xiao, Dr. Liuyun Li and all of students in Takarada's Laboratory for their support.

I gratefully acknowledge the financial support of this work by the Project of Prefecture Collaboration of Regional Entities for the Advancement of Technological Excellence, Japan Science and Technology Agency for one and half year. Greatly, I acknowledge the financial support of this work from Asian Jinzai project, Japan Government scholarship for one and half year. We would like to acknowledge Gunma University Faculty of Engineering and Hanoi University of Science and Technology for all their support throughout this research.

6. References

[1] Nordgreen, T., T. Li and K. Sjostrom; *Energy Fuels;* 20, 290–895 (2006)
[2] Maniatis, K. and A. A. C. M. Beenackers; *Biomass Bioenergy;* 18, 1–4 (2000)
[3] Milne T.A. and R.J. Evans. ; *US Department of Energy,* (1998)
[4] Morf P., P. Hasler, T.Nussbaumer; *Fuel;* 81, 843–853 (2002)
[5] Han J., and H.J. Kim; *Renewable and Sustainable Energy Reviews,* 12, 397–416 (2008)
[6] Gil J., Corella J., M. P. Aznar, and M. A.Caballero; *Biomass Bioenerg.,* 17, 389–403 (1999)
[7] Corma A., and A.V. Orchille´s; *Microporous and Mesoporous Materials;* 35–36, 21–30 (2000)
[8] Dayton D. A review of the literature on catalytic biomass tar destruction; USA: *National Renewable Energy Laboratory;* (2002)
[9] Davidian T., N. Guilhaume , H. Provendier, and C. Mirodatos; *Applied Catalysis A:* General 337, 111–120 (2008)
[10] Sutton D., B. Kelleher, J. R.H. and Ross; *Fuel Processing Technology;* 73, 155–173 (2001)
[11] Moulijn J.A., A.E. van Diepen, and F. Kapteijn; *Applied Catalysis A: General;* 212, 3–16 (2001)
[12] Asadullah, M., S. Ito, K. Kunimori, M. Yamada and K. Tomishige; *J. Catal.;* 208, 255–259 (2002)
[13] Zhang R, Brown RC, Suby A, and Cummer K.; *Energ Convers Manage;* 45, 995–1014 (2004)
[14] Furusawa, T., T. Sato, H. Sugito, Y. Miura, Y. Ishiyama, M. Sato, N. Itoh and N. Suzuki; *Int. J. Hydrogen Energy,* 32, 699–704 (2007)
[15] Srinakruang J., K. Sato, T. Vitidsant, and K. Fujimoto; *Fuel,* 85, *17–18,* 2419–2426 (2006)
[16] Srinakruang J., K. Sato, T. Vitidsant, and K. Fujimoto; *Catal. Communic.;* 6, 437–440 (2005)
[17] Sato K., K. Fujimoto; *Catalysis Communications,* 8, 1697–1701 (2007)
[18] Dou B, J. Gao, X. Sha, and S. W. Baek; *Appl Ther Eng;* 23, 2229–2239 (2003)
[19] Baker EG. *Ind Eng Chem Res;* 26, 1335–9 (1987)

[20] Miyazawa T., T. Kimura, J. Nishikawa, S. Kado, K. Kunimori, K. Tomishige; *Catalysis Today*; 115, 254–262 (2006)

[21] Tomita, A., K. Yoshida, Y. Nishiyama and Y. Tamai; *Carbon,* 10, 601–606 (1972)

[22] Tomita, A. and Y. Tamai; *Fuel,* 60, 992–994 (1981)

[23] Tomita, A., Y. Ohtsuka and Y. Tamai; *Fuel,* 62, 150–154 (1983)

[24] Tomita, A., Y. Watanabe, T. Takarada, Y. Ohtsuka and Y. Tamai; *Fuel,* 64, 795–800 (1985)

[25] Ohtsuka Y., and A. Tomita; *Fuel,* 65, 1653–1657 (1986)

[26] Takarada, T., T. Tonishi, Y. Fusegawa, K. Morishita, N. Nakagawa and K. Kato; *Fuel,* 72, 921–926 (1993)

[27] Sato M.; Master thesis; *Gunma Univer. Facul. of Eng.*; (2004)

[28] D. D. Le, X. Xiao, K. Morishita and T. Takarada; J. Chem. Engineering of Japan, 42, 51-57 (2009)

[29] Bodle, W. W., and Schora, F. C. *In* "Advances in Coal Utilization Technology," (K. S. Vorres and W. W. Waterman, eds.), p. 11. *Institute of Gas Technology*, Chicago (1979)

[30] Donald L. C. Academic Press, USA, (Chapter 3, 8, and 9) (1998)

[31] K. Aasberg–Petersen, J.–H. Bak Hansen, T.S. Christensen, I. Dybkjaer, P. Seier Christensen, C. Stub Nielsen, S.E.L. Winter Madsen, and J.R. Rostrup–Nielsen; *Applied Catalysis A: General*; 221, 379–387 (2001)

[32] Tomita, A., Y. Watanabe, T. Takarada, Y. Ohtsuka and Y. Tamai; *Fuel,* 64, 795–800 (1985)

[33] Tomita, A., K. Yoshida, Y. Nishiyama and Y. Tamai; *Carbon,* 10, 601–606 (1972)

Chapter 12

Characteristics of Animal Slurry as a Key Biomass for Biogas Production in Denmark

Jin M. Triolo, Alastair J. Ward, Lene Pedersen and Sven G. Sommer

Additional information is available at the end of the chapter

http://dx.doi.org/10.5772/54424

1. Introduction

Climate change has become an important global issue and animal manure has been pointed out as a major source of greenhouse gas (GHG) emissions. The Danish government targets animal manure as a key biomass with the aim of producing renewable fuels and reducing GHG emissions. Animal manure is a mixture of excreta and materials added during management. Apart from the major part of animal slurry which is feces and urine, animal slurry is composed of many materials, i.e., sand, water from cleaning, small branches and straw from the bedding materials. Thereby a wide variation of characteristics can be found depending on different management systems, animal type and diet, etc. which make for difficulties in the estimation of manure quality for biogas production.

There is no doubt that in the future the world's energy supply market will be dominated by renewable energy, since there is no alternative. While combustion is the most common method to gain energy from plant biomass such as wood and wood chip, the high content of water in animal slurry suits wet fermentation for conversion to energy, since direct combustion is not appropriate for most animal manures. Direct combustion dry matters (DM) content must be at least 45% [1]. Animal slurry is typically in a liquid form where DM typically contains 1-10% [2]. The production of energy through combustion can be made by enriching fiber fractions by separation technology. Fiber rich animal slurry through separation technology can potentially replace 3.6 PJ of coal energy, which corresponds to 4.3 ‰ of the yearly Danish energy consumption, if one third of the Danish manure is separated [3,4]. The European Commission made a considerable effort by making mandatory national targets for renewable energy shares of final energy consumption in 2020 with the goal: Increasing energy efficiency by 20% by 2020 and reducing GHG emissions at least 20% within the same period [5]. To commit to the targets of the European Commission, the Danish government targets animal slurry as a key element, setting an ambitious goal of increasing the utilization of animal manure for energy production from current levels (5%) up to 40% by 2020 [6].

Facing an "aggressive growth" of biogas production using animal slurry as prime feedstock, it is of great importance to understand critical barriers of characteristics of animal slurry on economic viability. Further, it is of current interest to find solid organic residues as co-substrate, in order to bring the best synergy by overcoming barriers of animal slurry. Biomass is the term given to all organic matter. Its production worldwide is estimated at 146 billion metric tons per year, composing mostly of wild plants [7-9]. The energy of biomass originates from solar energy through photosynthesis, which converts water and CO_2 into organic materials in plant biomass. It comprises i.e., plant, wood, energy crop, aquatic plants. Whereas plants store energy in the form of organic materials from solar energy directly, animals generate excreta through metabolizing and digesting. Hence, animal slurry has unique characteristics compared to other biomasses, since during digestion the relatively easily degradable organic matter is utilized while recalcitrant carbon concentrations are increased by animal digestion [10], which limits subsequent anaerobic degradability (BD) and biogas potential. Moreover, the quantity of organic polls in liquid slurry is often too small to perform economically viable operations [10,11].

Hence, the aim of this study is intensive investigation and identification of critical barriers in characteristics of animal slurry. The study was carried out using diverse animal slurry collected from 20 different farms in Denmark, firstly focusing on the Biochemical Methane Potential (BMP) of animal slurry with respect to the total feedstock fresh weight, organic fractions (VS) and DM. Physicochemical characteristics were determined to qualify animal slurries as prime substrates for biogas reactors, and the results were applied to construct algorithms to assess potential methane yield. This study finally highlights the characteristic digestibility of animal slurry compared to plant lignocellulosic biomass. The study further aims to improve our suggested model to predict BMP [10]. In accordance with the objective of the study, quantification of nutrients and characterization of indigestible organic pools of a wide range of animal slurry will be carried out.

2. Animal slurry as greenhouse gas source

Intensified livestock industry and increased consumption of meat and animal products are contributing to a surplus of animal by-products in Europe and other developed countries. In Europe more than 1500 million tons of animal slurry is produced every year [12]. Traditionally, slurry has been recycled as fertilizer, providing nitrogen (N) and phosphorous (P) source for plants and crops. However accumulation of carbon and leaching of N and P causes a serious and negative environmental impact (water, air and soil contamination). Thus, pathogens from improperly treated animal wastes often threaten public health. The emission of GHG during livestock slurry management has been widely ignored compared to the local environmental problem, as the impact itself is global and therefore indirect. It is not long ago that the climate changes became an important global issue, and animal slurry has been identified as a major source of GHG emissions in the agricultural sector.

The original solar energy stored in animal slurry is a form of organic material. The pathway of conversion of organic materials is of great importance to the ecological balance, as it determines the carbon flow. The principal of the conversion of organic materials is its oxidation either by oxygen in aerobic conditions or by transferring electrons when oxygen is not available (anaerobic condition). Degradation of organic materials in animal slurry in nature mostly occurs under anaerobic conditions that produce GHG, which breaks the carbon flow balance. To balance carbon flow, aerobic degradation must occur to bring the organic materials back to water and CO_2 which was spent for photosynthesis, however the oxygen in animal slurry is critical due to high contents of organic materials which consume the oxygen.

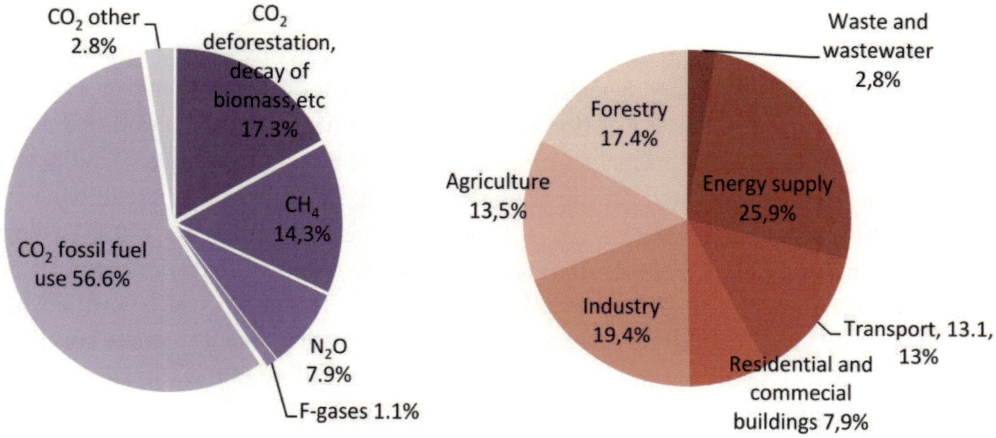

Figure 1. Share of anthropogenic greenhouse gas emissions: (a) Share of different anthropogenic GHGs of total emissions in 2004 in terms of carbon dioxide equivalents (CO2-eq). (b) Share of different sectors of total anthropogenic GHG emissions in 2004 in terms of CO2-eq. (Forestry includes deforestation.) [13].

Aerobic degradation may occur in the surface due to diffusion of oxygen but the amount is still insignificant. Hence, aerobic treatment of animal slurry often shows less environmental impact such as oxygen depletion of aquatic systems. The representative GHG in the agricultural sector are carbon dioxide (CO_2), methane (CH_4) and nitrous oxide (N_2O). In Denmark, animal manure accounts for about 40% of total CH_4 and 20% of total N_2O emissions [14]. CH_4 originates mainly from enteric fermentation in ruminant animals like cattle, whereas for pig production, slurry management is the primary source for CH_4 emission. Another important greenhouse gas is N_2O which is emitted from turnover of nitrogen in manures and in agricultural soils [15]. In comparison to CO_2, it is reported that the emission from CH_4 and N_2O is low [14], however their global warming potentials are 23 and 296 times higher than that of CO_2, respectively [2]. The distribution of GHG in total emissions is given in Figure 1, showing that the agricultural share of global emissions is 13.5% [13], while that of national emissions in Denmark is considerably higher at 18% [15].

3. Biogas production using animal slurry

Utilization of the energy from methane emitted by animal manure is of current ongoing interest. Biogas production is the technology that converts animal manure and other biomasses into viable fuel, recycling the carbon resource of animal slurry. Biogas production is known to be the most suitable technology to produce renewable fuels from wet biomass such as animal slurry.

Biogas can be produced from nearly all kinds of biomass, nevertheless, the largest resource represented is animal slurry. In an effort to obtain higher methane yields, co-digestion of livestock manure with industrial organic waste has been implemented successfully in large scale biogas plants in Denmark. Nevertheless, only a few biogas plants have generated economic profit in Denmark. Facing a 10-fold dramatic increase of Danish biogas production, the economic point of view should be integrated by ensuring the price of biogas being competitive in the energy market. This could be done either by increasing biogas yield or reducing operating costs per feedstock unit. The low profitability of biogas produced from animal slurry is due to the fact that quality and quantity of organic pools are critical. Low biodegradability (BD) of animal slurry is often caused by large amounts of indigestible fractions which are concentrated during animal digestion. The quantity of organic pools in slurry is often too small to perform economically viable operations [10,11]. Biogas productivity per unit of feedstock volume is inevitably related to its biochemical and physical composition. Hence, energy crop has been widely used as co-substrate to enhance biogas productivity particularly in Germany and Austria, using mostly maize, sunflower, grass and Sudan grass [16]. Meanwhile, in Denmark industrial organic waste is co-digested in most large scale biogas plants to increase methane yield. This results in limited availability of organic industrial waste, creating a setback of extending the biogas industry [11,17].

4. Methodology

4.1. Determination of methane potential

20 Animal manures from different farms were collected. The types of manure collected were dependent on the management of the farms. For pig manure, fattening pig, sow, piglet, and a mixture of sow and piglet were collected. Calf, dairy cow, cattle and mink manures were also included. Most of the samples collected are currently fed to biogas reactors except the calf manure.

The inoculum used for the BMP assay was collected from Fangel biogas plant in Denmark. Fangel biogas plant processes mixtures of pig manure and industrial organic waste (80:20 w/w) under mesophilic conditions (37°C). The BMPs of each subgroup were determined according to a standard protocol provided by VDI 4630 [18]. 1.1 liter batch infusion digesters were used for fermentation. 400mL of inoculum was used in each batch, with a 3:1 inoculum:substrate (I:S) ratio on a DM basis. A medium was added to ensure enough nutrients for bacterial growth and a standard pH buffer capacity following the recommendations of VDI 4630 [18] and ISO Standard 11734 [19] was also added. The composition of the medium used was shown in Table 1. The constituents were added to 1 L of distilled water containing less than 1 mg/L dissolved oxygen. The test medium prepared was flushed with nitrogen for 20 min to allow anaerobic conditions, and then 150 mL of the mixture of inoculum and substrate was added to each batch reactor.

Chemical compound	Molecular formula	g/L
Anhydrous potassium dihydrogen phosphate	KH_2PO_4	0.27
Disodium hydrogen phosphate didecahydrate	$Na_2HPO_4{\cdot}12H_2O$	1.12
Ammonium chloride	NH_4Cl	0.53
Calcium chloride dehydrate	$CaCl_2{\cdot}2H_2O$	0.02
Sodium sulphide nonahydrate	$Na_2S{\cdot}9H_2O$	0.1

Table 1. The composition of the medium used for this study

Digestion was carried out under mesophilic conditions (37°C) and terminated when daily biogas production per batch was less than 1% of cumulative gas production according to VDI 4630 [18]. On a daily basis each batch digester was mixed thoroughly by shaking to prevent dry layers and to encourage degassing. Gas volume was read off using a 500 ml syringe (Hamilton, Super syringe). Methane (CH_4) and carbon dioxide (CO_2) were determined by a gas chromatograph (HP 6890 series), equipped with a thermal conductivity detector and a 30 m × 0.32 mm column (J&W 113-4332). The carrier gas was helium (30 cm s^{-1}). Injector temperature was 110°C, and detector and oven temperatures were 250°C. Injection volume was 0.4 mL and the split rate was 1:100. Biogas production was given as gas volume of the gas flow and at STP conditions (273 K, 1.013 bar). Biomethane was quantified assuming that the dry biogas was composed of CO_2+CH_4, alone, consequently CH_4 production volume is calculated according to VDI 4630 [18] by multiplying the dry gas production by the ratio CH_4/ CO_2+CH_4. All the batch procedures and quantitative evaluation of biomethane production were similar. Blanks were measured in batches with inoculum to correct gas production. A control test was carried out using cellulose powder (Avicel PH-101 cellulose (Sigma Aldrich)) as a standard substrate. The BMP of cellulose was 386.7(±2.4) CH_4 NL (kg VS)$^{-1}$ and the ratio of BMP to theoretical BMP (TBMP) was 93.7%. TBMP of cellulose is 415 CH_4 NL (kg VS)$^{-1}$. The very low standard deviation (SD) indicates a high repeatability of results from batch fermentation of homogeneous substrate, and thus a good standard of the performed batch fermentations.

4.2. Physicochemical characterization and data analysis

DM, VS, Volatile Fatty Acids (VFA), total ammoniacal nitrogen (TAN), and total Kjeldahl nitrogen (TKN) were determined according to standard procedures [20]. Neutral detergent fibers (NDF) were determined by α-amylase neutral detergent extraction [21]. Acid detergent fiber (ADF) and acid detergent lignin (ADL) were determined ash free by acid detergent extraction as described in the ISO 13906 [22]. Organic nitrogen (N_{org}) was calculated as the difference between TKN and TAN. Crude protein was determined by multiplying N_{org} by 6.25 [10, 23]. Hemicellulose, cellulose and lignin were determined in accordance with Van Soest's characterization for fiber analysis [24,25]. The NDF was used to determine total cell wall components, including hemicelluloses, cellulose, lignin, and fiber-bound proteins, and it corresponds to lignocellulose [10]. The difference between VS and NDF is defined as neutral detergent soluble fraction (NDS) that corresponds to non-cell components. ADF consists of cellulose, lignin, and insoluble proteins. The difference between NDF and ADF can be identified with hemicellulose. ADL is identified with lignin, with the assumption that the fraction of lignin-bound nitrogen is insignificant. Thereby, the difference between ADF and ADL is defined as cellulose.

5. Results and discussion

Analysis of each compound gives a general view of the characteristics of each of the tested slurries.

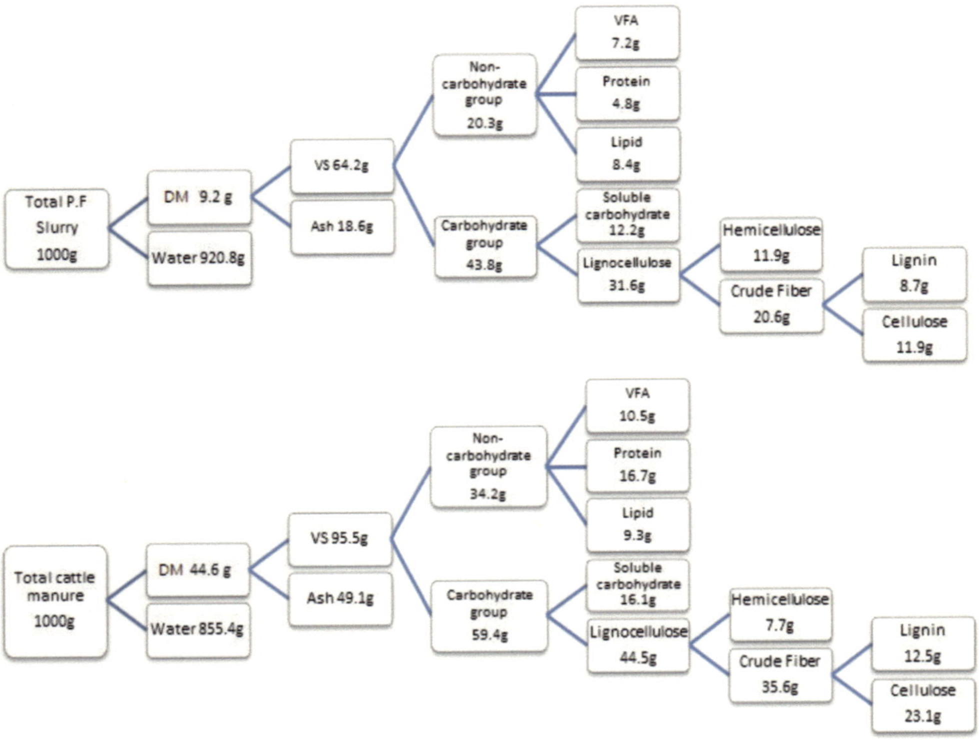

Figure 2. Distribution of each component in 1kg of pig fattening slurry (above) and of cattle manure (below); P.F: pig fattening.

The analysis is based on the measurement and simple mass balance calculation as follows:

- Water content: Total mass- DM(measurement)
- Ash: DM - Measurement of VS(measurement)
- Non-carbohydrate group = protein + VFA +lipid
- Carbohydrate group= VS- non carbohydrate
- Lignocellulose = hemicellulose + cellulose + lignin
- Crude fiber = cellulose + lignin Soluble carbohydrate

The overview of physicochemical characteristics using the distribution of component in cattle manure and pig fattening slurry is presented in Figure 2. Total slurry is separated as DM and water and DM is furthermore separated as VS and ash. VS is separated as a non-carbohydrate group and a carbohydrate group. The non-carbohydrate group is separated into VFA, protein and lipid. The carbohydrate group is separated into soluble carbohydrate and lignocellulose. Lignocellulose was separated into hemicellulose and crude fiber that composes of cellulose and lignin. As can be seen in Figure 2, the main characteristic of cattle manure is higher DM content which is the equivalent of approximately double DM concentration of pig fattening. The amount of DM is larger for cattle manure as well, as VS is a major fraction of DM. Since there is more VS in cattle manure than in pig fattening, the amount of each organic component including protein, lipid, etc., is larger in cattle manure as well. Nonetheless, the concentration of each organic component in VS is higher for pig fattening slurry than cattle manure.

5.1. Dry matters and organic matters

DM concentration is an important parameter to design the biogas reactor size and calculate capacity of a biogas plant such as an electrical power installation [2]. Too diluted animal slurry reduces economic viability but too high DM, for example higher than 15% DM, may cause a pumping problem. It is generally said that 10% DM is optimal.

The slurries included in this study had a wide range of DM contents (Table 2). It ranged between 34.1 (mink) to 238.6 kg^{-1} (calf). The highest DM was found in calf manure, since the majority was composed of straw bedding materials, but currently calf slurry is not used for biogas production in Denmark. DM concentration of all the tested samples was 9.7% of the mean value, close to the optimal DM concentration. However excluding the calf manure that is not used for biogas production, the mean DM concentration is much lower. Indeed, the DM concentration of the biogas reactor to which most of the manures tested were fed was 5.8%. As can be seen in Table 2, particularly piglet and mink manure have very low content of DM, which approximately amounts to 3-5% DM of total mass.

Slurry type	PH	DM	VS	
		(g kg^{-1})	(g kg^{-1})	% of DM
Piglet (n=4)	7.20(0.3)	54.3(31.0)	42.8(25.5)	77.4
Sow and piglet(n=3)	6.90(0.2)	66.5(18.9)	53.7(13.4)	81.7
Fattening pig (n=2)	7.53(0.3)	64.5(77.9)	52.9(67.5)	69.7
Sow (n=3)	7.74(0.5)	79.2(42.7)	64.2(36.8)	80.2
Dairy cow (n=3)	7.10(0.2)	94.1(12.1)	80.9(11.1)	85.9
Cattle (n=2)	7.42(0.2)	144.6(41.0)	95.6(1.8)	68.7
Calf (n=2)	NA	238.6(118.8)	218.8(108.1)	91.8
Mink (n=1)	7.28	34.1	27.0	79.2
Mean	7.31(0.3)	97.0(48.9)	79.47(37.5)	79.3

Table 2. The concentration of dry matters (DM) and organic materials (VS) of the slurry tested; given as mean values, standard errors in parentheses. n = number of samples included

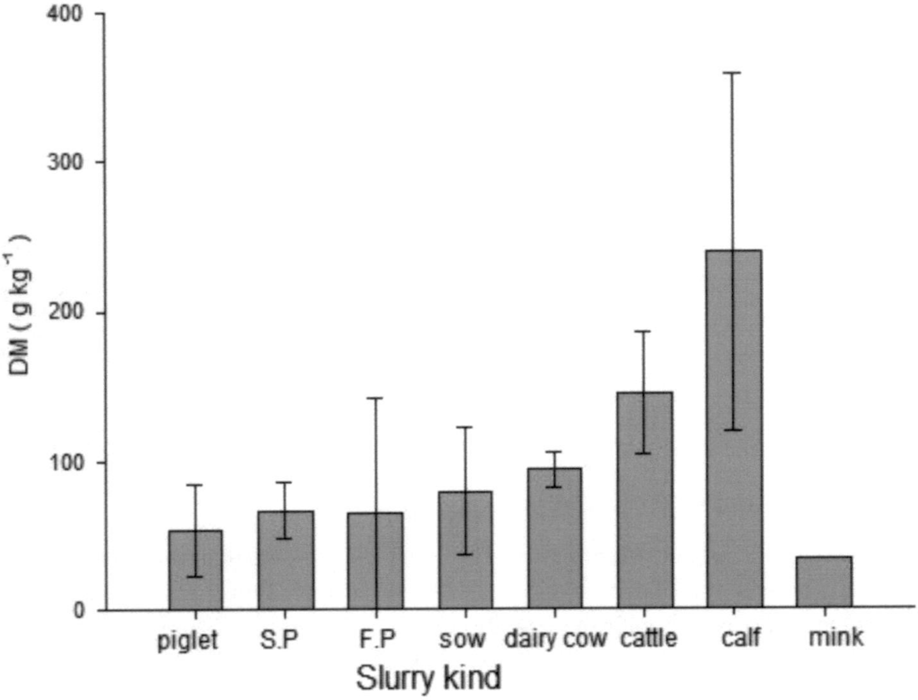

Figure 3. Comparison of dry matters (DM) depending on manure type; error bars show standard deviation; S.P: sow and piglet; F.P: fattening pig.

Compared to the large variation of DM concentration within and between manure groups (See Figure 3), the VS concentration (as a percentage of DM) varies much less. Table 2 shows that VS concentration varies between 70 to 90% of DM. The VS concentration is crucial to determine organic loading rate, and determines the methane yield. The variation of VS as a percentage of fresh weight is large, since VS is the organic fraction of DM.

5.2. Methane productivity

BMP is the maximum methane yield through anaerobic digestion, thereby BMP is identified with the cumulative methane yield at the end of a fermentation test. However, termination of fermentation is not clearly defined. Hence, the fermentation duration may vary from 7 to 365 days [26]. VDI 4630 [18] mentions that digestion should be terminated when daily biogas production per batch is less than 1% of the cumulative gas production, which is applied for our study. As BMP is the maximum methane yield, it is the most important parameter to evaluate the quality of feedstock for biogas production, and is used to design real scale biogas reactors. BMP is most frequently presented as being the unit of methane volume in terms of kg VS, hence, the BMP level varies depending on organic compositions in VS. Cumulative methane productions of the animal manures tested as a function of time are presented in Figure 4. As can be seen in Figure 4, the great majority of methane was produced in the first 2 weeks and thereafter only small amounts of gas were released. The cumulative methane curves generally follow first order kinetics, since the hydrolysis process is the rate limiting process [27,28].

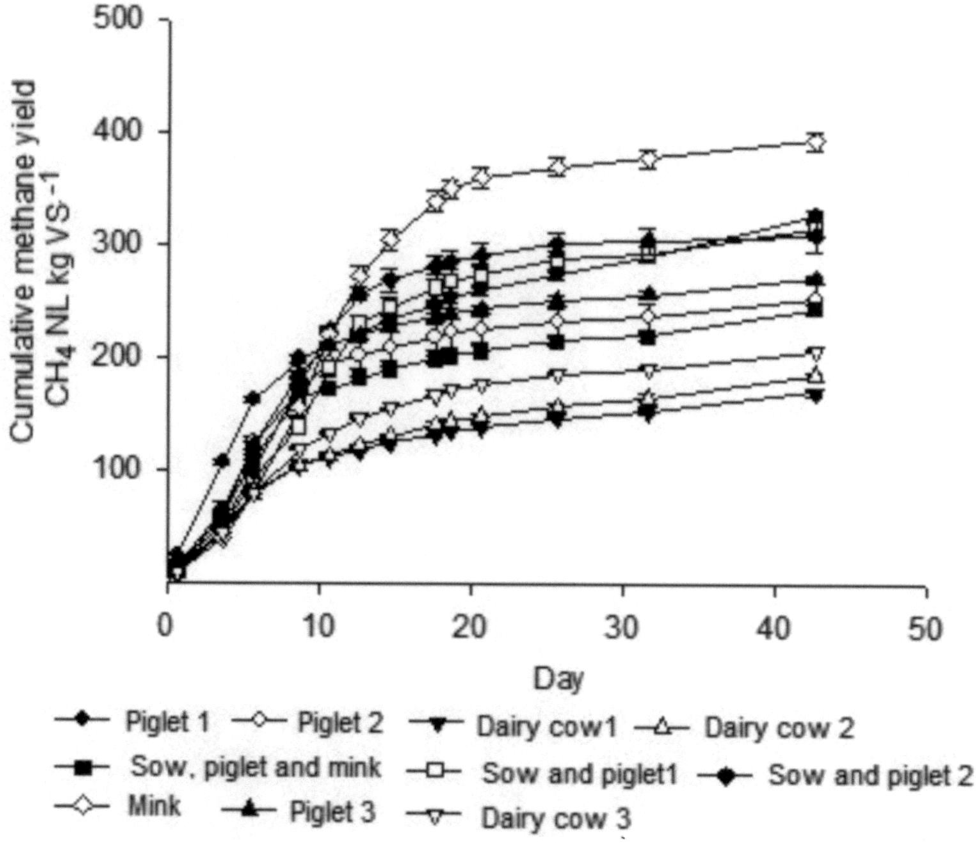

Figure 4. The cumulative methane yield curves from the biochemical methane potential determination test. Not all the data are present.

Whereas DM and VS are quantitative parameters for methane production potentials, BMP is the quality parameter that is reflected of bioconversion of organic compositions, which have dependency of methane potentials of each organic composition and its BD. Hence, the BMP value can be used as an index of the BD of substrates to biogas reactors [29].

Figure 5 gives the comparison of BMP results in terms of kg VS and of kg slurry of the animal slurry tested for this study. As can be seen in Figure 4, BMP of various animal slurries ranges between 170 – 400 CH_4 NL kg^{-1} VS. Most of the cow slurry is shown at the lowest level within the tested slurries, whereas high methane potential of pig slurries is found. This result has a good agreement with previous studies [10,23]. Mink slurry had the second highest BMP within the samples tested. BMP in terms of kg slurry had much larger variation in the range of 1.8 – 70 CH_4 NL kg^{-1} slurry of which two different terms of BMP were somewhat opposite, due to such a large variation of the DM concentration. Since the variation of the DM concentration in animal slurry is larger than methane potential per unit of VS, the results indicate that the water content of the animal slurry is the most significant parameter for methane productivity in reactors compared to BMP.

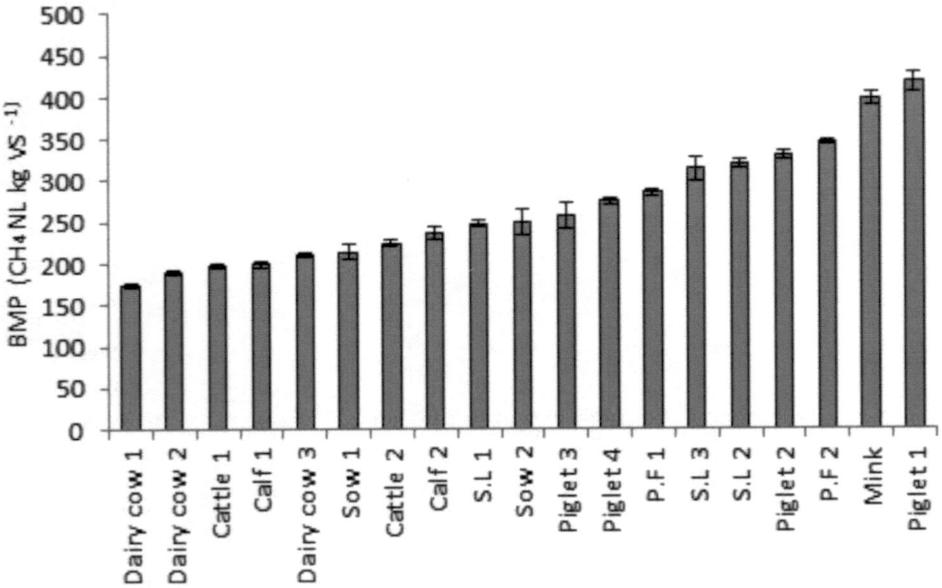

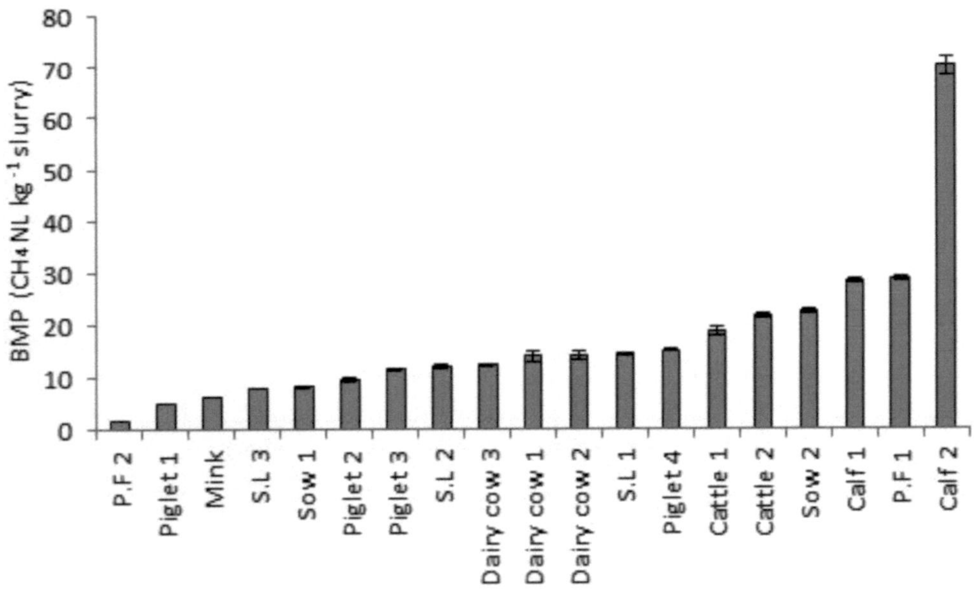

Figure 5. BMP results per kg of VS (above) and per kg of fresh weight of the animal manures tested for this study; vertical bars show standard deviations; S.P: sow and piglet; F.P: fattening pig.

Figure 5 indicates that control of the DM concentration is more crucial than control of BD of substrate with respect to increasing methane yield within the range that pumping is appropriate. Figure 6 shows a good linear correlation between DM concentration and biomethane potentials per kg slurry (R^2 =0.896). The results highlight the importance of a qualified control of water content in animal slurry. Controlling of DM could be achieved by co-digesting solid organic substrate such as energy crops, for this reason, energy crop has been widely used as co-substrate to enhance biogas productivity [16]. Sufficient water content is inevitable for the wet fermentation procedure, as too low concentrations of water decrease the biomethane production rate. However the 94.1% water content of effluent from the tested reactor indicates that there is need of optimizing it by codigesting solid organic residues. The high content of water was probably caused by spillage of cleaning water, which contributed to the lowest potential biomethane yield per unit of biogas reactor, in spite of high BMP results among the animal slurries included, as BMP is the methane potential in term of VS concentration.

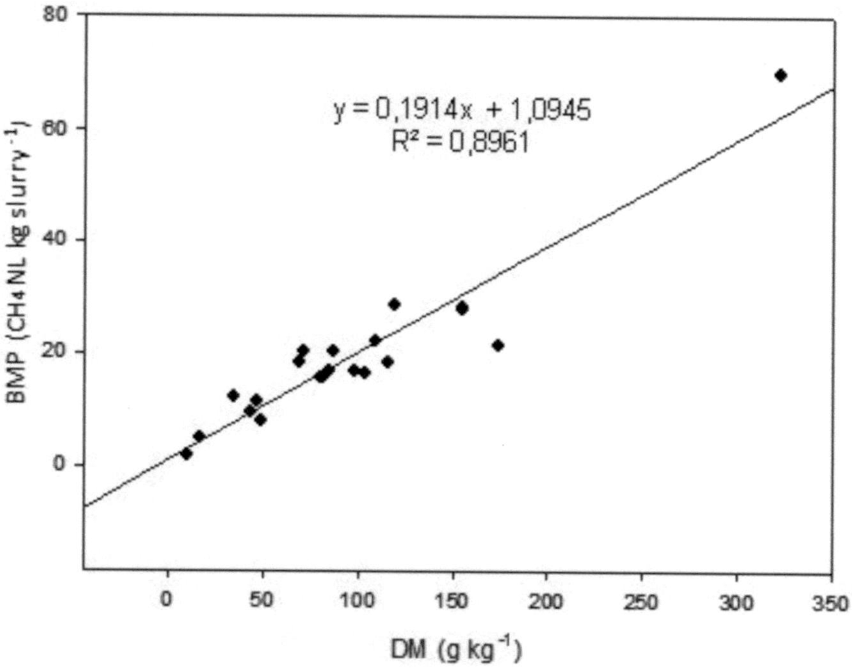

Figure 6. Relationship between the DM concentration and biomethane potential (BMP) per kg of slurry.

5.3. Lignocellulose in animal slurry

Lignocellulose is an element of the plant cell wall, and it majorly composes of hemicellulose, cellulose and lignin.

Lignin is a natural complex polymer and the chief noncarbohydrate constituent of wood binds to cellulose fibers providing mechanical strength and structural support of plants where it can be found extensively in the cell walls of all woody plants. Lignin is the most abundant natural source after cellulose, and between 40 and 50 million tons of lignin per annum are produced worldwide [30], constituting one-fourth to one-third of the total dry weight of trees. As the chemical composition of lignin has a certain variation, it is not possible to define the precise structure of lignin.

Due to the mechanical strength of lignocellulose supported by lignin, lignocellulose is known to be recalcitrant carbon pools. Lignocellulose is very slowly bioconvertible in anaerobic environments due to its rigid structure, as lignin is non-degradable [31] and the lignin suppresses degradation of lignocellulosic fibers such as hemicellulose and celluloses [10]. For this reason, it is often pointed out as the main cause of low BM in plant biomass. Many studies reported that lignin content and the efficiency of enzymatic hydrolysis have an inverse relationship. [23,29,32] In this text, pretreatment of substrate to increase biogas productivity usually focuses on improving hydrolysis by releasing lignocellulosic bindings, occasionally degrading lignin polymers. To the contrast of the critical role of lignin for anaerobic digestion, a larger amount of lignin is preferable to obtain energy from combustion, as higher heating values of biomass positively correlate with lignin content [33,34]. The higher heating value is the absolute value of the specific energy combustion, when solid biofuel burns in oxygen in a calorimetric bomb under specific conditions.

Lignocellulose is namely most abundant for plant biomass, likewise it's often called lignocellulosic biomass. High concentration of lignocellulose can also be found in animal slurry, since animals are fed plants i.e. grass, straw, etc. Bruni et al.[35] reported that the concentration of lignocellulose in DM ranged 40-50%. Lignocellulose in animal slurries has different characteristics compared to plants, whose structure is broken down during animal digestion. The concentrations of each lignocellulosic fibrous fraction are shown in Figure 7.

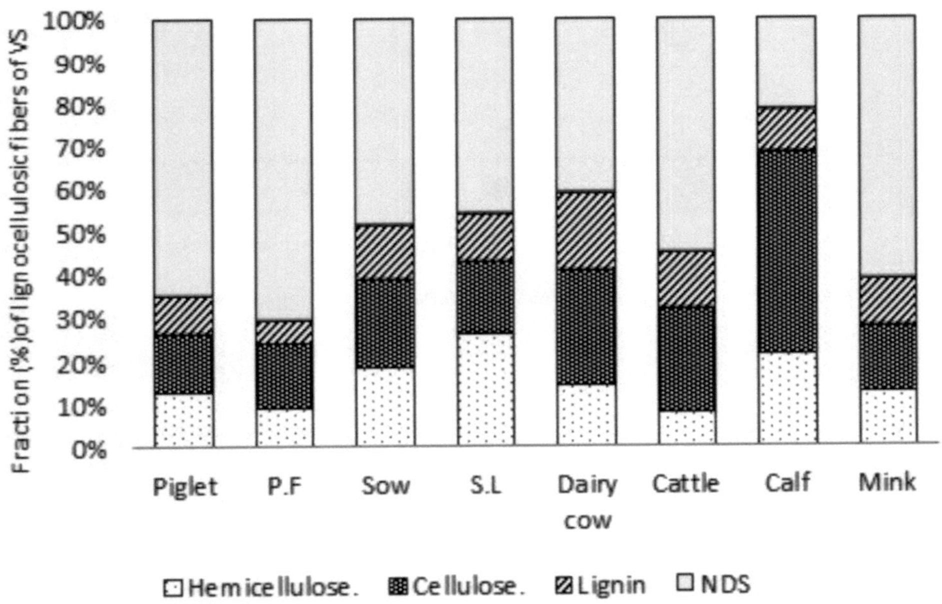

Figure 7. The concentrations of each lignocellulosic fibrous fraction.

Lignocellulose fraction in VS in animal slurry ranged 30 – 80%. Relatively lower lignocellulose was found in pig and mink slurry, whereas it was higher for cow slurry, which seems to be due to a different animal diet. The concentration of lignin in VS for most animal slurries was larger than 10% except for pig fattening slurry. Within the pig slurry, the lignin was highest for sow slurry. In detail, lignin was 8.6(±6.0)% for piglet, 4.8(±5.5)% for fattening pig, 12.5(±1.2)% for sow and 10.6(±1.1) for mixture of sow and piglet slurry, respectively. In case of cow manure, dairy cow had most abundant lignin at 18.0 (±2.1)%, whereas cattle and calf contains 13.1(±2.1)% and 10.1(±2.2)%, respectively. The concentration of lignin in mink slurry was 10.8%. The concentration of hemicellulose was similar to the lignin concentration, ranging 8.1% to 26.3%. However, the larger amounts were found in the slurry of young animals and in pig slurry, whereas high concentration of lignin was found cow in manure. The highest concentration of lignocellulose in VS with larger amount of cellulose and lignocellulose of calf slurry seems to be due to straw used for the bedding materials. In case of mink slurry, as can be seen in figure 6, the concentration of lignocellulose in VS and distribution of each fibrous fraction is similar to piglet slurry. The results of lignocellulose charaterisation and BMP clearly demonstrate that pattern of inverse relationship between lignin and BMP, which is in accordance to literatures [10,23,29,32].

In case of plant biomass, Triolo et al. [36] reported lignocellulose concentration in VS to be in the range of 49.0 - 82.8%, and lignin concentration in VS was 3.6 - 10.5% for grass and crop residues, whereas the concentration of lignin was larger for woody biomass, that is, 13.9 - 24.0%. In comparison with lignocellulosic characteristics of plant biomass from Triolo et al. [36], the concentrations of lignocellulose seem to be at approximately the same level, except pig fattening slurry. It is interesting that lignin of grass and pig slurries are relatively similar while the concentration of lignin in cow manure seems to be close to woody biomass to some extent. These results seem to be because lignin in straw and grass, which is cow diet, is up-concentrated up to the level of woody biomass, while relatively easily degradable organic pools are degraded. This result highlights that cow manure has critically high concentration of lignin that is the same level with woody biomass which is known as critical digestibility. Likewise, the difference between lignin and lignocellulose concentrations between pig and cow slurry seems to be more dependent on animal diet than management method, except calf manure.

5.4. Linear correlation between BMP and organic components.

VS is measured by burning dried materials for at least 2 hours at 525 °C, where the residues are defined as ash and the volatile fraction as VS. As each VS component has different stoichiometric methane potentials (TBMP) and different digestibility, knowing the composition of the VS component could be used to assess BMP alternatively instead of performing a fermentation test. Table 3 presents the TBMP of each organic component, where it shows that lipid and lignin is only preferable in respect to TBMP.

Whereas stoichiometric methane potential of each organic component is known relatively well, BD of it in animal slurry is poorly researched except VFA and Lignin. VFA is the intermediate during the procedure of digestion and the presence of VFA in animal slurry indicates the previous occurrence of hydrolysis. As hydrolysis decides degradation rate, we may hypothesise that the concentration of VFA in animal slurry may significantly correlate with digestibility, and that can further be correlated to BMP. For the lignin, Triolo et al. [10] confirmed that BD is significantly related to lignin concentration. Using the VFA results from the animal slurry used as independent variables against BMP, a reasonable correlation between VFA concentration and BMP was found (Figure 8). Furthermore, a fine correlation between lignin and BMP was also found.

	Formula	TBMP(*CH_4 L g^{-1} VS*)
VFA (mainly acetic acid)	C_2H_4O2	0.373
Protein	$C_5H_7O_2N$	0.496
VSED (Carbohydrate)	$C_6H_{10}O_5$	0.415
Lipid	$C_{57}H_{104}O_6$	1.014
Lignin	$C_{10}H_{13}O_3$	0.727

Table 3. Stoichiometric methane potential (TBMP) of each organic component

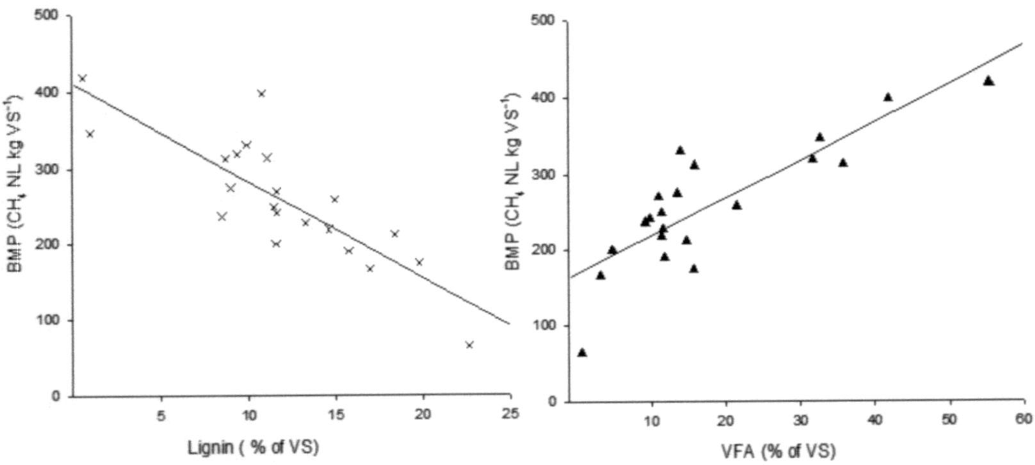

Figure 8. Relationship between VFA concentration (% of VS)(left) and BMP, and Lignin concentration (% of VS) and BMP (right) : as regression line for lignin ($y = -12.804x + 410.4$); for VFA ($y = 4.972x + 167.6$).

Statistical analysis showed that BMP significantly correlated with VFA, lignin and celluloses, though the correlation level of cellulose to BMP was quite weak. ($p<0.05$). On the other hand, it was not possible to find any correlation from other protein, hemicellulose, lipid, etc. The result of a simple linear regression test between BMP and organic components is given in Table 4, only showing significant models. Furthermore, multiple regression tests were performed using the significant variables, but excluding cellulose, since the model was not improved significantly including cellulose.

Due to the importance of BMP data, a large number of studies have proposed a BMP model based on the organic composition, since BMP is the reflection of destruction of organic materials (10, 37-43). Therefore we tested the precision of the algorithms obtained to test if the model could be used to predict BMP well enough.

Variable	R^2	p	RRMSE (%)	Algorithms
Lignin	0.698	<0.001	17.1	BMP = −12.804*lignin+410.4
VFA	0.701	<0.001	17.0	BMP = 4.972*VFA+167.6
Cellulose	0.249	<0.05	26.9	BMP = −3.574*cellulose +336.4
Lignin and VFA	0.766	<0.001	11.8	BMP = −7.807*lignin+3.057*VFA+295.5

Table 4. Summary of statistics results, algorithm obtained for BMP.

The precision of the model was evaluated by employing the relative root mean square error (RRMSE), which represents relative errors. As can be seen in Table 4, relative errors of the BMP model were similar for lignin and VFA, being 17% approximately, while relative error decreased to 11.8% when both of the variables were used for multiple regression tests.

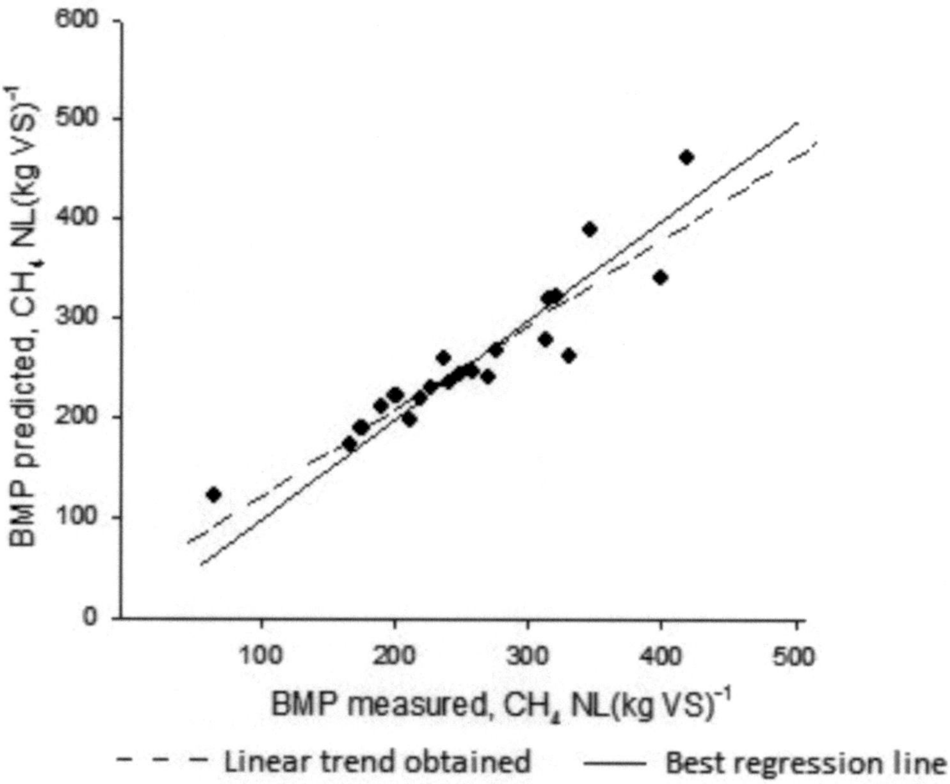

Figure 9. Measured BMP versus predicted BMP and the linear trend using the algorithm (BMP (CH_4 NL Kg VS^{-1}) = 295.5 + 3.057*VFA(% of lignin)-7.807*lignin(% of lignin)

Measured BMP versus predicted BMP using the model from multiple linear regression tests is plotted in figure 9, where it shows a good linear correlation. The slope of the best regression line and linear trend obtained was also very similar. The results indicate that the model predicted by cellulose is not preferable, whereas the BMP model using VFA and lignin could be useful for BMP assessment instead of time demanding fermentation tests.

5.5. New algorithm to predict potential biomethane yield

As it was commented above biomethane yield in terms of total slurry mass (BMP_{TM}) significantly correlated with DM concentrations. We tested the possibility of predicting BMP_{TM} using the concentration of DM, VS and the concentration of lignin and VFA, which were a significant variable for BMP. The results of the regression tests are shown in Table 5, where quite high correlations were found for all the models. However, critical relative errors using DM as an independent variable were found, that is, 62.1 %, which seems to be because the wide range of DM improved the correlation level. Hence, when assessing BMP_{TM}, only TS can be used when further characterisation is not possible. Apart from DM, relative errors were much lower when using VS and VS together with lignin and VFA, indicating a good potential of applying the model for prediction.

Variable	R^2	P	RRMSE (%)	Equation
DM (g kg $^{-1}$)	0.896	<0.001	62.1	$BMP_{TM} = -0.934+0.201*DM$
VS (g kg^{-1})	0.952	<0.001	19.8	$BMP_{TM} = 0.610+0.229*VS$
VS (g kg $^{-1}$), lignin (% of VS) and VFA (% of VS)	0.970	<0.001	15.6	$BMP_{TM} = 4.654+ 0.230*VS +0.009*VFA -0.360*lignin$

Table 5. Summary statistics results, algorithm obtained for BMP_{TM}.

6. Conclusion

The study highlights the critical quality of VS in cow manure and the critical quantity of VS in pig slurry which results in low viability of biogas production using animal slurry. The very high concentration of lignin in cattle and dairy cow manure indicates that there is a need of pretreatment either to reduce the influence of lignin by releasing lignocellosic bindings, or by depolymerizing lignin polymer. Whereas low digestibility of cow manure is problematic due to high concentration of lignin, lignin concentration of pig and mink slurry was relatively low. However despite of preferable digestibility of pig and mink slurry, the large amount of water and very low VS concentration in them indicates that there is a need of a qualified control of water content during management. Our study shows that control of DM concentration is more crucial than control of BD of substrate to enhance methane yield. Hence, the study highlights the importance of a qualified control of water content in feedstock by co-digesting solid organic substrates that can enrich VS concentrations prior to improvement of substrate digestibility by pretreatment.

Author details

Jin M. Triolo*, Lene Pedersen and Sven G. Sommer
University of Southern Denmark, Faculty of Engineering, Institute of Chemical Engineering, Biotechnology and Environmental Technology, Odense M, Denmark

Alastair J. Ward
Aarhus University, Dept. of Biosystems Engineering, AU Foulum, Tjele, Denmark

Acknowledgement

This study was supported by a grant from Energi Fyns Udviklingsfond.

7. References

[1] Al Seadi T., Rutz D., Prassl H., Köttner M., Finsterwalder T., Volk S., Janssen R. (2008) Biogas handbook. Published by University of Southern Denmark Esbjerg, Niels Bohrs Vej 9-10, DK-6700 Esbjerg, Denmark

[2] Sommer S G, Christensen K.V., Jensen L.S. (2008) Environmental Technology for Treatment and Management of Bio-waste. Compendium written by Sven G. Sommer and Knud Villy Christensen, University of Southern Denmark, Faculty of Engineering, Institute of Chemical Engineering, Biotechnology and Environmental Engineering & Lars Stoumann Jensen, University of Copenhagen, Faculty of Life Science, Plant and Soil Science Laboratory, Department of Agricultural Sciences.

[3] Statistics Denmark (2012), Available at http://statistikbanken.dk/statbank5a/default.asp?w=1466; accessed February 2012.

[4] Wittrup, S. (2010). Power plant also wants animal slurry (in Danish). Available at http://ing.dk/artikel/109248-kraftvaerker- vil-ogsaa-have-gylle; accessed February 2012.

[5] EREC (2010) Renewable energy in European: market, trends, and technologies (second ediction), European Renewable Energy Council.

[6] Klimakomission (2010) Grøn energi – vejen mod et dansk energisystem uden fossile brændsler, 28. september 2010, http://www.klimakommissionen.dk/daDK/OmKlimakommissionen/Klimakommissionensrapport/Sider/Forside.aspx

[7] Demirbas, A (2001) Biomass resource facilities and biomass conversion processing for fuels and chemicals. Energy Convers. Mgmt. 42:1357–1378.

[8] Demirbas, A. (2002) Electricity from biomass and hydroelectric development projects in Turkey.Energy Exploration and Exploitation 20:325–335.

[9] Demirbas, A., and Arin, G. (2002). An overview of biomass pyrolysis. Energy Sources 24:471–482

[10] Triolo, J.M., Sommer, S.G., Møller, H.B., Weisbjerg M.R., Jiang. X.Y., 2011. A new algorithm to characterize biodegradability of biomass during anaerobic digestion:

* Corresponding Author

Influence of lignin concentration on methane production potential. Bioresour. Technol. 102, 9395-9402.

[11] Møller, H.B., Nielsen, A.M., Nakakubo, R., Olsen, H.J., 2007. Process performance of biogas digesters incorporating pre-separated manure. Livestock Science, 112, 217-223.

[12] Holm-Nielsen, J.B., Seadi, B.P., Oleskowicz-Popiel, C., 2009. The future of anaerobic digestion and biogas utilization. Bioresour. Technol., 100, 5478–5484.

[13] IPCC, 2007. Intergovernmental Panel on Climate Change Assessment Reports. Intergovernmental Panel on Climate Change. Plenary XXVII (Valencia. Spain. 12-17 November 2007)

[14] Sommer, G.S., Peterson, S.O., Søgaard, H.T., 2000. Emission of green-house gases from stored cattle slurry and slurry fermented at a biogas plant. Journal of Environmental quality, 29,774-751.

[15] Olsen, H.J. (2006) Options for reducing the greenhouse gas emissions from agriculture. Department of Agroecology, Plantekongeres,93,3.

[16] Amon, T., Amon, B., Kryvoruchko, V., Zollitsch, W., Mayer, K., Gruber, L., 2007. Biogas production from maize and dairy cattle manure: influence of biomass composition on the methane yield. Bioresour. Technol. 100, 5777–5782.

[17] Raven M. Gregersen K.H., (2007). Biogas plants in Denmark: Successes and setbacks. Renewable and Sustainable Energ Revies 11(2007) 115-132.

[18] VDI (2006). VDI 4630: Fermentation of organic materials – Characterisation of the substrate, sampling, collection of material data, fermentation tests. In Verein Deutscher Ingenieure (VDI) (Ed.), VDI Handbuch Energietechnik. Berlin: Beuth Verlag GmbH.

[19] ISO (1995). ISO 11734: Water quality – evaluation of the 'ultimate' anaerobic biodegradability of organic compounds in digested sludge – method by measurements of the biogas production. Geneva: International Organization for Standardization

[20] APHA, 2005. Standard Methods for the Examination of Water and Wastewater (21st 11 ed.), American Public Health Assoc., Washington, DC (2005).

[21] Mertens, D. et al., 2002. Gravimetric determination of amylase-treated neutral detergent fiber in feeds with refluxing in beakers or crucibles: collaborative study. J. AOAC Int. 85 (6), 1217–1240.

[22] ISO (International Organization for Standardization), 2009. ISO 13906:2008. Animal Feeding Stuffs: Determination of Acid Detergent Fibre (ADF) and Acid Detergent Lignin (ADL) Contents. Available from: <http://www.iso.org/iso/ catalogue_detail.htm?csnumber=43032>.

[23] Møller, H.B., Sommer, S.G., Ahring, B.K., 2004. Methane productivity of manure, straw and solid fractions of manure. Biomass Bioenerg. 26, 485–495.

[24] Van Soest, P.J., 1963. Use of detergents in the analysis of fibrous feeds. II. A rapid method for the determination of fiber and lignin. J. Assoc. Offic. Agr. Chem. 46, 829–835.

[25] Goering, H.K., Van Soest, P.J., 1970. Forage Fiber Analysis (Apparatus, Reagents, Procedures and Some Applications). Agriculture Handbook No. 379. Agriculture Research Service, United States Department of Agriculture, Washington, DC.

[26] Raposo, F., Fernández-Cegrí, V., De la Rubia, M. A., Borja, R., Béline, F., Cavinato, C., Demirer, G., Fernández, B., Fernández-Polanco, M., Frigon, J. C., Ganesh, R., Kaparaju, P., Koubova, J., Méndez, R., Menin, G., Peene, A., Scherer, P., Torrijos, M., Uellendahl, H., Wierinck, I., & de Wilde, V. (2011). Biochemical methane potential (BMP) of solid organic substrates: evaluation of anaerobic biodegradability using data from an international inter-laboratory study. Journal of Chemical Technology and Biotechnology, 86(8), 1088-1098.

[27] Myinta M., Nirmalakhandanb N., Speecec R.E. (2007). Anaerobic fermentation of cattle manure: Modeling of hydrolysis and acidogenesis Water Research 323 – 332

[28] Bastone, D.J.; Keller, J.; Angelidaki, I.; Kalyuzhnyi, S.V.; Pavlostathis, S.G.; Rozzi, A.; Sanders, W.T.; Siegrist, H., Vavilin, V.A. (2002) Anaerobic digestion model no. 1. (ADM1). Water Science and Technology, vol. 45, no. 10, p. 65-73.

[29] Singh A., Yadav K., Sen A.K., (2012). Sal (Shorea Robusta) Leaves Lignin Epoxidation and Its Use in Epoxy Based Coatings. American Journal of Polymer Science, 2(1): 14-18

[30] Lesteur, M., Bellon-Maurel, V., Gonzalez, C., Latrille, E., Roger, J.M., Junqua, G., Steyer, J.P. (2010). Alternative methods for determining anaerobic biodegradability: a review. Process Biochem. 45 (4), 431–440.

[31] Mauseth, J.D., 1988. Plant Anatomy. The Benjamin/Cummings Publishing Company,CA.

[32] Buffiere P., Loisel D., Bernet N., Delgenes J.P. (2006) Towards new indicators for the prediction of solid waste anaerobic digestion properties. Water Sci Technol 53:233–41.

[33] Demirbas, A. (2000) Relationships between lignin contents and heating values of biomass. Energy Conversion & Management 42:183-188.

[34] Telmo, C., Lousada J., (2011) The explained variation by lignin and extractive contents on higher heating value of wood. Biomass and Bioenergy 35: 1663-1667.

[35] Bruni, E., Jensen, A.P., Angelidaki, I., 2010. Steam treatment of digested biofibers for increasing biogas production. Bioresource Technology 101. 7668–7671.

[36] Triolo, J.M., Pedersen L., Sommer, S.G.,(2012) Potential of plant waste for biomethane production: Characteristic biomethane production potential and anaearobic digestibility. Fourth International Symposium on energy from biomass and waste. (In press).

[37] Han, Y.W., Lee, J.S., Anderson, A.W., 1975. Chemical composition and digestibility of ryegrass straw. J.Agric. Food Chem. 23, 928–931.

[38] Chandler, J.A., Jewell, W.J., Gossett, J.M., Van Soest, P.J., Robertson, J.B., 1980. Predicting methane fermentation biodegradability. Biotechnol. Bioeng. Symp.10, 93–107.

[39] Bjorndal, K.A., Moore, J.E., 1985. Prediction of fermentability of biomass feedstocks from chemical characteristics. In: Smith, W.H. (Ed.), Biomass Energy Development. Plenum Press, New York, pp. 447–454.

[40] Tong, X., Smith, L.H., Mc Carty, P.L., 1990. Methane fermentation of selected lignocellulosic materials. Biomass 21, 239–255.

[41] Gunaseelan, V.N., 2007. Regression models of ultimate methane yields of fruits and vegetable solid wastes, sorghum and napiergrass on chemical composition. Bioresour. Technol. 98, 1270–1277.

[42] Schievano, A., Pognani, M., D'Imporzano, G., Adani, F., 2008. Predicting anaerobic biogasification potential of ingestates and digestates of a full-scale biogas plant using chemical and biological parameters. Bioresource Technology. 99, 8112–8117.

[43] Schievano A., Scaglia B., Imporzano G.D., Malagutti L., Gozzi A., Adani F., 2009. Prediction of biogas potentials using quick laboratory analyses: Upgrading previous models for application to heterogeneous organic matrices Bioresource Technology 100 (2009) 5777–5782.

Section 4

Biomass Processing

Chapter 13

A Real Story of Bioethanol from Biomass: Malaysia Perspective

K.L. Chin and P.S. H'ng

Additional information is available at the end of the chapter

http://dx.doi.org/10.5772/51198

1. Introduction

Rising fossil fuel prices associated with growing demand for energy, and environment concerns are the key factors driving strong interest in renewable energy sources, particular in biofuel. Biofuel refers to any type of fuel whose energy is derived from plant materials. Biofuel which includes solid biomass, liquid fuels and various biogases is among the most rapidly growing renewable energy technologies in recently. Biofuels are commonly divided into two groups based on the technology maturity which using the terms "conventional" and "advanced" for classification. Conventional biofuel technologies include well-established processes that are already producing biofuels on a commercial scale. These biofuels, commonly referred to as first-generation, include sugar- and starch-based ethanol, oil-crop based biodiesel and straight vegetable oil, as well as biogas derived through anaerobic digestion. First generation biofuel processes are useful but limited in most cases: there is a threshold above which they cannot produce enough biofuel without threatening food supplies and biodiversity. Whereas, advanced biofuel technologies are extensions from conventional technologies which some are still in the research and development (R&D), pilot or demonstration phase and they are commonly referred to as second- or third-generation. This category includes hydrotreated vegetable oil (HVO), which is based on animal fat and plant oil, as well as bioethanol based on lignocellulosic biomass, such as cellulosic-ethanol. Although there are wide varieties of advanced biofuels conversion technologies exists today, but they are not commercially available yet. Nevertheless, the most commercializable technology and most used biofuel on the global market is bioethanol.

2. Bioethanol

Bioethanol is chemically known as ethyl alcohol (C_2H_5OH) and produced from fermentation of fermentable sugars (i.e. glucose, sucrose, etc.) from plant sources using micro-organisms (yeasts or bacteria). Bioethanol is a clear colourless liquid, it is biodegradable, low in toxicity and causes little environmental pollution if spilt. In the 1970s, Brazil and the United States (US) started mass production of bioethanol grown from sugarcane and corn respectively. Current interest in bioethanol lies in production derived from lignocellulosic biomass. The most common usage of bioethanol is to power automobiles through mixed with petrol. It can be combined with gasoline in any concentration up to pure ethanol (E100). Anhydrous ethanol, that is, ethanol with at most 1% water, can be blended with gasoline in varying quantities to reduce consumption of petroleum fuels and in attempts reduce air pollution. Bioethanol burns to produce carbon dioxide (CO_2) and water. In addition to that, the use of bioethanol is generally CO_2 neutral. This is achieved because in the growing phase of the plant sources, CO_2 is absorbed by the plant and oxygen is released in the same volume that CO_2 is produced in the combustion of the fuel. This creates an obvious advantage over fossil fuels which only emit CO_2 as well as other poisonous emissions [1].

Blending bioethanol with gasoline help to reduce green house gases (GHG) emissions by oxygenate the fuel mixture so it burns more completely. On a life cycle basis, ethanol produced from corn results in about a 20 percent reduction in GHG emissions relative to gasoline. With improved efficiency and use of renewable energy, this reduction could be as much as 52 percent. In near future, bioethanol produced from cellulose has the potential to cut life cycle GHG emissions by up to 86 percent relative to gasoline as reported in EPA's Emission Facts [2].

3. Bioethanol in use

About 75% of bioethanol produced in the world being used to power automobiles, though it may be used for gasoline additives and other industries such as paints and cosmetics. Ethanol fuel blends are widely sold in the United States, Brazil, Europe and China. The most common blend is 10% ethanol and 90% petrol (E10). Vehicle engines require no modifications to run on E10 and vehicle warranties are unaffected also. However, only flexible fuel vehicles can run on up to 85% ethanol and 15% petrol blends (E85). Since 1976 the Brazilian government has made it mandatory to blend ethanol with gasoline with 5% ethanol and 95% petrol, and in 2007 the legal blend is around 25% ethanol and 75% gasoline (E25). Today, bioethanol contribute around 3% of total road transport fuel globally (on an energy basis) and considerably higher shares are achieved in certain countries [3]. The usage of bioethanol as transport fuel will be even more as the recent European Commission energy roadmap has set a target to increase the use of biofuels for transport from 5.75% from 2010 to 10% by 2020 under the Directive 2003/30/EC.

Bioethanol is also used as primarily gasoline additive and extender due to its high-octane rating. Bioethanol replacing lead as an oxygenate additive for traditional petrols in the form of Ethyl tertiary butyl ether (ETBE). The ethanol is mixed with isobutene (a non-renewable petroleum derivative) to form ETBE. At a 10% mixture, ethanol reduces the likelihood of engine knock, by raising the octane rating.

Beside the usage of bioethanol in fuel industry, bioethanol also can serve a wide range of uses in the pharmaceuticals, cosmetics, beverages and medical sectors as well as for industrial uses. The market potential for bioethanol is therefore not just limited to transport fuel or energy production but has potential to supply the existing chemicals industry. These include for use in acetaldehyde (raw material for other chemicals e.g. binding agent for paints and dyes), acetic acid (raw material for plastics, bleaching agent, preservation), ethylacetate (paints, dyes, plastics, and rubber), detergents, thermol (cold medium for refrigeration units and heat pumps), solvent for spirits industry, cosmetics, print colours and varnish, isopropyl alcohol (IPA), ethyl acetate (EAC), WABCO-antifreeze (disinfectant, cleaning agent for electronic devices, solvents) and vinasse, potassium sulphate (feeding stuffs, fertilizer).

4. Bioethanol technology

Bioethanol can be produced either from conventional or advance biofuel technologies depending on the state of sugars polymerization. The predominant technology for producing bioethanol is through fermentation of sucrose from sugar crops such as sugarcane, sugar beet and sweet sorghum. Bioethanol produced from sugar or starchy materials is categorize under the conventional technology and the bioethanol so called first generation bioethanol. Whereas, at present, much focus is on the bioethanol produced from biomass that possesses lignocellulosic content. This second generation bioethanol or cellulosic ethanol could be produced from abundant low-value material, including wood chips, grasses, crop residues, and municipal waste.

Regardless of the bioethanol technologies used to produce bioethanol, the bioethanol process have to undergo several treatment steps in which normally involves pre-treatment, extraction of fermentable sugars and fermentation. Pre-treatment process mainly deals with the preparation of the feedstock into smaller size (higher surface to volume ratio) for ease of sugars extraction. Whereas, extraction process with the aim of transforming the various sugars polymer chains into simple fermentable sugars. Fermentation process is a biological process in which fermentable sugars are converted into cellular energy and thereby produce ethanol and carbon dioxide as metabolic waste products in the absence of oxygen (anaerobic process) using *Saccharomyces cerevisiae*. The theoretical yield of bioethanol is 0.51 g per one gram of glucose consumed during fermentation.

5. Bioethanol conversion yield

Commercial production of bioethanol deals with the biotechnological production from different feedstock. The selection of the most appropriate feedstock for ethanol production strongly depends on the local conditions. Due to the agro-ecological conditions, North American and European countries have based their ethanol industry on the starchy materials. In Brazil, sugarcane is the main feedstock for bioethanol production. World production of ethanol (all grades) in 2010 was nearly 70 billion litres (IEA, 2010). Although many countries produce ethanol from a variety of feedstocks, Brazil and the United States are the major producers of ethanol in the world, each accounting for approximately 35 percent of global production [4].

The theoretical yield of ethanol from sucrose is 163 gallons of ethanol per tonne of sucrose. Factoring in maximum obtainable yield and realistic plant operations, the expected actual recovery would be about 141 gallons per tonne of sucrose [5]. Using [6],[7] and [8] reports, average sugar recovery rates, one tonne of sugarcane would be expected to yield 70 L of ethanol and one tonne of sugar beets would be expected to yield 100 L of ethanol. One tonne of molasses, a byproduct of sugarcane and sugar beet processing, would yield about 260 L of ethanol. Corn had the highest ethanol yield per tonne feedstock (403 L/t), followed by wheat with 350 L/t [9]. A lower ethanol yield per tonne of feedstock was obtained for cassava compared to corn. The ethanol yield from starchy materials were basically higher than sugar containing material because of the higher amount of fermentable sugars (glucose) that may be released from the original starchy material [10].

The conversion of sugar containing material into bioethanol is easier compared to starchy materials and lignocellulosic biomass because previous hydrolysis of the feedstock is not required since this disaccharide can be broken down directly by the yeast cells [11]. Therefore, using raw sugar as a feedstock, one tonne would yield 500 L of ethanol while refined sugar would yield 530 L ethanol. Molasses, from either sugarcane or sugar beets, was found to be the most cost competitive feedstock. The table below summarizes the estimated ethanol production yield and conversion efficiency from starchy and sugar containing materials from all over the world, as well as research ethanol yield produced from lignocellulosic biomasses.

Bioethanol is currently produced from raw materials such as sugar cane, or beet or starch from cereals. Recent interest was on the low cost and abundant availability of lignocellulosic biomass as the potential feedstock for bioethanol production. Lignocellulosic biomass which includes agricultural and forestry residues and waste materials, has the advantage of providing a greater choice of potential feedstock that does not conflict with land-use for food production, and that will be cheaper than conventional bioethanol sources. Many researchers from around the world are now working on transforming lignocellulosic biomass such as straw, and other plant wastes, into "green" gold - cellulosic ethanol. Cellulosic ethanol, a fuel produced from the stalks and stems of plants (rather than only from sugars and starches, as with corn ethanol), is starting to take root in the United States.

The bioconversion of lignocellulosic biomass to monomeric sugars is harder to accomplish than the conversion of starch, presently used for bioethanol production. However, many countries are making efforts to utilize these lignocellulosic biomasses into ethanol; Sweden, Australia, Canada and Japan are planning to invest into lignocellulosic ethanol mill [21]. The highest ethanol yield from lignocellulosic materials was obtained using switchgrass, 201 L/t with 80% conversion efficiency. Ballesteros *et. al* [20] studied on ethanol conversion using woody material such as *Populus nigra* and *Eucalyptus globule* found that the yield of 145 L/t and 137 L/t feedstock and conversion efficiency ranging 59% - 64% was observed. The conversion efficiency for lignocellulosic materials was lower than the conversion efficiency obtained from sugar-containing material and starchy material.

Feedstock		Sugar convertible materials (%)	EtOH yield (L/t)		Conversion efficiency (%)	Source
			Actual ethanol yield	Theoretical ethanol yield		
Sugar containing materials	Sugar cane juice(80% MC)	12	70	78	90	[6]; [7]
	Sugar beet (75% MC)	18	100	116	86	[8]
Starchy materials	Cassava (40% MC)	32	178	207	86	[12]
	Sweet sorghum (14% MC)	15	80	97	82	[13]
	Wheat (14% MC)	66	350	427	82	[14]
	Corn (15% MC)	70	403	452	89	[15]
Lignocellulosic biomass*	Cane bagasse	33	140	213	66	[16]
	Wheat straw	36	140	233	60	[17]
	Corn stalk	35	130	226	63	[18]
	Switchgrass	39	201	252	80	[19]
	Populus nigra	35	151	226	64	[20]
	Eucalyptus globulus	36	138	232	59	[20]
	Brassica carinata	33	128	213	60	[20]

* Note: Sugar convertible materials are referred as cellulose content.

Table 1. Comparative indexes for three main types of bioethanol feedstocks

The selection of the feedstock is in concordance with the interests of each country based on their availability and low cost. Because feedstocks typically account for greater than one-third of the production costs, maximizing the bioethanol yield is imperative [22].

6. Bioethanol from lignocellulosic biomass

Second generation bioethanol which made from lignocellulosic biomass or woody crops, agricultural residues or waste is considered a future replacement for the food crops that are currently used as feedstock for bioethanol production. Technology for producing bioethanol from biomass is moving out of the laboratory and into the commercial place. Breakthroughs in bioethanol technology in the past decade has lead to commercialization of biomass conversion technology. In U.S alone, Six companies were listed by the U.S Environmental

Protection Agency (EPA) as cellulosic ethanol producers, and their combined anticipated production volume is 8 million ethanol-equivalent gallons for coming years [23]. The six companies are DuPont Danisco Cellulosic Ethanol LLC, Fiberight LLC, Fulcrum Bioenergy Inc., Ineos Bio, KL Energy Corp. and ZeaChem Inc. In April 2011, Mossi & Ghisolfi Group (M&G) (Chemtex) commenced construction of a commercial-scale 13 million gallons/year (50 million liters) cellulosic ethanol production facility in Crescentino, Italy. Beside that, there is Abengoa Company, which has a 5m litre/year demonstration plant at Salamanca, Spain. In October 2010, Norway-based cellulosic ethanol technology developer Weyland commenced production at its 200,000 liter (approximately 53,000 gallon) pilot-scale facility in Bergen, Norway. In Asia, Nippon Oil Corporation and other Japanese manufacturers including Toyota Motor Corporation plan to set up a research body to develop cellulose-derived biofuels. The consortium plans to produce 250,000 kilolitres (1.6 million barrels) per year of bioethanol by March 2014. In China, cellulosic ethanol plant engineered by SunOpta Inc. and owned and operated by China Resources Alcohol Corporation that is currently producing cellulosic ethanol from corn stover (stalks and leaves) on a continuous, 24-hour per day basis.

6.1. Process

Various process configurations are possible for the production of bioethanol from lignocellulosic biomass, the most common method for bioethanol conversion technology from lignocellulosic biomass involves three key steps:

Pre-treatment : During biomass pre-treatment lignocellulosic biomass is pre-treated with acids or enzymes in order to reduce the size of the feedstock and to open up the plant structure. Normally, the structure of cellulosic biomass is altered; lignin seal is broken, hemicelluloses is reduced to sugar monomers, and cellulose is made more accessible to the hydrolysis that convert the carbohydrates polymers into fermentable sugars.

Hydrolysis: This is a chemical reaction that releases sugars, which are normally linked together in complex chains. In early biomass conversion processes, acids were used to accomplish this. Recent research has focused on enzyme catalysts called "cellulases" that can attack these chains more efficiently, leading to very high yields of fermentable sugars. Although the decomposition of the material into fermentable sugars is more complicated, the fermentation process step is basically identical for bioethanol from either food crops or lignocellulosic biomass.

Fermentation : Microorganisms that ferment sugars to ethanol include yeasts and bacteria. Research has focused on expanding the range and efficiency of the organisms used to convert sugar to ethanol.

6.1.1. Pre-treatment

The aim of the pretreatment is to break down the lignin structure and disrupt the crystalline structure of cellulose for enhancing acid or enzymes accessibility to the cellulose during hydrolysis step [24],[25]. Lignocellulosic biomass consists of three major components; Cellulose, hemicellulose and lignin and are in the form of highly complex lignocellulosic matrix. Depending on type of lignocellulosic biomass, the lignin content varies from about 10 – 25%, the hemicelluloses content from about 20 – 35% and the cellulose content from about 35 – 50%. Lignin is a polymer of phenyl propanoid units interlinked through a variety of non-hydrolysable C - C and C-O-C bonds. It therefore is a complex molecule with no clear chemical definition as its structure varies with plant species. Hemicellulose is an amorphous heterogenous group of branched polysaccharides. Its structure is characterised by a long linear backbone of one repeating sugar type with short branched side chains composed of acetate and sugars. Cellulose is a linear molecule consisting of repeating cellobiose units held together by Beta- glycosidic linkages. Cellulose is more homogeneous than hemicellulose but is also highly crystalline and highly resistant to depolymerisation. The three components of lignin, hemicellulose and cellulose are tightly bound to each other in the biomass. In fact hemicellulose acts as a bonding agent between cellulose and lignin. In order to convert this biomass to fuel ethanol, the biomass has to be broken up into the individual components first before the molecular chains within each component can be broken up further into simpler molecules.

6.1.2. Hydrolysis

Once the celluloses disconnect from the lignin, acid or enzymes will be used to hydrolyze the newly freed celluloses into simple monosaccharides (mainly glucose). There are three principle methods of extracting sugars from sugars. These are concentrated acid hydrolysis, dilute acid hydrolysis and enzymatic hydrolysis.

6.1.2.1. Concentrated acid hydrolysis process

The primary advantage of the concentrated acid process is the potential for high sugar recovery efficiency [18]. It has been reported that a glucose yield of 72-82% can be achieved from mixed wood chips using such a concentrated acid hydrolysis process [26]. In general, concentrated acid hydrolysis is much more effective than dilute acid hydrolysis [27]. Furthermore, the concentrated-acid processes can operate at low temperature (e.g. 40ºC), which is a clear advantage compared to dilute acid processes. However, the concentration of acid used is very high in this method (e.g. 30-70%), and dilution and heating of the concentrated acid during the hydrolysis process make it extremely corrosive. Therefore, the process requires either expensive alloys or specialized non-metallic constructions, such as ceramic or carbon-brick lining. The acid recovery is an energy-demanding process.

Despite the disadvantages, the concentrated acid process is still of interest. The concentrated acid process offers more potential for cost reductions than the dilute sulfuric acid process [28]. The concentrated acid hydrolysis process works by adding 70-77% sulfuric acid to the pre-treated lignocellulosic biomass. The acid is added in the ratio of 1.25 to 1.5 acid to 1 lignocellulosic biomass and the temperature is controlled at 40-60ºC. Water is then added to dilute the acid to 20-30% and the mixture is again heated to 100ºC for 1 hour. The gel produced from this mixture is then pressed to release an acid sugar mixture. The acid is then recovered partly by anion membranes and partly in the form of H_2S from anaerobic waste water treatment. The process was claimed to have a low overall cost for the ethanol produced [29].

6.1.2.2. Dilute acid hydrolysis

Dilute acid hydrolysis process is similar to the concentrated acid hydrolysis except using very low concentration of sulfuric acid at higher cooking temperature. Biomass is treated with dilute acid at relatively mild conditions which the hemicelluose fraction is hydrolyzed and normally higher temperature is carried out for depolymerisation of cellulose into glucose. The highest yield of hemicellulose derived sugars were found at a temperature of 190°C, and a reaction time of 5 – 10 min, whereas in second stage hydrolysis considerably higher temperature (230 °C) was found for hydrolysis of cellulose [30].

6.1.2.3. Enzymatic hydrolysis

The enzymatic hydrolysis reaction is carried out by means of enzymes that act as catalysts to break the glycosidic bonds. Instead of using acid to hydrolyse the freed cellulose into glucose, enzymes are use to break down the cellulose in a similar way. Bacteria and fungi are the good sources of cellulases, hemicellulases that could be used for the hydrolysis of pretreated lignocellulosics. The enzymatic cocktails are usually mixtures of several hydrolytic enzymes comprising of cellulases, xylanases, hemicellulases and mannanases.

6.1.3. Fermentation process

The hydrolysis process breaks down the cellulostic part of the biomass into glucose solutions that can then be fermented into bioethanol. Yeast *Saccharomyces cerevisiae* is added to the solution, which is then heated at 32°C. The yeast contains an enzyme called zymase, which acts as a catalyst and helps to convert the glucose into bioethanol and carbon dioxide. Fermentation can be performed as a batch, fed batch or continuous process. For batch process, the fermentation process might takes around three days to complete. The choice of most suitable process will depend upon the kinetic properties of microorganisms and type of lignocellulosic hydrolysate in addition to process economics aspects.

The chemical reaction is shown below:

$$\underset{(Glucose)}{C6H12O6} \xrightarrow[Catalyst]{Zymase} \underset{(Bioethanol)}{2C2H5OH} + \underset{Carbon\ dioxide}{2CO2}$$

6.2. Current development in cellulosic bioethanol

At present, much focus is on the development of methods to produce higher recovery yield bioethanol from lignocellulosic biomass. This can be done through two methods; (1) use of pre-treatment to increase the readiness of lignocellulosic biomass for hydrolysis. (2) increase the conversion yield of lignocellulosic biomass into bioethanol through simultaneous fermentation of glucose and xylose into bioethanol.

As mentioned, one barrier to the production of bioethanol from biomass is that the sugars necessary for fermentation are trapped inside the lignocellulosic biomass. Lignocellulosic biomass has evolved to resist degradation and to confer hydrolytic stability and structural robustness to the cell walls of the plants. This robustness is attributable to the crosslinking between the polysaccharides (cellulose and hemicellulose) and the lignin via ester and ether linkages. Ester linkages arise between oxidized sugars, the uronic acids, and the phenols and phenylpropanols functionalities of the lignin. The cellulose fraction can be only hydrolysed to glucose after a pre-treatment aiming at hydrolytic cleavage of its partially crystalline structure. A number of pre-treatment methods are now available – steam explosion, dilute acid pre-treatment [31] and hydrothermal treatment [32]. Hydrothermal treatment prevent the degradation of cellulose content inside the lignocellulosic biomass during pre-treatment because hydrothermal can be performed without addition of chemicals and oxygen to the lignocellulosic biomass. Hydrothermal treatment involves two process where during the first process, lignocelluosic biomass was soaked in water at 80 °C to soften it before being treated in the second process with higher temperature at 190–200°C.

Another way to increase the recovery yield of bioethanol from lignocellulosic biomass is to convert every bit of biomass into bioethanol. This means using all the available sugars from cellulose and hemicelluose and fermented into bioethanol. Lignocellulosic biomass have high percentage of pentoses in the hemicellulose, such as xylose, or wood sugar, arabinose, mannose, glucose and galactose with majority sugar in hemicelluloses is xylose which account more than 90% present. Unlike glucose, xylose is difficult to ferment. This meant that as much as 25% of the sugars in biomass were out of bounds as far as ethanol production was concerned. At the moment, research shows that steam explosion or mild acid treatment performed under adequate temperature and time of incubation, render soluble the biomass hemicellulose part with the formation of oligomers and C5 sugars that are easily extracted from the biomass. The C5 sugar stream can be individually fermented to ethanol by microorganisms such as *E.coli, Pichia stipitis* and *Pachysolen* , that are able to metabolise xylose, or be used as carbon source in a variety of other fermentative processes [33].

7. Bioethanol from lignocellulosic biomass - Malaysia perspective

Malaysia formulated the National Biofuel Policy with envisions to put the biofuel as one of the five energy sources for Malaysia, enhancing the nation's prosperity and well being. This is in line with nation's Five-Fuel Diversification Policy, a national policy to promote renewable energy (RE) as the fifth fuel along with fossil fuels and hydropower. The National Biofuel Policy was implemented in March 2006 to encourage the production of Biofuels, particularly biodiesel from palm oil, for local use and for export. However, in 2007, the Government has announced that the implementation of the whole biodiesel project has been put on hold indefinitely owing to the current high price of refined, bleached and deodorized palm olein.

Recently, the Government of Malaysia launched new strategy to promote the biofuel through the National Biomass Strategy 2020 on year 2011. The aim of National Biomass

Strategy 2020 is to create higher value-added biomass economic activities that contribute towards Malaysia's gross national income (GNI) and creating high value jobs for the benefit of Malaysians. This Strategy outline the production of bioethanol produced from lignocellulosic biomass particularly the oil palm biomass as a starting point with extended to include biomass from other sources such as wood waste. The palm oil sector correspondingly generates the largest amount of biomass, around 80 million dry tonnes in 2010. This is expected to increase to about 100 million dry tonnes by 2020, primarily driven by increases in plantation area. A conservative estimation of utilising an addition 20 million tonnes of oil palm biomass for bioethanol has the potential to contribute significantly to the nation's economy while at the same time reduce the green house gasses emission.

The National Biomass Strategy 2020 proposes a mandate of bioethanol blending of 10 percent in petrol fuel in Malaysia by 2020 to cut down the green house gasses emissions. This would generate a domestic demand for one million tonnes of bioethanol per annum with the first bioethanol from lignocellulosic biomass plant is expected to be commercially viable between 2013 and 2015 [34]. As a result, much attention has been focuses on generating bioethanol from oil palm biomass and wood waste.

As mentioned early, bioethanol utilization as automobile fuel is especially promising as the United States, Brazil and Europe has introduced. However, low-cost supply associated with high bioethanol yield of the bioethanol is indispensable for its wide use. The discussion of economic feasibility of bioethanol production from lignocellulosic biomass in Malaysia in this paper was based on the experimental data through laboratory worked done by [35] and [36] and comparison was made with sugarcane and corn.

7.1. Experiment data

Optimum cellulose conversion to glucose with the hydrolysis efficiency of 82%, 67% and 66% for oil palm trunk, rubberwood and mixed hardwood, respectively obtained using two-stage concentrated sulfuric acid hydrolysis at elevated temperature using 60% sulfuric acid treated in a water bath with a temperature of 60°C for 30 min at the first stage hydrolysis and subsequently subjected to 30% sulfuric acid at 80°C for 60 min at the second stage [36]. As stated in the study by [35], optimum fermentation parameters for lignocellulosic hydrolysates using *Saccharomyces cerevisae* was obtained using 33.2°C and pH 5.3 with the fermentation efficiency of 80%, 85% and 90% for oil palm trunk, rubberwood and mixed hardwood, respectively. The optimum cellulose conversion and fermentation efficiency were used to calculate the actual ethanol yield per tonne (L/t) and the conversion efficiency of lignocellulosic biomass. The conversion efficiency was calculated in percentage of actual yield over the theoretical yield. The theoretical yield was calculated in assumptions that all the cellulose was converted to glucose and further converted to ethanol theoretical yield (51%) in 100% conversion rate. Using the cellulose conversion and fermentation efficiencies, the actual ethanol yields per tonne lignocellulosic biomass can be calculated for lignocellulosic biomass as the equation below:

$$\text{Ethanol yield in liter per tonne of feedstock (L/t)} = [1000\,(\text{kg}) \times \text{cellulose content} \times \text{actual hydrolysis efficiency} \times \text{ethanol theoretical yield}\,(0.51) \times \text{actual fermentation efficiency}] / 0.789$$

(Note: Ethanol has a density of 0.789 kg/L)

The results of bioethanol yield per tonne for oil palm trunk, rubberwood and mixed hardwood and their conversion efficiencies were presented in Table 2.

	Oil palm trunk	Rubberwood	Mixed hardwood
Celulose content	0.48	0.56	0.56
Hydrolysis efficiency	0.82	0.67	0.66
Ethanol theoretical yield at 100% fermentation efficiency	0.51	0.51	0.51
Actual fermentation efficiency	0.80	0.85	0.90
Actual Ethanol Yield/tonne of dried raw materials	204 L	206 L	215 L
Theoretical Ethanol Yield/tonne of dried raw materials	310 L	362 L	361 L
Total Ethanol Conversion efficiency	66%	57%	60%

Table 2. Ethanol Yield Per Tonne of Feedstock And The Ethanol Conversion Efficiency

As shown from the Table 2, using the same amount of feedstock, mixed hardwood produced slightly higher in volume of bioethanol (215 L/t) compared to oil palm trunk and rubberwood with the ethanol yield per tonne of 204 L/t and 206L/t, respectively. The volume of bioethanol produced using oil palm trunk, rubberwood and mixed hardwood per metric tones of dry weight basically were higher than those reported by [20] as shown in Table 1. The highest conversion efficiency was obtained from oil palm trunk (66%), followed by mixed hardwood (60%) and rubberwood (57%).

If bioethanol yield per tonne feedstock values are taken into consideration, the three lignocellulosic biomass studied was higher than most of the comparing feedstock. The three lignocellulosic biomass ethanol yields per tonne of feedstock were much higher than sugarcane, sugarbeet and cassava. This could be explain by the high moisture content of the sugarcane, sugarbeet and cassava implies the use of a greater amount of feedstock to reach the same sugar content that may released from the lignocellulosic material. However, lower bioethanol yield per tonne feedstock of the studied lignocellulosic biomass was found to be lowered than those wheat and corn feedstocks. This is due to the higher glucose convertible substance in the wheat and corn which contributed to higher ethanol yield. Overall, the conversion efficiency for the studied lignocellulosic biomass was lower than sugar containing material and starchy material. This showed how critical the hydrolysis and fermentation efficiency of the lignocellulosic biomass contributed to a higher ethanol yield to make it comparative with these commercial feedstocks. The three lignocellulosic biomass used in this study in terms of ethanol yield per tonne feedstock were found to be comparable with the results obtained from the lignocellulosic biomass obtained from other studies and conversion efficiency (Table 1). The studied lignocellulosic biomass contained higher amount of cellulose as the glucose convertible material. Therefore, this may contributed to higher ethanol yield per tonne of feedstock.

7.2. Economic feasibility of bioethanol

The cost of biethanol per litre presented here mainly calculated from the cost of raw materials used; i.e. lignocellulosic biomass and sulfuric acid and processing cost. Fixed operating costs are excluded from this calculation. Fixed operating costs including labour and various overhead items are fully incurred regardless of the operating production capacity and their contribution to the total cost of bioethanol is estimated at 15 to 18%. [37] stated that cost of biomass contribute almost 60% to the total production cost which is the highest contributor to the cost of bioethanol. Therefore, the main focus here is to estimate the effect of raw materials price on the cost of bioethanol.

7.2.1. Cost of lignocellulosic biomass

Assessing the various costs of mobilising lignocellulosic biomass today which include harvesting, collection, pre-processing, substitution and transportation to a downstream hub, the order of biomass can be mobilised at globally competitive costs, i.e., at a cost of less than RM 250 per dry-weight tonne. The distance of transportation should be less than 100km in radius from the collection area.

7.2.2. Cost of sulfuric acid and recovery charge

The sulfuric acid is sells at RM 264 per tonne. By far, sulfuric acid is the largest expenditure of raw materials in the process of making bioethanol from lignocellulosic biomass. Nonetheless, the current technology enable the acid-sugar solution from hydrolysis separated into acid and sugar components by means of chromatographic separation using commercial available ion exchange resins to separate the components without diluting the sugar. The separated sulfuric acid is recirculated and reconcentrated to the level required by the decrystallization and hydrolysis steps. Using this technology almost up to 100% of the sulfuric acid can be recovered from the process.

7.2.3. State of art scenario

The state of art scenario presented here makes use of the conversion rates from the experiment data (Table 2). Approximately, 200 L of bioethanol yields per dry tones of lignocellulosic biomass and anticipated prices of RM 250 per dry tones of lignocellulosic biomass and RM 264 per tones of 60% concentrated sulfuric acid. The feedstock cost for one litre of bioethanol produced using either from oil palm trunk or wood wastes is estimated at about RM 1.25/litre. The production cost for one litre bioethanol from lignocellulosic biomass is estimated at RM 0.26 with the hydrolysis cost contributed RM 0.20 based on the sulfuric acid is added at a ratio of 5:1 (acid: dry weight of biomass) with acid lost in the sugar stream is not more than 3% during recovery (97% recoverable). Fermentation cost contributed RM 0.06 with the yeast would be grown at the site without cultivation process [38]. Therefore, the total cost per litre of bioethanol produced is RM 1.51 excluding capital and fixed variable costs. However, without the recovery of sulfuric acid during hydrolysis, the cost of bioethanol from lignocellulosic biomass would be rose up to RM7.85, excluding capital and fixed variable costs. With ethanol prices now at RM 2.10 per litre, it is possible for the Malaysia to produce the bioethanol from oil palm trunk and wood wastes, yet it would be not profitable to produce ethanol from lignocellulosic biomass without using the recovery system for sulfuric acid during hydrolysis.

The table below shows different scenario on the biomass feedstock and bioethanol yield that might affect the cost of bioethanol in Malaysia. The scenarios were based on 97% sulfuric acid recovered during hydrolysis and no change on the cost of fermentation production.

Scenario Analysis :

The economic feasibility of bioethanol production in Malaysia from lignocellulosic biomass is highly dependent on the feedstock cost and recovery yield. The cost of feedstock contributed approximately 80% (excluding capital and fixed variable costs) to the total bioethanol cost when the feedstock price estimated at RM 250 per dry weight ton. As the feedstock price increase 5% to 15% per dry ton, the cost of bioethanol increased from as low as 4% up to almost 13%. Higher recovery yield from the bioethanol process will surely reduce the cost of bioethanol produced per litre when the cost of feedstock remains the same. However, as the conversion yield of bioethanol decrease from 200 L per dry weight ton of biomass, the cost of biothenol per litre increase from 5% up to 17%.

Like corn in the United States and sugarcane in Brazil, the relatively low feedstock cost will only makes this process economically competitive. The cost of producing ethanol from sugarcane in Brazil is estimated at about RM 0.60 per litre, excluding capital costs. U.S. ethanol conversion rates utilizing corn as the feedstock are estimated at approximately 2.65

gallons of ethanol per bushel for a wet mill process and 2.75 gallons per bushel for a dry mill process. Net feedstock costs for a wet mill plant are estimated at about RM 0.30 per litre with total ethanol production costs estimated at RM 0.76 per litre. Net feedstock costs for a dry mill plant are estimated at RM 0.38 per litre with total ethanol production costs at RM 0.76 per litre. Molasses, from either sugarcane or sugar beets, was found to be the most cost competitive feedstock beside the lignocellulosic biomass. Estimated ethanol production costs using molasses were approximately RM 0.92 per litre with a RM 0.66 per litre feedstock cost [39].

	Bioethanol yield (L/T)	Feedstock Price per ton (RM)	Price of Sulfuric Acid per ton (RM)	Cost of Feedstock per litre of bioethanol (RM)	Cost of Production per litre (RM)	Cost of bioethanol per litre (RM)
Laboratory worked	200	250.00	264.00	1.25	0.26	1.51
Scenario 1: Reducing in conversion yield	-5% -10% -15%	Remain Remain Remain	Remain Remain Remain	1.31 1.39 1.47	0.27 0.28 0.29	1.58 (+4.6%) 1.67 (+10.6%) 1.76 (+16.6%)
Scenario 2: Increase of feedstock cost	Remain Remain Remain	+5% +10% +15%	Remain Remain Remain	1.31 1.38 1.44	Remain Remain Remain	1.57 (4.0%) 1.64 (+8.6%) 1.70 (+12.6%)
Scenario 3: Increase of sulfuric acid cost	Remain Remain Remain	Remain Remain Remain	+5% +10% +15%	Remain Remain Remain	0.27 0.28 0.29	1.52 (+0.6%) 1.53 (+1.3%) 1.54 (+2.0%)

Table 3. Cost of bioethanol per litre with different scenario on cost of raw materials and conversion yield

8. Conclusion

The studied lignocellulosic biomass has a higher bioethanol yield per tonne feedstock (L/t) than most of the commercialized bioethanol feedstock. However, improvement had to be made on the conversion efficiency to obtained higher ethanol yield to make it more comparable with the sugar containing and starchy material. The composition of substance that can be converted to glucose played a big influence on the ethanol yield per tonne feedstock. With the large amount of glucose convertible material and abundant availability, these lignocellulosic biomasses are potential feedstock for bioethanol production.

Author details

K.L. Chin[1] and P.S. H'ng
Faculty of Forestry, Universiti Putra Malaysia, UPM Serdang, Selangor, Malaysia

9. References

[1] Wang M, Saricks C, and Santini D (1999) Effects of Fuel Ethanol Use on Fuel-Cycle Energy and Greenhouse Gas Emissions. Argonne, IL.: USDOE Argonne National laboratory, Center for Transportation Research.

[2] EPA (2007) Emission Facts; Greenhouse Gas Impacts of Expanded Renewable and Alternative Fuels Use. Emission Facts Report (EPA420-F-07-035). Office of Transportation and Air Quality, EPA. US.

[3] IEA (2010) Medium Term Oil and Gas Markets 2010, OECD/IEA, Paris.

[4] F.O. Lights (2011) World Ethanol and Biofuels Report. F.O. Lights. Vol. 10, No. 6.

[5] Bolling C, Suarez NR (2001) The Brazilian Sugar Industry: Recent Developments. Sugar and Sweetener Situation and Outlook. U.S. Department of Agriculture Economic Research Service, 222-232, Septemper 2001.

[6] Baucum LE, Rice RW, Schuneman TG (2006) ,. Electronic Data Information Source (EDIS) SC032. Agronomy Department, University of Florida, Gainesville, FL.

[7] Coelho S (2005) Brazilian Sugarcane Ethanol: Lessons Learned. The Brazilian Reference Center on Biomass. STAP Workshop on Liquid Biofuel, New Delhi, India, 29 August-1 September, 2005.

[8] Berg C (2001) World Fuel Ethanol. Analysis and Outlook. F.O. Lichts.

[9] Pongsawatmanit R, Temsiripong T, Suwonsichon T (2007) Thermal And Rheological Properties Of Tapioca Starch And Xyloglucan Mixtures In The Presence Of Sucrose. Food research international. 40: 239–248 June.

---

[1] Corresponding Author

[10] Yoosin S, Sorapipatana C (2007) A Study Of Ethanol Production Cost For Gasoline Substitution In Thailand And Its Competitiveness. Thammasat int. j. sc. tech. 12:69-80.

[11] Cardona CA, Sanchez OJ (2007) Fuel Ethanol Production: Process Design Trends and Integration Opportunities. Bioresour. technol. 98: 2415-2457.

[12] Wang W (2002) Cassava Production For Industrial Utilization In China - Present And Future Perspective. In: Cassava research and development in Asia: exploring new opportunities for an ancient crop. Seventh regional cassava workshop. Bangkok, Thailand, October 28–November 1, 2002. pp.33–38.

[13] Rao DB, Ratnavathi CV, Karthikeyan K, Biswas PK, Rao SS, Vijay Kumar BS, Seetharama N (2004) Sweet Sorghum Cane For Bio-Fuel Production: A SWOT Analysis In Indian Context. National Research Centre for Sorghum. pp. 20.

[14] Agu R, Bringhurst T, Brosnan J (2006) Production of grain whisky and ethanol from wheat, maize and other cereals. Journal of the Institute of Brewing. 112(4): 314-323.

[15] Baker A., Zahniser S. (2006) Ethanol Reshapes the Corn Market. Amber Waves. Economic Research Service/USDA. 4(2): pp 30-35. Available: http://www.bioenergypro.com/bioenergy-articles/ethanol/ethanol-reshapes-the-corn-market. Accessed 2012 March 20.

[16] Moreira JS (2000) Sugarcane for Energy – Recent Results and Progress In Brazil. Energy for sustainable development. 6(3): 43–54.

[17] Ballesteros I, Negro MJ, Oliva JM, Cabañas A, Manzanares P, Ballesteros (2006) Ethanol production from steam-explosion pretreated wheat straw. Appl Biochem Biotechnol.129-132:496-508.

[18] Demirbas A (2005) Bioethanol From Cellulosic Materials: A Renewable Motor Fuel From Biomass. Energy sources. 27: 327–337.

[19] Bakker RR, Gosselink RJA, Maas RHW, de Vrije T, de Jong E, van Groenestijn JW, Hazewinkel JHO (2004) Biofuel Production From Acid-Impregnated Willow And Switchgrass. In 2nd World Conference on Biomass for Energy, Industry and Climate Protection. 10–14 May 2004, Rome, Italy. pp. 1467–1470.

[20] Ballesteros M, Oliva JM, Negro MJ, Manzanares PI (2004) Ethanol From Lignocellulosic Materials By A Simultaneous Saccharification And Fermentation Process (SFS) With *Kluyveromyces marxianus* CECT 10875. Process Biochemistry. 39(12):1843-1848.

[21] F.O. Lights (2008) World Ethanol and Biofuel Report. F.O. Lights. Vol.4, No.7. pp.71-92.

[22] Sánchez OJ, Cardona CA (2008) Trends In Biotechnological Production Of Fuel Ethanol From Different Feedstocks. Bioresour. technol. 99: 5270-5295.

[23] Bevill K (2001) Ethanol Producer Magazine, BBI International.

[24] Mosier N, Wyman CE, Dale BD, Elander RT, Lee YY, Holtzapple M, Ladisch CM (2005) Features Of Promising Technologies For Pretreatment Of Lignocellulosic Biomass. Bioresour. technol. 96: 673–686.

[25] Puls J, Schuseil J (1993) Chemistry Of Hemicelluloses: Relationship Between Hemicellulose Structure And Enzymes Required For Hydrolysis. In: M.P. Coughlan, G.P. Hazlewood, editors. Hemicellulose and Hemicellulases, Portland Press Research Monograph. pp. 1–27.

[26] Iranmahboob J, Nadim F, Moemi S (2002) Optimizing Acid-Hydrolysis: A Critical Step For Production Of Ethanol From Mixed Wood Chips. Biomass and bioenergy. 22: 401–404.

[27] Grahmann K, Torget R, Himmel M (1985) Optimization Of Dilute Acid Pretreatment Of Biomass. Biotechnology and bioengineering symposium. 15: 59–80.

[28] Farooqi R, Sam AG (2004) Ethanol as A Transportation Fuel. Centre for Applied Business Research in Energy and the Environment (CARBREE) Climate Change Inititive. University of Alberta, Canada.

[29] Van Groenestijn J, Hazewinkel O, Bakker R (2006) Pretreatment Of Lignocellulose With Biological Acid Recycling (Biosulfurol Process). Zuckerindustrie. 131(9): 639-641.

[30] Chandel AK, Chan ES, Rudravaram R, Narasu ML, Rao LV, Ravindra P (2007) Economics and Environmental Impact of Bioethanol Production Technologies: An Appraisal. Biotecnology and molecular biology review. 2(1): 14-32.

[31] Schell DJ, Farmer J, Newman M and McMillan JD (2003) Dilute-Sulfuric Acid Pretreatment of Corn Stover in Pilot-Scale Reactor: Investigation of Yields, Kinetics, and Enzymatic Digestibilities of Solids. Applied biochemistry and biotechnology. 105-108: 69-86.

[32] Thomsen MH, Thygesen A, Jorgensen H, Larsen J, Christensen BH, Thomsen AB (2006) Preliminary Results On Optimizing Of Pilot Scale Pretreatment Of Wheat Straw Used In Coproduction Of Bioethanol And Electricity. Applied Biochemistry and Biotechnology. 129–132: 448–460.

[33] Bon EPS, Ferrara MA (2007) The Role of Agricultural Biotechnologies for Production of Bioenergy in Developing Countries. FAO seminar, Rome, 12 October 2007.

[34] AIM, 2011. National Biomass Strategy 2020: New wealth creation for Malaysia's palm oil industry. Agensi Inovasi Malaysia. Putrajaya.

[35] Chin KL, H'ng PS, Wong LJ, Tey BT, Paridah MT (2010) Optimization Study Of Ethanolic Fermentation From Oil Palm Trunk, Rubberwood and Mixed Hardwood Hydrolysates Using *Saccharomyces cerevisiae*. Bioresour. technol. 101: 3287-3291.

[36] Chin KL, H'ng PS, Wong LJ, Tey BT, Paridah MT (2011) Production Of Glucose From Oil Palm Trunk and Sawdust Of Rubberwood and Mixed Hardwood. Applied energy. 88: 4222-4228.

[37] Wang L, Sharifzadeh M, Templer R, Murphy RJ (2012) Technology Performance And Economic Feasibility Of Bioethanol Production From Various Waste Papers. Energy environ. sci. 5: 5717–5730.

[38] Farone WA, Cuzens JE (1996) Method of Separating Acids and sugars Resulting from Strong Acid Hydrolysis. U.S. Patent No. 5,580,389. December 3.

[39] Shapouri H, Salassi M, Fairbanks JN (2006) The Economic Feasibility Of Ethanol Production From Sugar In The United States. USDA's Renewable Energy. United State Department of Agriculture, Office of the Chief Economist, Office of Energy Policy and New Uses.

Chapter 14

Lignocelluloses Feedstock Biorefinery as Petrorefinery Substitutes

Hongbin Cheng and Lei Wang

Additional information is available at the end of the chapter

http://dx.doi.org/10.5772/51491

1. Introduction

1.1. Lignocelluloses feedstock (LCF) biorefinery

1.1.1. Background

The material needs from our society are reaching the crisis point, as the demand for resources will soon exceed the capacity of the present fossil resource based infrastructure [1]. Currently, fossil-based energy resources, such as petroleum, coal, and natural gas, are responsible for about three quarters of the primary energy consumption in our world. While decreasing crude-oil reserves, enhanced demand for fuels worldwide, increased climate concerns about the use of fossil-based energy carriers, and political commitment, the focus has recently turned to develop the utilization of renewable energy resources [2]. Gullón *et al.* [3] described the variety of problems on present social, economic and technological situation, which including: the fear for a shortening of the supplies of basic resources, as the population growth; the increasing *per capita* demands of the developing economies for goods and energy, derived from the increasing purchase power of the population; environmental challenges, especially those related to effects of greenhouse gas emissions (emphasis on CO_2) on the global climate; the national security issues surrounding reliance on imported oil [4].

On our market, nowadays, there are more than 2500 different oil-based products. The petroleum crisis of the 1970s resulted in a shift from total reliance on fossil resources and simultaneously triggered research into biomass based technologies. As a result of the oil crisis, renewable resources became a popular phrase [5]. Currently, the most of energy requirements in the world are still met by fossil fuels. The limited deposits of these fossil fuels coupled with environmental problems have prompted people to look for sustainable resources as alternatives to meet the increasing energy demand. Bio-energy production has the advantage of forming smaller amounts of greenhouse gases compared to the conversion of fossil fuels, as the carbon dioxide generated during the energy conversion is consumed during subsequent biomass re-growth [6]. However, simply providing sustainable and non-polluting energy will not be enough. In our life, clothes, shelter, tools, medications and so on are all, to a greater or lesser degree, dependent on organic carbon. As fossil-based resources will be replaced, new sources of organic carbon have be found or alternate applications and processing of existing sources must be developed. The challenge is to find replacements not only for current usage, but also for the even future greater energy consumption, with a likely concomitant increase in biomass demand for manufacturing [7].

1.1.2. What is biorefinery?

The core aim for biorefineries is to produce both high-volume liquid fuels and high-value chemicals. As petroleum refinery uses petroleum as the major input and processes it into many different products, the term 'biorefinery' has been coined to describe the processing complexes that will use biomass as feedstocks to produce a wide spectrum of chemicals, fuels and bio-based materials, that can be used as industrial intermediates or sold directly to consumers [1, 8, 9]. Biorefineries have been considered as the key for access to an integrated production of chemicals, materials, goods, fuels and energy of the future [10]. As oil prices continue to rise and biorefining technology matures, biorefineries are playing an increasingly major role in the global economic system, with the potential to ultimately replace petroleum refineries as the world's principal method of fuel generation.

1.1.3. Lignocelluloses feedstock (LCF) biorefinery

The largest organic carbon reservoir in our world is the biomass - plants and algae. Each year, plants fix approximately 90 billion tons of CO_2, most of this as wood [11]. Lignocelluloses are the natural combination of cellulose, hemicelluloses and lignin. It's the raw material for potential conversion to energy fuels and chemical feedstock for manufacturing. LCF biorefinery has been drfined as one of the so-called phase-III biorefinery concepts which are characterized by the ability to use a variety of resources by different routes to generate multiple products [12].

A LCF biorefinery uses lignocellulosic biomass, including forestry residue, agricultural residue, yard waste, wood products, animal wastes, etc. Initially, plant material is cleaned and broken down into the three main fractions (hemicellulose, cellulose, and lignin) by chemical digestion or enzymatic hydrolysis. Hemicellulose and cellulose can be produced by alkaline and acid. Lignin can also be further broken down with enzymes. The hemicellulose and cellulose are sugar polymers, which can be converted to their component sugars through hydrolysis. A hemicellulose is a polymer that contains five-carbon sugars (usually D-xylose and L-arabinose), six-carbon sugars (D-galactose, D-glucose, and D-mannose), and uronic acid. Cellulose is a polymer of only glucose. The hydrolysis process of hemicelluloses and cellulose result in the aforementioned sugars [13].

The LCF Biorefinery is a promising alternative due to the abundance and variety of available raw materials and the good position of the conversion products on the market [14]. Its profitability is also dependent on the technology employed to alter the structure of lignocellulosic biomass in order to produce high value co-products from its three main fractions *i.e.* cellulose, hemicellulose, and lignin [15].

Currently the main feedstock for biorefineries is still based on starch. The practiced technologies in fuel ethanol industry are primarily based on the fermentation of sugars derived from starch and sugar crops, which are quite mature with little possibility of process improvements. However, using starch and sugar crops to produce ethanol also has been questioned since it draws its feedstock from a food stream. Lignocellulosic biomass is a more promising renewable resource as it is available in large quantities and does not compete with food or feed. Lignocellulosic biomass is a renewable resource that stores energy from sunlight in its chemical bonds, with great potentials for the production of affordable fuel ethanol [16, 17]. Its main obstacle for a major breakthrough is the high production costs for bioenergy products.

On the other hand, lignocellulosic biomass-derived products can significantly reduce green house gas emissions, compared to fossil-based products. Also, many common petrochemicals could be obtained with lower green house gas emissions from bio-based feedstocks. The maturity and economics of the conversion processes and logistics is a major challenge for lignocellulosic biomass [18].

1.1.4. The main goal of Biorefinery

With, implementing innovative, environmentally sound and cost-effective production technologies for a variety of products, the integrated biorefinery is increasing the availability and use of bioenergy and bio-based products. The main objective of a biorefinery is to produce high-value low volume and low-value high-volume products by a series of producing processes. The processes are designed to maximize the valued products while minimizing the waste streams by converting low-value high-volume intermediates into energy. The high-value products can enhance the profitability, and the high-volume fuels will help to meet the global energy demand. The power produced from a biorefinery can also help to reduce the overall cost. Figure 1 shows the elements of a biorefinery, in which biomass is used to produce various useful products such as fuel, power, and chemicals by biological and chemical conversion processes [13].

Traditionally, the matured biorefinery pathways include bioconversion (aerobic and anaerobic digestion) and chemical conversion (bio-pulping). There are two most promising emerging biorefinery platforms. One is the sugar platform and the other is the thermo-chemical platform (syngas platform). In sugar biorefineries platform, biomass will be broken down into different types of component sugars for fermentation or other biological processing into various fuels and chemicals. In thermo-chemical biorefineries platform, biomass will be synthesized hydrogen and carbon monoxide or pyrolysis oil, the various components of which could be directly used as fuel [19].

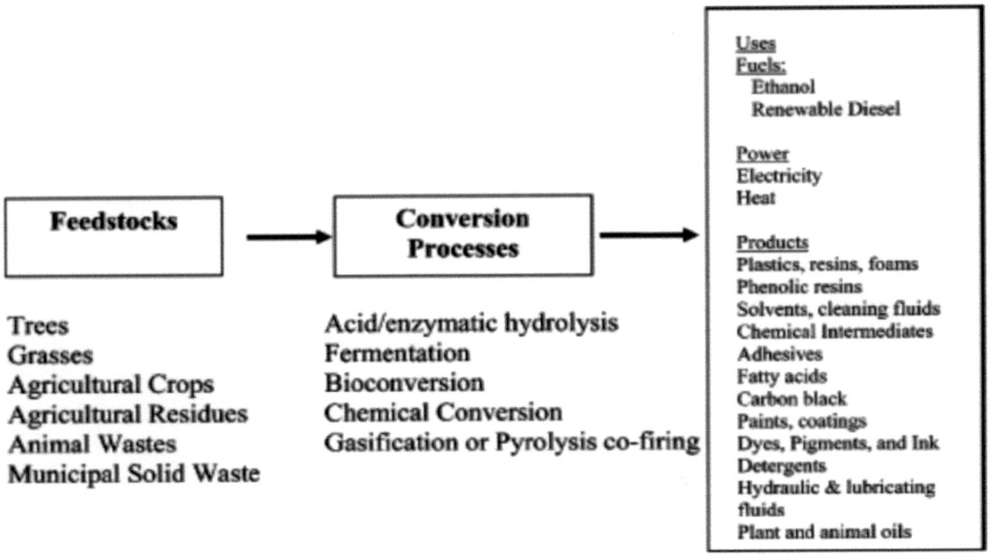

Figure 1. Simple procedure for three-step biomass-process-products [13]

1.1.5. Disadvantages

It is very important to increase the reaction rates, as slow reactions rates is one of the main disadvantages for biological conversions in biorefinery processes. Another disadvantage is the often low product concentrations, which means the high product recovery costs with existing technology. The lower yields of targeted products is often found in some multiple products systems [20]. Therefore the biorefinery processes to become an actual alternative to fossil fuels and petroleum-derived products, biorefinery processes must be competitive and cost-effective [21].

2. Biofuels

As an important category of bioenergy, biofuel is a type of fuel which is biologically derived from biomass. The biofuels, which include liquid, solid biofuels and various biogases, can replace the conventional petroleum or petroleum derived products. Many biological reactions involved in biofuels production are at mild conditions, can offer relatively high products yields and generally result in low levels of contamination to the environment. The modern application of biological transformations, known as biotechnology, is also an evolving field that has great promise for substantial improvements and significant cost reductions. In this section, several liquid and gases biofuels are introduced *e.g.* (1) bioethanol, biobutanol, and biodiesel which can replace the gasoline used as transportation fuels; and (2) biogas, which is produced from anaerobic digestion of biomass as a substitute for natural gas either for industrial applications or for transportation.

2.1. Bioethanol

Bioethanol is a promising transport fuel alternative to gasoline because it has higher oxygen content and no sulphur or nitrogen when compared with gasoline [22]. Currently, the blends E5 and E10 that consist of 5% (v/v) and 10% (v/v) ethanol respectively, have a widespread usage since these blends can supply the existing vehicular fleet without major changes to engines. High bioethanol blends (E100, E95 and E85) require modified or dedicated vehicles.

Bioethanol can be produced from three types of raw materials: sugars (from sugarcane, sugar beet, molasses, and fruits), starch (from corn, cassava, potatoes, and root crops), and cellulose (from wood, agricultural residues, waste sulphite liquor from pulp and paper mills). Among the three main types of raw materials, cellulose contained in lignocellulosic biomass represents the most abundant global source of biomass, which can be utilised for bioethanol production [23]. There are also two approaches for producing bioethanol from lignocellulosic biomass through (1) Biochemical (2) Thermochemical processes.

2.1.1. Biochemical production of bioethanol

Figure 2 illustrates the high level technologies for producing bioethanol from these various biomass feedstocks. Typically, the common steps for biologically producing bioethanol from different feedstocks are fermentation and distillation. For the first generation (1G) bioethanol production, the sugar extracted from sugar-rich crops and that from starch digestion by amylases or acids is directly fermented to bioethanol. To convert lignocellulosic biomass into second generation (2G) bioethanol, an additional step of pre-treatment is usually required.

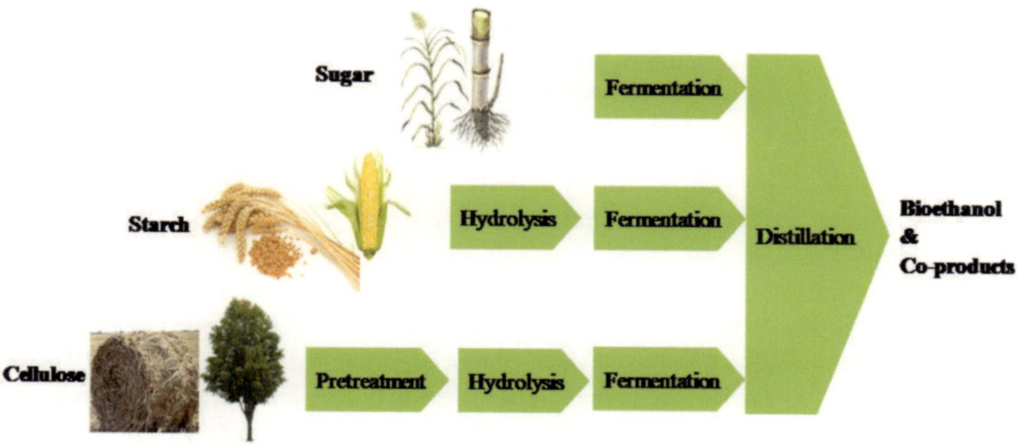

Figure 2. Technologies required producing bioethanol from biomass. [24]

A wide variety of lignocellulosic feedstocks are potentially available for bioethanol production such as wood, grass, agricultural waste and MSW (municipal solid waste). Their physical structures and chemical compositions are different; therefore technologies applied for bioethanol production can be diverse. In addition to the main product bioethanol, co-products are also usually produced, such as heat and electricity generated by burning lignin-rich residue from fermentation and also, potentially, a wide range of high value-added chemicals like acetic acid, furfural and hemicellulose sugar syrup and the low molecular weight lignin.

General technologies required for biologically producing 2G bioethanol include (1) pre-treatment, (2) enzymatic hydrolysis, (3) fermentation, and (4) distillation.

Pre-treatment is applied to enhance the accessibility of enzyme to biomass by increasing available biomass particle surface area for enzyme to attack. This can be achieved by partially removing lignin and/or hemicellulose, changing the structure of biomass fibres to decrease cellulose crystallinity and its degree of polymerization. The current available pre-treatment methods can be classified as mechanical, chemical and biological. Table1 summarised some typical pre-treatment methods and their characterisations. Pre-treatment has been viewed as the most expensive step in the biologically production of bioethanol. Therefore, it is important to assess the economic feasibility of the pre-treatment method in addition to its technology performance. More information about each pre-treatment method can be found in Section 5.

Enzymatic hydrolysis is carried out under mild conditions with potentially high sugar yields and relatively low maintenance costs. Nevertheless, major challenges for cost-effective commercialisation remain, such as the high cost of enzymes, the slow rate of enzymatic reaction and potential inhibition by sugar degradation products from pre-treatments [48]. In enzymatic hydrolysis, cellulose is hydrolysed by a suite of enzymes, including cellulase and β-glucosidase crudely purified from lignocellulose-degrading fungi such as *Trichoderma reesi*, *Trichoderma viride* and *Aspergillus niger*. Cellulase refers to a class of enzymes including endocellulase breaking internal bonds of cellulose, exocellulase cleaving from the free ends of chains produced by endocellulase to form cellobiose (a dimer of glucose), and cellobiase (β-glucosidase) then hydrolysing cellobiose to produce glucose monomers. In addition, most of cellulase mixtures contain hemicellulase that facilitates hemicellulose hydrolysis to assist with the overall effectiveness of enzymatic hydrolysis.

After the enzymatic hydrolysis, sugar monomers can then be fermented to ethanol by micro-organisms (*e.g. Saccharomyces cerevisiae* and *Zymomonas mobilis*). Fermentation has been commercialised in brewery and food manufacturing for centuries and itself is not a complex and expensive process. The challenges regarding fermentation for the bioethanol industry are: (1) to convert pentose (C5 sugar) which cannot be fermented by the conventional yeast efficiently, and (2) to prevent inhibition caused by sugar degradation products from pre-treatments. Research has shown the feasibility of construction and application of genetically engineered yeasts capable of converting both pentose and hexose to ethanol [49]. Further potential lies in using bacteria with the metabolic pathways necessary to ferment all sugars available from lignocellulosic biomass. *Z. mobilis* has shown to be capable of metabolising 95% of glucose, 80% of xylose and 40% of other sugars in corn stover hydrolysate [50]. Metabolic engineered *Geobacillus thermoglucosidasius* has demonstrated an ethanol yield of over 90% of theoretical at temperatures in excess of 60°C [51].

Pre-treatment method	Process and conditions	Possible changes in biomass	Disadvantages	Reference
Steam explosion	No agent temperature:160-260°C,20-50 bar , 2-5 minutes	Dissolve hemicelluloses Low sugar degradation	Partially degrade hemicellulose	[25-27]
Ammonia fibre explosion (AFEX)	Ammonia as agent, 65-90°C, 0.5-3 hours	Change biomass physical structure Enhancing hemicelluloses hydrolysis	Limited effects on soft and hardwood	[28, 29]
SO_2/H_2SO_4 explosion	SO_2 as agent, 160-220°C, < 2 minutes	Dissolve hemicelluloses effectively for hardwood and agricultural residues	Degradation of hemicelluloses, less effective for softwood	[30, 31]
CO_2 explosion	CO_2 as agent, 35°C, 56.2 bar, 10-60 minutes	Interrupt crystalline structure of cellulose	Inefficient for softwood and high capital cost	[32, 33]
Hot liquid water	Water as agent, 190-230°C, 45 seconds-4 minutes	Effectively dissolve hemicelluloses Very low degradation	Water recycling prohibitively expensive	[34-36]
Dilute acid	H_2SO_4 as agent , over 160°C, 2-10 minutes	Effectively dissolve hemicelluloses	Needs neutralisation, significant formation of fermentation inhibitors	[37-39]
Alkaline	NaOH/ $Ca(OH)_2$ /Ammonia as agent, 70-120°C, 20-60 minutes	Removal of lignin Low hemicelluloses degradation	Costs of reagents and wastewater treatment are high	[40-42]
Oxidation	$Ca(OH)_2$+O_2/H_2O_2 as agent, 140°C, 3 hours	Removal of lignin Low hemicelluloses degradation	Costs of reagents and wastewater treatment are high	[43, 44]
Organic solvent	Ethanol as agent, 140-200°C, 30-150 minutes	Removal of lignin	Cost of solvent recovery is high	[45, 46]
Ionic liquid	Ionic liquid as agent, 120°C, 22 hours	Remove of lignin and hemicellulose	Costs of reagents and long treatment time	[47]

Table 1. Chemical pre-treatment methods for lignocellulosic biomass.

Bioconversion process configurations, including Separate Hydrolysis and Fermentation (SHF), Simultaneously Saccharification and Fermentation (SSF), Simultaneously Saccharification and Co-Fermentation (SSCF), and Consolidated Bioprocessing (CBP). The SHF has many advantages, such as allowing both enzyme and micro-organisms to operate at their optimum conditions. Also, any accidental failure of enzymatic hydrolysis and fermentation would not affect the other steps. Alternatively the enzymatic hydrolysis may also be combined with fermentation and can thus be carried out simultaneously in a same reactor - this being known as the simultaneous saccharification and fermentation (SSF). During enzymatic hydrolysis, the cellulases are strongly inhibited by hydrolysis products: glucose and short cellulose chains ('end-point' inhibition). SSF can overcome this inhibition by fermenting the glucose to ethanol as soon as it appears in solution. However, ethanol itself inhibits the action of fermenting micro-organisms and cellulase although ethanol accumulation is less inhibitory than high concentrations of hydrolysis products [52]. Nevertheless, SSF operating at the compromised temperature (37-40 °C) has some drawbacks caused by the different optimal temperatures for the action of cellulases (45-50° C) and the growth of microorganisms (typically 28-35°C). One method to overcome this disadvantage is the utilisation of thermo-tolerant fermenting organisms. SSCF is a promising SSF process where the micro-organism co-ferment pentose and hexose to bioethanol. CBP currently becomes the focus of most research efforts to date; it integrates cellulase production, cellulose hydrolysis and fermentation in one step by using an engineered strain [53]. Many studies have been reported in CBP technologies developments recently [54-56].

Nevertheless, other significant efforts are also required to enable future integrated biorefinery. They include (1) promising process designs to integrate energy consumption and minimise the water footprint (2) producing a range of high value added by products, *e.g.* power, chemicals, and lignin-derived products *etc.*

2.1.2. Thermo-chemical production of bioethanol

The thermo-chemical bioethanol production refers to a series of processes including biomass indirect gasification, alcohol synthesis and alcohol separation as shown in Figure 3.

The biomass is processed and dried by flue gas before being fed to biomass gasifier. The biomass is chemically converted to a mixture of syngas components (*i.e.* CO, CH_4, CO and H_2 *etc*), tars, and a solid char which is the fixed carbon residual from the biomass. The heat required for endothermic gasification reactions is supplied by circulating hot synthetic olivine 'sand' between the gasifier and combustor. The solid char and 'sand' from the gasifier are separated by cyclones and then sent to a char combustor where the char is oxidised by oxygen injected. The heat released from the oxidation of the char reheats the 'sand' over 980 °C. The hot 'sand' is then sent to the gasifier to provide heat required by gasification reactions. The ash from the char combustor and sand particles captured are sent to landfill after being cooled and moistened. The tar produced in the gasifier is reformed to CO and H_2 with the presence of catalyst in a bubbling fluidized bed reactor. The syngas generated in the biomass gasifier goes through a cooling and clean-up process to remove CO_2 and H_2S. During this process, the tar is reformed in an isothermal fluidized bed reactor and the catalyst is regenerated. The cleaned syngas is then converted to alcohols in a fixed bed reactor. The produced alcohol stream is depressurised in preparation of dehydration and separation afterwards. The evolved syngas in alcohol stream is recycled to the Gas Cleanup & Conditioning section. Finally, the alcohol mix is separated to methanol, ethanol and other higher molecular weight alcohols. The heat required for the gasifier and reformer operations and electricity for internal power requirements is provided by a conventional steam cycle. The steam cycle produces steam by recovering heat from the hot process streams throughout the plant.

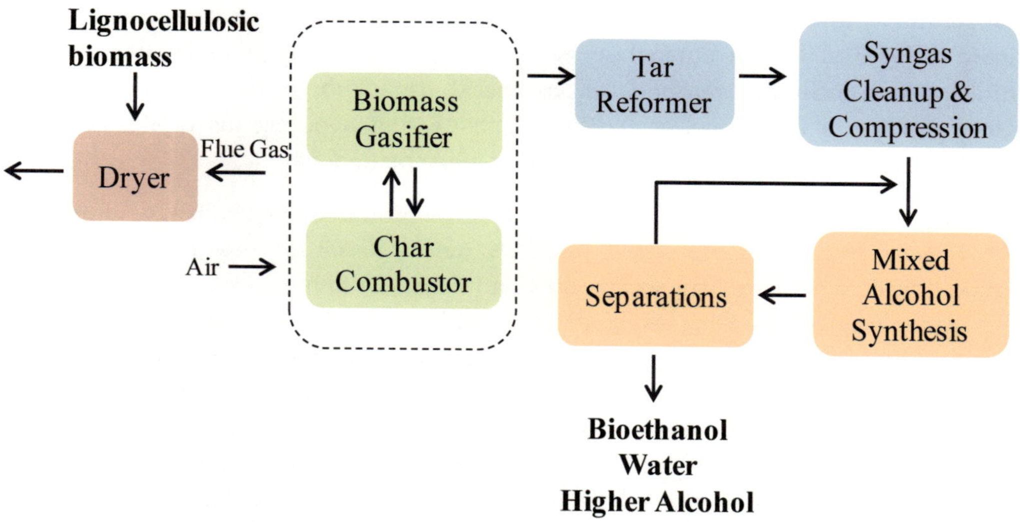

Figure 3. The schematic of a thermo-chemical cellulosic ethanol production process [57]

To compare these two approaches (biochemical *vs.* thermo-chemical) for producing bioethanol from economic point of view, process simulation and economic analysis are usually performed to calculate the minimum ethanol selling price (MESP) calculated from the discounted cash flow method. The MESP is defined as the selling price of bioethanol that makes the net present value of the biomass to bioethanol process equal to zero with a certain discounted cash flow rate with in a return period over the life of the plant [37]. In other words, it refers to the ethanol price at the break-even point which means annual costs and income are equal at this price. Several studies suggested that the estimated prices for 2G bioethanol produced biochemically is in the range of 2.16 to 4.44 USD $/gallon, depending on the type of biomass feedstock, technologies applied and the reference year based on [37, 58-61]. On the other hand, NREL (National Renewable Energy Laboratory) reported a relatively low MESP for bioethanol produced thermo-chemically as 1.07 USD $/gallon. Nevertheless, raw materials cost (including biomass feedstock and catalyst or enzyme) is the main contributor to the MESP. For example, the cost of corn stover accounts for 40% and 43% of the MESP for bioethanol biochemically and thermo-chemically produced respectively [37, 57].

From environmentally point of view, a comparative LCA study showed that biochemical approach offers a slightly better performance on greenhouse gas emission and fossil fuel consumption impact categories, but the thermo-chemical pathway has significantly less water consumption [62].

2.2. Butanol

Butanol is another attention attracted alternative fuel to gasoline besides ethanol because of its properties with respect to gasoline blending, distribution and refuelling, and end use in existing vehicles. For instance, butanol has relatively high energy content which is 30% higher than ethanol and is closer to gasoline. Additionally, butanol has low vapor pressure, low sensitivity to water and it is less volatile, and less flammable when compared with other liquid fuels [63]. Therefore butanol can be handled conventionally in the existing petroleum infrastructure, including transport *via* pipeline. It also can be blended, at any ratio, with either gasoline or diesel fuel at existing refineries, thus avoiding the capital investment associated with plant revamps and the need for major operational, *etc.*

Similarly to bioethanol, butanol can be biochemically produced from both agricultural crops and lignocellulosic biomass using *Clostridium acetobutylicum* or *C. beijerinckii* to ferment lignocellulosic hydrolysate sugars (hexoses and pentoses) to butanol. Traditionally, sugar-rich agricultural crops such as corn, cane molasses and whey permeate have been successfully used as feedstocks in the commercial production of butanol for decades. However, the cost for these food crops rises significantly nowadays; therefore, lignocellulosic biomass becomes more popular as substrates for butanol production. In similar ways of producing bioethanol, pre-treatments are required prior to enzymatic hydrolysis (using cellulase and cellobiose). However, one of technology challenges is the inhibition caused by by-products in pre-treatments such as furfural, HMF, acetic acid, and ferulic acid generated in dilute acid pre-treatments *etc.* Among these by-products, ferulic and ϱ-coumaric acids were found can significantly inhibit fermentation but furfural and HMW were surprisingly stimulating to the cell culture [64].

The resulted lignocellulosic hydrolysate is then fermented by microorganisms *via* Acetone-butanol-ethanol (ABE) fermentation (Figure 4). The main challenge in the ABE fermentation is the product butanol itself is toxic to the fermenting microorganisms. In order to overcome this drawback, focused research efforts are to (1) improve the fermentation strategies to minimise the level of inhibitors accumulated such as simultaneously removing butanol and (2) to develop or genetically improve butanol – producing cultures.

However, biobutanol has several potential shortcomings. It is more toxic to humans and animals in the short term than ethanol or gasoline (although some components of gasoline, such as benzene, are more toxic and/or carcinogenic). And it is not clear whether butanol will degrade the materials commonly used in automobiles that can come into contact with motor fuels; building evidence suggests that it will not cause problems, but there has been no definitive testing on the wide range of potentially affected polymers and metals [65].

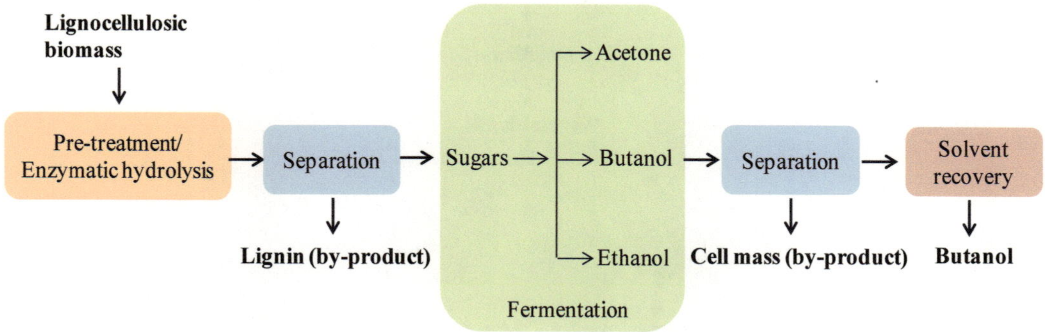

Figure 4. Phases of ABE fermentation for producing butanol

Additionally, butanol is reported cannot deliver a better economic feasibility and a more sustainable environmental performance when compared with bioethanol under the current level of technology [66]. The relatively low yield of solvents out of glucose (mixture of acetone, ethanol and butanol), which is in the range of 33% - 45% (wt), is the main cause for the high cost of butanol. This economic study argued that butanol perhaps can be sold as chemicals rather than transport fuel unless the technology would be improved to make butanol production economically competitive with bioethanol.

2.3. Biodiesel

Biodiesel refers to a liquid fuel alternative to petroleum diesel which can be used alone or blended with petroleum diesel. Similarly to bioethanol blends, blends of 20% biodiesel (B20) or lower can be used in diesel equipment without or with only minor modifications. Biodiesel can be produced from animal fat or oil from plants such as soybean and *Jatropha,* or from microalgae and fungi.

2.3.1. Biodiesel from vegetable oil

Conventionally, the biodiesel is produced from vegetable oil with the presence of alcohol/alkaline/acid catalyst. This process is known as transesterificaiton or alcoholysis as shown in Figure 5 [67].

The vegetable oil is converted to esters and glycerol by reacting with an alcohol which can be ethanol, methanol or butanol. During this reaction, catalysts (*e.g.* alkalis, acids or enzymes) are required to improve the reaction rate and yield. Alkalis including NaOH, KOH and carbonates *etc.* are usually used as catalyst when feedstock containing less than 4% fatty acids. Acids, which are normally used when feedstocks contain more than 4% free fatty acids, include sulfuric acid, hydrochloric acid and sulphonic acids *etc.* Lipase, an enzyme that catalyses the hydrolysis of fats, can be used as a biocatalyst [68].

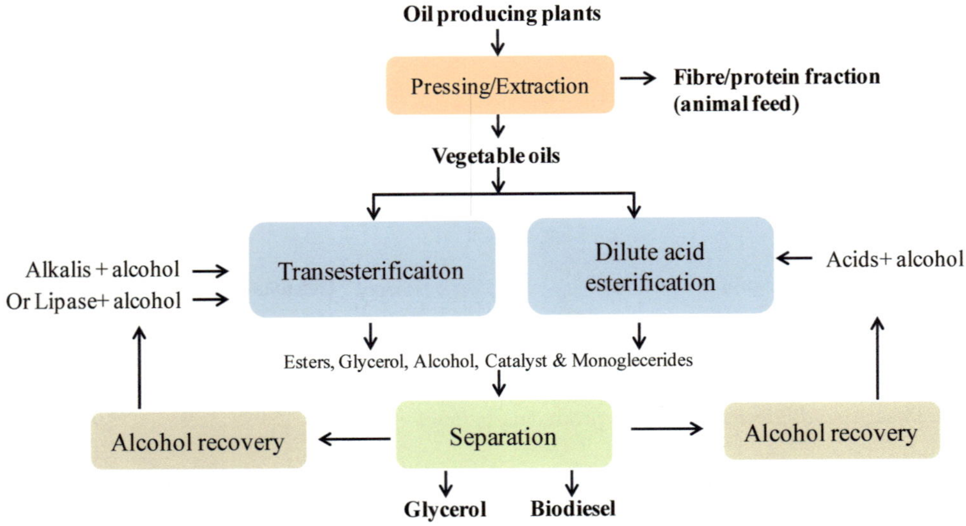

Figure 5. The schematic of biodiesel production [67]

A review by Ma and Hanna [69] summarized the parameters significantly influencing the rate of transesterificaiton reaction which include temperature, ratio of alcohol to oil, type of catalyst and catalyst concentration. The ester yield is increased by rising the transesterificaiton temperature; however, it will increase the risk of forming methanol bubbles when the temperature is close to methanol's boiling point. The ratio of alcohol to oil depends on the type of catalyst used which is approximately 6:1 for alkali catalyst and 30:1 for acid catalyst [70]. Enzyme used as a catalyst is becoming more attractive nowadays because it tolerates free fatty acid and water contents in the oil to avoid soap formation and thus results in an easier purification of biodiesel and glycerol [68]. However, the relatively high price of enzyme catalyst makes its utilization in the commercial production of biodiesel challenging.

Nowadays, 90% of U.S. biodiesel is made from soybean oil. The price relationship between vegetable oils and petroleum diesel is key influential factor to the profitability of biodiesel industry. Because of the increasing price of vegetable oils, biodiesel industry is suffering uncomfortable situations [71]. As a result, alternative non-food feedstocks and the associated technologies are becoming the focused research in biodiesel area.

Jatropha curcas is an agro-forestry crop growing in tropical and sub-tropical countries, such as India, Sahara Africa, South East Asia and China. This crop grows rapidly and takes 2-3 years to reach maturity with economic yields [72]. Lu *et al.* reported a higher than 98% biodiesel yield by a pre-esterification using solid acid followed by a transesterificaiton using KOH [73]. A high yield of 98% (wt) is also reported by Shah *et al.* [74] which is obtained from *Jatropha* oil using *Pseudomonas cepacia* lipase. Kumari *et al.* [75] also documented a relatively high yield of 94% (wt) biodiesel yield from *Jatropha* oil using lipase from *Enterobacter aerogene*. They also reported negligible loss in lipase activity even after repeated use for several cycles.

2.3.2. *Biodiesel from microalgae*

Due to biodiesel produced from oil crops, waste cooking oil and animal fat cannot meet the high demand for renewable transport fuels, another biomass feedstock microalgae becomes attractive. This is because (1) microalgae are sunlight-driven cells, (2) grow rapidly with biomass double time of 24 hours, (3) require less high quality land used compared to other feedstock, (4) many are exceedingly rich in oil and (5) biodiesel produced from microalgae is 'carbon neutral' [76] (see Figure 6). However, several challenges need to be tackled in order to produce biodiesel from microalgae commercially. Scott *et al.* [77] provides a comprehensive review discussing these challenges and potential tackles.

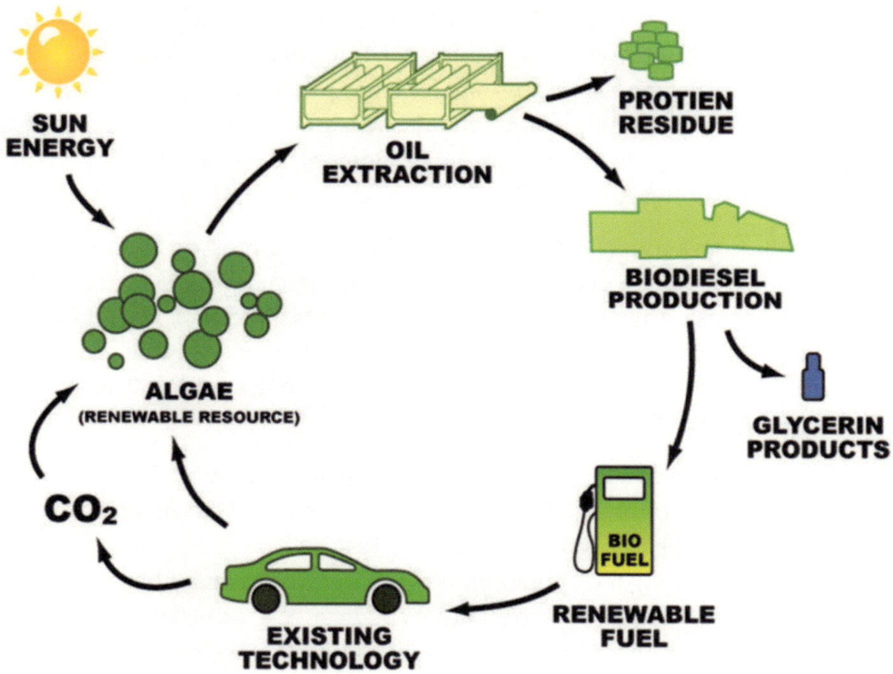

Figure 6. Life cycle of biodiesel produced from microalgae

There are estimated 300 000 species in algal strain. After screening, typical species including *Botryococcus braunii, Nannochloropsis sp., Neochloris oleoabundans, Nitzschia sp.,* and *Schizochytrium sp.* have up to 77% (dry wt) oil content [76]. Microalgal biomass is produced with the presence of light, fed carbon dioxide and essential inorganic elements including nitrogen (N), phosphorus (P), iron and in some cases silicon. Biomass is then harvested and extracted to obtain oil for biodiesel production using transesterificaiton with methanol. Nutrients and spent biomass are recycled in the downstream process.

Factors involved in these phases are all important to be considered and optimized to maximize the biomass yield and minimize the production cost. First of all, the light level needs to be manipulated to deliver an optimal light to all of the algae cells within the culture. The excess light level not only can results in less efficient use of absorbed light energy but also can cause biochemical damage to the photosynthetic machinery [77]. Secondly, though minimal nutrients requirement can be estimated according to the approximate molecular formula of microalgae which is $CO_{0.48}\ H_{1.83}\ N_{0.11}\ P_{0.01}$[78], nutrients such as phosphorous must be supplied in excess. In order to minimise the nutrient cost, sea water supplement with commercial nitrate and phosphate fertilisers can be used for growing microalgae [76]. Thirdly, the choice of facility (open raceway ponds or closed photobioreactor) is important since the scale-up of biomass production is largely depending on the surface area rather than volume because light only penetrate a few centimeters [77]. The former raceway pond is an open-top close loop recirculation channel with a typical depth of 0.3 m. It is relatively cheap to build and has been operated with extensive experience for decades. However, the drawbacks for this type of facility are (1) it is difficult to avoid microbial contamination, (2) it requires for extensive areas of land for ht raceways and substantial cost regarding harvesting, and (3) it has poorly mixed therefore has optically dark zone [76, 77]. The photobioreactor a tubular reactors consists of an array of glass or plastic transparent tubes. It requires a large amount of energy for pumping and compressing air for sparging culture [77].

The biomass broth from production phase is harvested and processed to remove water and residual nutrients which are recycled. The concentrated biomass paste is then extracted to obtain oil and lipids using water and extraction solvent (*e.g.* hexane) [79]. It is difficult to release lipids from microalgae intracellular location using an energy-efficient way because of the large amount of solvent required. Also it is key to avoid significant contamination by other cellular components such as DNA [77].

The efforts in academic research and industrial commercialization of biodiesel production from microalgae include: (1) integration of production process such as energy integration, water and nutrient recycling; (2) improvement of microalgae biology *via* genetic and metabolic engineering such as enhancing their photosynthetic efficiency, increasing biomass yield and oil content and improving temperature tolerance to reduce cost associated with cooling; (3) improving photobioreactors regarding their capacity and operational ability [76].

2.4. Biogas from anaerobic digestion

Anaerobic digestion (AD) has been used to treat biodegradable solid waste such as MSW, industrial waste and sewage sludge over decades. Biogas containing methane and carbon dioxide is the main product form AD digester. Generally, biogas is collected in the gas tank and they can be directly exported to national gas grid or sent to combustion in the CHP system to generate electricity (with a yield in the range of 0.7 – 2.0 kwh/m^3 biogas) and heat.

AD process is a dynamic complex system involving microbiological, biochemical, and physical-chemical processes though which the biodegradable waste are turned into biogas. Among biological waste treatment methods, AD has been identified as the most environmentally sustainable option for treating biowaste since it offers a unique technology which enables not only diverting biodegradable from landfill but also producing bio-energy and potential by-products such as a beneficial soil conditioner [80].

AD systems generally have four classifications [80]:

- Mesophilic (30 - 40 °C) or thermophilic (50 - 65°C) according to temperature
- Wet digestion (< 15% dry solid) or dry digestion (between 20% - 40% dry solid) according to the solid content in feedstock
- Single step (one vessel) or multiple step digestions (normally two-step digestion i.e. hydrolysis and methanogenesis)
- Batch digestion (loading feedstock in the beginning and remove products at the end of process) or continuous digestion (loading feedstock and withdraw products continuously)

Generally, five trophic groups are considered to be relevant to the process such as hydrolysing bacteria, acidogenic bacteria, acetogenic bacteria, aceticlastic and hydrogentrophic methanogens. They are involved in a series of digestion steps which are described as following and in Figure 7 [81] :

1. Carbohydrates, lipids, proteins *etc.* are broken down through hydrolysis to sugars, long – chain fatty acids and amino acids by extracellular enzymes released by hydrolytic bacteria;
2. Then these molecules are converted into volatile fatty acids, alcohols, CO_2 and H_2 in acidogenesis step;
3. These molecules are then further converted by acetogenic bacteria mainly into acetic acid, H_2 and CO_2;
4. Finally, all these intermediate products are turned into CH_4, CO_2 and water in the last step where methanogenic bacteria are involved. Three biochemical pathways are used by methanogens to produce methane gas:

   a. Acetotrophic methanogenesis: $4\ CH_3COOH \rightarrow 4\ CO_2 + 4\ CH_4$

   b. Hydrogenotrophic methanogenesis: $CO_2 + 4\ H_2 \rightarrow CH_4 + 2\ H_2O$

   c. Methylotrophic methanogenesis: $4\ CH_3OH + 6\ H_2 \rightarrow 3\ CH_4 + 2\ H_2O$

Due to for different substances, biological consortia and digestion conditions, the overall biogas yield and methane content will vary. Typically, the methane content of biogas is in the range of 40-70 % (v/v) [82].

Several key factors influence the Ad performance. They include pH, temperature, organic loading rate (OLR), the ratio of inoculum to substance (I/S) and the presence of inhibitory substances. Generally, mesophilic AD (35 - 37 °C) is more preferred than thermophilic AD (50 - 60°C) since the latter one offers less methane yield and it is more sensitive to environment change [81]. The pH range suggested for AD process is in the range of 6.8 -7.2 [80]. In addition, anaerobic digestion requires attention to the loading of nutrients for bacteria including carbon and nitrogen. The proper ratio of these two components (C/N) depends on the digestibility of the carbon and nitrogen sources between 20: 1 and 30:1. Other nutrients such as S, Mg, K, P, Ca, Fe, Zn, Al, Ni, Co, Cu and vitamin B12 are necessary [80]. However, these components are generally contained in the Organic Fraction of MSW (OFMSW) while they are added in the laboratory scale AD systems.

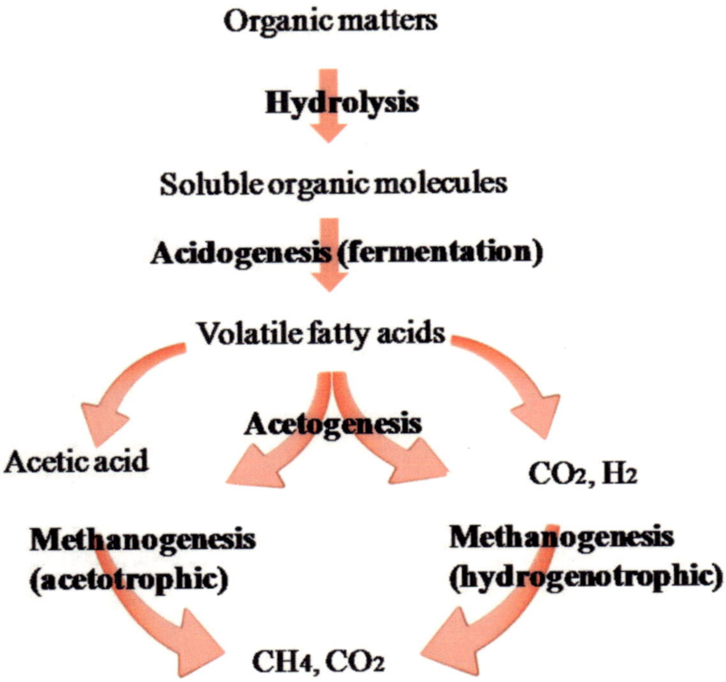

Figure 7. Anaerobic digestion biochemical conversion pathways

Regarding the AD process operation, I/S ratio is considered as one the most important parameter. It is suggested to be approx 1 by Raposo *et al.*[83] who found that biogas production was inversely proportional to the I/S ratio in the range of 1 to 3. two stage AD is more preferred because it provides optimum environmental conditions for each bacteria group, offers accelerated digestion rates, better stability and thus increased methane yield [80]. Another process parameter is retention time which includes hydraulic retention time and solid retention time. The former refers to the mean time that any portion of liquid feed remains in a digestion system; the latter is defined as the mean time for any portion of solid feed or microbal biomass remains in the digester. These two retention times are the same in a single stage digester; while in a two stage digestion system the longer solid retention time is, the higher degradation rates and biogas yields are obtained [80]. In addition to the above process parameters discussed, the organics loading rate (OLR) is also critical which is measured as volatile solid (VS) or chemical oxygen demand (COD) of feed to a unit volume of digester per unit time [80]. The range for OLR is suggested in the range of 6 - 9.7 kg VS/day/m^3 which is varied with the biodegradability of feedstock and AD systems [81].

Furthermore, the quality of OFMSW treated in biogas plants is also crucial for balanced performance of the biogas process, the technical feasibility of process and the use of residual/effluent as agricultural soil conditioner. Therefore, the costs associated with waste collection, sorting and pre-treatment should be considered [84].

Currently, most of MSW in the U.S. are sent to composting as an alternative to landfill. This is because it is more difficult to treat OFMSW than treating wastewater or manure. In addition, the AD of OFMSW requires a large amount of investment and technological experience as well as a higher capital and operating cost than compositing and landfilling [82]. The relatively low gate fees for landfill in the U.S. and relatively low energy prices make AD difficult to be commercialized in the U.S compared to those in Europe [82]. However, in the UK, there is currently very little waste treatment using AD apart from the use of AD to treat sewage sludge and wastewaters [85].

However, LCA studies have shown that AD of MSW reduces environmental impacts and is more cost - effective (in Europe) on a whole system basis than composting or landfilling options [86, 87].

3. Commodity chemicals and materials

Today, only a small numbers of chemicals are produced from lignocellulosic feedstocks via fermentation. Much less attention has been given to biomass as a feedstock for organic chemicals, while there has been a strong political and technical focus on using biomass to produce transportation fuels. However, replacement of petroleum-derived chemicals with those from biomass will play a key role in sustaining the growth of the chemical industry [88]. One way to replace petroleum is through biological conversion of lignocellulosic resources into products now derived from petroleum. The current developments especially in fermentation technologies, membrane technologies and genetic manipulation open new possibilities for the biotechnological production of market relevant chemicals from renewable resources [5].

In lignocellulosic feedstocks biorefinery processes, the sugars or some of the fermentation products can be chemically converted into a variety of chemicals, which could be used to form biological materials, such as protein polymers, xantham gum, and polyhydroxybutyrate. The lignin as remaining fraction from lignocellulosic feedstocks, could be converted through hydrogenation processes into materials, such as phenols, aromatics, and olefins, or simply burned as a boiler fuel for cost efficiency of the overall process. Currently, conventional chemicals include acetaldehyde, acetic acid, acetone, n-butanol, ethylene, and isopropanol can simply be derived from LCF. Appropriate organisms could then convert the sugars into the desirable products and co-products for this process. The advantage to such products is that the market is already established, and minimal effort is required to integrate these products into existing markets. However, co-product markets might be limited, and caution must be taken in considering their impact on overall economics, especially for large-scale implementation. A sequence of processes comparable to those employed for cellulosic ethanol production would be used to pre-treat the lignocellulosic biomass to open its structure for the weight of the feedstock. Therefore, lignocellulosic biomass might be expected as the low cost of raw materials could be converted to a variety of commodity chemicals and materials [20].

3.1. Present promising commodity chemicals and materials from LCF biorefinery

3.1.1. Lactic acid

Lactic acid represents a chemical with a small world market, and the market for traditional applications of lactic acid is estimated to be growing at about 3–5% annually. New products based on lactic acid may increase the world market share significantly, which includes the use of derivatives such as ethyl esters to replace hazardous solvents like chlorinated hydrocarbon solvents in certain industrial applications. In theory, one mole of glucose results in almost two moles of lactic acid. The recovery process for lactic acid is much more sophisticated than that of the ethanol fermentations, involving various precipitations, chromatographic and distillation steps [5].

Lactic acid can be converted to methyl lactate, lactide, and polylactic acid (PLA) by fermentation [89]. The PLA is a biodegradable polymer used as environmentally friendly biodegradable plastic, which can be the replacement for polyethylene terephthalates (PETs) [90]. Recently, attempts have been made to produce PLA homopolymer and its copolymer by direct fermentation by metabolically engineered [91], shows a great potential for utilizing lignocellulosic feedstock for the key biodegradable polymers. Efforts are also under way to develop efficient processes for converting biologically produced lactic and hydroxypropionic acids to methacrylic and acrylic acids [88].

Lactic acid can be produced either chemically or by microbial fermentation. A major disadvantage for chemical synthesis is the racemic mixture of lactic acid. Microbial fermentation offers both utilization of renewable carbohydrates and production of pure L- or D-lactic acid depending on the strain selected. Currently, most of lactic acid production is produced mainly from corn starch. However, the use of lignocellulosic feedstock for lactic acid production appears to be more attractive because they do not impact the food chain for humans. But the process for converting lignocellulosic feedstock into lactic acid is not cost efficient due to the high cost of cellulase enzymes involved in cellulose hydrolysis [92; 93]. In addition, the main bottleneck during the hydrolysis of lignocellulosic feedstock by cellulases is the inhibition on cellulase by glucose and cellobiose, which remarkably slows down the rate of lignocellulosic feedstock hydrolysis [94]. Economic improvements on the process are mainly focused on increasing the lactic acid tolerance, reducing the requirements for complex and cost intensive growth supplements and products recovery [95].

3.1.2. Acetone–butanol–ethanol (ABE)

An acetone – butanol – ethanol blend (in a ratio of 3-6-1) may serve as an excellent car fuel, which can be easily mixed not only with petrol but also with diesel. ABE as a fuel additive has the advantage of a similar heat of combustion to hydrocarbons, and perfect miscibility with hydrocarbons, even when water is present. The fermentative production of ABE used to be the second largest industrial fermentation after ethanol production [5]. Product inhibition caused principally by butanol is the main problem that hindering commercial development of the fermentation process. One way to overcome this inhibition problem would be to couple the fermentation process to a continuous product removal technique, so that inhibitory product concentrations are never reached. However, even with continuous product removal, product formation in these systems does not proceed indefinitely, because of the inhibition caused by the accumulation of mineral salts in the reactor [96]. Due to the shortage of raw materials, namely corn and molasses, and to decreasing prices of oil, ABE fermentation is not profitable when compared to the production of these solvents from petroleum. During the 1950s and 1960s, ABE fermentation was replaced by petroleum chemical plants.

Currently, the production of mixtures of acetone, butanol and ethanol (ABE) by sugars derived from lignocellulosic feedstocks continues to receive attention because of its potential commercial significance. The traditional fermentative production of acetone–butanol– ethanol is batch anaerobic bacteria fermentation with Clostridia. The substrate consists of molasses, and phosphate and nitrogen sources. Instead of molasses other sugar sources like sugar from lignocellulosic feedstock can also serve as a raw material for fermentation [97].

3.2. Xylan

As one of main polysaccharides in lignocellulosic biomass, xylan has a variety of applications in our everyday life and affects our well-being. For example, (1) xylans are important functional ingredients in baked products [98]; (2) xylans can be potentially used for producing hydrogels as biodegradable coatings and also encapsulation matrices in many industrial applications; (3) xyl, the main constituent from xylans, can be converted to xylitol which is used as a natural food sweetener and a sugar substitute [99]; (4) xylans can be used for clarification of juices and improvement in the consistency of beer [100]; (5) xylans are also important for livestock industry as they are critical factors for silage digestibility; (6) xylans are major constituents in non-nutritional animal feed [101]; (7) xylans can be converted to sugars and then further to fuels and chemicals; (8) enzymes that degrade xylan can facilitate paper pulping and biobleaching of pulp [100].

Xylans, the main component in hemicellulose, are heteropolysaccharides with homopolymeric backbone chains of 1,4 linked β-d-xylopyranose units. In addition to xylose, xylans may also contain arabinose, glucuronic acid or its 4-*O*- methyl ether, acetic, ferulic, and *p*-coumaric acids. Xylans can be categorized as linear homoxylan, arabinoxylan, glucuronoxylan, and glucuronoarabinoxylan. Depends on the different sources of xylan (i.e. soft- and hard- wood, grasses, and cereals), the composition of xylans differs [100].

Hemicellulose can be derived via chemical treatment or enzymatic hydrolysis. As discussed in Section 2.1.1, several pre-treatments listed in Table 1 are available to fractionate, solubilize and hydrolyze and separate hemicellulose from cellulose and lignin components. Generally, hemicelluloses are solublized by either high temperature and short residence time (270°C, 1 min) or lower temperature and longer residence time (190 °C, 10 min) [102]. However, some of chemical treatment result in hemicellulose degradation by-products such as furfural and

5-hydroxymethyl furfural (HMF) which are inhibitors for microorganisms involved in downstream fermentation if applicable.

Biodegradation of xylan requires enzymes including endo-β-1,4-xylanase, β-xylosidase, and several accessory enzymes, such as α-L-arabinofuranosidase, α-glucuronidase, acetylxylan esterase, ferulic acid esterase, and *p*-coumaric acid esterase, which are necessary for hydrolyzing various substituted xylans. The endo-xylanase attacks the main chains of xylans while β-xylosidase breaks xylooligosaccharides to monomeric sugar xylose. The α-arabinofuranosidase and α-glucuronidase remove the arabinose and 4-*O*-methyl glucuronic acid substituents from the xylan backbone [100]. The esterases hydrolyze the ester linkages between xylose units of the xylan and acetic acid (acetylxylan esterase) or between arabinose side chain residues and phenolic acids , for example ferulic acid (ferulic acid esterase) and *p*-coumaric acid (*p*-coumaric acid esterase) [100].

Hemicellulose hydrolysates from lignocellulosic biomass either obtained by chemical treatment or enzymatic hydrolysis are attractive feedstock for producing bioethanol, 2,3-butanediol or xylitol. Other value added products from hemicellulose hydrolysate include (1) ferulic acid, and (2) lactic acid which can be used in the food, pharmaceutical, and cosmetic industries [100].

3.3. Other main chemicals and materials from lignocellulosic feedstock

Acetic acid, at present, most demand of the commercial acetic acid is met synthetically. The production involves fermentation by a species of Acetobacter, which converts ethanol to acetic acid with a small final concentrations percentage (4–6%), using almost exclusively for vinegar production. In commercial practice, the actual yield roughly 75–80% of the theoretical yield [5].

Ferulic acid, as a precursor for numerous aromatic chemicals used in the chemistry industry, can be produced from lignocellulosic feedstock [88].

Levulinic Acid, Formic Acid and Furfural, their biorefinery process usually involves the use of dilute acid as a catalyst but it differs from other dilute-acid lignocellulosic-fractionating processes in that free monomer sugars are not the product. Instead, these monosaccharides are converted into the platform chemicals levulinic acid and furfural as the final products by multiple acid-catalysed reactions [103].

3.4. Opportunities and challenges

New products from lignocellulosic feedstock including new adhesives, biodegradable plastics, degradable surfactants, and various plastics and polymers could also be derived through the unique biotechnologies. The products with desirable properties that are not easily matched by petrochemical processing are particularly promising targets. Therefore, less price pressure would exist initially for such new products. However, to have a substantial impact on petroleum consumption, it is necessary to ensure that large markets have to be eventually resulted [20].

Even today, the potential of microorganisms for the production of bulk chemicals is far from being fully exploited. The cost of feedstocks still remains one of the crucial points if biotechnological processes are to succeed. The transition of industrial chemical production from petrochemical to biomass feedstock faces real hurdles. Biorefinery processes do not require the high pressures and temperatures compared with most non-biological chemical processes, thus have the potential to reduce costs. However, current non-biological chemical processes (often continuous, and well integrated) for production of commodity chemicals have become highly efficient by evolved through considerable investment. Therefore biorefinery processes for production of commodity chemicals must rapidly approach similar levels of efficiency and productivity. Nevertheless, available technologies, economic opportunities, and environmental imperatives make the use of lignocellulosic feedstock and biorefinery for industrial chemical production not only feasible but highly attractive from multiple perspectives [88].

Simple criteria have been devised to allow rapid screening of potential chemicals and materials from lignocellulosic feedstock for their economic merit. We now need to identify products that have economic potential and improve the technology to a point where these technologies can be applied in a cost-effective way [20].

4. Fractionation of lignocellulosic feedstock

4.1. Definition

Conversion of lignocellulosic materials to higher value products requires fractionation of the material into its components: lignin, cellulose, and hemicellulose, which convert to fuels, and chemicals for the production of most of our synthetic plastics, fibres, and rubbers is technically feasible. Liquefaction of LCF might serve as feedstocks for cracking to chemicals in the similar way that crude oil is presently used. Currently commercial products of LCF fractionation include levulinic acid, xylitol, and alcohols [104]. The ultimate goal of LCF fractionation is the efficient conversion of lignocellulose materials into multiple streams that contain value-added compounds in concentrations that make purification, utilization, and/or recovery economically feasible [15].

Fractionation of LCF is being developed as a means to improve the overall biomass utilization. Hemicellulose when separated from the LCF may find broader use for chemicals, fuel, and food application. The lignin separated in the process can be used as a fuel [105]. Unlike the lignin generated from pulping process, lignin fractionated from biomass by our approach is relatively clean, free of sulphur or sodium.

Fractionation of lignocellulosic materials is very difficult to accomplish efficiently, because of their complex composition and structure [106, 107]. However, fractionation of lignocellulosic materials is essential for some important applications, for example, paper-making, and in their conversion into basic chemical feedstocks or liquid fuels.

Figure 8 shows that fractionation of lignocellulosic biomass into its three major components, cellulose, hemicelluloses and lignin. It has been proposed as the first step of LCF refining to high value-added products [108]. Achieving high fractionation yields and maintaining the integrity of the macromolecular fractionation products are of major importance regarding the effectiveness of the whole refining process [109].

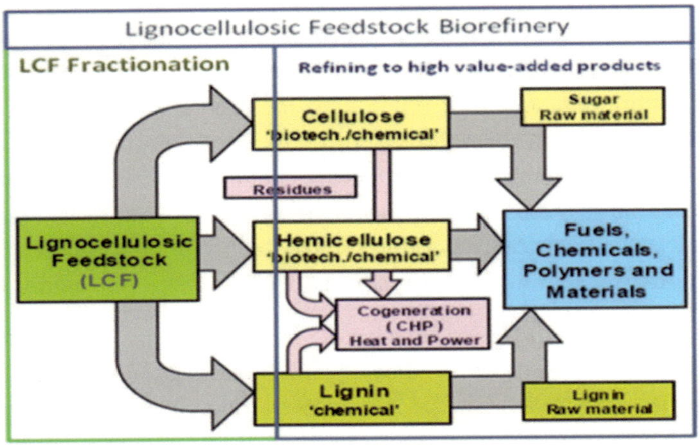

Figure 8. Lignocellulosic Feedstock Biorefinery [110]

4.2. Organosolv fractionation

The organosolv process is a unique and promising LCF fractionation. Using organosolv, lignocellulosic biomass can be converted into cellulosic fibres, hemicellulose sugars and low molecular weight lignin fractions in one-step fractionation [111-113]. Organosolv fractionation is the process to using organic solvents or their aqueous solutions to remove or decompose the network of lignin from lignocellulosic feedstocks with varying simultaneous hemicellulose solubilisation [114]. In this process, an organic or aqueous organic solvent mixture with or without an acid or alkali catalysts is used to dissolve the lignin and part of the hemicellulose, leaving reactive cellulose in the solid phase [106, 115-117]. Usually, the presence of catalyst can increase the solubilisation of hemicellulose and the digestibility of substrate is also further enhanced [118]. Comparing to other chemical pre-treatments the main advantage of organosolv process is that relatively pure, low molecular weight lignin is recovered as a by-product [119]. Organic solvents are always easy to recover by distillation and recycled for fractionation; the chemical recovery in organosolv fractionation processes can separate lignin as a solid material and carbohydrates as syrup, both of which can be used as chemical feedstocks [112, 120, 121]. A variety of organic solvents have been used in the organosolv process such as ethanol, methanol, acetone, ethylene glycol, triethylene glycol, tetrahydrofurfuryl alcohol, glycerol, aqueous phenol, aqueous n-butanol, esters, ketones, organic acids, *etc* [117, 119, 122]. For economic reasons, among all possible solvents, the use of low-molecular-weight alcohols with lower boiling points such as ethanol and methanol has been favoured [123].

Organic solvents are costly and their use requires high-pressure equipment due to their high volatility. The applied solvents should be separated from the system is necessary because the residual solvents may be inhibitors to enzymatic hydrolysis and fermentation [106], and they should be recycled to reduce operational costs. Otherwise organic solvents are always expensive, so it should be recovered as much as possible, but this causes increase of energy consumption.

The organosolv fractionation seems more feasible for biorefinery of lignocellulosic biomass, as it considers the utilization of all the biomass components. However, there are inherent drawbacks to the organosolv fractionation. In order to avoid the re-precipitation of dissolved lignin, the fractionated solids have to be washed with organic solvent previous water washing, the cumbersome washing processes means more cost. In addition, organosolv fractionation must be performed under extremely tight and efficient control due to the volatility of organic solvents. No digester leaks can be tolerated because of inherent fire and explosion hazard [121]. Its successful commercialization will depend on the development of high-value co-products from lignin and hemicelluloses [124].

4.3. Ionic liquids fractionation

The ionic liquids (ILs) is a group of promising green solvents for the efficient fractionation of lignocellulosic materials. This technology has been used for delignification of lignocellulosic materials in paper-making [125]. Moreover, by fractionating lignocelluloses with ionic liquids it is possible to extract cellulose cleanly, which establishes a platform for the development of cellulose composites and derivatives.

ILs are liquid salts exist at relatively low temperatures (often at room temperature), which typically composed of large organic cations and small inorganic anions. By adjusting the anion and the alkyl constituents of the cation, ILs' solvent properties can be varied. The solvent properties include chemical and thermal stability, non-flammability, low vapour pressures and a tendency to remain liquid in a wide range of temperatures [126]. ILs are called "green" solvents, as no toxic or explosive gases are formed.

Most ILs are nonflammable and recyclable solvents with very low volatility and high thermal stability. Carbohydrates and lignin can be simultaneously dissolved in ILs, and the intricate network of non-covalent interactions between biomass polymers of cellulose, hemicellulose, and lignin is effectively disrupted while minimizing formation of degradation products [127-129].

ILs can dissolve large amounts of cellulose at considerable mild conditions and feasibility of recovering nearly 100% of the used ILs to their initial purity makes them attractive [130]. ILs as cellulose solvents, comparing with regular volatile organic solvents of biodegradability, possesses several advantages including low toxicity, broad selection of anion and cation combinations, low hydrophobicity, low viscosity, enhanced electrochemical stability, thermal stability, high reaction rates, low volatility with potentially minimal environmental impact, and non-flammable property.

However, ILs fractionation using ionic liquids faces many challenges in putting these potential applications into industrial scale., for example, the high cost of ILs, regeneration requirement [16]. Their toxicity toward enzymes and microorganisms must also be established before ILs can be considered as a real option for LCF pre-treatment [129].

Other main challenges are the recovery of ionic liquids and the recovery of hemi-cellulose and lignin from the ionic liquids after extraction of cellulose [126].

4.4. Liquid hot water (LHW) fractionation

Liquid hot water fractionation does not employ any catalyst or chemicals. Pressure is utilized to maintain water in the liquid state at elevated temperatures (160–240 °C) and provoke alterations in the structure of the lignocelluloses [131-133]. LCF in LHW undergoes high temperature cooking in water with high pressure. LHW pre-treatment has been reported to have the potential to enhance cellulose digestibility, sugar extraction, and pentose recovery, with the advantage of producing hydrolysates containing little or no inhibitor of sugar fermentation [134].

Water is an abundant, non-toxic, environmentally benign and inexpensive solvent. LHW is the part range of sub-critical water that near its critical point (374 ºC, 22.1 MPa), Sub-critical water (SCW) possesses marvellous properties which are very different from that of ambient liquid water [135-138]. In SCW, dielectric constant, surface tension, and viscosity decrease dramatically with increasing temperature, which enhances the solubility of organic compounds. Sub-critical water is more like non-polar organic solvent (similar with acetone), thus it can substitute for some of organic solvents, and become a clean medium for chemical reactions. SCW is a tunable reaction medium for conducting ionic/free radical reactions, and an effective medium for energy and mass transfer. The ionic product of SCW is larger by three orders of magnitude than that of ambient water, which means concentrations of hydrogen and hydroxide ions are much higher. Therefore, in addition to the increase in kinetic rates with temperature, both acid and base catalyses by water are enhanced in SCW, which can be a solvent or reactant participated in chemical reaction. And without any pollution, hydrolysis in SCW is an environment-friendly technology

The objective of the liquid hot water is to solubilise mainly the hemicellulose to make the cellulose more accessible and to avoid the formation of inhibitors. By keeping the pH between 4 and 7 the autocatalytic formation of fermentation inhibitors are avoided during the fractionation [34, 139, 140]. If catalytic degradation of sugars occurs it results in a series of reactions that are difficult to control and result in undesirable side products.

The slurry generated after pre-treatment can be filtered to obtain two fractions: one solid cellulose-enriched fraction and a liquid fraction rich in hemicellulose derived sugars [34]. Lignin is partially depolymerised and solubilised as well during hot water fractionation but complete delignification is not possible using hot water alone, because of the re-condensation of soluble components originating from lignin.

Water under high pressure can penetrate into the LCF, hydrate cellulose, and remove hemicellulose and part of lignin. The major advantages are no addition of chemicals and no requirement of corrosion-resistant materials for hydrolysis reactors in this process. Liquid hot water pre-treatments are attractive from no catalyst requirement and low-corrosion potential. Liquid hot water has the major advantage that the solubilised hemicellulose and lignin products are present in lower concentrations, due to higher water input and subsequently concentration of degradation products like furfural and the condensation and precipitation of lignin compounds is reduced. However, water demanding in the process and energetic requirement are higher and it is not developed at commercial scale [141].

4.5. Combined technology for LCF fractionation

The efficiency of lignocelluloses utilization can be significantly improved by fractionation [40]. Fractionation of lignocellulosic materials may be achieved by various physical, chemical and biological methods. Combination of different methods may lead to more efficient fractionation processes of lignocellulosic materials [5].

The most promising combined technology for LCF fractionation is the combination of liquid hot water (LHW) with the assisted technologies, which usually are performed before or during the LHW fractionation, including steam explosion, CO2 explosion, Ammonia fibre explosion (AFEX), acid or alkaline pre-treatment, High energy radiation pre-treatment, Wet oxidation and Ozonolysis *etc.*

4.5.1. Combination with steam explosion

Steam explosion is the most widely employed physical-chemical pre-treatment for lignocellulosic biomass. It is a hydrothermal pre-treatment in which the biomass is subjected to pressurised steam for a period of time ranging from seconds to several minutes, and then the pressure is suddenly reduced and makes the materials undergo an explosive decompression. The treatment leads to the disruption of the structure of the material due to the rapid expansion of the water vaporized inside it. The temperatures involved are higher than, or close to, the glass transition temperature of hemicellulose, lignin and cellulose impregnated with water [142, 143], so that the internal cohesion of lignocelluloses is weakened and disaggregation and defibration of the material are facilitated. This pre-treatment combines mechanical forces and chemical effects due to the hydrolysis (auto-hydrolysis) of acetyl groups present in hemicelluloses.

Hydrolytic treatments of lignocellulosic biomass by saturated steam, with (un-catalyzed) and without (catalyzed) addition of small amounts of mineral acids, have been widely studied as a method to weaken the lignocellulosic structure and increase its chemical reactivity and enzyme accessibility [144, 145].

Un-catalyzed steam-explosion is one of only a very limited number of cost-effective pre-treatment technologies that have been advanced to pilot scale demonstration and commercialized application [16]. Autohydrolysis takes place when high temperatures promote the formation of acetic acid from acetyl groups; furthermore, water can also act as an acid at high temperatures. The mechanical effects are caused because the pressure is suddenly reduced and fibres are separated owing to the explosive decompression. In combination with the partial hemicellulose hydrolysis and solubilisation, the lignin is redistributed and to some extent removed from the material [146]. Catalyzed steam-explosion is very similar to un-catalyzed steam-explosion on their action modes, except that some acidic chemicals (gases and liquids), primarily including SO_2, H_2SO_4, CO_2, oxalic acid, etc. are used as catalysts to impregnate the LCF prior to steam-explosion, to improve recovering both cellulose and hemicellulose fractions [147]. It is recognized as one of the most cost-effective pre-treatment processes [148, 149]. Compared to un-catalyzed steam explosion, catalyzed steam-explosion has more complete hemicellulose removal leading to more increased enzymatic digestibility of LCF with less generation of inhibitory compounds [150]. A steam-explosion/separation process offers several attractive features when compared to the alternative hydrolysis and pulping processes. These include the potential for significantly lower environmental impact, lower capital investment, more potential for energy efficiency, less hazardous process chemicals and conditions [151]. Steam-explosion allows the recovery of all constitutive LCF components without the destructive degradation of any one component in favour of any other [152]. The process is generally followed by fractionation steps in order to separate the various components.

4.5.2. Combination with CO_2 explosion

Carbon dioxide explosion can also be used for lignocellulosic biomass pre-treatment. The method is based on the utilization of CO_2 as a supercritical fluid, which refers to a fluid that is in a gaseous form but is compressed at temperatures above its critical point to a liquid like density. Supercritical carbon dioxide has been used as an extraction solvent for non-extractive purposes, due to some advantages such as availability at relatively low cost, non-toxicity, non- flammability, easy recovery after extraction, and environmental friendly [153]. Besides a liquid-like solvating power, supercritical carbon dioxide displays gas-like mass transfer properties [154].

Supercritical pre-treatment conditions can effectively increase substrate digestibility by removing lignin. Addition of co-solvents such ethanol can improve delignification. Supercritical carbon dioxide has been mostly used as an extraction solvent but it is being considered for non-extractive purposes due to its many advantages [155]. CO_2 molecules are comparable in size to water and ammonia and they can penetrate in the same way the small pores of lignocelluloses. This mechanism is facilitated by high pressure. After CO2 explosive, pressure released, disruption of cellulose and hemicellulose structure is observed and consequently accessible surface area of the substrate to enzymatic attack increases [141].

4.5.3. Combination with ammonia fibre explosion (AFEX)

Similar to steam explosion, AFEX is one of the alkaline physical-chemical pre-treatment processes. Here the biomass is exposed to liquid ammonia under high pressure for a period time, and then the pressure is suddenly released, resulting in a rapid expansion of the ammonia gas that causes swelling and physical disruption of LCF fibres and partial decrystallization of cellulose. This swift reduction of pressure opens up the structure of lignocellulosic biomass leading to increased digestibility of biomass.

One of the main advantages of AFEX pre-treatment is no formation of some types of inhibitory by-products, which are produced during the other pre-treatment methods, such as furans in steam explosion pre-treatment.

AFEX has been studied for decreasing cellulose crystallinity and disrupt lignin–carbohydrates linkages [156]. Ammonia recovery and recycle is feasible despite of its high volatility [157] but the associated complexity and costs of ammonia recovery may be significant regarding industrial scale using of the AFEX pre-treatment [34, 158].

There are some disadvantages in using the AFEX process compared to some other processes. AFEX simultaneously de-lignify and solubilize some hemicellulose while decrystallizing cellulose, but does not significantly solubilize hemicellulose as acid and acid-catalyzed steam-explosion pre-treatments [159-161]. The AFEX produces only a pre-treated solid fraction, while steam explosion produces a slurry that can be separated in a solid and a liquid fractions [15]. Furthermore, ammonia must be recycled after the pre-treatment to reduce the cost and protect the environment [106, 158].

4.5.4. Combination with acid or alkaline treatment

A way to improve the effect of LHW fractionation is to add an external acid or alkali, which can catalyze the solubilisation of the hemicellulose, reduce the optimal pre-treatment temperature and gives a better enzymatic hydrolysable substrate [162-164].

Acid pre-treatments can be performed with concentrated or diluted acid. However utilization of concentrated acid is less attractive for ethanol production due to the formation of inhibiting compounds, and high acid concentration (e.g. 30-70%) in the concentrated-acid process makes it extremely corrosive and dangerous [165, 166]. Diluted acid pre-treatment appears as more favourable method for industrial applications and have been studied for fractionation wide range of lignocellulosic feedstocks, including softwood, hardwood, herbaceous crops, agricultural residues, wastepaper, and municipal solid waste. It performed well on most biomass materials, mainly xylan, but also converting solubilised hemicellulose to fermentable sugars. Of all acid-based pre-treatment methods, sulphuric acid has been most extensively studied since it is inexpensive and effective. Organic acids such as fumaric or maleic acids are appearing as alternatives to pre-treat LCF for fractionation. Organic acids also can pre-treat lignocellulosic materials with high efficiency although fumaric acid was less effective than maleic acid. Furthermore, less amount of furfural was formed in the maleic and fumaric acid pre-treatments than with sulphuric acid [167]. Phosphoric acid, hydrochloric acid and nitric acid have also been tested [34].

Alkali pre-treatment refers to remove lignin and a part of the hemicellulose, by use of alkaline solutions such as NaOH and Ca(OH)2, and efficiently increase the accessibility of enzyme to the cellulose. Alkali pre-treatment can be used at room temperature and times ranging from seconds to days. It is reported to cause less sugar degradation than acid pre-treatment. It is basically a delignification process, in which a significant amount of hemicellulose is solubilised as well. Alkaline pre-treatment of lignocellulosic materials causes swelling, increasing the internal surface of cellulose and decreasing the degree of polymerization and crystallinity, which provokes lignin structure disruption, and separation of structural linkages between lignin and carbohydrates [117]. In general, alkaline pre-treatment is more effective on hardwood, herbaceous crops, and agricultural residues with low lignin content than on softwood with high lignin content [168]. Alkali pre-treatment was shown to be more effective on agricultural residues than on wood materials [169]. Addition of an oxidant agent (oxygen/H_2O_2) to alkaline pre-treatment (NaOH/$Ca(OH)_2$) can favour lignin removal to improve the performance [170].

4.5.5. Combination with ammonia and carbon dioxide solution

The aim of combination is to enhance alkaline or acidic intensity of liquid hot water by ammonia or carbon dioxide for lignocelluloses fractionation.

Ammonia is an extremely important widely used bulk chemical. The polarity of Ammonia molecules and their ability to form hydrogen bonds explains to some extent the high solubility of ammonia in water. In aqueous solution, ammonia acts as a base, acquiring hydrogen ions from H2O to yield ammonium and hydroxide ions.

$$NH_3(aq) + H2O(l) \approx NH_4+(aq) + OH\text{-}(aq)$$

The production of hydroxide ions when ammonia dissolved in water gives aqueous solutions of ammonia the characteristics of alkaline properties.

Carbon dioxide can be considered as an ideal solvent for the treatment of natural products, because of the relatively low critical pressure (73.8 atm) and critical temperature (31.1 °C), it. In contrast with organic solvent, Super-critical carbon dioxide is non-toxic, non-flammable, non corrosive, cheap and readily available in large quantities with high purity [171].

Carbon dioxide dissolves in water becomes acidic due to the formation and dissociation of carbonic acid:

$$CO_2 + H_2O \approx H_2CO_3 \approx H^+ + HCO_3^-$$

Over the temperature range 25-70 °C and pressure range 70-200 atm, the pH of solution ranged between 2.80 and 2.95, and increases with increasing temperature and decreases with increasing pressure [172]. It was shown that in the presence of water, supercritical CO_2 can efficiently improve the enzymatic digestibility of lignocellulosic materials [32].

4.5.6. Combination with high energy radiation treatment

Digestibility of lignocellulosic materials can be enhanced by the application of high energy radiation methods, such as microwave heating [173-175] and ultrasound [176, 177]. The treatments can cause hydrolysis of hemicellulose, and partial depolymerization of lignin, the increase of specific surface area, decrease of the degrees of polymerization and crystallinity of cellulose.

Microwave treatment is a physical-chemical process involving both thermal and non-thermal effects. Treatments can be carried out by immersing the biomass in dilute chemical reagents and exposing the slurry to microwave radiation for a period of time [178]. The treatment of ultrasound on lignocellulosic biomass have been used for extracting hemicelluloses, cellulose and lignin [179]. Some researchers have also shown that saccharification of cellulose is enhanced efficiently by ultrasonic pre-treatment [180]. The efficiency of ultrasound in the treatment of vegetal materials has been already proved [181]. The well known benefits from ultrasounds, such as swelling of vegetal cells and fragmentation due to the cavitational effect associated to the ultrasonic treatment. Furthermore, mechanical impacts produced by the collapse of cavitation bubbles, give an important benefit of opening up the solid substrates surface for enzymatic hydrolysis [180].

However, the high energy radiation methods are usually energy-intensive and prohibitively expensive; appear to be strongly substrate-specific. The current estimation of overall cost from high energy radiation techniques looks too high, lack commercial appeal.

4.5.7. Combination with oxidative treatment

Wet oxidation

Wet oxidation is an oxidative pre-treatment method which employs oxygen or air as catalyst, and can be operated at relatively low temperatures and short reaction times [182]. It is an exothermic process, therefore self-supporting with respect to heat while the reaction is started [183]. Wet oxidation of the hemicellulose fraction is a balance between solubilisation and degradation. Wet oxidation has been proven to be an efficient method for separating the cellulosic fraction from lignin and hemicellulose [184], and also been widely used for ethanol production followed by SSF [185]. Wet oxidation pre-treatment mainly causes the formation of acids from hydrolytic processes, as well as oxidative reactions. The hemicelluloses are extensively cleaved to monomer sugars, cellulose is partly degraded, and the lignins undergo both cleavage and oxidation in wet oxidation pre-treatment. Therefore lignin produced by wet oxidation cannot be used as a fuel [186]. In general, low formation of inhibitors and efficient removal of lignin can be achieved with wet oxidation pre-treatment.

Ozonolysis

Ozone is a powerful oxidant that shows high delignification efficiency [106]. This method can effectively degrade lignin and part of hemicellulose. The pre-treatment is usually carried out at room temperature, and does not lead to inhibitory compounds [187]. It is usually performed at room temperature and normal pressure and does not lead to the formation of inhibitory compounds that can affect the subsequent hydrolysis and fermentation. However, ozonolysis might be expensive since a large amount of ozone is required, which can make the process economically unviable [106].

5. Other bioconversion technologies

5.1. Landfill gas (LFG) production

As discussed in Section 2.4, anaerobic digester is a suitable waste treatment method to deal with wastewater, sewage sludge and animal mature since the high solid content of other types waste would challenge the anaerobic digester operation technologies. Currently most of biodegradable waste is sent to landfill where landfill gas (LFG) is generated.

Because the wastes sent to landfill include not only biodegradable components but also other hazard wastes, the LFG produced contains approx 40 - 60% methane, CO_2, and varying amounts of nitrogen, oxygen, water vapour, volatile organics (VOC), H_2S and other contaminates (also known as non-methane organic compounds NMOCs). Some other inorganic contaminants, for example, heavy metals are found present in the LFG. Therefore, the direct release of the landfill gas to atmosphere will cause serious greenhouse gas emissions and pollutions. LFG produced from landfill site has to be monitored and managed appropriately. The general LFG managing options are: flaring (burn without energy recovery), boiler (produces heat), internal combustion (producing electricity), gas turbine (producing electricity), fuel cell (producing electricity), convert the methane to methyl alcohol, or sent to natural gas lines after cleaning process [188].

5.2. Biopulping and wood utilization

Biopulping, also known as biological pulping, refers a type of industrial biotechnology using fungus to convert wood chips to paper pulp. This technology has the potential to improve the quality of paper pulp, reduce energy consumption and environmental impacts when compare with the traditional chemical pulping technologies [189].

The aim of pulping is to extract cellulose from plant material. The traditional approaches are mechanical and chemical pulping. The former method is generally accomplished by refining grinding or thermo-mechanical pulping. The latter way is to dissolve lignin from the cellulose and hemicellulose fibers via chemical treatment, such as kraft pulping in which wood chips are cooled in a solution containing sodium hydroxide and sodium sulfide [190]. These traditional pulping technologies have several drawbacks: (1) high energy demand; (2) low cellulose yield, especially from chemical pulping due to partial degradation of cellulose; (3) potential hazards chemicals emitted to the environment [189].

Lignin is a complex polymer which serves as a structural component of higher plants and is highly resistant towards chemical degradation [191]. White-rot and brown-rot fungi are two classifications of wood-rotting basidiomycetes. White-rot basidimycetes have been reported enable to, selectively or simultaneously with cellulose, degrade lignin in different types of wood [191, 192]. Brown-rot basidiomycetes, which grow mainly on softwood, can degrade wood polysaccharides but cause only a partial modification of lignin. Besides white- and brown- rot basidimycetes, some scomycetes so-called soft-rot fungi which can degrade wood under extreme environmental conditions such as high or low water potential that prohibit the activity of other fungi [191].

The fungal treatment process fits in a paper mill operation well. After wood is debarked, chipped and screened, wood chips are briefly steamed to reduce natural chip microorganisms, cooled with air, and inoculated with the biopulping fungus for 1 to 4 weeks prior to further processing. The biopulping has been indicated as a technology technologically feasible and economically beneficial [193].

This biological treatment of wood using fungi has also been studied and used as a pre-treatment approach prior to enzymatic hydrolysis for biofuel production [194-196]. However, more research are required to understand the mechanism of wood degradation, structural changes of wood cell wall caused by these wood decay fungus and to improve the treatment technologies [197, 198].

6. Conclusions

The concept of 'biorefinery' has emerged since the potential of lignocellulosic based products substituting fossil fuel derived products has been discovered. Biorefienries may play a major role in tackling climate change by reducing the demand on fossil fuel energy and providing sustainable energy, chemicals and materials, potentially aiding energy security, and creating opportunities and market. This paper reviewed a wide range of such lignocellulosic derived products and current available biorefinery technologies. Some of these technologies have been or being close to the industrialization and others are still at the early stage of development. However, more research efforts are required to improve the technologies and integrate the biorefinery system in order to achieve the maximum outputs and to make biorefinery work at scale.

Author details

Hongbin Cheng*
Department of Process Engineering, Stellenbosch University, South Africa
New China Times Technology Ltd, China

Lei Wang*
Department of Life Science, Imperial College London, UK
New China Times Technology Ltd, China

7. References

[1] Clark, J.H., *Green chemistry for the second generation biorefinery—sustainable chemical manufacturing based on biomass.* Journal of Chemical Technology & Biotechnology, 2007. 82(7): p. 603-609.

* The two authors contribute equally to this study

[2] Stöcker, M., *Biofuels and Biomass-To-Liquid Fuels in the Biorefinery: Catalytic Conversion of Lignocellulosic Biomass using Porous Materials.* Angewandte Chemie International Edition, 2008. 47(48): p. 9200-9211.
[3] Gullón, P., et al., *Selected Process Alternatives for Biomass Refining: A Review.* The Open Agricultrure Journal 2010. 4: p. 135-144.
[4] Kelley, S.S., *Lignocellulosic Biorefineries: Reality, Hype, or Something in Between?*, in *Materials, Chemicals, and Energy from Forest Biomass*. 2007, American Chemical Society. p. 31-47.
[5] Danner, H. and R. Braun, *ChemInform Abstract: Biotechnology for the Production of Commodity Chemicals from Biomass.* ChemInform, 2000. 31(6): p. no-no.
[6] Chheda, J.N., G.W. Huber, and J.A. Dumesic, *Liquid-Phase Catalytic Processing of Biomass-Derived Oxygenated Hydrocarbons to Fuels and Chemicals.* Angewandte Chemie International Edition, 2007. 46(38): p. 7164-7183.
[7] Rowlands, W.N., A. Masters, and T. Maschmeyer, *The Biorefinery—Challenges, Opportunities, and an Australian Perspective.* Bulletin of Science, Technology & Society, 2008. 28(2): p. 149-158.
[8] Sanders, J.P.M., B. Annevelink, and D.A.v.d. Hoeven, *The development of biocommodities and the role of North West European ports in biomass chains.* Biofuels Bioproducts and Biorefining, 2009. 3(3): p. 395-409.
[9] Kamm, B., *Production of Platform Chemicals and Synthesis Gas from Biomass.* Angewandte Chemie International Edition, 2007. 46(27): p. 5056-5058.
[10] Kamm, B., P. Schönicke, and M. Kamm, *Biorefining of Green Biomass – Technical and Energetic Considerations.* CLEAN – Soil, Air, Water, 2009. 37(1): p. 27-30.
[11] Li, L., S. Lu, and V. Chiang, *A Genomic and Molecular View of Wood Formation.* Critical Reviews in Plant Sciences, 2006. 25(3): p. 215-233.
[12] Kamm, B., et al., *Biorefinery Systems – An Overview*, in *Biorefineries-Industrial Processes and Products*. 2008, Wiley-VCH Verlag GmbH. p. 1-40.
[13] Fernando, S., et al., *Biorefineries: Current Status, Challenges, and Future Direction.* Energy & Fuels, 2006. 20(4): p. 1727-1737.
[14] Kamm, B. and M. Kamm, *Biorefinery - Systems.* Chemical and Biochemical Engineering Quarterly, 2004. 18(1): p. 1-6.
[15] Mosier, N., et al., *Features of promising technologies for pretreatment of lignocellulosic biomass.* Bioresource Technology, 2005. 96(6): p. 673-686.
[16] Zheng, Y., Z. Pan, and R. Zhang, *Overview of biomass pretreatment for cellulosic ethanol production.* International Journal of Agricultural and Biological Engineering, 2009. 2(3): p. 51-68.
[17] McKendry, P., *Energy production from biomass (part 1): overview of biomass.* Bioresource Technology, 2002. 83(1): p. 37-46.
[18] Kristian, M. and M. Hurme, *Lignocellulosic biorefinery economic evaluation.* Cellulose Chemistry and Technology, 2011. 45(7-8): p. 443-454.
[19] Lynn, W., *Biomass Energy Data Book: Edition 1.* 2006.
[20] Wyman, C. and B. Goodman, *Biotechnology for production of fuels, chemicals, and materials from biomass.* Applied Biochemistry and Biotechnology, 1993. 39-40(1): p. 41-59.

[21] García, A., et al., *Biorefining of lignocellulosic residues using ethanol organosolv process.* Chemical Engineering Transactions, 2009. 18: p. 911-916.

[22] Puppan, D., *Environmental evaluation of biofuels.* Periodica Polytechnica Ser. Soc. Man. Sci., 2002. 10(1): p. 95-116.

[23] Lin, Y. and S. Tanaka, *Ethanol fermentation from biomass resources: current state and prospects.* Applied Microbiology and Biotechnology, 2006. 69(6): p. 627-642.

[24] Slade, R.B., *Prospects for cellulosic ethanol supply-chains in Europe: a techno-economic and environmental assessment*, in *Centre for Process Systems Engineering and Centre for Environmental Policy*. 2009, Univiersity of London. p. 170.

[25] Josefsson, T., H. Lennholm, and G. Gellerstedt, *Steam Explosion of Aspen Wood. Characterisation of Reaction Products.* Holzforschung, 2002. 56(3): p. 289-297.

[26] Ballesteros, I., et al., *Effect of chip size on steam explosion pretreatment of softwood.* Applied Biochemistry and Biotechnology, 2000. 84-86(1): p. 97-110.

[27] Ruiz, E., et al., *Evaluation of steam explosion pre-treatment for enzymatic hydrolysis of sunflower stalks.* Enzyme and Microbial Technology, 2008. 42(2): p. 160-166.

[28] Holtzapple, M., et al., *Pretreatment of lignocellulosic municipal solid waste by ammonia fiber explosion (AFEX).* Applied Biochemistry and Biotechnology, 1992. 34-35(1): p. 5-21.

[29] Alizadeh, H., et al., *Pretreatment of switchgrass by ammonia fiber explosion (AFEX).* Applied Biochemistry and Biotechnology, 2005. 124(1): p. 1133-1141.

[30] Tengborg, C., et al., *Comparison of SO_2 and H_2SO_4 impregnation of softwood prior to steam pretreatment on ethanol production.* Applied Biochemistry and Biotechnology, 1998. 70-72(1): p. 3-15.

[31] Öhgren, K., M. Galbe, and G. Zacchi, *Optimization of Steam Pretreatment of SO_2;-Impregnated Corn Stover for Fuel Ethanol Production*, in *Twenty-Sixth Symposium on Biotechnology for Fuels and Chemicals*, B.H. Davison, et al., Editors. 2005, Humana Press. p. 1055-1067.

[32] Kim, K.H. and J. Hong, *Supercritical CO2 pretreatment of lignocellulose enhances enzymatic cellulose hydrolysis.* Bioresource Technology, 2001. 77(2): p. 139-144.

[33] Zheng, Y., H.M. Lin, and G.T. Tsao, *Pretreatment for Cellulose Hydrolysis by Carbon Dioxide Explosion.* Biotechnology Progress, 1998. 14(6): p. 890-896.

[34] Mosier, N., et al., *Optimization of pH controlled liquid hot water pretreatment of corn stover.* Bioresource Technology, 2005. 96(18): p. 1986-1993.

[35] Laser, M., et al., *A comparison of liquid hot water and steam pretreatments of sugar cane bagasse for bioconversion to ethanol.* Bioresource Technology, 2002. 81(1): p. 33-44.

[36] Allen, S.G., et al., *A Comparison between Hot Liquid Water and Steam Fractionation of Corn Fiber.* Industrial & Engineering Chemistry Research, 2001. 40(13): p. 2934-2941.

[37] Aden, A., Ruth, M., Ibsen, K., Jechura, J., Neeves, K., Sheehan, J., Wallace, B., Montague, L., Slayton, A., Lukas, J, *Lignocellulosic Biomass to Ethanol Process Design and Economics Utilizing Co-Current Dilute Acid Prehydrolysis and Enzymatic Hydrolysis for Corn Stover.* 2002, National Renewable Energy Laboratory (NREL). p. 95. NREL/TP-510-32438.

[38] Torget, R., M.E. Himmel, and K. Grohmann, *Dilute sulfuric acid pretreatment of hardwood bark.* Bioresource Technology, 1991. 35(3): p. 239-246.

[39] Schell, D.J., et al., *Dilute-sulfuric acid pretreatment of corn stover in pilot-scale reactor - Investigation of yields, kinetics, and enzymatic digestibilities of solids.* Applied Biochemistry and Biotechnology, 2003. 105: p. 69-85.

[40] Kim, T.H. and Y.Y. Lee, *Fractionation of corn stover by hot-water and aqueous ammonia treatment.* Bioresource Technology, 2006. 97(2): p. 224-232.

[41] Vaccarino, C., et al., *Effect of SO2, NaOH and Na2CO3 pretreatments on the degradability and cellulase digestibility of grape marc.* Biological Wastes, 1987. 20(2): p. 79-88.

[42] Chander Kuhad, R., et al., *Fed batch enzymatic saccharification of newspaper cellulosics improves the sugar content in the hydrolysates and eventually the ethanol fermentation by Saccharomyces cerevisiae.* Biomass and Bioenergy, 2010. 34(8): p. 1189-1194.

[43] Sierra, R., L.A. Garcia, and M.T. Holtzapple, *Selectivity and delignification kinetics for oxidative short-term lime pretreatment of poplar wood. Part I: Constant-pressure.* Biotechnology Progress, 2011. 27(4): p. 976-985.

[44] Chang, V., et al., *Oxidative lime pretreatment of high-lignin biomass.* Applied Biochemistry and Biotechnology, 2001. 94(1): p. 1-28.

[45] Pasquini, D., et al., *Extraction of lignin from sugar cane bagasse and Pinus taeda wood chips using ethanol–water mixtures and carbon dioxide at high pressures.* The Journal of Supercritical Fluids, 2005. 36(1): p. 31-39.

[46] Hallac, B.B., et al., *Effect of Ethanol Organosolv Pretreatment on Enzymatic Hydrolysis of Buddleja davidii Stem Biomass.* Industrial & Engineering Chemistry Research, 2010. 49(4): p. 1467-1472.

[47] Brandt, A., et al., *Ionic liquid pretreatment of lignocellulosic biomass with ionic liquid-water mixtures.* Green Chemistry, 2011. 13(9): p. 2489-2499.

[48] Hamelinck, C.N., G.v. Hooijdonk, and A.P.C. Faaij, *Ethanol from lignocellulosic biomass: techno-economic performance in short-, middle- and long-term.* Biomass and Bioenergy, 2005. 28(4): p. 384-410.

[49] Sedlak, M. and N. Ho, *Production of ethanol from cellulosic biomass hydrolysates using genetically engineered saccharomyces yeast capable of cofermenting glucose and xylose.* Applied biochemistry and biotechnology, 2004. 114(1): p. 403-416.

[50] Humbird, D. and A. Aden, *Biochemical Production of Ethanol from Corn Stover: 2008 State of Technology Model.* 2009, National Renewable Energy Laboratory (NREL). NREL/TP-510-46214.

[51] Cripps, R.E., et al., *Metabolic engineering of Geobacillus thermoglucosidasius for high yield ethanol production.* Metabolic Engineering, 2009. 11(6): p. 398-408.

[52] Bezerra, R. and A. Dias, *Enzymatic kinetic of cellulose hydrolysis: Inhibition by ethanol and cellobiose.* Applied Biochemistry and Biotechnology, 2005. 126(1): p. 49-59.

[53] Lynd, L.R., et al., *Consolidated bioprocessing of cellulosic biomass: an update.* Current Opinion in Biotechnology, 2005. 16: p. 577-583.

[54] van Zyl, W., et al., *Consolidated Bioprocessing for Bioethanol Production Using Biofuels*, L. Olsson, Editor. 2007, Springer Berlin / Heidelberg. p. 205-235.

[55] Jin, M., et al., *Consolidated bioprocessing (CBP) performance of Clostridium phytofermentans on AFEX-treated corn stover for ethanol production.* Biotechnology and Bioengineering, 2011. 108(6): p. 1290-1297.

[56] Olson, D.G., et al., *Recent progress in consolidated bioprocessing.* Current Opinion in Biotechnology, (0).

[57] Phillips, S., et al., *Thermochemical Ethanol via Indirect Gasification and Mixed Alcohol Synthesis of Lignocellulosic Biomass.* 2007, National Renewable Energy Laboratory (NREL). NREL/TP-510-41168.

[58] Dutta, A., et al., *An economic comparison of different fermentation configurations to convert corn stover to ethanol using Z. mobilis and Saccharomyces.* Biotechnology Progress, 2010. 26(1): p. 64-72.

[59] Stephenson, A.L., et al., *The environmental and economic sustainability of potential bioethanol from willow in the UK.* Bioresource Technology, 2010. 101(24): p. 9612-9623.

[60] Sassner, P., M. Galbe, and G. Zacchi, *Techno-economic evaluation of bioethanol production from three different lignocellulosic materials.* Biomass and Bioenergy, 2008. 32(5): p. 422-430.

[61] Wang, L., et al., *Technology performance and economic feasibility of bioethanol production from various waste papers.* Energy & Environmental Science 2012. 5(2): p. 5717-5730.

[62] Mu, D., et al., *Comparative Life Cycle Assessment of Lignocellulosic Ethanol Production: Biochemical Versus Thermochemical Conversion.* Environmental Management, 2010. 46(4): p. 565-578.

[63] Qureshi, N. and T.C. Ezeji, *Butanol, 'a superior biofuel' production from agricultural residues (renewable biomass): recent progress in technology.* Biofuels, Bioproducts and Biorefining, 2008. 2(4): p. 319-330.

[64] Ezeji, T., N. Qureshi, and H.P. Blaschek, *Butanol production from agricultural residues: Impact of degradation products on Clostridium beijerinckii growth and butanol fermentation.* Biotechnology and Bioengineering, 2007. 97(6): p. 1460-1469.

[65] Cascone, R., *Biobutanol – a replacement for bioethanol?* Chemical Engineering Progressing, 2008. 104: p. S4-S9.

[66] Pfromm, P.H., et al., *Bio-butanol vs. bio-ethanol: A technical and economic assessment for corn and switchgrass fermented by yeast or Clostridium acetobutylicum.* Biomass and Bioenergy, 2010. 34(4): p. 515-524.

[67] Shereena, K.M. and T. Thangaraj, *Biodiesel: an alternative fuel produced from vegetable oils by transesterification.* Electronic Journal of Biology, 2009. 5(3): p. 67-74.

[68] Zabeti, M., W.M.A. Wan Daud, and M.K. Aroua, *Activity of solid catalysts for biodiesel production: A review.* Fuel Processing Technology, 2009. 90(6): p. 770-777.

[69] Ma, F. and M.A. Hanna, *Biodiesel production: a review.* Bioresource Technology, 1999. 70(1): p. 1-15.

[70] Marchetti, J.M., V.U. Miguel, and A.F. Errazu, *Possible methods for biodiesel production.* Renewable and Sustainable Energy Reviews, 2007. 11(6): p. 1300-1311.

[71] Schnepf, R., *Agriculture-Based Biofuels: Overview and Emerging Issues.* 2010, Congressional Research Service. R41282.

[72] Tamalampudi, S., et al., *Enzymatic production of biodiesel from Jatropha oil: A comparative study of immobilized-whole cell and commercial lipases as a biocatalyst.* Biochemical Engineering Journal, 2008. 39(1): p. 185-189.

[73] Lu, H., et al., *Production of biodiesel from Jatropha curcas L. oil.* Computers & Chemical Engineering, 2009. 33(5): p. 1091-1096.

[74] Shah, S. and M.N. Gupta, *Lipase catalyzed preparation of biodiesel from Jatropha oil in a solvent free system.* Process Biochemistry, 2007. 42(3): p. 409-414.

[75] Kumari, A., et al., *Enzymatic transesterification of Jatropha oil.* Biotechnology for Biofuels, 2009. 2(1): p. 1.

[76] Chisti, Y., *Biodiesel from microalgae.* Biotechnology Advances, 2007. 25: p. 294-306.

[77] Scott, S.A., et al., *Biodiesel from algae: challenges and prospects.* Current Opinion in Biotechnology, 2010. 21(3): p. 277-286.

[78] Grobbelaar, J., *Turbulence in mass algal cultures and the role of light/dark fluctuations.* Journal of Applied Phycology, 1994. 6(3): p. 331-335.

[79] Chisti, Y., *Biodiesel from microalgae beats bioethanol.* Trends in Biotechnology, 2008. 26(3): p. 126-131.

[80] Monson, K.D., et al., *Anaerobic digestion of biodegrabale municiple solid wastes: A review.* 2007, University of Glamorgan

[81] Trzcinski, A.P., *Anaerobic membrane bioreactor technology for solid waste stabilization,* in *Chemical Engineering and Chemical Technology*. 2009, Imperial College London: London.

[82] Rapport, J., et al., *Current anaerobic digestion technologies used for treatment of municipal organic solid waste*. 2008, Department of Biological and Agricultural Engineering, University of California Davis, CA

[83] Raposo, F., et al., *Influence of inoculum to substrate ratio on the biochemical methane potential of maize in batch tests.* Process Biochemistry, 2006. 41(6): p. 1444-1450.

[84] Hartmann, H., H.B. Moller, and B.K. Ahring, *Efficiency of the anaerobic treatment of the organic fraction of municipal solid waste: collection and pretreatment.* Waste Management & Research, 2004. 22: p. 35-41.

[85] WRAP, *Anaerobic digestate.* 2008, Waste & Resources Action Programme (WRAP)

[86] Edelmann, W., K. Schleiss, and A. Joss, *Ecological, energetic and economic comparison of anaerobic digestion with different competing technologies to treat biogenic wastes.* Water Science and Technology, 2000. 4(3): p. 263-273.

[87] Edelmann, W., U. Baier, and H. Engeli, *Environmental aspects of the anaerobic digestion of the organic fraction of municipal solid wastes and of solid agricultural wastes.* Water Science and Technology, 2004. 52(1-2): p. 203-208.

[88] Dodds, D.R. and R.A. Gross, *Chemicals from Biomass.* Science, 2007. 318(5854): p. 1250-1251.

[89] Carlson, T.L. and E.M. Peters. 2006. Patent application US2006/094093 A1.

[90] Lorenz, P. and H. Zinke, *White biotechnology: differences in US and EU approaches?* Trends in Biotechnology, 2005. 23(12): p. 570-574.

[91] Jung, Y.K., et al., *Metabolic engineering of Escherichia coli for the production of polylactic acid and its copolymers.* Biotechnology and Bioengineering, 2010. 105(1): p. 161-171.

[92] Wyman, C.E., et al., *Coordinated development of leading biomass pretreatment technologies.* Bioresource Technology, 2005. 96(18): p. 1959-1966.

[93] Yáñez, R., et al., *Production of D(−)-lactic acid from cellulose by simultaneous saccharification and fermentation using <i>Lactobacillus coryniformis</i> subsp. <i>torquens</i>.* Biotechnology Letters, 2003. 25(14): p. 1161-1164.

[94] Adsul, M.G., et al., *Development of biocatalysts for production of commodity chemicals from lignocellulosic biomass.* Bioresource Technology, 2011. 102(6): p. 4304-4312.

[95] Danner, H., et al., *Bacillus stearothermophilus for thermophilic production of l-lactic acid.* Applied Biochemistry and Biotechnology, 1998. 70-72(1): p. 895-903.

[96] Maddox, I.S., N. Qureshi, and K. Roberts-Thomson, *Production of acetone-butanol-ethanol from concentrated substrate using clostridium acetobutylicum in an integrated fermentation-product removal process.* Process Biochemistry, 1995. 30(3): p. 209-215.

[97] Jones, D.T. and D.R. Woods, *Acetone-butanol fermentation revisited.* Microbiology and Moclecular Biology Reviews, 1986. 50(4): p. 484-524.

[98] Faik, A., *Xylan biosysthesis: News from the grass.* Plant Physiology, 2010. 153: p. 396-402.

[99] Granström, T., K. Izumori, and M. Leisola, *A rare sugar xylitol. Part II: biotechnological production and future applications of xylitol.* Applied Microbiology and Biotechnology, 2007. 74(2): p. 273-276.

[100] Saha, B., *Hemicellulose bioconversion.* Journal of Industrial Microbiology & Biotechnology, 2003. 30(5): p. 279-291.

[101] Choct, M. and G. Annison, *Anti-nutritive activity of wheat pentosans in broiler diets.* British Poultry Science, 1990. 31(4): p. 811-821.

[102] Duff, S.J.B. and W.D. Murray, *Bioconversion of forest products industry waste cellulosics to fuel ethanol: a review.* Bioresource Technology, 1996. 55: p. 1-33.

[103] Hayes, D.J., et al., *The Biofine Process – Production of Levulinic Acid, Furfural, and Formic Acid from Lignocellulosic Feedstocks*, in *Biorefineries-Industrial Processes and Products*. 2008, Wiley-VCH Verlag GmbH. p. 139-164.

[104] Fatih Demirbas, M., *Biorefineries for biofuel upgrading: A critical review.* Applied Energy, 2009. 86, Supplement 1(0): p. S151-S161.

[105] Saddler, J.N., *Bioconversion of Forest and Agricultural Plant Residues*. Biotechnology in agriculture. 1993: CABI, Wallingford, UK.

[106] Sun, Y. and J.Y. Cheng, *Hydrolysis of lignocellulosic materials for ethanol production: a review.* Bioresource Technology, 2002. 83(1): p. 1-11.

[107] Zhu, S., et al., *FED-BATCH SIMULTANEOUS SACCHARIFICATION AND FERMENTATION OF MICROWAVE/ACID/ALKALI/H2O2 PRETREATED RICE STRAW FOR PRODUCTION OF ETHANOL.* Chemical Engineering Communications, 2006. 193(5): p. 639-648.

[108] Koukios, E.G. and G.N. Valkanas, *Process for chemical separation of the three main components of lignocellulosic biomass.* Industrial & Engineering Chemistry Product Research and Development, 1982. 21(2): p. 309-314.

[109] Koukios, E.G., *Biomass refining: a non-waste approach.*, in *Economics and Ecosystem Management*, D.O. Hall, N. Myers, and N.S. Margaris, Editors. 1985, Dr W. Junk Publishers: Dordrecht, The Netherlands. p. 233-244.

[110] Kamm, B., P.R. Gruber, and M. Kamm, *Biorefineries – Industrial Processes and Products*, in *Ullmann's Encyclopedia of Industrial Chemistry*. 2000, Wiley-VCH Verlag GmbH & Co. KGaA.

[111] Kleinert, T.N., *Organosolv pulping with aqueous alcohol.* Tappi Journal, 1974. 57: p. 99-102.

[112] Lora, J.H. and S. Aziz, *Organosolv pulping — a versatile approach to wood refining.* Tappi Journal, 1985. 68(6): p. 94-97.

[113] Sarkanen, K.V., *Chemistry of solvent pulping.* Tappi Journal, 1990: p. 215-219.

[114] Pan, X., et al., *Bioconversion of hybrid poplar to ethanol and co-products using an organosolv fractionation process: Optimization of process yields.* Biotechnology and Bioengineering, 2006. 94(5): p. 851-861.

[115] Pan, X., et al., *Effect of organosolv ethanol pretreatment variables on physical characteristics of hybrid poplar substrates.* Applied Biochemistry and Biotechnology, 2007. 137-140(1): p. 367-377.

[116] Pan, X., et al., *The bioconversion of mountain pine beetle-killed lodgepole pine to fuel ethanol using the organosolv process.* Biotechnology and Bioengineering, 2008. 101(1): p. 39-48.

[117] Taherzadeh, M.J. and K. Karimi, *Pretreatment of lignocellulosic wastes to improve ethanol and biogas production: A review.* International Journal of Molecular Sciences, 2008. 9(9): p. 1621-1651.

[118] Chum, H.L., et al., *Organosolv pretreatment for enzymatic hydrolysis of poplars: I. Enzyme hydrolysis of cellulosic residues.* Biotechnology Bioengineering, 1988. 31(7).

[119] Zhao, X., K. Cheng, and D. Liu, *Organosolv pretreatment of lignocellulosic biomass for enzymatic hydrolysis.* Applied Microbiology and Biotechnology, 2009. 82(5): p. 815-827.

[120] Johansson, A., O. Aaltonen, and P. Ylinen, *Organosolv pulping — methods and pulp properties.* Biomass, 1987. 13(1): p. 45-65.

[121] Aziz, S. and K.V. Sarkanen, *Organosolv pulping—a review.* Tappi Journal, 1989. 72: p. 169-175.

[122] Thring, R.W., E. Chornet, and R.P. Overend, *Recovery of a solvolytic lignin: Effects of spent liquor/acid volume ratio, acid concentration and temperature.* Biomass, 1990. 23(4): p. 289-305.

[123] Sidiras, D. and E. Koukios, *Simulation of acid-catalysed organosolv fractionation of wheat straw.* Bioresource Technology, 2004. 94(1): p. 91-98.

[124] Zhu, J.Y. and X.J. Pan, *Woody biomass pretreatment for cellulosic ethanol production: Technology and energy consumption evaluation.* Bioresource Technology, 2010. 101(13): p. 4992-5002.

[125] Myllymaki, V. and R. Aksela. 2005. Dissolution and delignification of lignocellulosic materials with ionic liquid solvent under microwave irradiation. WO patent 2005/017001

[126] Hayes, D.J., *An examination of biorefining processes, catalysts and challenges.* Catalysis Today, 2009. 145(1–2): p. 138-151.

[127] Dadi, A.P., S. Varanasi, and C.A. Schall, *Enhancement of cellulose saccharification kinetics using an ionic liquid pretreatment step.* Biotechnology and Bioengineering, 2006. 95(5): p. 904-910.

[128] Zhu, S., *Use of ionic liquids for the efficient utilization of lignocellulosic materials.* Journal of Chemical Technology & Biotechnology, 2008. 83(6): p. 777-779.

[129] Reichert, W.M., et al., *Derivatization of chitin in room temperature ionic liquids,* in *222 ACS National Meeting*. 2001: Chicago, IL, United States.

[130] Heinze, T., K. Schwikal, and S. Barthel, *Ionic Liquids as Reaction Medium in Cellulose Functionalization.* Macromolecular Bioscience, 2005. 5(6): p. 520-525.

[131] Dien, B.S., et al., *Enzymatic saccharification of hot-water pretreated corn fiber for production of monosaccharides.* Enzyme and Microbial Technology, 2006. 39(5): p. 1137-1144.

[132] Negro, M., et al., *Hydrothermal pretreatment conditions to enhance ethanol production from poplar biomass.* Applied Biochemistry and Biotechnology, 2003. 105(1): p. 87-100.

[133] Rogalinski, T., T. Ingram, and G. Brunner, *Hydrolysis of lignocellulosic biomass in water under elevated temperatures and pressures.* The Journal of Supercritical Fluids, 2008. 47(1): p. 54-63.

[134] van Walsum, G., et al., *Conversion of lignocellulosics pretreated with liquid hot water to ethanol.* Applied Biochemistry and Biotechnology, 1996. 57-58(1): p. 157-170.

[135] Franck, E.U., *Fluids at high pressures and temperatures.* Chemical thermodynamics, 1987. 19: p. 225-242.

[136] Shaw, R.W., et al., *Supercritical water. A medium for chemistry.* Chemical Engineering News, 1991. 23: p. 26-39.

[137] Savage, P.E., et al., *Reactions at supercritical conditions: {A}pplications and fundamentals.* AIChE Journal, 1995. 41(7): p. 1723-1778.

[138] Katritzky, A.R., et al., *Reactions in High-Temperature Aqueous Media.* Chemical Reviews, 2001. 101(4): p. 837-892.

[139] Kohlmann, K.L., et al., *Enhanced Enzyme Activities on Hydrated Lignocellulosic Substrates,* in *Enzymatic Degradation of Insoluble Carbohydrates.* 1996, American Chemical Society. p. 237-255.

[140] Weil, J., et al., *Continuous pH monitoring during pretreatment of yellow poplar wood sawdust by pressure cooking in water.* Applied Biochemistry and Biotechnology, 1998. 70-72(1): p. 99-111.

[141] Alvira, P., et al., *Pretreatment technologies for an efficient bioethanol production process based on enzymatic hydrolysis: A review.* Bioresource Technology, 2010. 101(13): p. 4851-4861.

[142] Goring, D.A.I., *Thermal softening of lignin, hemicelluloses and cellulose.* 1963: Pulp and Paper Research Institute of Canada.

[143] Overend, R.P., E. Chornet, and J.A. Gascoigne, *Fractionation of Lignocellulosics by Steam-Aqueous Pretreatments [and Discussion].* Philosophical Transactions of the Royal Society of London. Series A, Mathematical and Physical Sciences, 1987. 321(1561): p. 523-536.

[144] Mason, W.H. 1929. Apparatus and process of explosion defibration of lignocellulosic material. U.S. Patent No. 1655618

[145] Montane, D., et al., *Application of steam explosion to the fractionation and rapid vapor-phase alkaline pulping of wheat straw.* Biomass and Bioenergy, 1998. 14(3): p. 261-276.

[146] Pan, X., et al., *Strategies to enhance the enzymatic hydrolysis of pretreated softwood with high residual lignin content.* Applied Biochemistry and Biotechnology, 2005. 121-124: p. 1069-1079.

[147] Eklund, R., M. Galbe, and G. Zacchi, *The influence of SO2 and H2SO4 impregnation of willow prior to steam pretreatment.* Bioresource Technology, 1995. 52(3): p. 225-229.

[148] Fein, J.E., D. Potts, and D. Good, *Development of an optimal wood-to-fuel ethanol process utilizing best available technology*. Energy Biomass and Waste, 1991. 15(745-765).

[149] Ropars, M., et al., *Large-scale enzymatic hydrolysis of agricultural lignocellulosic biomass. Part 1: Pretreatment procedures*. Bioresource Technology, 1992. 42(3): p. 197-204.

[150] Morjanoff, P.J. and P.P. Gray, *Optimization of steam explosion as method for increasing susceptibility of sugarcane bagasse to enzymatic saccharification*. Biotechnology Bioengineering, 1987. 29(6): p. 733-741.

[151] Focher, B., A. Marzetti, and V. Crescenzi, *Steam Explosion Techniques, Fundamentals and Industrial Applications*. 1991, Gordon and Breach Publishers: Philadelphia. p. 412.

[152] Avellar, B.K. and W.G. Glasser, *Steam-assisted biomass fractionation. I. Process considerations and economic evaluation*. Biomass and Bioenergy, 1998. 14(3): p. 205-218.

[153] Zheng, Y. and G.T. Tsao, *Avicel hydrolysis by cellulase enzyme in supercritical CO2*. Biotechnology Letters, 1996. 18(4): p. 451-454.

[154] Zheng, Y., et al., *Supercritical carbon dioxide explosion as a pretreatment for cellulose hydrolysis*. Biotechnology Letters, 1995. 17(8): p. 845-850.

[155] Schacht, C., C. Zetzl, and G. Brunner, *From plant materials to ethanol by means of supercritical fluid technology*. The Journal of Supercritical Fluids, 2008. 46(3): p. 299-321.

[156] Laureano-Perez, L., et al., *Understanding Factors that Limit Enzymatic Hydrolysis of Biomass*

[157] *nty-Sixth Symposium on Biotechnology for Fuels and Chemicals*, B.H. Davison, et al., Editors. 2005, Humana Press. p. 1081-1099.

[158] Teymouri, F., et al., *Optimization of the ammonia fiber explosion (AFEX) treatment parameters for enzymatic hydrolysis of corn stover*. Bioresource Technology, 2005. 96(18): p. 2014-2018.

[159] Eggeman, T. and R.T. Elander, *Process and economic analysis of pretreatment technologies*. Bioresource Technology, 2005. 96(18): p. 2019-2025.

[160] Ghosh, S., et al., *Pilot-scale gasification of municipal solid wastes by high-rate and two-phase anaerobic digestion (TPAD)*. Water Science and Technology 2000. 41(3): p. 101-110.

[161] Beccari, M., et al., *Enhancement of anaerobic treatability of olive oil mill effluents by addition of Ca(OH)2 and bentonite without intermediate solid/liquid separation*. Water Science and Technology, 2001. 43(11): p. 275-282.

[162] Tanaka, S., et al., *Effects of thermochemical pretreatment on the anaerobic digestion of waste activated sludge*. Water Science and Technology, 1997. 35(8): p. 209-215.

[163] Brownell, H.H., E.K.C. Yu, and J.N. Saddler, *Steam-explosion pretreatment of wood: Effect of chip size, acid, moisture content and pressure drop*. Biotechnology and Bioengineering, 1986. 28(6): p. 792-801.

[164] Gregg, D. and J. Saddler, *A techno-economic assessment of the pretreatment and fractionation steps of a biomass-to-ethanol process*. Applied Biochemistry and Biotechnology, 1996. 57-58(1): p. 711-727.

[165] Chang, V.S., et al., *Simultaneous saccharification and fermentation of lime-treated biomass*. Biotechnology Letters, 2001. 23(16): p. 1327-1333.

[166] Sun, X.F., et al., *Characteristics of degraded cellulose obtained from steam-exploded wheat straw*. Carbohydrate Research, 2005. 340(1): p. 97-106.

[167] Jones, J.L. and K.T. Semrau, *Wood hydrolysis for ethanol production — previous experience and the economics of selected processes.* Biomass, 1984. 5(2): p. 109-135.

[168] Kootstra, A.M.J., et al., *Comparison of dilute mineral and organic acid pretreatment for enzymatic hydrolysis of wheat straw.* Biochemical Engineering Journal, 2009. 46(2): p. 126-131.

[169] Bjerre, A.B., et al., *Pretreatment of wheat straw using combined wet oxidation and alkaline hydrolysis resulting in convertible cellulose and hemicellulose.* Biotechnology and Bioengineering, 1996. 49(5): p. 568-577.

[170] Kumar, P., et al., *Methods for Pretreatment of Lignocellulosic Biomass for Efficient Hydrolysis and Biofuel Production.* Industrial & Engineering Chemistry Research, 2009. 48(8): p. 3713-3729.

[171] Carvalheiro, F., L.C. Duarte, and F.M. Gírio, *Hemicellulose biorefineries: a review on biomass pretreatments.* Scientific & Industrial Research, 2008. 67: p. 849-864.

[172] Molero Gómez, A., C. Pereyra López, and E. Martinez de la Ossa, *Recovery of grape seed oil by liquid and supercritical carbon dioxide extraction: a comparison with conventional solvent extraction.* The Chemical Engineering Journal and the Biochemical Engineering Journal, 1996. 61(3): p. 227-231.

[173] Toews, K.L., et al., *pH-Defining Equilibrium between Water and Supercritical CO2. Influence on SFE of Organics and Metal Chelates.* Analytical Chemistry, 1995. 67(22): p. 4040-4043.

[174] Intanakul, P., M. Krairiksh, and P. Kitchaiya, *Enhancement of Enzymatic Hydrolysis of Lignocellulosic Wastes by Microwave Pretreatment Under Atmospheric Pressure.* Journal of Wood Chemistry and Technology, 2003. 23(2): p. 217-225.

[175] Saha, B.C., A. Biswas, and M.A. Cotta, *Microwave Pretreatment, Enzymatic Saccharification and Fermentation of Wheat Straw to Ethanol.* Journal of Biobased Materials and Bioenergy, 2008. 2(3): p. 210-217.

[176] Ma, H., et al., *Enhanced enzymatic saccharification of rice straw by microwave pretreatment.* Bioresource Technology, 2009. 100(3): p. 1279-1284.

[177] Imai, M., K. Ikari, and I. Suzuki, *High-performance hydrolysis of cellulose using mixed cellulase species and ultrasonication pretreatment.* Biochemical Engineering Journal, 2004. 17(2): p. 79-83.

[178] Nitayavardhana, S., et al., *Ultrasound pretreatment of cassava chip slurry to enhance sugar release for subsequent ethanol production.* Biotechnology and Bioengineering, 2008. 101(3): p. 487-496.

[179] Keshwani, D.R., *Microwave Pretreatment of Switchgrass for Bioethanol Production.* 2009, North Carolina State University.

[180] Sun, R.C. and J. Tomkinson, *Characterization of hemicelluloses obtained by classical and ultrasonically assisted extractions from wheat straw.* Carbohydrate Polymers, 2002. 50(3): p. 263-271.

[181] Yachmenev, V., et al., *Acceleration of the enzymatic hydrolysis of corn stover and sugar cane bagasse celluloses by low intensity uniform ultrasound.* Biobased Material Bioenergy, 2009. 3: p. 25-31.

[182] Vinatoru, M.e.a., *Ultrasonically assisted extraction of bioactive principles from plants and their constituents*, in *Advances in Sonochemistry*, T.J. Mason, Editor. 1999, JAI Press. p. 209-248.
[183] Palonen, H., et al., *Evaluation of wet oxidation pretreatment for enzymatic hydrolysis of softwood.* Applied Biochemistry and Biotechnology, 2004. 117(1): p. 1-17.
[184] Schmidt, A.S. and A.B. Thomsen, *Optimization of wet oxidation pretreatment of wheat straw.* Bioresource Technology, 1998. 64(2): p. 139-151.
[185] Chum, H.L., et al., *Evaluation of pretreatments of biomass for enzymatic hydrolysis of cellulose.* 1985, Solar Energy Research Institute: Golden, Colorado. p. 64
[186] Martín, C., et al., *Wet oxidation pretreatment, enzymatic hydrolysis and simultaneous saccharification and fermentation of clover–ryegrass mixtures.* Bioresource Technology, 2008. 99(18): p. 8777-8782.
[187] Galbe, M. and G. Zacchi, *A review of the production of ethanol from softwood.* Appl Microbiol Biotechnol, 2002. 59: p. 618 - 628.
[188] Vidal, P., *Ozonolysis of Lignin – Improvement of in vitro digestibility of poplar sawdust.* Biomass, 1988. 16(1): p. 1-17.
[189] Energy Information Administration, *Chapter 10 Growth of the Landfill Gas Industry*, in *Renewable Energy Annual 1996*. 1997.DOE/EIA-0603(96)
[190] Breen, A. and F.L. Singleton, *Fungi in lignocellulose breakdown and biopulping.* Current Opinion in Biotechnology, 1999. 10(3): p. 252-258.
[191] Hischier, R., *Life Cycle Inventories of Packaging and Graphical Papers*, in *Ecoinvent Report No.11*. 2007, Swiss Centre for Life Cycyle Inventories: Dübendorf
[192] Martínez, Á.T., et al., *Biodegradation of lignocellulosics: microbial, chemical, and enzymatic aspects of the fungal attack of lignin.* International micorbiology, 2005. 8: p. 195-204.
[193] Maijala, P., et al., *Biomechanical pulping of softwood with enzymes and white-rot fungus Physisporinus rivulosus.* Enzyme and Microbial Technology, 2008. 43(2): p. 169-177.
[194] Shukla, O.P., U.N. Rai, and S.V. Subramanyam. *Biopulping and Biobleaching: An Energy and envioronment Saving Technology for Indian Pulp and Paper Industry.* 2004 [cited 2012 April]; Available from: http://isebindia.com/01_04/04-04-3.html.
[195] Ray, M.J., et al., *Brown rot fungal early stage decay mechanism as a biological pretreatment for softwood biomass in biofuel production.* Biomass and Bioenergy, 2010. 34(8): p. 1257-1262.
[196] Wan, C. and Y. Li, *Fungal pretreatment of lignocellulosic biomass.* Biotechnology Advances, (0).
[197] Yu, H., et al., *The effect of biological pretreatment with the selective white-rot fungus Echinodontium taxodii on enzymatic hydrolysis of softwoods and hardwoods.* Bioresource Technology, 2009. 100(21): p. 5170-5175.
[198] Monrroy, M., et al., *Structural change in wood by brown rot fungi and effect on enzymatic hydrolysis.* Enzyme and Microbial Technology, 2011. 49(5): p. 472-477.

Chapter 15

Biomass Extraction Methods

Adina-Elena Segneanu, Florentina Cziple, Paulina Vlazan,
Paula Sfirloaga, Ioan Grozescu and Vasile Daniel Gherman

Additional information is available at the end of the chapter

http://dx.doi.org/10.5772/55338

1. Introduction

Biomass represents an extremely valuable potential to obtain new clean energy sources and natural structurally complex bioactive compounds. Renewable energy can be produced from any biological feedstock, that contains appreciable amounts of sugar or materials that can be converted into sugar (e.g. starch or cellulose). Lignocellulose's biomass–dendromass and phytomass is natural based material consisting of complex of heterogenic macromolecules with cell structure (celluloses, hemicelluloses and lignin) as well as numerous organic and inorganic structures with low molecule weight (Sun, 2002).

Long-term economic and environmental concerns have resulted in a great amount of research in the past couple of decades on renewable sources of liquid fuels to replace fossil fuels. Producing of cellulose and alcohol from biomass is important technological process. Conversion of abundant lignocellulosic biomass to biofuels as transportation fuels presents a viable option for improving energy security and reducing greenhouse emissions. Lignocellulosic materials such as agricultural residues (e.g., wheat straw, sugarcane bagasse, corn stover), forest products (hardwood and softwood), and dedicated crops (switchgrass, salix) are renewable sources of energy. These raw materials are sufficiently abundant and generate very low net greenhouse emissions. The use of biomass with low economic value, the waste from agriculture, forestry and wild flora as sources of clean energy, is a viable way to avoid potential conflicts with the biomass production for food, which represent the main concern of UE regarding the biofuels production from biomass.

The presence of lignin in lignocelluloses leads to a protective barrier that prevents plant cell destruction by fungi and bacteria for conversion to fuel. For the conversion of biomass to fuel, the cellulose and hemicellulose must be broken. The digestibility of cellulose present in lignocellulosic biomass is hindered by many physicochemical, structural, and compositional factors. The lignocellulosic biomasses need to be treated prior to fuel production to expose cellulose. In present, there is many different type of pretreatment of lignocelluloses materials. Pretreatment uses various techniques, including ammonia fiber explosion, chemical treatment, biological treatment, and steam explosion, to alter the structure of cellulosic biomass to make cellulose more accessible. The purpose of the pretreatment is to remove lignin and hemicellulose, reduce cellulose crystallinity, and increase the porosity of the materials. Then, acids or enzymes can be used to break down the cellulose into its constituent sugars. Enzyme hydrolysis is widely used to break down cellulose into its constituent sugars. Pretreatment can be the most expensive process in biomass-to-fuels conversion but it has great potential for improvements in efficiency and lowering of costs through further research and development. Cellulose chains can also be broken down into individual glucose sugar molecules by enzymes known as cellulose. Cellulose refers to a class of enzymes produced by fungi, bacteria, and protozoans that catalyze the hydrolysis of cellulose. But, one of the main drawn back of convention chemical methods used in ethanol formation process is degradation of carbohydrates and formation of undesirable by-products, which severely inhibition of ethanol during the fermentation process: furfural, 5-hydroxymethylfurfural, uronic acid, levulinic acid, acetic acid, formic acid, hydroxybenzoic acid, vanillin, phenol, cinnamaldehyde, formaldehyde, and so (Nenkova et.al., 2011). Some inhibitors such as terpene compounds are present in the biomass–dendromass.

Lignin is a complex reticulated phenolic polymer that occurs in xylem of most terrestrial plants and is the second most abundant biopolymer in nature, corresponding to around 30% of the biosphere organic carbon. This macromolecule is one of the biggest wood components and also one of the most important. Even the lignin has a significant role in technology, in the bioethanol production process valuable chemical properties and functions from lignin and hemicelluloses are not fully recovery, the black liquor result from process being using specially for energy recovery. About half of wood components are dissolved into this black liquor. The dissolved organic compounds consist mainly in degraded lignin and also hemicelluloses and cellulose degradation products. Also, phenols derived from biomass are valuable and useful chemicals, due to their pharmacological properties including antiviral inhibitor (anti-HIV). These compounds with good antioxidant activity can be used to preserve food from lipid peroxidation and oxidative damage occurring in living systems (Martínez et.al., 1996; Mahugo Santana et.al., 2009; Nenkova, et.al.2011). Antioxidants can also prevent the loss of food color, flavor and active vitamins content, providing the stabilization of the molecules involved in such characteristics. They can also be used for the production of adhesives and for the synthesis of polymer.

It is well known that, biomass also contains many other natural products: waxes and fatty acids, polyacetylenes, terpenoids (e.g., monoterpenoids, iridoids, sesquiterpenoids, diterpenoids, triterpenoids), steroids, essential oils, phenolics, flavonoids, tannins, anthocyanins, quinones, coumarins, lignans, alkaloids, and glycosidic derivatives (e.g., saponins, glycosides, flavonoid glycosides) (Alonso et.al., 1998; Japón-Luján et.al., 2006; Faustino, 2010; Fang et.al., 2009; Gallo, 2010; Carro, 1997; Kojima, 2004). In this regards, are needed more studies to recover these important compounds from biomass for use in pharmaceutical industry, food industry, and so.

2. Extraction techniques

Actually, there are known many different techniques used for biomass extraction: liquid-solid extraction, liquid-liquid extraction, partitioning, acid-base extractions, ultrasound extraction (UE), microwave assisted extraction (MAE). The capability of a number of extraction techniques have been investigated, such as solvent extraction (J.A. Saunders, D.E. Blume, 1981) and enzyme-assisted extraction (B.B. Li, B. Smith, M.M. Hossain, 2006). However, these extraction methods have drawbacks to some degree.

The choice of extraction procedure depends on the nature of the natural material and the components to be isolated. The main conventional extraction procedures are liquid-liquid extraction and liquid-solid extraction. For liquid-liquid extraction is using two different solvents, one of which is always water, (water-dichloromethane, water-hexane, and so). Some of the disadvantages of this method are: cost, toxicity and flammability (Kaufmann 2002; McCabe, 1956; Perry, 1988; Sarker et. al., 2006).

Solid-phase extraction (SPE) can be used to isolate analytes dissolved or suspended in a liquid mixture are separated from a wide variety of matrices according to their physical and chemical properties. Conventional methods include: soxhlet extraction, maceration, percolation, extraction under reflux and steam distillation, turbo-extraction (high speed mixing) and sonication. Although these techniques are widely used, have several shortcomings: are very often time-consuming and require relatively large quantities of polluting solvents, the influence of temperature which can lead to the degradation of thermo labile metabolites (Kaufmann 2002; McCabe, 1956; Sarker et. al., 2006; Routray, 2012).

Supercritical fluid extraction (SFE), microwave-assisted extraction (MAE) and pressurised solvent extraction (PSE) are fast and efficient unconventional extraction methods developed for extracting analytes from solid matrixes.

2.1. Supercritical fluid extraction

Supercritical fluid extraction (SFE) is one of the relatively new efficient separation method for the extraction of essential oils from different plant materials. The new products, extracts, can be used as

a good base for the production of pharmaceutical drugs and additives in the perfume, cosmetic, and food industries. Use of SFE under different conditions can allow selecting the extraction of different constituents. The main reason for the interest in SFE was the possibility of carrying out extractions at

temperature near to ambient, thus preventing the substance of interest from incurring in thermal denaturation.

Supercritical fluid extraction has proved effective in the separation of essential oils and its derivatives for use in the food, cosmetics, pharmaceutical and other related industries, producing high-quality essential oils with commercially more satisfactory compositions (lower monoterpenes) than obtained with conventional hydro-distillation (Ehlers et al., 2001; Diaz-Maroto et al., 2002; Ozer et al., 1996). Also, extraction with supercritical fluids requires higher investment but can be highly selective and more suitable for food products. This plays a mechanistic role in supercritical fluid chromatography (SFC), where it contributes to the separation of the solutes that are injected into the chromatographic system.

Supercritical fluid extraction is an interesting technique for the extraction of flavouring compounds from vegetable material. It can constitute an industrial alternative to solvent extraction and steam distillation processes (Stahl, E. and Gerard, D. 1985). SFE allows a continuous modification of solvent power and selectivity by changing the solvent density (Nykanen, I.et al., 1991). Nevertheless, the simple SFE process, consisting of supercritical CO_2 extraction and a one-stage subcritical separation, in many cases does not allow a selective extraction because of the simultaneous extraction of many unwanted compounds.

2.2. Ultrasound-assisted solvent extraction

Ultrasound assisted extraction is very efficient extraction procedure. Sonication induces cavitation, the process in which bubbles with a negative pressure are formed, grown, oscillated, and may split and implode. By this process different chemical compounds and particles can be removed from the matrix surface by the shock waves generated when the cavitation bubbles collapse. The implosion of the cavities creates microenvironments with high temperatures and pressures. Schock waves and powerful liquid micro jets generated by collapsing cavitation bubbles near or at the surface of the sample accelerate the extraction (R. Kellner et al., 2004). Ultrasonic assisted extraction has many advantages since it can be used for both liquid and solid samples, and for the extraction of either inorganic or organic compounds (S.L. Harper et al., 1983). If extracted from solid samples, different problems can occur: there is a possibility of the decomposition of the analyte which could be trapped inside of the collapsing cavitational bubbles. The ultrasound extraction system can be also applied as a dynamic system in which the analytes are removed as soon as they are transferred from the solid matrix to the solvent. In this process, furthermore, the sample is continuously exposed to the solvent (I. Rezic´ et al., 2008).

This is a modified maceration method where the extraction is facilitated by the use of ultrasound (high-frequency pulses, 20 kHz). Ultrasound is used to induce a mechanical stress on the cells of biomass solid samples through the production of cavitations in the sample. The cellular breakdown increases the solubilization of metabolites in the solvent and improves extraction yields. The efficiency of the extraction depends on the instrument frequency, and length and temperature of sonication. Ultrasonification is rarely applied to large-scale extraction; it is mostly used for the initial extraction of a small amount of material. It is commonly applied to facilitate the extraction of intracellular metabolites from plant cell cultures (Kaufmann, 2002; Sarker, 2006).

2.3. Pressurized Solvent Extraction (PSE)

Pressurized solvent extraction or "accelerated solvent extraction," employs temperatures that are higher than those used in other methods of extraction, and requires high pressures to maintain the solvent in a liquid state at high temperatures. It is best suited for the rapid and reproducible initial extraction of a number of samples. The solid biomass sample is loaded into an extraction cell, which is placed in an oven. The solvent is then pumped from a reservoir to fill the cell, which is heated and pressurized at programmed levels for a set period of time. The cell is flushed with nitrogen gas, and the extract, which is automatically filtered, is collected in a flask. Fresh solvent is used to rinse the cell and to solubilize the remaining components. A final purge with nitrogen gas is performed to dry the material. High temperatures and pressures increase the penetration of solvent into the material and improve metabolite solubilization, enhancing extraction speed and yield. Moreover, with low solvent requirements, pressurized solvent extraction offers a more economical and environment-friendly alternative to conventional approaches

As the material is dried thoroughly after extraction, it is possible to perform repeated extractions with the same solvent or successive extractions with solvents of increasing polarity. An additional advantage is that the technique can be programmable, which will offer increased reproducibility. However, variable factors, e.g., the optimal extraction temperature, extraction time, and most suitable solvent, have to be determined for each sample (Kaufmann, 2002; Tsubaki, 2010; Sarker, 2006).

Microwave-assisted extraction (MAE) or simply microwave extraction is a relatively new extraction technique that combines microwave and traditional solvent extraction. The microwave energy has been investigated and widely applied in analytical chemistry to accelerate sample digestion, to extract analytes from matrices and in chemical reactions. Application of microwaves for heating the solvents and plant tissues in extraction process, which increases the kinetic of extraction, is called microwave-assisted extraction. Microwave energy is a non-ionizing radiation that causes molecular motion by migration of ions and rotation of dipoles, without changing the molecular structures if the temperature is not too high. Nonpolar solvents, such as hexane and toluene, are not affected by microwave energy and, therefore, it is necessary to add polar additives. Microwave-assisted extraction (MAE) is an efficient extraction technique for solid samples which is applicable to thermally stable compounds accepted as a potential and powerful alternative to conventional extraction techniques in the extraction of organic compounds from materials. The microwave-assisted extraction technique offers some advantages over conventional extraction methods.

Compared to conventional solvent extraction methods, the microwave-assisted extraction (MAE) technique offers advantages such as improved stability of products and marker compounds, increased purity of crude extracts, the possibility to use less toxic solvents, reduced processing costs, reduced energy and solvent consumption, increased recovery and purity of marker compounds, and very rapid extraction rates.

The use of MAE in natural products extraction started in the late 1980s, and through the technological developments, it has now become one of the popular and cost-effective extraction methods available today, and several advanced MAE instrumentations and methodologies have become available, e.g., pressurized microwave-assisted extraction (PMAE) and solvent-free microwave-assisted extraction (SFMAE).

Comparison between conventional and MAE extraction method

This technique has been used successfully for separation of phenolic compounds from types of biomass, polyphenols derivates, pyrimidine glycosides, alkaloids, terpenes, and so.

In most cases, the results obtained suggested that the microwave assisted method was more convenient even compared to the ultrasound extraction method.

Pyrimidine glycosides

The studies regarding the microwave extraction of vicine and convicine (toxic pyrimidine glycosides) from *Vicia faba* using a methanol: water mixture (1:1 v/v) involves two successive microwave irradiations (30 s each) with a cooling step in between. No degradation could be observed under these conditions,but further irradiation was found to decrease the yield of vicine and convicine. The yield obtained was 20% higher than with the conventional Soxhlet extraction method.

Alkaloids Sparteine, a lupine alkaloid, was extracted from *Lupinus mutabilis*, with methanol: acetic acid (99:1, v/v) in a common microwave oven and the microwave irradiation program used one to five cycles of 30 s with a cooling step in between and conduct to 20% more sparteine than was obtained with a shaken-flask extraction using the same solvent mixture for 20 min.

Terpenes Five terpenic compounds: linalool, terpineol, citronellol, nerol and geraniol, associated with grape (*Vitis vinifera*) aroma was extracted from must samples by MAE (Carro et al.,1997). Was investigated the influence of the parameters: extracting solvent volume, extraction temperature, and amount of sample and extraction time. Several conditions were fixed, such as the extraction time (10 min) and the applied power (475 W). The solvent volume appeared to be the only statistically significant factor, but was limited to 15 mL by the cell size. The highest extraction yield was obtained with both the solvent volume and the temperature at their maximum tested values. In contrast, the sample amount had to be minimized in order to obtain the best recoveries. The final optimized extraction conditions were as follows: 5 mL sample amounts extracted with 10 mL of dichloromethane at a temperature of 90°C for 10 min with the microwave power set at 50% (475 W).

Steroids Recently, was demonstrated that only 30–40 s were sufficient to extract ergosterol quantitatively by MAE using 2 mL methanol and 0.5 mL 2 M sodium hydroxide. Microwave irradiation was applied at 375W for 35 s and the samples were cooled for 15 min before neutralization with 1 M hydrochloric acid followed by pentane extraction. The yield was similar to or even higher than that obtained with the traditional methanolic extraction followed by alkaline saponification and pentane extraction.

Alkaloids The extraction of two alkaloids cocaine and benzoylecgonine by focused MAE was optimized by taking into account several parameters such as the nature of the extracting solvent, particle size distribution, sample moisture, applied microwave power and radiation time. MAE was found to generate similar extracts to those obtained by conventional SLE but in a more efficient manner. Indeed, 30s were sufficient to extract cocaine quantitatively from leaves, using methanol as solvent and a microwave power of 125 W. (Kaufmann, 2002).

Phenolic compounds

In recent years, synthetic antioxidants were reported to have the adverse effects such as toxicity and carcinogenicity and this situation has forced scientists to search for new natural antioxidants from herbs or the other materials. Phenolic compounds, the most important bioactive compounds from plant sources, are among the most potent and therapeutically useful bioactive substances, providing health benefits associated with reduced risk of chronic and degenerative disease (Luthria, 2006; Tsubaki et al., 2010; Proestos, 2008).

Extraction is one of the most imperative steps in the evaluation of phenolic compounds from plant. Often is done a saponification prior to the extraction step because is necessary to cleave the ester linkage to the cell walls (Robbins, 2003).

The capability of a number of extraction techniques have been investigated, such as solvent extraction and enzyme-assisted extraction. However, these extraction methods have drawbacks to some degree. For example, solvent extraction is time consuming and enzyme in enzyme assisted extraction is easy to denature. In the case of Soxhlet extraction, the extraction time vary from 1 minute to 6 h. Ultrasonic is one of the most industrially used methods to enhance mass transfer phenomena (Japón-Luján et.al. 2006; Luthria, 2006; Pérez-Serradilla, 2007). Meanwhile, microwave assisted extraction heats the extracts quickly and significantly accelerates the extraction process (Martínez, 1996; Kojima, 2004; Patsias, 2009). Simultaneous ultrasonic/microwave assisted extraction (UMAE) coupled the advantage of microwave and ultrasonic, presenting many advantages (Kojima, 2004).

Extraction of phenolic compounds from solid samples is usually carried out by stirring (Luthria, 2006; Nepote, 2005), although the use of auxiliary energies has demonstrated to accelerate the process (Japón-Luján, et.al.2006; Pérez-Serradilla, 2007). Microwave-assisted extraction (MAE) is the process by which microwave energy is used to heat polar solvents in contact with solid samples and to partition compounds of interest between the sample and the solvent, reducing both extraction time and solvent consumption.

The conventional liquid–solid extraction techniques, such as heat reflux extraction (HRE), ultrasonic extraction (UE) and maceration extraction (ME), are discommodious, laborious, time-consuming and require large volumes of toxic organic solvents. So increasing attention is paid to the development of more efficient extraction methods for the rapid extraction of active compounds from materials.

The current analytical methods used to extract phenolic compounds from liquid samples are based on *liquid-liquid extraction* (LLE). Although this technique offers efficient and precise results, it is relatively time-consuming, possibly harmful due the use of large volume of organic solvents (frequently toxic) and highly expensive. For these reasons, there is an increasing tendency to replace LLE by solid-phase extraction (SPE) for liquid samples. SPE was developed in the 1980s, and has emerged as a powerful tool for chemical isolation and purification. This methodology is an alternative extraction to LLE due to it reduces organic solvents consumption, the length of analysis and it can be automated (Martínez, et. al., 1996; Kojima, 2004; Patsias, et.al., 2009).

Although most attention has been focused on the determination of phenolic compounds in aqueous samples, more substituted phenols, such as pentachlorophenol, show limited transport in water and they are more likely absorbed in sediments and soils. This fact contributes to the persistent of these compounds in the environment and it results in high concentrations of them that could affect aquatic and earth organism. For extraction, Soxhlet extraction is one of the most popular techniques for isolating phenolic compounds from solid samples, due to its simplicity, inexpensive extraction apparatus. Despite the good results obtained with this methodology, Soxhlet extraction makes the analysis procedure excessive time consuming. Moreover, it requires large amount of hazardous organic solvents.

Ultrasonic extraction is another conventional technique to extract analytes from solid samples. Although sonication is faster than Soxhlet extraction, it also requires large volumes of toxic and expensive organic solvents.

The studies show that the compounds are extracted more effectively when the energy provided by microwave is employed (Perez-Serradilla, 2011).

3. Experimental studies

The efficiencies of different solvents (water, acid and alcohol) in the extraction of caffeine and phenols from leaves of white, black, green and red tea in different solvents: ethanol, isopropanol, methanol and water. Extraction was performed comparative by ultrasonic and by MAE. Determination of the total amount of phenolic compounds was studied comparative using different extraction times 5, 15 and respectively 30 minutes. The microwave irradiation shortens time necessary to extract phenols and caffeine from tea samples (between 30 and 50 seconds). The results of the comparison investigation are presented in the figure 1.

4. Conclusion

Chromatographic determination of phenolic compounds isolated from the tea samples by ultrasonic and MAE extraction is comparable. The difference between the two methods of extraction consists in extraction time and amount of solvents used. Also, the yield for MAE was about 20% is 20% higher than that of the ultrasonic extraction.

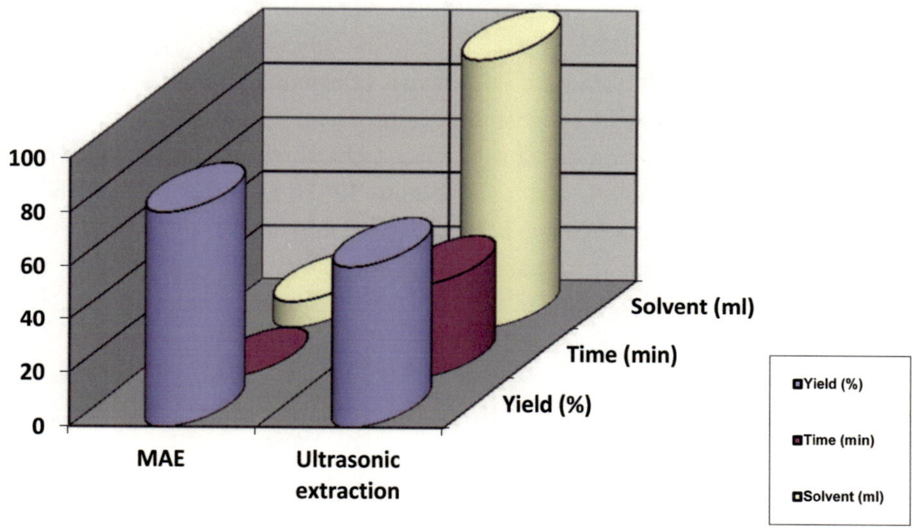

Figure 1. Comparison between the two extraction methods

Author details

Adina-Elena Segneanu, Paulina Vlazan,
Paula Sfirloaga and Ioan Grozescu
National Institute of Research and Development for Electrochemistry and Condensed Matter – INCEMC Timisoara, Romania

Florentina Cziple
Eftimie Murgu University, Resita, Romania

Vasile Daniel Gherman
Politehnica University of Timisoara, Romania

5. References

Alonso M.C., Puig D., Silgoner I., Grasserbauer M., Barcelo′ D., (1998), Determination of priority phenolic compounds in soil samples by various extraction methods followed by liquid chromatography–atmospheric pressure chemical ionisation mass spectrometry, *Journal of Chromatography A*, 823, 231–239;

Carro,N., GarciaC.M., Cela R., (1997), Microwave-assisted extraction of monoterpenols in must samples, *Analyst*, 122 (4), 325–330;

Diaz-Maroto MC, Perez-Coello MS, Cabezudo MD. Supercritical carbon dioxide extraction of volatiles from spices – comparison with simultaneous distillation – extraction. J. of Chromatography A, 947, 23-29, 2002.

Ehlers D, Nguyen T, Quirin KW, Gerard D. Anaylsis of essential basil oils-CO_2 extracts and steam-distilled oils. Deutsche Lebensmittel-Rundschau, 97, 245-250, 2001.

Fang X., Wang J., Zhou H., Jiang X., Zhu L., Gao X.(2009), Microwave-assisted extraction with water for fast extraction and simultaneous RP-HPLC determination of phenolic acids in Radix Salviae Miltiorrhizae, *J. Sep. Sci.*, 32, 2455 – 2461;

Faustino H., Gil N., Baptista C., Duarte A.P., (2010), Antioxidant Activity of Lignin Phenolic Compounds Extracted from Kraft and Sulphite Black Liquors, *Molecules*, 15, 9308-9322;

Gallo M., Ferracane R., Graziani G., Ritieni A. Fogliano V. (2010), Microwave Assisted Extraction of Phenolic Compounds from Four Different Spices, *Molecules*, 15, 6365-6374;

Harper S.LWalling., J.F., Holland D.M., Pranger L.J., Anal. Chem. 55 (1983) 1553.

Japón-Luján, R., Luque-Rodríguez, J.M., & Luque de Castro, M. D. (2006), Multivariate optimization of the microwave-assisted extraction of oleuropein and related biophenols from olive leaves, *Analytical and Bioanalytical Chemistry*, 385, 753–759;

Kaufmann B., Christen P., (2002), Recent extraction techniques for natural products: microwave-assisted extraction and pressurised solvent extraction, *Phytochemical Analysis*, Vol.13, 2, 105–113;

Kellner R., Mermet J., Otto M., Valcarel M., Widmer H.M., Analytical Chemistry, second ed., Wiley-VCH, Germany, 2004.

Kojima, M., Tsunoi, S., Tanaka, M., (2004), High performance solid-phase analytical derivatization of phenols for gas chromatography–mass spectrometry, *J. Chromatogr. A 1042*,1-7;

Li X.; Zeng, Z.; Zhou, J. (2004), High thermal-stable sol–gel-coated calix[4]arene fiber for solid-phase microextraction of chlorophenols, *Anal. Chim. Acta 509*, 27-37;

Li B.B., Smith B. Hossain M.M., Separation and Purification Technology 48 (2006) 189.

Luthria, D.L., Mukhopadhyay, S., Kwansa, A.L. (2006), A systematic approach for extraction of phenolic compounds using parsley (Petroselinum crispum) flakes as a model substrate, *Journal of the Science of Food and Agriculture,* 86, 1350–1358;

Mahugo Santana C., Sosa Ferrera Z., Torres Padrón M. E., Santana Rodríguez J.J., (2009), Methodologies for the Extraction of Phenolic Compounds from Environmental Samples: New Approaches, *Molecules, 14,* 298-320;

McCabe, W. L., and J. C. Smith, *Unit Operations of Chemical Engineering,* McGraw- Hill, 1956;

Martínez, D., Pocurull, E., Marcé, R.M., Borrull, F., Calull M., (1996), Comparative study of the use of high-performance liquid chromatography and capillary electrophoresis for determination of phenolic compounds in water samples, *Chromatographia, 43,* 619-624;

Nenkova S., Radoykova T., Stanulov K., (2011), Preparation and antioxidant properties of biomass low molecular phenolic compounds (review), *Journal of the University of Chemical Technology and Metallurgy, 46, 2, 109-120;*

Nepote, V, Grosso, N.R., Guzman, C.A. (2005), Optimization of extraction of phenolic antioxidants from peanut skins, *Journal of the Science of Food and Agriculture,* 85, 33–38;

Nykanen, I., Nykanen, L. and Alkio, M. (1991) J. Essential Oil Res. 3, 229.

Ozer EO, Platin S, Akman U, Hortasçsu O. Supercritical Carbon Dioxide Extraction of Spearmint Oil from Mint-Plant Leaves. Can. J. Chem. Eng., 74, 920-928, 1996.

Patsias, J., Papadopoulou-Mourkidou, E. (2009), Development of an automated on-line solid-phase extraction–high-performance liquid chromatographic method for the analysis of aniline, phenol, *Molecules, 14,* 315;

Perry, R. H., and D. Green, *Perry's Chemical Engineers' Handbook,* 6th Edition, McGraw-Hill, 1988;

Pérez-Serradilla, J.A., Japón-Luján, R., Luque de Castro, M.D. (2007), Simultaneous microwave-assisted solid–liquid extraction of polar and nonpolar compounds from Alperuja, *Analytica Chimica Acta,* 602, 82–88;

Proestos C., Komaitis M., (2008), Application of microwave-assisted extraction to the fast extraction of plant phenolic compounds, *LWT* 41, 652–659;

Rezic I., Krstic D., Bokic Lj., Ultrasonic extraction of resins from an historic textile, Ultrasonics Sonochemistry 15 (2008) 21–24

Robbins, R. J. (2003), Phenolic Acids in Foods: An Overview of Analytical Methodology, *J. Agric. Food Chem., 51,* 2866-2887;

Routray W., Orsat V., (2012), Microwave-Assisted Extraction of Flavonoids:A Review, *Food Bioprocess Technol.* 5, 409–424;

Sarker S.D., Latif Z., Gray A. I.,(2006), Natural Products Isolation Second Edition, *Humana Press* Inc.;

Saunders J.A., Blume D.E., Journal of Chromatography A 205 (1981) 147.

Stahl, E. and Gerard, D. (1985) Perfumer Flavorist 10, 29.

Sun,Y., Cheng, J. (2002), Hydrolysis of lignocellulosic materials for ethanol production: a review. *Bioresource Technology,* 83, 1–11;

Tsubaki S., Sakamoto M., Azuma J., (2010), Microwave-assisted extraction of phenolic compounds from tea residues under autohydrolytic conditions, *Food Chemistry,* 123,1255–1258;

Chapter 16

High-Efficiency Separation of Bio-Oil

Shurong Wang

Additional information is available at the end of the chapter

http://dx.doi.org/10.5772/51423

1. Introduction

1.1. What is fast pyrolysis?

Biomass is a CO_2-neutral energy source that has considerable reserve. It can replace fossil feedstock in the production of heat, electricity, transportation fuels, chemicals, and various materials. Liquid bio-fuels, which are considered to be substitutes for traditional petrol liquid fuels, can be produced from biomass in different ways, such as high-pressure liquefaction, hydrothermal pyrolysis, and fast pyrolysis.

Fast pyrolysis is a technology that can efficiently convert biomass feedstock into liquid biofuels. The liquid obtained from fast pyrolysis, which is also called crude bio-oil, may be used as burning oil in boilers or even as a transportation fuel after upgrading. Fast pyrolysis is a process in which lignocellulosic molecules of biomass are rapidly decomposed to short-chain molecules in the absence of oxygen. Under conditions of high heating rate, short residence time, and moderate pyrolysis temperature, pyrolysis vapor and some char are generated. After condensation of the pyrolysis vapor, liquid product can be collected in a yield of up to 70 wt% on a dry weight basis (Bridgwater et al., 1999; Lu et al., 2009). The obvious advantages of the process are as follows:

1. Low-grade biomass feedstock can be transformed into liquid biofuels with relatively higher heating value, thus making storage and transportation more convenient.
2. The by-products are char and gas, which can be used to provide the heat required in the process or be collected for sale.
3. For waste treatment, fast pyrolysis offers a method that can avoid hazards such as heavy metal elements in the char and reduce pollution of the environment.

Many researchers have focused on the techniques of fast pyrolysis, and various configurations of reactor have been developed to satisfy the requirements of high heating rate, moderate reaction temperature, and short vapor residence time for maximizing bio-oil production. During the past decades, many types of reactor have been designed to promote the large-scale and commercial utilization of biomass fast pyrolysis, such as the fluidized bed reactor (Luo et al., 2004; Wang et al., 2002), the ablative reactor (Peacocke & Bridgwater, 1994), the rotating cone reactor (Muggen, 2010; Peacocke; Wagenaar, 1994) and Vacuum reactor (Bridgwater, 1999; Yang et al., 2001).

1.2. The composition and properties of bio-oil

The chemical composition of bio-oil is significantly different from that of petroleum fuels. It consists of different compounds derived from decomposition reactions of cellulose, hemicellulose, and lignin. The chemical composition of bio-oil varies depending on the type of biomass feedstock and the operating parameters. Generally speaking, bio-oil is a mixture of water and complex oxygen-rich organic compounds, including almost all such kinds of organic compounds, that is, alcohols, organic acids, ethers, esters, aldehydes, ketones, phenols, etc. Normally, the component distribution of bio-oil may be measured by GC-MS analysis.

Crude bio-oil derived from lignocellulose is a dark-brown, viscous, yet free-flowing liquid with a pungent odor. Crude bio-oil has an oxygen content of 30–50 wt%, resulting in instability and a low heating value (Oasmaa & C., 2001). The water content of bio-oil ranges from 15 to 50 wt%. The high water content of bio-oil derives from water in the feedstock and dehydration reactions during biomass pyrolysis (Bridgwater, 2012). Heating value is an important indicator for fuel oils. The heating value of bio-oil is usually lower than 20 MJ/kg, much lower than that of fuel oil. The high water content and oxygen content are two factors responsible for its low heating value. The density of bio-oil derived from fast pyrolysis is within the range 1100–1300 kg/m^3(Adjaye et al., 1992). The pH value of bio-oil is usually in the range 2–3 owing to the presence of carboxylic acids such as formic acid and acetic acid. The strong acidity can corrode pipework and burner components. Measurements of the corrosiveness of bio-oil have shown that it can induce an apparent mass loss of carbon steel and the breakdown of a diesel engine burner (Wright et al., 2010).

Fresh bio-oil is a homogeneous liquid containing a certain amount of solid particles. After long-term storage, it may separate into two layers and heavy components may be deposited at the bottom. As mentioned above, the high content of oxygen and volatile organic compounds are conducive to the ageing problems of bio-oil. The aldol condensation of aldehydes and alcohols and self-aggregation of aldehydes to oligomers are two of the most likely reactions to take place. Coke and inorganic components in the bio-oil may also have a catalytic effect, thereby enhancing the ageing process (Rick & Vix, 1991).

1.3. The utilization of bio-oil

The oxygenated compounds in bio-oil can lead to several problems in its direct combustion, such as instability, low heating value, and high corrosiveness. Although higher water content can improve the flow properties and reduce NOx emissions in the fuel combustion process, it causes many more problems. It not only decreases the heating value of the fuel, but also increases the corrosion of the combustor and can result in flame-out. The low pH value of bio-oil also aggravates corrosiveness problems, which may lead to higher storage and transportation costs. Many researchers have tested the combustion of bio-oil in gas boiler systems, diesel engines, and gas turbines (Czernik & Bridgwater, 2004).

Fresh bio-oil from different feedstocks can generally achieve stable combustion in a boiler system. One problem, however, is the difficulty of ignition. The high water content of bio-oil not only decreases its heating value, but also consumes a large amount of latent heat of vaporization (Bridgwater & Cottam, 1992). Thus, the direct ignition of bio-oil in a cold furnace is not easy, and an external energy source is needed for ignition and pre-heating of the furnace. The combustion of bio-oil in diesel engines is more challenging. Its long ignition delay time, short burn duration, and lower peak heat release have limited its combustion properties (Vitolo & Ghetti, 1994). Experiments employing bio-oil in gas turbines have proved largely unsuccessful. The high viscosity and high ash content of bio-oil result in severe blocking and attrition problems in the injection system. Moreover, acid in the bio-oil is harmful to the mechanical components of the gas turbine.

Even though many combustion tests of bio-oil have shown its combustion performances to be inferior to those of fossil fuels, the environmental advantages of bio-oil utilization cannot be ignored. Comparative tests have shown that the SO_2 emissions from bio-oil combustion are much lower than those from fossil fuel combustion.

Bio-oil is a mixture of many organic chemicals, such as acetic acid, turpentine, methanol, etc. Many compounds in bio-oil are important chemicals, such as phenols used in the resins industry, volatile organic acids used to produce de-icers, levoglucosan, hydroxyacetaldehyde, and some agents applied in the pharmaceutical, synthetic fiber, and fertilizer industries, as well as flavoring agents for food products (Radlein, 1999). Besides, bio-oil can also be used in a process that converts traditional lime into bio-lime (Dynamotive Corporation, 1995).

2. Separation of bio-oil for upgrading or refinement

2.1. The importance of separation technology

Bio-oil cannot be directly applied as a high-grade fuel because of its inferior properties, such as high water and oxygen contents, acidity, and low heating value. Thus, it is necessary to upgrade bio-oil to produce a high-grade liquid fuel that can be used in engines (Bridgwater, 1996; Czernik & Bridgwater, 2004; Mortensen et al., 2011).

In view of its molecular structure and functional groups, and using existing chemical processes for reference, such as hydrodesulfurization, catalytic cracking, and natural gas steam reforming, several generic bio-oil upgrading technologies have been developed, including hydrogenation, cracking, esterification, emulsification, and steam reforming.

Components with unsaturated bonds, such as aldehydes, ketones, and alkenyl compounds, influence the storage stability of bio-oil, and hydrogenation could be used to improve its overall saturation (Yao et al., 2008). Hydrogenation can achieve a degree of deoxygenation of about 80%, and transform bio-oil into high-quality liquid fuel (Venderbosch et al., 2010; Wildschut et al., 2009). This process requires a high pressure of hydrogen, which increases both the complexity and cost of the operation. Alcohol hydroxyl, carbonyl, and carboxyl groups were easily hydrodeoxygenated, and phenol hydroxyl and ether groups were also reactive, while furans, having a cyclic structure, were more difficult to convert (Furimsky, 2000). After the separation of bio-oil, the components with alcohol hydroxyl, carbonyl, carboxyl, phenol hydroxyl, and ether groups can be efficiently hydrodeoxygenated at a low hydrogen pressure, while the hydrodeoxygenation of more complex components, such as ethers and furans, may be achieved by developing special catalysts.

Catalytic cracking of bio-oil refers to the reaction whereby oxygen is removed in the form of CO, CO_2, and H_2O, in the presence of a solid acid catalyst, such as zeolite, yielding a hydrocarbon-rich high-grade liquid fuel. In the process of cracking, oxygenated compounds in bio-oil are thought to undergo initial deoxygenation to form light olefins, which are then cyclized to form aromatics or undergo some other reactions to produce hydrocarbons (Adjaye & Bakhshi, 1995a). Since bio-oil has a relatively low H/C ratio, and dehydration is accompanied by the loss of hydrogen, the H/C ratio of the final product is generally low, and carbon deposits with large aromatic structures tend to be formed, which can lead to deactivation of the catalyst (Guo et al., 2009a). The cracking of crude bio-oil is always terminated in a short time, with a coke yield of about 20% (Adjaye & Bakhshi, 1995b; Vitolo et al., 1999). Alcohols, ketones, and carboxylic acids are efficiently converted into aromatic hydrocarbons, while aldehydes tend to condense to form carbon deposits (Gayubo et al., 2004b). Phenols also show low reactivity and coking occurs readily (Gayubo et al., 2004a). Besides, some thermally sensitive compounds, such as pyrolitic lignin, might undergo aggregation to form a precipitate, which would block the reactor and lead to deactivation of the catalyst. Consequently, efforts have been made to avoid this phenomenon by separating these compounds through thermal pre-treatment (Valle et al., 2010). Therefore, to maintain the stability and high performance of the cracking process, it is necessary to obtain fractions suitable for cracking by separation of bio-oil, to achieve the partial conversion of bio-oil into hydrocarbon fuels.

Bio-oil has a high content of carboxylic acids, so catalytic esterification is used to neutralize these acids. Both solid acid and base catalysts display high activity for the conversion of carboxylic acids into the corresponding esters, and the heating value of the upgraded oil is thereby increased markedly (Zhang et al., 2006). Since this method is more suitable for the transformation of carboxylic acids, which constitute a relatively small proportion of crude bio-oil, an ester fuel with a high heating value can be expected to be produced from the esterification of a fraction enriched with carboxylic acids obtained from the separation.

The emulsion fuel obtained from bio-oil and diesel is homogeneous and stable, and can be burned in existing engines. Research on the production of emulsions from crude bio-oil and diesel suggested that the emulsion produced was more stable than crude bio-oil. Subsequent tests of these emulsions in different diesel engines showed that because of the presence of carboxylic acids, the injector nozzle was corroded, and this corrosion was accelerated by the high-velocity turbulent flow in the spray channels (Chiaramonti et al., 2003a; Chiaramonti et al., 2003b). Besides corrosion, the high water content of bio-oil will lower the heating value of the emulsion as a fuel, and some high molecular weight components such as sugar oligomers and pyrolitic lignin will increase the density and reduce the volatility of the emulsion. Thus, it is beneficial to study the emulsification of the separated fractions that contain less water and fewer high molecular weight components.

Catalytic steam reforming of bio-oil is also an important upgrading technology for converting it into hydrogen. Research on the steam reforming of acetic acid and ethanol is now comparatively mature, with high conversion of reactants, hydrogen yields, and stability of the catalysts (Hu & Lu, 2007). However, some oxygenated compounds in bio-oil show inferior reforming behavior. Phenol cannot be completely converted even at a high steam-to-carbon ratio, while m-cresol and glucose not only show low reactivity, but are also easily coked (Constantinou et al., 2009; Hu & Lu, 2009). To improve the reforming process, some further investigations of steam reforming based on other separating methods are needed.

Therefore, it is necessary to combine crude bio-oil utilization with the current upgrading technologies. Taking advantage of efficient bio-oil separation to achieve the enrichment of compounds in the same family or the components that are suitable for the same upgrading method is a significant strategy for the future utilization of high-grade bio-oil.

2.2. Conventional separation technologies

The efficient separation of bio-oil establishes a solid foundation for its upgrading. Currently, conventional methods for bio-oil separation include column chromatography, solvent extraction, and distillation.

2.2.1. Solvent extraction

The solvents for extraction include water, ethyl acetate, paraffins, ethers, ketones, and alkaline solutions. In recent years, some special solvents, such as supercritical CO_2, have also been used for extraction or other research. By selecting appropriate solvents for extraction of the desired products, good separation of bio-oil can be achieved.

Some researchers have used non-polar solvents for the primary separation of bio-oil, such as toluene and n-hexane, and then proceeded to extract the solvent-insoluble fraction with water; finally, the water-soluble and water-insoluble fractions were further extracted with diethyl ether and dichloromethane, respectively (Garcia-Perez et al., 2007; Oasmaa et al., 2003). A lot of organic solvents are consumed during the process. Considering the cost of these solvents and the difficulty of the recovery process, the operating costs are unacceptable, which hinders its industrialization.

Supercritical fluid extraction is based on the different dissolving abilities of supercritical solvents under different conditions. Supercritical fluid extraction at low temperatures contributes to preventing undesirable reactions of thermally sensitive components. Researchers usually use CO_2 as the supercritical solvent. In a supercritical CO_2 extraction, compounds of low polarity (aldehydes, ketones, phenols, etc.) are selectively extracted, while acids and water remain in the residue phase (Cui et al., 2010).

2.2.2. *Column chromatography*

The principle of column chromatography is that substances are separated based on their different adsorption capabilities on a stationary phase. In general, highly polar molecules are easily adsorbed on a stationary phase, while weakly polar molecules are not. Thus, the process of column chromatography involves adsorption, desorption, re-adsorption, and re-desorption. Silica gel is commonly used as the stationary phase, and an eluent is selected according to the polarity of the components. Paraffin eluents, such as hexane and pentane, are used to separate aliphatic compounds. Aromatic compounds are usually eluted with benzene or toluene. Some other polar compounds are obtained by elution with methanol or other polar solvents (Ertas & Alma, 2010; Onay et al., 2006; Putun et al., 1999).

2.2.3. *Distillation*

Distillation is a common separating technology in the chemical industry. This method separates the components successively according to their different volatilities, and it is essential for the separation of liquid mixtures. Atmospheric pressure distillation, vacuum distillation, steam distillation, and some other types of distillation have been applied in bio-oil separation.

Due to its complex composition, the boiling of bio-oil starts below 100 °C under atmospheric pressure, and then the distillation continues up to 250–280 °C, whereupon 35–50% of residue is left (Czernik & Bridgwater, 2004).

The thermal sensitivity of bio-oil limits the operating temperature of distillation. In view of the unsatisfactory results obtained by atmospheric pressure distillation, researchers have employed vacuum distillation to lower the boiling points of components, and bio-oil could thereby be separated at a low temperature. Characterization of the distilled organic fraction showed that it had a much better quality than the crude bio-oil, containing little water and fewer oxygenated compounds, and having a higher heating value.

Steam distillation is performed by introducing steam into the distilling vessel, to heat the bio-oil and decrease its viscosity, and finally the volatile components are expelled by the steam. In a study combining steam distillation with reduced pressure distillation, bio-oil was first steam distilled to recover 14.9% of a volatile fraction. The recovered fraction was then further distilled by reduced pressure distillation to recover 16 sub-fractions (Murwanashyaka et al., 2001). In this process, a syringol-containing fraction was separated and syringol with a purity of 92.3% was obtained.

Due to its thermal sensitivity, it is difficult to efficiently separate bio-oil by conventional distillation methods. Molecular distillation seems to offer a potential means of realizing bio-oil separation, because it has the advantages of low operating temperature, short heating time, and high separation efficiency.

2.3. Molecular distillation

There are forces between molecules, which can be either repulsive or attractive depending on intermolecular spacing. When molecules are close together, the repulsive force is dominant. When molecules are not very close to each other, the forces acting between them are attractive in nature, and there should be no intermolecular forces if the distance between molecules is very large. Since the distances between gas molecules are large, the intermolecular forces are negligible, except when molecules collide with each other. The distance between collisions with another molecule is called its free path.

The mean free path of an ideal gas molecule can be described by Eq. (1):

$$\lambda_m = \frac{k}{\sqrt{2}\pi}\frac{T}{d^2 p} \tag{1}$$

Where T (°C) is the local temperature; λ_m (m) refers to the mean free path; d (m) is the effective diameter of the molecule; P (Pa) is the local pressure; and k is the Boltzmann constant.

As is apparent from Eq. (1), the molecular mean free path is inversely proportional to the pressure and the square of the effective molecular diameter. Under certain conditions, that is, if the temperature and pressure are fixed, the mean free path is a function of the effective molecular diameter. Apparently, a smaller molecule has a shorter mean free path than a larger molecule. Furthermore, molecular mean free path will increase with increasing temperature or decreasing pressure.

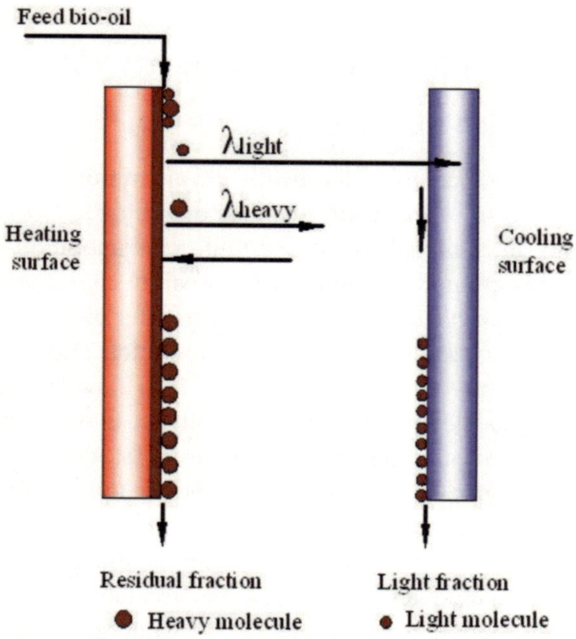

Figure 1. Schematic representation of molecular distillation.

Molecules will move more rapidly when the liquid mixture is heated. Surface molecules will overcome intermolecular forces and escape as gas molecules when they obtain sufficient energy. With an increased amount of gas molecules above the liquid surface, some molecules will return to the surface. Under certain conditions, the molecular motion will achieve dynamic equilibrium, which is manifested as equilibrium on a macroscopic scale.

Traditional distillation technology separates components by differences in their boiling points. However, molecular distillation (or short-path distillation) is quite different and precisely relies on the various mean free paths of different substances. As shown in Fig. 1, the distance between the cooling and heating surfaces is less than the mean free path for a light molecule, but greater than that for a heavy molecule. Therefore, the light molecules escaping from the heating surface can easily reach the cooling surface and be condensed. The dynamic balance is thereby broken, and the light molecules are continuously released from the liquid phase. On the contrary, the heavy molecules are not released and return to the liquid phase. In this way, the light and heavy molecules are effectively separated.

Molecular distillation technology has been widely used in the chemical, pharmaceutical, and foodstuff industries, as well as in scientific research to concentrate and purify organic chemicals. It is a feasible process for the separation of thermally unstable materials, taking into account that it only takes a few seconds to complete the separation process. Bio-oil is a complex mixture of many compounds with a wide range of boiling points. It is thermally sensitive and easily undergoes reactions such as decomposition, polymerization, and oxygenation. Additionally, most of the compounds are present in low concentrations. Molecular distillation is not limited by these unfavorable properties and is suitable for the separation of bio-oil to facilitate analysis and quantification of its constituent compounds.

3. High-efficiency separation of bio-oil at Zhejiang University

3.1. A molecular distillation apparatus

Fig. 2 shows a KDL-5 wiped-film molecular distillation apparatus used for bio-oil separation research at Zhejiang University, which was manufactured by UIC Corporation in Germany. It consists of four main units, namely a feeding unit, an evaporation unit, a condensation unit, and a reduced pressure unit. The feeding unit mainly comprises a graduated dosing funnel with a double jacket, which is filled with heat-transfer oil to control the temperature and to ensure free flowing of the feedstock. The evaporation unit comprises a cylindrical evaporator with a surface area of 0.048 m^2, encased in a double jacket containing heat-transfer oil to maintain good temperature homogeneity. It is worth noting that all of the temperatures of these sections are independent. The condensation unit has two cold traps. The first cold trap (or internal condenser) is located in the center of evaporator, and condenses the volatile compounds reaching the cooling surface. There is another cold trap to prevent uncondensed volatile organic compounds from entering the pump. In the reduced pressure unit, the condensation temperature is usually set at –25 °C. The evaporation temperature ranges from room temperature to 250 °C, while the operating pressure can be as low as 5 Pa.

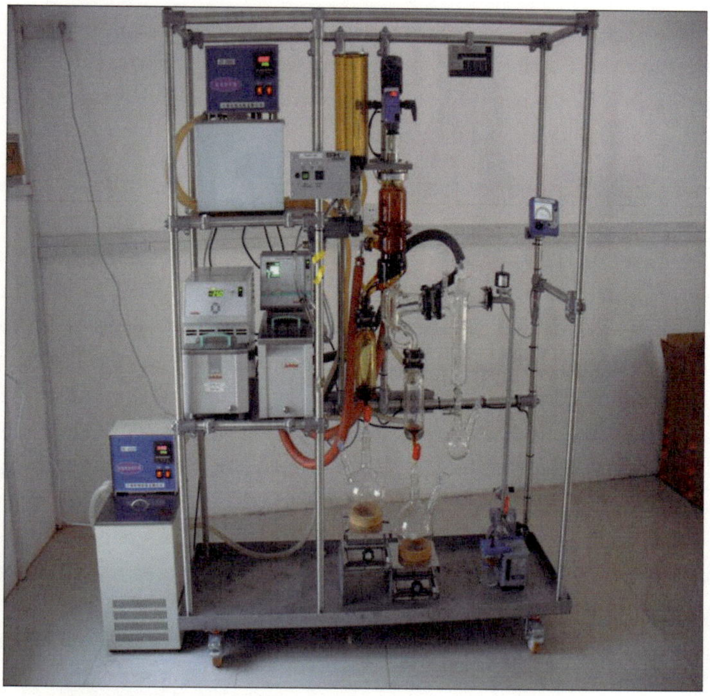

Figure 2. KDL-5 molecular distillation apparatus.

The bio-oil used at Zhejiang University was produced from a bench-scale fluidized bed fast pyrolysis reactor (Wang et al., 2008). Crude bio-oil often contains some solid particles, which would abrade the evaporator surface and block the orifice of the dosing funnel, so it is necessary to perform some pre-treatments. Centrifugation and filtration are usually used to remove the solid particles, and traditional reduced pressure distillation can also be used to remove water and volatile compounds. The pre-treated bio-oil is placed in the funnel and then the separation process starts. The volatile components released from the thin liquid film are condensed by the internal condenser to form the distilled fraction, while the heavy compounds that are not vaporized flow along the evaporator surface and are collected as the residual fraction.

Because of the short residence time of the feed material at the evaporation temperature, this gentle distillation process only puts a low thermal load on the materials to be distilled. It is therefore appropriate for the separation of bio-oil, which is thermally unstable.

3.2. Single separation process under different operating conditions

3.2.1. Physical characteristics of samples

Bio-oil used in the single separation process was produced by the pyrolysis of Mongolian pine sawdust (Wang et al., 2008). Wang et al. (Guo et al., 2009b; Wang et al., 2009) carried out experimental research on molecular separation of the bio-oil, which was pre-treated by centrifugation and filtration to remove solid particles. Molecular distillation of the bio-oil at

50, 70, 100, and 130 °C, respectively, was investigated under a fixed pressure of 60 Pa. Under all of the tested conditions, the light fraction collected by the second condenser placed before vacuum pump was designated as LF, the middle fraction condensed by the internal condenser as MF, and the heavy fraction as HF.

The color of the distilled fractions becomes lighter while the residual fractions become darker. Under the four conditions, water was concentrated in the LFs, which had water contents of about 70 wt%. The LFs could not be burned because of their high water contents. The pH values of the LFs were in the range 2.13–2.17 as a result of their carboxylic acid contents. On the other hand, the HFs had the highest heating values and the lowest water contents, resulting in good ignitability but inferior fluidity. At a distillation temperature of 70 °C, the water content of the MF was as low as 2 wt%. The total mass of the bio-oil distillation fractions amounted to more than 97% of the bio-oil feed. With increasing temperature, the yield of the LF increased without any coking or polymerization problem. Water and volatile carboxylic acids were evaporated from the feedstock in the temperature range 50–130 °C under low pressure, and more carboxylic acids escaped from the liquid at higher temperature. However, on further increasing the temperature, this phenomenon was not so pronounced, due to more and more molecules of higher boiling point also being distilled. The yield of the distilled fraction increased with increasing distillation temperature. However, too high temperature may lead to decomposition of some chemical compounds in the crude bio-oil. Hence, there must be an optimum temperature to realize reasonable separation.

3.2.2. Distribution of acidic compounds in bio-oil fractions

The high content of carboxylic acids in bio-oil is one of the main reasons for its corrosiveness, which damages storage tanks, boilers, and gas turbines. As a consequence, detailed research on the separation of acidic compounds has been carried out under the condition of distillation at 50 °C.

The carboxylic acid content in the refined bio-oil was used to estimate the separation efficiency. Guo et al. (2009b) chose five major acids in bio-oil and studied their separation characteristics. As shown in Fig. 3, the amount of acetic acid, the most abundant acid in bio-oil, was reduced to 1.9 wt% and 0.96 wt% in the MF and HF, respectively. The results indicated that acidic compounds could be effectively separated from the crude bio-oil by means of molecular distillation technology. The LF, which was rich in water and carboxylic acids, was valuable for further catalytic esterification of bio-oil acidic compounds. Both MF and HF could be further upgraded to produce high-quality fuels.

3.2.3. Distribution characteristics of several chemicals in three fractions

Fig. 4 illustrates the distributions of selected compounds in bio-oil, MF, and HF. Six chemicals were selected as being representative of ketones, aldehydes, phenols, and sugars, respectively. 1-Hydroxy-2-propanone, the most abundant ketone in bio-oil, could not be detected in the MF or HF after separation, indicating that it was extremely enriched in the

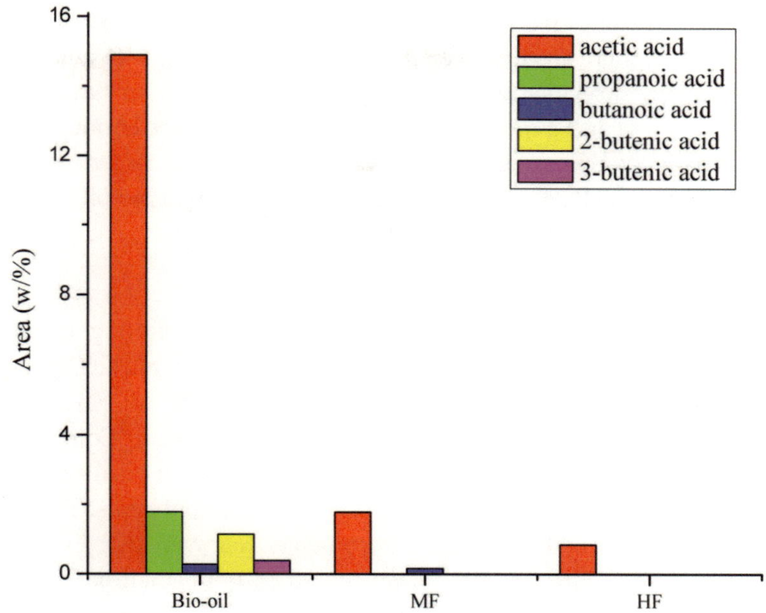

Figure 3. Contents of acidic compounds in three samples obtained at 50 °C (Guo et al., 2009b).

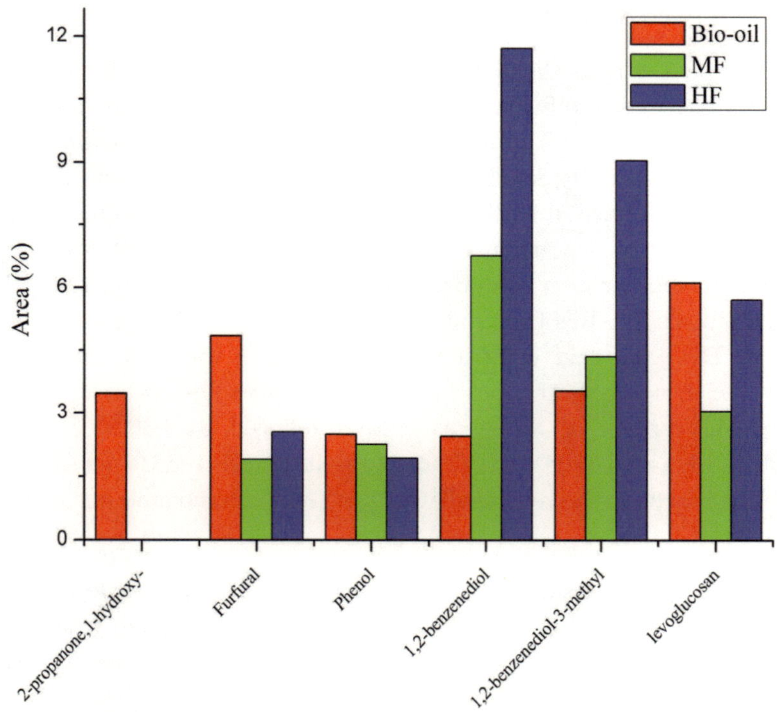

Figure 4. Distributions of selected compounds obtained at 50 °C in three fractions.

LF. The content of furfural in the HF was just a little higher than that in the MF, but much lower than that in bio-oil. The distributions of these two compounds reflected the enrichment of small ketone and aldehyde molecules in the LF. Phenol appeared to be difficult to separate as there were similar distributions in bio-oil, MF, and HF. In contrast, compounds of higher molecular weight tended to be enriched in the MF and HF. For example, 1,2-benzenediol and 3-methyl-1,2-benzenediol were more abundant in the MF and HF than in the bio-oil before separation. In particular, the relative content of 1,2-benzenediol amounted to 11.73 wt% in HF, about five times higher than that in bio-oil (2.45 wt%).

3.2.4. Statistical method to evaluate the separation of bio-oil

As the composition of the bio-oil and the effects of the operating conditions on the distribution of each fraction are both complicated, Wang et al. (2009) put forward a statistical method to directly evaluate the separation level of bio-oil by molecular distillation. The separation coefficients of four groups, "Complete Isolation", "Nonvaporization", "Enrichment", and "Even Distribution", were calculated from the ratios of relative peak of a single component with respect to total components. The results showed that "Complete Isolation" had the largest percentage, followed by "Even Distribution", "Non-vaporization", and "Enrichment" which contained only small parts. Meanwhile, the temperature had a significant effect on the distributions of the compounds.

3.3. Multiple molecular distillation for bio-oil separation

Based on the above single distillation experiments, a multiple molecular distillation experiment was carried out to further evaluate the separation characteristics of bio-oil (Guo et al., 2010b). The feed bio-oil, which was pre-treated by centrifugation, filtration, and vacuum distillation, was firstly distilled at 80 °C and 1600 Pa to obtain the distilled fraction 1 (DF-1) and the residual fraction 1 (RF-1). A part of RF-1 was then further distilled at 340 Pa to obtain DF-2 and RF-2 fractions. In the multiple distillation process, the distilled fraction yield of each distillation process was about 26 wt%. The amounts of water in RF-1 and RF-2 were greatly reduced. The RFs from the two processes had higher heating values than the feed bio-oil or DFs. The acid content was 11.37 wt% in the feed bio-oil, while it was 17.36 wt% for DF-1, nearly four times higher than that in RF-1 (4.56 wt%). In the second process, the acid content of RF-2 was further reduced to 1.38 wt%. The content of monophenols in RF-1 was 36.24 wt%, about twice that in DF-1 (18.02 wt%). Sugars showed non-distillable character in the two distillation processes, and no amounts could be detected in the DF.

In order to gain a deeper insight into the bio-oil distillation properties, Guo (Guo et al., 2010b) proposed a separation factor to evaluate the separation characteristics. The separation factors of acetic acid and 1-hydroxy-2-propanone were approximately 0.9, implying that they could be mostly distilled off. 2-Methoxyphenol, phenol, 2(5H)-furanone, and 2-methoxy-4-methylphenol, the separation factors of which ranged from 0.61 to 0.74, proved to be difficult to separate effectively. Higher molecular weight compounds, such as 3-methoxy-1,2-benzenediol, 4-methoxy-1,2-benzenediol, and 1,2-benzenediol, were very difficult to distil, having separation factors close to zero.

3.4. The joint distillation system at Zhejiang University

Based on the operation experiences gained with the KDL5 molecular distillation apparatus, a larger-scale joint reduced pressure and molecular distillation set-up was established in the State Key Laboratory of Clean Energy Utilization, Zhejiang University. The flow diagram of this joint distillation system is illustrated in Fig. 5. The processing capacities of the reduced pressure distillation and molecular distillation units were both 8–10 kg/h, and they could be run at temperatures up to 300 °C and pressures down to 50 Pa. The reduced pressure distillation unit could be operated separately to remove the water from bio-oil as well as to obtain bio-oil fractions. When these two units were assigned to run together, the pre-treated bio-oil from the first reduced pressure distillation unit could be pumped directly into the molecular distillation unit.

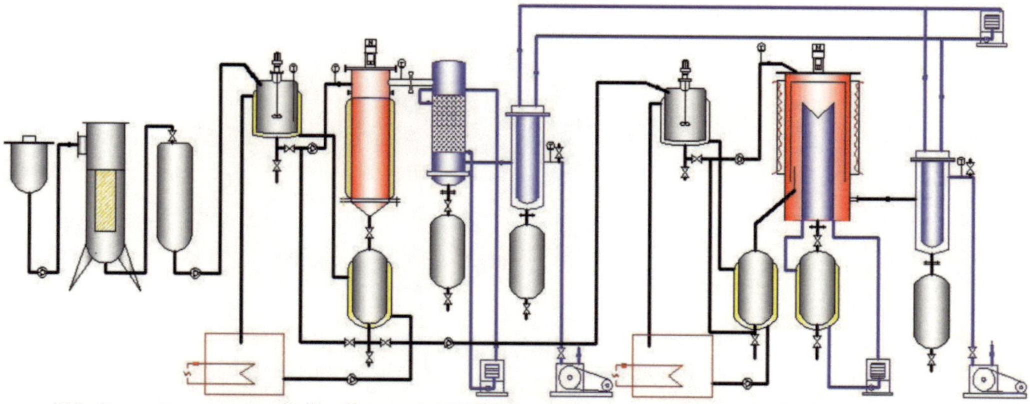

Figure 5. Schematic diagram of the joint distillation system.

3.5. Further research on the distilled fractions

Based on the molecular distillation results, a scheme of the process combining molecular distillation separation with bio-oil upgrading is proposed. The light fraction rich in carboxylic acids and other light components could be used for esterification, catalytic cracking, and steam reforming, to produce ester fuel, hydrocarbons, and hydrogen, respectively. For the middle fraction, steam reforming at high temperature or hydrodeoxygenation at high pressure could efficiently convert this fraction into hydrogen or hydrocarbons. The heavy fraction, which consisted mainly of pyrolytic lignin and sugar oligomers, could be emulsified with diesel to obtain emulsion fuel with a relatively high heating value. On the other hand, the extraction of some valuable chemicals can benefit the overall economy of this process.

Recently, some further research has been performed, aiming at investigating some characteristics of the distilled fractions and devising more promising upgrading methods. Thermal decomposition processes and the pyrolysis products of crude bio-oil and distilled fractions were investigated by means of TG-FTIR by Guo (Guo et al., 2010a). The light fraction (LF) was completely evaporated at 30–150 °C, with the maximum weight loss rate at about 100 °C due to the volatilization of water and compounds of lower boiling point. The middle fraction (MF) and heavy fraction (HF) contained more lignin-derived compounds, and these decomposed continuously over a wide temperature range of 30–600 °C, leaving a final residue yield of 25–30%. Upgrading of the distilled fraction rich in carboxylic acids and ketones was carried out by Guo (Guo et al., 2011). Carboxylic acids accounted for 18.39% of the initial fraction, with acetic acid being the most abundant. After upgrading, the carboxylic acid content decreased to 2.70%, with a conversion yield of 85.3%. The content of esters in the upgraded fraction increased dramatically from 0.72% to 31.1%. The conversion of corrosive carboxylic acids into neutral esters reduced the corrosivity of the bio-oil fraction.

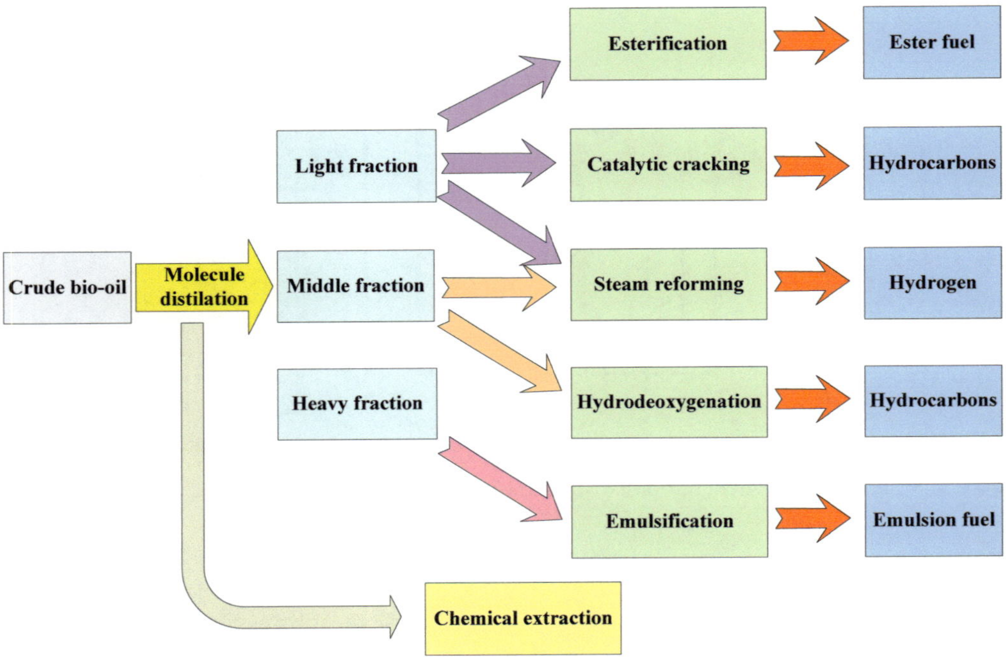

Figure 6. A scheme of the process combining molecular distillation separation with bio-oil upgrading.

Author details

Shurong Wang
Zhejiang University, China

Acknowledgement

The author acknowledges the financial support from the Program for New Century Excellent Talents in University, the International Science & Technology Cooperation Program of China (2009DFA61050), Zhejiang Provincial Natural Science Foundation of China (R1110089), the Research Fund for the Doctoral Program of Higher Education of

China (20090101110034), the National Natural Science Foundation of China (50676085) and the National High Technology Research and Development Program of China (2009AA05Z407). The author also highly appreciates the kind support from Mr. Zuogang Guo, Mr. Qinjie Cai, Mr. Long Guo and Miss Yurong Wang, who have been involved in the experimental research and the preparation of this chapter.

4. References

Adjaye, J.D. & Bakhshi, N.N. (1995a). Catalytic Conversion Of A Biomass-Derived Oil To Fuels And Chemicals .1. Model-Compound Studies And Reaction Pathways. *Biomass & Bioenergy*, Vol.8, No. 3, pp. 131-149, ISSN 0961-9534.

Adjaye, J.D. & Bakhshi, N.N. (1995b). Production Of Hydrocarbons By Catalytic Upgrading Of A Fast Pyrolysis Bio-Oil .1. Conversion Over Various Catalysts. *Fuel Processing Technology*, Vol.45, No.3, pp. 161-183, ISSN 0378-3820.

Adjaye, J.D., Sharma, R.K. & Bakhshi, N.N. (1992). Characterization And Stability Analysis Of Wood-Derived Bio-Oil. *Fuel Processing Technology*, Vol.31, No.3, pp. 241-256, ISSN 0378-3820.

Bridgwater, A.V. (1999). Principles and practice of biomass fast pyrolysis processes for liquids. *Journal of Analytical and Applied Pyrolysis*, Vol.51, No.1-2, pp. 3-22, ISSN 0165-2370.

Bridgwater, A.V. (1996). Production of high grade fuels and chemicals from catalytic pyrolysis of biomass. *Catalysis Today*, Vol.29, No.1-4, pp. 285-295, ISSN 0920-5861.

Bridgwater, A.V. (2012). Review of fast pyrolysis of biomass and product upgrading. *Biomass and Bioenergy*, Vol.38, pp. 68-94, ISSN 0961-9534.

Bridgwater, A.V. & Cottam, M.L. (1992). Opportunities For Biomass Pyrolysis Liquids Production And Upgrading. *Energy & Fuels*, Vol.6, No.2, pp. 113-120, ISSN 0887-0624.

Bridgwater, A.V., Meier, D. & Radlein, D. (1999). An overview of fast pyrolysis of biomass. *Organic Geochemistry*, Vol.30, No.12, pp. 1479-1493, ISSN 1479-1493.

Chiaramonti, D., Bonini, A. & Fratini, E. (2003a). Development of emulsions from biomass pyrolysis liquid and diesel and their use in engines - Part 1: emulsion production. *Biomass & Bioenergy*, Vol.25, No.1, pp. 85-99, ISSN 0961-9534.

Chiaramonti, D., Bonini, A. & Fratini, E. (2003b). Development of emulsions from biomass pyrolysis liquid and diesel and their use in engines - Part 2: tests in diesel engines. *Biomass & Bioenergy*, Vol.25, No.1, pp. 101-111, ISSN 0961-9534.

Constantinou, D.A., Fierro, J.L.G. & Efstathiou, A.M. (2009). The phenol steam reforming reaction towards H_2 production on natural calcite. *Applied Catalysis B-Environmental*, Vol.90, No.3-4, pp. 347-359, ISSN 0926-3373.

Cui, H., Wang, J. & Wei, S. (2010). Supercritical CO2 extraction of bio-oil. *Journal Of Shandong Unirersity Of Technology (Science And Technology)*, Vol.24, No.6, pp. 1-5, 10, ISSN 1672-6197.

Czernik, S. & Bridgwater, A.V. (2004). Overview of applications of biomass fast pyrolysis oil. *Energy & Fuels*, Vol.18, No.2, pp. 590-598, ISSN 0887-0624.

Dynamotive Corporation. (1995). *Acid emission reduction.U.S. Patent 5458803.*

Ertas, M. & Alma, M.H. (2010). Pyrolysis of laurel (Laurus nobilis L.) extraction residues in a fixed-bed reactor: Characterization of bio-oil and bio-char. *Journal of Analytical and Applied Pyrolysis*, Vol.88, No.1, pp. 22-29, ISSN 0165-2370.

Furimsky, E. (2000). Catalytic hydrodeoxygenation. *Applied Catalysis A-General*, Vol.199, No.2, pp. 147-190, ISSN 0926-860X.

Garcia-Perez, M., Chaala, A. & Pakdel, H. (2007). Characterization of bio-oils in chemical families. *Biomass & Bioenergy*, Vol.31, No.4, pp. 222-242, ISSN 0961-9534.

Gayubo, A.G., Aguayo, A.T. & Atutxa, A. (2004a). Transformation of oxygenate components of biomass pyrolysis oil on a HZSM-5 zeolite. I. Alcohols and phenols. *Industrial & Engineering Chemistry Research*, Vol.43, No.11, pp. 2610-2618, ISSN 0888-5885.

Gayubo, A.G., Aguayo, A.T. & Atutxa, A. (2004b). Transformation of oxygenate components of biomass pyrolysis oil on a HZSM-5 zeolite. II. Aldehydes, ketones, and acids. *Industrial & Engineering Chemistry Research*, Vol.43, No.11, pp. 2619-2626, ISSN 0888-5885.

Guo, X.J., Wang, S.R. & Guo, Z.G. (2010a). Pyrolysis characteristics of bio-oil fractions separated by molecular distillation. *Applied Energy*, Vol.87, No.9, pp. 2892-2898, ISSN 0306-2619.

Guo, X.Y., Zheng, Y. & Zhang, B.H. (2009a). Analysis of coke precursor on catalyst and study on regeneration of catalyst in upgrading of bio-oil. *Biomass & Bioenergy*, Vol.33, No.10, pp. 1469-1473, ISSN 0961-9534.

Guo, Z.G., Wang, S.R. & Gu, Y.L. (2010b). Separation characteristics of biomass pyrolysis oil in molecular distillation. *Separation and Purification Technology*, Vol.76, No.1, pp. 52-57, ISSN 1383-5866.

Guo, Z.G., Wang, S.R. & Xu, G.H. (2011). Upgrading Of Bio-Oil Molecular Distillation Fraction With Solid Acid Catalyst. *Bioresources*, Vol.6, No.3, pp. 2539-2550, ISSN 1930-2126.

Guo, Z.G., Wang, S.R. & Zhu, Y.Y. (2009b). Sepatation of acid compounds for refining biomass pyrolysis oil. *Journal of Fuel Chemistry and Technology*, Vol.37, No.1, pp. 49-52, ISSN 1872-5813.

Hu, X. & Lu, G.X. (2007). Investigation of steam reforming of acetic acid to hydrogen over Ni-Co metal catalyst. *Journal of Molecular Catalysis A-Chemical*, Vol.261, No.1, pp. 43-48, ISSN 1381-1169.

Hu, X. & Lu, G.X. (2009). Investigation of the steam reforming of a series of model compounds derived from bio-oil for hydrogen production. *Applied Catalysis B-Environmental*, Vol.88, No.3-4, pp. 376-385, ISSN 0926-3373.

Lu, Q., Li, W.Z. & Zhu, X.F. (2009). Overview of fuel properties of biomass fast pyrolysis oils. *Energy Conversion and Management*, Vol.50, No.5, pp. 1376-1383, ISSN 0196-8904.

Lu, Q., Yang, X.L. & Zhu, X.F. (2008). Analysis on chemical and physical properties of bio-oil pyrolyzed from rice husk. *Journal of Analytical and Applied Pyrolysis*, Vol.82, No.2, pp. 191-198, ISSN 0165-2370.

Luo, Z.Y., Wang, S.R. & Liao, Y.F. (2004). Research on biomass fast pyrolysis for liquid fuel. *Biomass & Bioenergy*, Vol.26, No.5, pp. 455-462, ISSN 0961-9534.

Muggen, G. 2010. Empyro Project Summary. in: *PyNe newsletter*, Vol. 27, pp. 3-5.

Murwanashyaka, J.N., Pakdel, H. & Roy, C. (2001). Seperation of syringol from birch wood-derived vacuum pyrolysis oil. *Separation and Purification Technology*, Vol.24, No.1-2, pp. 155-165, ISSN 1383-5866.

Oasmaa, A., Kuoppala, E. & Solantausta, Y. (2003). Fast pyrolysis of forestry residue. 2. Physicochemical composition of product liquid. *Energy & Fuels*, Vol.17, No.2, pp. 433-443, ISSN 0887-0624.

Onay, O., Gaines, A.F. & Kockar, O.M. (2006). Comparison of the generation of oil by the extraction and the hydropyrolysis of biomass. *Fuel*, Vol.85, No.3, pp. 382-392, ISSN 0016-2361.

Peacocke, G.V.C. Techno-economic assessment of power production from the Wellman and BTG fast pyrolysis. *Science in Thermal and Chemical Biomass Conversion, 2006, Vol.2, pp. 1785-1902*

Peacocke, G.V.C. & Bridgwater, A.V. (1994). Ablative Plate Pyrolysis Of Biomass For Liquids. *Biomass & Bioenergy*, Vol.7, No.1-6, pp. 147-154, ISSN 0961-9534.

Putun, A.E., Ozcan, A. & Putun, E. (1999). Pyrolysis of hazelnut shells in a fixed-bed tubular reactor: yields and structural analysis of bio-oil. *Journal of Analytical and Applied Pyrolysis*, Vol.52, No.1, pp. 33-49, ISSN 0165-2370.

Radlein, D. 1999. *The production of Chemicals from Fast Pyrolysis Bio-oils. Fast Pyrolysis of Biomass: A handbook.* CPL Press, Newbury.

Rick, F., Vix, U. 1991. Product standards for pyrolysis products for use as fuel in industrial firing plant. in: *Biomass pyrolysis liquids upgrading and utilization*, Elsevier applied science, pp. 177-218.

Valle, B., Gayubo, A.G. & Aguayo, A.T. (2010). Selective Production of Aromatics by Crude Bio-oil Valorization with a Nickel-Modified HZSM-5 Zeolite Catalyst. *Energy & Fuels*, Vol.24, pp. 2060-2070, ISSN 0887-0624.

Venderbosch, R.H., Ardiyanti, A.R. & Wildschut, J. (2010). Stabilization of biomass-derived pyrolysis oils. *Journal of Chemical Technology and Biotechnology*, Vol.85, No.5, pp. 674-686, ISSN 0268-2575.

Vitolo, S. & Ghetti, P. (1994). Physical And Combustion Characterization Of Pyrolytic Oils Derived From Biomass Material Upgraded By Catalytic-Hydrogenation. *Fuel*, Vol.73, No.11, pp. 1810-1812, ISSN 0016-2361.

Vitolo, S., Seggiani, M. & Frediani, P. (1999). Catalytic upgrading of pyrolytic oils to fuel over different zeolites. *Fuel*, Vol.78, No.10, pp. 1147-1159, ISSN 0016-2361.

Wagenaar, B.M. 1994. The rotating cone reactor for rapid thermal solids processing, PhD, University of Twente.

Wang, Q., Liu, Q. & He, B. (2008). Experimental research on biomass flash pyrolysis for bio-oil in a fluidized bed reactor. *Journal of Engineering Thermophysics*, Vol.29, pp. 885-888, ISSN 0253-231X.

Wang, S.R., Gu, Y.L. & Liu, Q. (2009). Separation of bio-oil by molecular distillation. *Fuel Processing Technology*, Vol.90, No.5, pp. 738-745, ISSN 0378-3820.

Wang, S.R., Luo, Z.Y. & Dong, L.J. (2002). Flash pyrolysis of biomass for bio-oil in a fluidized bed reactor. *Acta Energiae Solaris Sinica*, Vol.1, pp. 185-188, ISSN 0254-0096.

Wildschut, J., Mahfud, F.H. & Venderbosch, R.H. (2009). Hydrotreatment of Fast Pyrolysis Oil Using Heterogeneous Noble-Metal Catalysts. *Industrial & Engineering Chemistry Research*, Vol.48, No. 23, pp. 10324-10334, ISSN 0888-5885.

Wright, M.M., Satrio, J.A., Brown, R.C. 2010. Techno-Economic Analysis of Biomass Fast Pyrolysis to Transportation Fuels. NREL.

Yang, J., Blanchette, D., B, D.C., Roy, C. 2001. *Progress in thermochemical biomass conversion.*

Yao, Y., Wang, S.R. & Luo, Z.Y. (2008). Experimental research on catalytic hydrogenation of light fraction of bio-oil. *Journal of Engineering Thermophysics*, Vol.29, No.4, pp. 715-719, ISSN 0253-231X.

Zhang, Q., Chang, J. & Wang, T.J. (2006). Upgrading bio-oil over different solid catalysts. *Energy & Fuels*, Vol.20, No.6, pp. 2717-2720, ISSN 0887-0624.

Section 5

Biomass Production

Chapter 17

Short-Rotation Coppice of Willows for the Production of Biomass in Eastern Canada

Werther Guidi, Frédéric E. Pitre and Michel Labrecque

Additional information is available at the end of the chapter

http://dx.doi.org/10.5772/51111

1. Introduction

The production of energy by burning biomass (*i.e.* bioenergy), either directly or through transformation, is one of the most promising alternative sources of sustainable energy. Contrary to fossil fuels, bioenergy does not necessarily result in a net long-term increase in atmospheric greenhouse gases, particularly when production methods take this concern into account. Converting forests, peatlands, or grasslands to production of food-crop based biofuels may release up to 400 times more CO_2 than the annual greenhouse gas (GHG) reductions that these biofuels would provide by displacing fossil fuels. On the other hand, biofuels from biomass grown on degraded and abandoned agricultural lands planted with perennials do not have a negative effect on carbon emissions [1]. In addition, when properly managed, bioenergy can enhance both agricultural and rural development by increasing agricultural productivity, creating new opportunities for revenue and employment, and improving access to modern energy services in rural areas, both in developed and developing countries [2].

Biofuels constitute a very broad category of materials that can be derived from sources including municipal by-products, food crops (*e.g.* maize, sugar cane etc.), agricultural and forestry by-products (straws, stalks, sawdust, etc.) or from specifically-conceived fuel crops. Our analysis focuses on agricultural biofuel crops that can be grown in temperate regions. These crops can be divided into four main categories (Table 1).

Oilseed crops have long been grown in rotation with wheat and barley to produce oil for human, animal or industrial use. Today, these crops primarily provide feedstock for biodiesel. Biodiesel is produced by chemically reacting a vegetable oil with an alcohol such as methanol or ethanol, a process called transesterification. Cereals and starch crops, whose main economical use is for food and fodder, can also be transformed to produce biofuels. For example, the starch in the grains of maize (*Zea mays* L.), wheat (*Triticum aestivum* L.) and

Category	Common name	Botanical name	Habit	Crop life cycle	Main destination
Oil crops	Camelina	*Camelina sativa* (L.) Crantz	Herbaceous	Annual	Biodiesel
	Castor	*Ricinus communis* (L.)		Mostly annual	
	Field mustard	*Sinapis alba* (L.)		Annual	
	Groundnut	*Arachis hypogaea* (L.)			
	Hemp	*Cannabis sativa* (L.)			
	Linseed	*Linum usitatissimum* (L.)			
	Oilseed rape	*Brassica napus* (L.)			
	Safflower	*Carthamus tinctorius* (Mohler)			
	Soybean	*Glycine max* (L.) Merr.			
	Sunflower	*Helianthus annuus* (L.)			
Cereals	Barley	*Hordeum vulgare* (L.)	Herbaceous	Annual	1st gen. ethanol / Solid biofuel
	Maize	*Zea mays* (L.)			
	Oats	*Avena sativa* (L.)			
	Rye	*Secale cereale* (L.)			
	Wheat	*Triticum aestivum* (L.)			
Starch crops	Jerusalem artichoke	*Helianthus tuberosus* (L.)	Herbaceous	Perennial	1st gen. ethanol
	Potato	*Solanum tuberosum* (L.)		Annual	
	Sugar beet	*Beta vulgaris* (L.)		Biennial	
	Sugarcane	*Saccharum officinarum* (L.)		Perennial	
Dedicated bioenergy crops	Kenaf	*Hibiscus cannabinus* (L.)	Herbaceous	Annual	Solid biofuel / 2nd gen. ethanol
	Sorghum	*Sorghum bicolor* (L.) Moench			
	Cardoon	*Cynara cardunculus* (L.)	Herbaceous	Perennial	
	Giant reed	*Arundo donax* (L.)			
	Miscanthus	*Miscanthus* spp.			
	Reed canary grass	*Phalaris arundinacea* (L.)			
	Switchgrass	*Panicum virgatum* (L.)			
	Short-Rotation Coppice	*Eucalyptus* spp. *Populus* spp. *Salix* spp.	Woody	Perennial	

Table 1. The main bioenergy crops for regions with a temperate climate.

sorghum (*Sorghum bicolor* (L.) Moench) can be converted to sugars and then to ethanol by traditional fermentation methods for use in transportation and other fuels (*e.g.* bioethanol). These crops may also be used to produce biogas, composed principally of methane and carbon dioxide produced by anaerobic digestion of biomass. These energy crops have the advantage of being relatively easy to grow. Most are traditional agricultural crops and are easy to introduce at the farm level since they do not require particularly cutting-edge technological equipment. However, using food crops as a source of bioenergy raises serious issues related to food supply and costs, and consequently has been under increasing criticism from the scientific community and society. In particular, the use of these crops for bioenergy competes directly with their use as food. In addition, since many of these crops are annuals, they require large energy inputs and fertilizer for establishment, growth and management, and thus in the end result in minimal energy gains. For such reasons, these crops may not be efficient either for achieving energy balances or for reducing greenhouse gas emissions.

The category of dedicated energy crops notably includes all lignocellulosic (mostly perennial) crops grown specifically for their biomass and used to produce energy. Such crops include herbaceous (*e.g.* miscanthus, switchgrass, reed canary grass, etc.) and woody (willow, poplar, eucalyptus) species that have been selected over the past decades for their high biomass yield, high soil and climate adaptability, and high biomass quality. In addition, especially if grown on marginal arable lands, they do not compete directly for use for food [3], do not require large amounts of inputs in terms of annual cultivation and fertilizer applications [4], nor involve the destruction of native forests with severe negative effects on carbon sequestration [5] and biodiversity [6-7].

We shall limit our description to woody species, because they constitute the focus of our research.

Woody crops for energy production include several silvicultural species notably sharing the following characteristics: fast growth and high biomass yield, potential to be managed as a coppice and high management intensity (highly specific needs with regard to fertilization, irrigation, etc).

A recent review of the literature revealed that about ten different terms are used to refer to the silvicultural practice of cultivating woody crops for energy production: short-rotation woody crops, short-rotation intensive culture, short-rotation forestry, short-rotation coppice, intensive culture of forest crops, intensive plantation culture, biomass and/or bioenergy plantation culture, biofuels feedstock production system, energy forestry, short-rotation fiber production system, mini-rotation forestry, silage sycamore, wood grass [8]. The same author suggested adoption of standard terminology based on an earlier work [9] that had defined this cropping system as *"a silvicultural system based upon short clear-felling cycles, generally between one and 15 years, employing intensive cultural techniques such as fertilization, irrigation and weed control, and utilizing genetically superior planting material"*, to which he proposed to add *"and often relying on coppice regeneration"*, since most species used are able to sprout following harvest. The term coppice refers to a silvicultural practice in which the stem of a tree is cut back at ground level, allowing new shoots to regenerate from the stump.

The early growth rate of coppice sprouts is much greater than that of seedlings or cuttings and in this way trees managed as coppice are characterized by remarkably fast growth and high biomass yield [10-11]. The main species under this cultivation regime in temperate climates are poplar (*Populus* spp) [12], willow (*Salix* spp) [13] and eucalyptus (*Eucalyptus* spp.) [14], and to a lesser extent, black locust (*Robinia pseudoacacia* L.) [15] and alder (*Alnus* spp.) [16]. All of these species, which are cultivated for biomass production in a specific region, are fast-growing under local conditions, cultivated in dense stands (to take maximum advantage of available nutrients and light, resulting in maximum growth), harvested after short rotation periods (usually between 2-8 years), and coppicable (thus reducing establishment costs). In addition, willows and poplars demonstrate ease of vegetative propagation from dormant hardwood cuttings, a broad genetic base and ease of breeding. These characteristics make them ideal for growing in biomass systems and facilitate clonal selection and ensure great environmental adaptability [17].

2. Willow short-rotation coppice in Quebec

2.1. A brief history

Scientific interest in short-rotation bioenergy willows in Canada dates back to the mid-1970s' oil crisis, which stimulated the use of biomass for energy production. The Federal government's 1978 ENFOR (ENergy from the FORest) program, coordinated by the Canadian Forest Service was part of a federal interdepartmental initiative on energy research and development to promote projects in the forest bioenergy sector. Scientists from the Faculty of Forestry at the University of Toronto pioneered the investigation of willow's potential for bioenergy in Canada, convinced that willows could produce high annual yields in temperate zones [18-19] Louis Zsuffa's (1927-2003) work on selection and breeding of poplars and willows through genetic trials on small surfaces inspired the next generation of researchers, including one of his graduate students, Andrew Kenney, who implemented short-rotation intensive culture technology on the first prototype energy plantations in Canada [20]. As well, Gilles Vallée, of the Quebec ministry of Natural Resources, investigated the genetic improvement of hybrid poplar and willow with the aim of developing clones adapted to the shorter growing seasons of boreal forest locations. Our own *Institut de recherche en biologie végétale* (Plant Biology Research Institute), located at the Montreal Botanical Garden, grew out of the ENFOR program in the early 1990'. Our research team initially set out to identify willow species and clones well-adapted to short-rotation coppice in southern Quebec (Eastern Canada). Our experiments showed that Quebec's climate and soil are very favourable for growing various willow clones in short rotation, and that wastewater sludge can be an effective low-cost and environmentally-friendly fertilizer [21]. Researchers from Federal and provincial ministries also initiated diverse willow projects during the 1980s and 1990s, including the genetic improvement of hybrid poplar and willow clones adapted to the short growing seasons of boreal forests [22]. Simultaneously, Natural Resources Canada, a federal ministry, collaborated with several committees, including the International Energy Agency, to improve cooperation and information exchange between countries that have national programs in bioenergy research.

From the early 1990s to the present, dedicated, continuous research on willows in the Canadian context has been concentrated at the Montréal Botanical Garden. As a result of these extensive research efforts, approximately 300 ha of willows have been established on marginal agricultural lands in Quebec over the last 20 years.

2.2. Site selection

Several environmental factors can potentially influence a willow short-rotation coppice plantation and all should be evaluated prior to plantation establishment to maximize success. Ecologically, the majority of willow species are common in cold temperate regions and are adapted to mesic-hydric habitats. However, most riparian species require well-aerated substrate and flowing moisture, whereas non-riparian species have less exacting soil aeration requirements [23]. Moisture availability is an important factor determining native distribution in natural environments, successful plant establishment and high biomass yield. On average, willow coppice requires more water for growth than conventional agricultural crops [24] and consequently highly moisture retentive soil is an essential prerequisite. The lower St. Lawrence Valley, where most willow plantations in Quebec have been successfully established over the past two decades, is characterized by a temperate and humid climate with an annual average temperature of 6.4°C, average growing season (May-October) temperature of 15.8°C and a mean total annual precipitation of 970 mm. The period without freezing is on average 182 days and the total number of growing degree-days (above 5°C) is 2100.

Soil composition is another important factor for ensuring willow crop establishment and yield. In general, willow can be grown on many types of agricultural land. However, since this species is more water-dependent than other crops, particularly dry land should be avoided. On the other hand, although willow has been shown to be a rather flood-tolerant species compared to other woody energy crops [25], permanently submerged soils also constitute unsuitable sites. Ideally, willows should be grown on a medium textured soil that is aerated but still retains a good supply of moisture. Most willows grow best in loamy soils, with a pH ranging from 5.5 –7.0, although to a certain extent suitable soil types may range from fine sands to more compact clay soils. Several studies have shown that heavy clay soils are not very suitable for willows [26]. Most abandoned agricultural lands in Quebec are thus highly suited to growing willows, being situated in temperate regions and often adequately fertile. Other pre-establishment considerations are linked to the location of the plantation. Economical (and ecological) benefits can be maximized when high production levels of willows are achieved in combination with low input requirements, which result in high-energy efficiency and low environmental impact. For this reason, choosing the right location is crucial for achieving a sustainable energy production system. Normally, the plantation should be situated as close as possible to the end utilisation point (*e.g.* within 50-100 km from a power plant or transformation industry, etc.) and in any case should be established in proximity to main roads, highways or railroads. For the same reasons, the shape of willow fields should be as regular as possible to avoid loss of time and energy during management and harvest operations. For practical reasons (mainly linked to tillage and harvest) land with an elevated slope (>15%) should be avoided. Ideal sites are flat or with a slope not exceeding 7-8%.

2.3. Choice of planting material

Willow yield varies greatly depending on both environmental and genetic factors. The genus *Salix*, to which willows belong, comprises 330 to 500 species worldwide of deciduous or, rarely, semi-evergreen trees and shrubs [27] and the number and variety of species along with the ease of breeding have facilitated clonal selection adapted to several goals (ornamental, silvicultural, environmental applications, etc.). However, a large number of willow species are not suitable for biomass production because of their slower growth rate. Nowadays, the exploitation of the wide biological diversity within the genus *Salix* is focused primarily on a few species (*S. viminalis, S. purpurea, S. triandra, S. dasyclados, S. eriocephala, S. miyabeana, S. purpurea, S. schwerinii,* and *S. sachalinensis*), whereas there has been a recent increase in the number of selected intra- and interspecific hybrid cultivars offering higher yields, improved disease resistance and tolerance of a higher planting density (Table 2).

In Quebec, the first trials for evaluating willow biomass potential began on small plots in the early 1990s with two species, one indigenous (*S. discolor*) and the other a European cultivar (*S. viminalis* 5027). Two growing seasons after establishment, their total aboveground biomass yield was very similar – between 15 and 20 t ha^{-1} of dry-matter per year, confirming the high potential of these two species under Quebec's agro-ecological conditions [28]. A subsequent trial aimed at evaluating these two species comparatively with *S. petiolaris* Smith; both the first-tested species were shown superior to the latter in terms of biomass productivity [21]. However, since after a number of years this *S. viminalis* cultivar showed sensitivity to insect attacks, particularly to the potato leaf hopper, and since the risk of epidemic diseases increases as the plantation area expands, a new set of selected clones was investigated. These experiments showed that in contrast to *S. viminalis'* poor performance due to high sensitivity to pests and diseases, other willow cultivars (*S. miyabeana* SX64 and S. *sachalinensis* SX61) could achieve high biomass yields [29]. Now, 10 years later, *S. miyabeana* (SX64) and *S. sachalinensis* (SX61) cultivars still provide the highest biomass yield and greatest growth in diameter and height among willows in the Upper St. Lawrence region. However, selected cultivars from indigenous (*i.e.* North-American) willow species, especially *S. eriocephala* (cultivars S25 and S546) and *S. discolor* (cultivar S 365), perform well and only slightly below SX64, thus making them preferable for use on large-scale plantations in Quebec due to their less rigorous maintenance requirements and sensitivity to insect and pest attacks.

New selected planting material has also been made extensively available by several willow growers interested in development of willow cultivation in Quebec and operating jointly with researchers. Agro Énergie (www.agroenergie.ca) was the first large-scale commercial nursery in Quebec to produce diverse varieties of willow and has continued to expand its willow plantations across Eastern Canada. For the joint project between our research team and Agro Énergie, we provide scientific expertise in terms of plantation layout, species selection, cultivation methods and management practices. The 100 hectares of land provided by Agro Énergie represent an opportunity to scale up experimental technology, perfect techniques and evaluate costs and yield, using the high performance agricultural equipment necessary for large-scale commercial production.

Taxon	English common name	Origin	Comercial varieties and hybrids
S. nigra Marshall	Black willow	North America	S05*
S. triandra L.	Almond-leaved willow	Eurasia	Noir de Villaines+, P6010+,
S. alba L.	White willow	Europe, Africa, & west Asia	S44*
S. eriocephala Michx.	Heart-leaved willow	North America	S25*, S546*
S. discolor Muhl.	American pussy willow	North America	S365*¥
S. dasyclados Wimm.	Wooly-stemmed willow	Eurasia	SV1*¥
S. schwerinii Wolf	Schwerin willow	East Asia	
S. udensis (sin S. *sachalinensis*)Trautv.		East Asia	SX61*
S. viminalis L.	Common osier or basket willow	Eurasia	SVQ*, S33*,5027*, Jorr+
S. miyabeana Seemen	Miyabe willow	East Asia	SX64*, SX67*
S. purpurea L.	Purple willow or purple osier	Northern Africa & Europe	Fish Creek*
S. acutifolia Willd.	Pointed-leaf willow	Eastern Europe	S54*
S. sachalinensis x *S. miyabeana*			Sherburne*, Canastota*
S. purpurea x *S. miyabeana*			Millbrook*
S. eriocephala x S. interior			S625*
S. viminalis x S. schwerinii			Bjorn+, Tora+, Torhild+, Sven+, Olof+

Table 2. Most common *Salix* taxa and corresponding commercial varieties for biofuel production in Quebec (* Selected in North America; + Selected in Europe;¥ Its identity is currently under study).

2.4. Land preparation and weed control

Appropriate soil preparation is essential to ensure good plant establishment and vigorous growth. This is particularly true when willows are to be established on soil with low fertility or marginal land. The main goal of any land preparation operation should be to eliminate weeds, aerate soil and create a uniform soil surface for planting. Once the planting site has been chosen, the first operation to be performed is preparation of the land much as for any other agricultural crop. The productivity of trees under short-rotation intensive culture is strongly influenced by herbaceous competition. One of the first trials conducted by our research team in the early 1990s showed that weed suppression was essential to willow establishment [30]. On Quebec's generally well-drained lands, the most common weeds are broad-leaved annuals such as white goosefoot (*Chenopudium album* L.) and redroot pig-weed (*Amaranthus retroflexus* L.), whereas on poorly drained lands, annual grasses, barnyard grass (*Echinochloa crusgalli* L.) and perennials such as Canada thistle (*Cirsium arvense* L.) and quack grass (*Agropyron repens* (L.) Beauv.) are more common [30]. In the case of abandoned agricultural lands or in the presence of a high concentration of weeds, one or two applications of a systemic herbicide (*e.g.* glyphosate 2- 4 L/ha) during the summer of the year prior to planting are strongly recommended to promote establishment. A few weeks later, the destroyed plant mass should be incorporated into the soil using a rotating plough. In Quebec, a first ploughing should be performed in the fall prior to planting. Autumn ploughing allows the soil to break down over the winter, and also increases the amount of moisture in the planting bed. Suitable equipment includes any common mouldboard, chisel or disc plough (20 – 30cm depth), following usual agronomical practices for other crops (*e.g.* maize). Power harrowing (15- 18 cm depth) or cross disking of the site should be carried out in the spring immediately prior to planting to ensure a flat, regular planting bed.

2.5. Plantation design and planting

Willows can be planted according to two different layouts. In most North European countries (Sweden, UK, Denmark) and in the US, the most frequent planting scheme is the double row design with 0.75 m distance between the double rows and 1.5 m to the next double row, and a distance between plants ranging from 1 m to 0.4 m, corresponding to an initial planting density of 10,000 - 25,000 plants ha^{-1}. The most common plantation density in these countries is currently around 15,000 (1.5 x 0.75 x 0.59 m) plants ha^{-1} [31]. This rectangular planting arrangement is used to facilitate field machine manoeuvres through the plantation site. Tractors overlap the double row and the wheels run in the wider strips between those rows [32]. In Quebec, a simpler willow planting design, similar to that used for poplar in short rotations, has been in use since initial trials with only minimal modifications. It consists of a single row design ranging from 0.33 m between plants on a row and 1.5 m between rows (20,000 plants ha^{-1}) in the very first plantations, to 0.30 m on the row and 1.80 m between rows (18,000 plants ha^{-1}) in newer willow plantations. Theoretically, this design facilitates weed control during the establishment phase (the first three years), and consequently willow rooting and growth. In fact, the design choice depends mostly on machinery available for planting and harvesting, since it has been clearly demonstrated that planting design has less impact than plant density and cutting cycle on the yield of *Salix* plants, due to their ability to take advantage of the space available to each stool [32]. The choice of planting density must take into account other ecological factors as well. On sites with appropriate water supply, plantation establishment and subsequent biomass production depend largely on agronomic considerations such as plant spacing and harvesting cycles. Many studies have reported a correlation between spacing and harvesting cycles. In general, maximum yields are achieved early in dense willow plantations, but wider-spaced plantations ensure the highest long-term biomass yield [33-34]. On the other hand, under short harvesting cycles, willow stands have a shorter duration, as they are likely to be more exposed to pathogens [35]. At present, most willow short-rotation stands in Quebec have a plantation density of about 16,000 to 17,000 cuttings ha^{-1} and are harvested every two to three years.

Planting material consists of dormant willow stem sections, either rods or cuttings, depending on the planting machinery to be adopted. In some countries, for example in the UK and in the USA, 'step planters' are the most commonly used machines. Willow rods 1.5-2.5 m long are fed into the planter by two or more operators, depending on the number of rows being planted. The machine cuts the rods into 18-20 cm lengths, inserts these cuttings vertically into the soil and firms the soil around each cutting. Step planters have been calculated to cover 0.6 ha/hr in a UK study. [31]. In Quebec, the most common planting machine is a cutting planter that uses woody cuttings (20-25 cm long) and may operate on 3 rows simultaneously (Figure 1).

Figure 1. Willow planting machine operating on 3 rows simultaneously

Normally, a cutting planter inserts cuttings into the soil at a depth of about 18 cm. Based on empirical experience, this equipment can plant 3,600-4,000 cuttings per hour (1 ha of willow every 3-4 hours), although the duration of this operation may vary depending on several factors (site topography, soil type, plot shape, etc.). Planting material in Quebec is prepared by harvesting one-year-old stems (about 3 m long) in the autumn (*i.e.* when plants are dormant) of the year prior to planting. This material is wrapped in plastic film to avoid moisture loss, and stored in a refrigerator at -2 to -4°C. In spring, two to three weeks prior to planting, healthy willow rods 1-2 cm in diameter (with no symptoms of disease on bark or wood) are selected to prepare cuttings. Tips of stems bearing flower buds are first discarded. Then the rest of the whip is cut into 20-25 cm lengths using an adapted rotary saw and stored in boxes, ready to be planted (Figure 2).

Figure 2. Willow cuttings before planting

If cuttings are left in temperatures above 0°C, a break in their dormancy will occur, adventitious roots will develop and the buds may burst. This will lead to a reduction in water and nutrient content and consequently reduced viability. Thus, it is very important to plan the planting operation carefully in advance, calculating the number of cuttings that can be planted.

The time of planting varies according to meteorological and soil conditions. Planting should be undertaken as soon as possible in the spring, to allow plants to benefit from the high soil water content following snowmelt, and then to establish quickly and take maximum advantage of a long growing season. In addition, a late willow planting is also more subject to failure due to drought if a dry summer should occur. However, there are several additional factors that play an important role in determining the planting date. In order for soil preparation (*e.g.* harrowing) to begin in the spring, soil should be free from snow but not so muddy that soil structure could easily be damaged by tractors. The date at which such conditions are met vary considerably from year to year, but in southern Quebec, it usually falls during May, although late planting (up to mid-June) is possible and, in our experience, does not result in serious problems in plant establishment. Planting willow in the colder, northernmost regions of Quebec (*e.g.* Abitibi) may take place up to the beginning of July. In all of these situations, rapid colonisation by highly competitive weed species occurs on fertile sites, thus the use of appropriate residual herbicides is essential to maximize plant survival and early growth. Pre-emergence residual herbicide should be applied immediately upon completion of planting (within a maximum delay of 3-5 days). A mixture of two herbicides (2.30 kg Devrinol and 0.37 kg Simazine per hectare) has been effective on most of our plantations. Since the treatment must reach the zone of weed seed germination, most pre-emergence herbicides require mechanical incorporation (such as by a power tiller) as well as adequate irrigation or natural moisture (rainfall or snow) for best results. More recently, a new herbicde (SureGuard, a.i. flumioxazin) has received approval for pre-emergent use at the time of planting on poplar and willow (including planting stock production in the field, on both stoolbeds and bareroot beds).

2.6. Crop management

2.6.1. *Establishment year*

All operations carried out in a willow stand during the first year are aimed at promoting plant establishment and a high survival rate, thereby ensuring the on-going productive life of the plantation. Weeds are the main problem encountered in willow crop, and they may still colonise fields despite pre-emergence treatments. It was established decades ago that during the first year after planting, vigorous weeds reduce willow growth by between 50% and 90% [36]. Most of these invasive species have higher growth rates than young willow shoots, and compete with them mainly for light [37], and to a lesser extent for water and nutriments, leading to high plant mortality within the first few months. Hence, great care should be taken to control weed development in the field in the weeks following planting. On most willow plantations in Quebec, one to three passes with a rotary tiller cultivator between rows are needed to control weeds during the establishment year. In case of a severe weed problem, manual weeding may be required between plants within each row.

2.6.2. *Cutback*

There is much evidence that most newly-established willow plantations profit immensely from being cut back at the end of the first growing season (Figure 3).

Figure 3. After cutback willows sprout vigorously from the stumps

Not only does cutback encourage established cuttings to produce vigorous multiple shoots the following spring, it also helps reduce competition by weeds, thereby reducing the need for continued chemical weed control [38]. Furthermore, cutback facilitates entering the field at the beginning of the second growing season to fertilize and till soil between rows. Cutback is normally performed in the fall by cutting all newly-formed shoots at ground level using conventional agricultural equipment, such as reciprocating mowers for large surfaces or a trimmer/brush-cutter for small plots.

2.6.3. Fertilization

For many reasons, fertilization is a controversial aspect of short-rotation plantation, subject to fluctuations in practice. Our review of the historical evolution of willow short-rotation forestry in different countries suggests that the initially highly favourable attitude toward using chemical fertilizers has tended to attenuate over time, mainly because other issues beyond the biomass yield (both economical and environmental) have arisen. Different perspectives on this topic have also arisen out of legislation that in some countries has favored more environmental-friendly management (*e.g.* by reducing mineral fertilization and enhancing the application of biosolids and waste materials) of bioenergy cropping systems.

However, it is an irremediable fact that, due to high biomass yields, most willow energy crops grown in short-rotation and intensively managed and harvested remove nutrients at a high rate, though evidence varies somewhat (Table 3).

Annual nutrient removal (kg tDM^{-1})					
N	P	K	Ca	Mg	Reference
20.6	6.9	13.7	-	-	[39]
13.6	1.5	8.5	-	-	[40]
13.0	1.6	8.3	-	-	
6.3	1.0	7.5	-	-	[41]
5.7	1.0	3.0	3.0	1.0	[42]
5.3	0.9	3	7.2	0.7	
7.5	0.6	1.8	4.2	0.4	[43]
5.0	0.7	1.8	3.5	0.3	
3.9	0.5	1.5	3.6	0.2	[44]
3.5	0.5	2.5	-	-	[45]

Table 3. Average mass of nutrient removal (kg) per oven dry ton of aboveground willow biomass

Some authors have highlighted that N fertilization in willow plantations at the beginning of the cutting-cycle, excluding the year of planting, is generally a very efficient way to enhance plant growth [45-46]. On the other hand, willow nutrient requirements are relatively low, due to efficient recycling of N from litter and the relatively low nutrient content retained in biomass (stem). Therefore, much less nitrogen fertilizer should be applied than is typical with agricultural crops, although dosage should also be based on formal soil chemical analyses performed prior to plant establishment. Several authors have indicated that no nitrogen is required in the planting year for short-rotation coppice [39-47]. This also reduces the competitiveness of weeds that would take advantage of fertilizer application. Economical considerations are yet another factor to consider when determining the dose of fertilizer to be used, since fertilizer constitutes a significant percentage of the financial cost involved in the production of willow biomass crops. A recent study conducted in New York

State showed that fertilizer represents up to 10–20% of the cost of production over several rotations [48]. The average dose generally recommended in Quebec ranges from the equivalent of 100-150 kg N, 15 kg – 40 kg P and around 40 kg K per hectare per year after the establishment year. Because it is not possible to introduce heavy equipment into the field after plantation establishment, fertilizer application is normally performed one year after planting and after any harvest, when tractors can circulate freely in the field.

An interesting alternative to mineral fertilizers are biosolids and other industrial and agricultural byproducts, which have been tested in many countries since the early 1990s. These include municipal wastewater [49], wastewater from the dairy industry, landfill leachate [50], diverted human urine [51], industrial wastewaters such as log-yard runoff [52], as well as solid wastes like digested or granulated sludge [53] and pig slurry [54]. In fact, the majority of these products contain high levels of nitrogen and phosphorous, elements that might constitute a source of pollution for the environment but at the same time represent a source of nutrients for the plant. Thus there are many advantages to using such products in willow plantations:

1. recycling of nutrients, thereby reducing the need for farmers to invest in chemical fertilizer;
2. conservation of water;
3. prevention of river pollution, canals and other surface water, into which wastewater and sewage sludge would otherwise be discharged;
4. low-cost, hygienic disposal of municipal wastewater and sludge.

Willow cultivated in short rotation is a very suitable crop for fertilization with these products for several reasons. First, it has been determined, both by measured and estimated models, that this crop has high evapotranspiration rates and thereby consumes water quantities as high as any other vegetation cover, which allows significant wastewater disposal over each growing season [24-55-56]. Furthermore, willow short-rotation stands have been shown to be able to uptake large amounts of nutrients present in this waste [57]. Last but not least, willow coppice is a no food no fodder crop and, if properly handled, any possible source of human or environmental contamination is strongly reduced [58]. In some early trials carried out in Quebec to test the possibility of using sludge in willow short-rotation culture, it was found that a moderate dose of dried and palletized sludge (100-150 kg of "available" N ha^{-1}) might constitute a good fertilizer during the establishment of willows, especially on clay sites [53-59]. Today, the recommended dose of derived wastewater sludge fertilizer in Quebec ranges between 18-21 t $ha^{-1}$of dried material, which corresponds to 100-120 kg available nitrogen per hectare. Fertilization is performed in spring of the second year after planting with ordinary manure spreading machines. Another recent project investigated the effect of the use of pig slurry as fertilizer on the productivity of willow in short-rotation coppice (Figure 4).

The results showed that pig slurry is good fertilizer for willow plantations [54]. In fact, very high biomass yields were obtained over two years, and even made it possible to predict that typical three-year rotation cycles could be reduced to two years, under the proper

Figure 4. Pig slurry application to a willow plantation

production conditions. This means that even though nitrogen in slurry may be less efficient than that in a mineral fertilizer, a significant reduction in the production costs of willow-based biomass as well as recycling of a greater quantity of slurry can be achieved simultaneously [54].

2.7. Pests and diseases

Although there are a great number of insects feeding on willows, three main species are of concerns for willow short rotation coppice in Quebec. The first is the willow leaf beetle (*Plagiodera versicolora* Laicharteg.), one of the most common insects found on willows. The willow leaf beetle is a small (4 - 6 mm long), metallic-blue beetle widely distributed around the world. In Quebec, adults emerge from their overwintering quarters under the loose bark and feed on young willow foliage in spring. Egg laying begins in mid-June. Females lay yellow eggs grouped on the undersides of the leaves. The young larvae emerge a few days later and begin feeding on both sides of the leaves and eating the tissue between the veins, thus skeletonizing the leaves and, depending on the extent of the attack, in all probability leading to a reduction of plant growth. In Quebec, this insect has been frequently observed feeding on leaves of clones of *Salix viminalis* and to a much lesser extent on most common commercial varieties of *S. miyabeana* (SX64 and SX67) and *S. sachalinensis* (SX61). To date, the reported threshold of damage caused by this insect has never been high enough to justify any type of control. However, in case of severe attack, non-toxic products based on *Bacillus thuringiensis*, shown to be effective in eliminating this pathogen, can be used [60].

The other predominant insects found feeding on willow trees and shrubs are two aphid species: the giant willow aphid, *Tuberolachnus salignus* (Gmelin) and the black willow aphid, *Pterocomma salicis* (L) [61].

The giant willow aphid. is one of the largest aphids ever recorded, measuring up to 5.8 mm in length [62]. It feeds almost exclusively on willow, but has very occasionally been recorded on poplar (*Populus* spp.). The species is strongly aggregative, forming vast colonies on infested trees. These colonies can cover a significant portion of the 1-3 year old stem surface of a willow tree. Laboratory experiments with willows grown in soil and in hydroponic culture have revealed that this species can reduce the above-ground yield of biomass willows, have severe negative effects on the roots and reduce the survival of both newly planted and established trees [63]. Other preliminary studies carried out in the UK have shown that this insect's feeding behavior is affected by chemical cues from the host. Researchers found that one of its most preferred willows was *S. viminalis* [64]. Although large colonies of this insect have recently been found on several willow varieties in Quebec, it is not yet possible to estimate its threat to willow plantations in this region (Figure 5).

Figure 5. Giant aphids feeding on willow. This insect is often found forming large colonies at base of the stem.

The black willow aphid, *Pterocomma salicis* (L) may actually pose a threat only if severe, frequent attacks occur. Several studies have shown that this species is less damaging than the giant willow aphid, with a less persistent negative impact on willow growth. In Quebec, high density populations of this species have recently been found at the end of June on a willow plantation in the upper St. Lawrence River valley (Huntingdon), mainly on *S. miyabeana* (SX67 and SX64); it did not seem to feed on *S. viminalis.*

Other less damaging insects have been found on willow plantations in Quebec. *Calligrapha multipunctata bigsbyana* adults and larvae may feed on willow leaves without destroying leaf veins, with consequences quite similar to those of *Plagiodera versicolora.* Willow flea beetles of the genus *Crepidodera* (*C. nana* and *C. decora*also feed on *Salicaceae* leaves [65], and are easy to recognize by their brilliant metallic and bicoloured upper surface; blue or green head and pronotum tinged with strong bronze, copper or violet; and unicolorous blue or green elytra. This beetle feeds on either the upper or lower leaf surface, consuming the epidermis and tissue below, but not on the opposite side. After desiccating, the tissue falls out, resulting in a leaf with a bullet-hole appearance. Varieties of willows developed in Europe, based on pedigrees with *Salix viminalis* or *S. viminalis* x *S. schwerinii,* are susceptible to potato leafhopper (*Empoasca fabae* Harris), which causes serious damage to this species and its cultivars or hybrids. Willow shoot sawfly (*Janus abbreviates* Say) larvae have recently been found in Quebec, carving deep tunnels on young willow *S. miyabeana* SX64 shoots where they cause wilting, change of colour (brown or black) and eventually drooping of shoot tips. It has been observed that in some cases 30% of individuals of SX64 in Huntingdon showed at least one shoot affected by this insect. However, only repeated and severe attacks in young willow plantations may adversely affect tree growth.

Willow can be injured by several diseases [66]. Willow leaves may be sensitive to *Alternaria* spp., *Melamsora* spp. and *Venturia* spp., whereas *Cryptodiaporthe* spp., *Glomerella* spp. and *Valsa* spp. are found to affect stems and branches and *Armillaria* spp., *Fusarium* spp. and *Verticilium* spp. roots [67]. However, the most widespread, frequent and damaging disease in willow plantations is leaf rust, caused by *Melampsora* spp. In northern Europe, leaf rust is considered a major factor limiting growth of short-rotation coppice willow [68]. It can cause premature defoliation, poor cold acclimation, premature leaf senescence, and a predisposition to abiotic stress (*e.g.*, competition and drought) in host trees, along with secondary disease organisms, and it may reduce yields by as much as 40% [69]. One of the main alternative solutions to spraying fungicides proposed in northern Europe is growing willow in inter- and intra-species mixtures [70]. If a variety dies out of a mixture due to disease, competition or some other factor, the remaining varieties can compensate for the loss [71]. In some willow plantations in Quebec, severe attacks of *Melamsora* spp. have been detected mainly on a specific commercial clone S301 (*S. interior* 62 x *S. eriocepala* 276), which seemed to be more vulnerable to rust than any other clone studied in the area [29]. Few rust attacks have been reported for most commercial clones, however, chemical or biological disease control is generally not required.

2.8. Harvesting and yields

Willow should be harvested at the end of each rotation cycle (2-5 years), normally in fall, after leaf shedding. All willow stems should be cut at a height of 5 - 10 cm above the soil surface in order to leave a stump from which new buds will form sprouts the following spring. Essentially, there are three ways to harvest willows, the choice largely depending on the final destination of biomass and the equipment available. When willows are grown to produce rods to be used in environmental engineering structures such as sound barriers, snow fences and wind breaks along highways and streets [72-73] or to produce new cuttings, plants are harvested with trimmer brush-cutters. Whole willow rods can also be stored in heaps at the edge of the field and chipped after drying.

Another option involves the use of direct-chip harvesting machines (*e.g.* Class Jaguar and Austoft). This technique uses modified forage harvesters specifically designed to harvest and direct chip willow stems: the stems are cut, chipped and dropped into a trailer either driven parallel to the harvester or connected directly to it. Although this harvest model is very economically efficient and recommended in many countries, it also presents several disadvantages that should be carefully evaluated. Willow biomass has a moisture content of 50-55% (wet basis) at harvest. Consequently, storage and drying of the freshly chipped wood may cause problems. It has been shown that stored, fresh wood chip in piles can heat up to 60°C within 24 hours and start to decompose. Biomass piles require careful management because internal fermentation can cause combustion and the high level of fungi spore production can lead to health problems for operators. Decomposition processes cause a loss of biomass of up to 20% and a significant reduction in calorific value (*i.e.* energy value) of the biomass [74]. Thus, this type of harvest system requires infrastructures to mechanically dry the biomass (*e.g.* ventilation, heating, mixing machinery) and these post-harvest operations increase the production cost. Alternatively, the freshly chipped material should be delivered to heating plants as soon as possible.

A third harvest system recently developed in Canada, mainly adapted to willow short-rotation coppice, is a cutter-shredder-baler machine that performs light shredding and bales willow stems [22], producing up to 40 bales hr^{-1} (20 t hr^{-1}) on willow plantations (Figure 6).

Figure 6. Willow cutter-shredder-baler harvester operating in Quebec

The main advantage is that, since bales can be left to dry before being chipped, the risks linked to handling wet biomass are reduced [75]. In Quebec, willow biomass harvest is usually done in fall after leaf shedding.

As with any other agricultural crop, biomass yield of willow short-rotation coppice depends on many co-occurring factors including cultivar, site, climate and management operations. Soil type, water availability, and pest and weed control also affect yield. Data from existing commercial sites in the UK suggest that average yields of around 8-10 odt $ha^{-1}yr^{-1}$ are representative of plantations using older cultivars, whereas biomass yields as high as 15-18 odt $ha^{-1}yr^{-1}$ can be obtained by using selected genetic material [31]. In other northern European countries, an average annual growth of 15–20 odt $ha^{-1}yr^{-1}$ has been observed in early experiments [76], although more recent figures suggest that an average of 10 odt $ha^{-1}yr^{-1}$ is more realistic [77]. Experimental yields of short-rotation willow ranging from 24 to 30 oven dry tonnes (odt) ha^{-1} yr^{-1} have been measured in the US and Canada [43-44], although typical yields are more often in the range of 10 to 12 odt ha^{-1} yr^{-1} [78].

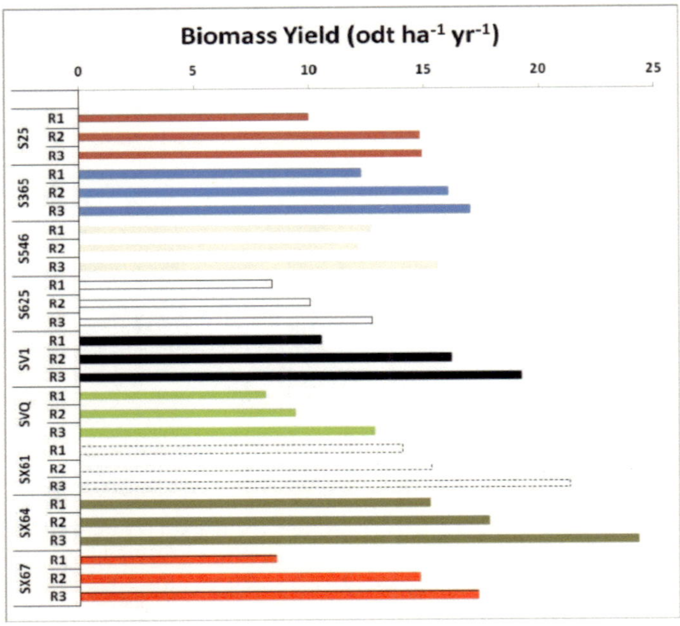

Figure 7. Average biomass yield for nine willow cultivars during three successive rotations (10 years) in the Upper St. Lawrence region (Quebec) on former farmlands. Clones SX64 and SX61 along with some indigenous species (S25, S365, S546) are the most productive and thus are considered to be very suitable for short-rotation forestry in southern Quebec.

Long-term trials show that under southern Quebec's pedoclimatic conditions, short-rotation willow coppice can provide high biomass yields over many years, although results vary according to variety. In one clonal test for instance, at the end of the third (3-years) rotation cycle, the most productive willow cultivars were SX64 (19 Odt ha^{-1} yr^{-1}) and SX61 (17 Odt ha^{-1} yr^{-1}) (Figure 7). Also, indigenous (*i.e.* North-American) willow cultivars, especially *S. eriocephala* (S25 and S546) and *S. discolor* (S 365) cultivars, show high biomass potential (13 - 15 Odt ha^{-1} yr^{-1}). A scientific follow up of an old willow plantation established in Huntingdon in southern Quebec (Canada), showed that willows were still able to maintain a high level of productivity after five coppicing cycles. Plants can remain vigorous and produce high yields (14 Odt ha^{-1} yr^{-1}) even after 18 years of cultivation (Table 4). This represents a very important demonstration of the viability of long-term economic exploitation of willows.

Rotation	Average biomass yield Total (Odt ha^{-1})	Annual (Odt $ha^{-1} yr^{-1}$)
First (1195-1997)	45.3	15.1
Second (1998-2001)	88.1	22
Third (2002-2004)	51.7	17.2
Fourth (2005-2008)	67.4	16.9
Fifth (2009-2011)	42	14

Table 4. Average biomass yield for *Salix viminalis* L. (clone 5027) achieved during five successive rotations in southern Quebec (Canada)

3. Perspectives for future research: The use of willows in phytoremediation

In Canada, it is estimated that millions of hectares of arable land lie uncultivated. These so-called marginal lands tend to be less productive, less accessible, poorly drained, or even contaminated [79]. Willows have been successfully used to capture leached nutrient and heavy metals from soils [54, 59, 80, 81]. The various species of *Salix* have been shown to establish well on these marginal and contaminated soils, which provides new research opportunities for future applications.

3.1. Phytoremediation

The main types of contaminants found in Quebec soils are petroleum products and heavy metals [82]. In many urban areas, past industrial activities have resulted in thousands of contaminated sites that require decontamination prior to any further utilization. Estimates by the province's ministry of environment have shown that, in the region of Montreal alone, there are over 1350 contaminated sites of which only 54% are in the process of being rehabilitated by traditional methods [83]. Current decontamination methods imply the excavation of the contaminated soils, transport to a landfill treatment facility followed by chemical cleaning, vitrification, incineration or dumping; these steps are extremely expensive [84]. Plant-based *in situ* decontamination technologies, *i.e.* phytoremediation, represent a cost-effective alternative [84]. Plants have the capacity to accumulate, translocate, concentrate, or degrade contaminants in their tissues. Phytoremediation takes advantage of the microbial communities (bacteria and fungi) present in soils to increase the potential of plants to uptake pollutants from the soil matrix. Willows are among the species most widely used for phytoremediation, given their diversity and tolerance of high levels of contaminants [85]. Also, willows develop an extensive root system that stimulates rich and diverse microbial communities that are involved in the degradation of organic pollutants, These characteristics, combined with exceptionally high biomass production, make them very suitable for phytoremediation [86].

Phytoremediation using willows is becoming an increasingly popular alternative approach to decontamination, and several studies and pilot projects are underway. Willows have been used successfully to treat highly toxic organic contaminants such as PCBs, PAHs, and nitro-aromatic explosives [87]. Similarly, willows, in particular *S. viminalis* and *S. miyabeana*, have been shown to accumulate Cd and Zn in their stems and leaves while sequestering Cu, Cr, Ni and Pb in their roots [85,88,89,90]. In previous studies, the efficiency of willows in short-rotation intensive plantation for the elimination of heavy metals contained in wastewater sludge has been investigated [28, 59, 90]. We have also found that willow may be useful for improving sites polluted by mixed organic-inorganic pollution [91] (Figure 8).

Figure 8. Phytoremediation using willows on a former oil refinery around Montreal

Although the fast-growing perennial habits of short-rotation coppice willow planted at high densities result in a low concentration of metals accumulated in biomass after one year of growth, the high biomass production of *Salix* spp. over several harvesting cycles (2-3 years) allows them to accumulate large quantities of metals over the long-term, suggesting great potential as a phytoremediation tool.

3.2. Genetic improvement of willow for phytoremediation

Historically, most genetic selection to improve willow germplasm has been oriented toward increased capacity for biomass production [92], adapted to temperate climates and resistant to pathogens. However, in the context of phytoremediation, the ideal willow genotype must also: i) be adapted to specific pedo-climatic conditions; ii) be fast growing; ii) produce a large root biomass; iv) be resistant to a variety of contaminants; v) have a high concentration factor of contaminants; vi) be easy to establish, maintain and collect. The exceptional diversity of the genus *Salix* makes it an ideal candidate for breeding programs seeking to develop cultivars more efficient at phytoremediation.

To our knowledge, one of the rare efforts to understand the genetic and genomic bases underlying the potential of willow for phytoremediation is the three-year Genorem project (www.genorem.ca) launched by research teams at the Université de Montréal and McGill University (Project Leaders Dr. B. Franz Lang and Dr. Mohamed Hijri, both of the *Université de Montréal*) and involving over thirty scientists, students and staff. The project integrates traditional field and molecular biology experiments, employing recently developed life science technologies: genomics, proteomics, metabolomics and bioinformatics. GenoRem's objectives include the development of guidelines for phytoremediation procedures respectful of the environment that will ultimately be useful to both government and corporate sectors. The transcriptomes of 11 willow genotypes will be sequenced, resulting in basic molecular information about the genes activated in willow when in presence of soil contaminants. GenoRem will also investigate the close relationship established between the willow cultivars studied and the associated soil microorganisms. Ultimately, project results will provide willow breeders with gene markers linked with increased phytoremediation potential.

Phytoremediation as a decontamination technology can be applied to large surface areas, causes less environmental disturbances and represents a significantly cheaper approach than traditional methods. However, treatment is lengthy (several years), and the methodologies appropriate for each type of contamination require refinement. While the biomass produced in the context of a phytoremediation project may potentially be contaminated, this does not affect its utilization as a product outside the food chain. Moreover, the highly concentrated ashes resulting from conversion of the biomass to fuel facilitate disposal and treatment of the contaminant, particularly for a large, diluted volume of contaminated soil. Hence the decontamination by means of phytoremediation is a less intensive technique.

4. Conclusions

Eastern Canada is one region where willow short-rotation coppice has been the focus of numerous research projects over the last 15-20 years. Most experimental data published during this period concerning Quebec have found a high biomass potential, due to a combination of several factors, including the very high biomass yield of certain willow varieties, favourable pedoclimatic conditions and the very low incidence of severe pests and diseases. These high biomass yields have encouraged some growers to choose willows as an alternative agricultural crop, leading to a dramatic expansion of land devoted to willow short-rotation coppice in the province, especially over the last five years. However, the future evolution of this crop's production will most certainly be influenced by the development of an active market for such biomass, which would encourage farmers to grow willow over a much larger surface area. In particular, developments in the technology of feedstock transformation and marketing issues related to product potential both merit further study. The high potential of willow for bioenergy production and environmental applications, including phytoremediation, in the Quebec context has been clearly demonstrated.

Author details

Werther Guidi, Frédéric E. Pitre and Michel Labrecque
Institut de Recherche en Biologie Végétale (IRBV – Plant Biology Research Institute) – Université de Montréal – The Montreal Botanical Garden, Montréal, Canada

5. References

[1] Fargione, J., Hill, J., Tilman, D., Polasky, S., Hawthorne, P., 2008. Land Clearing and the Biofuel Carbon Debt. Science 319, 1235-1238

[2] FAO, 2012. Impacts of Bioenergy on Food Security – Guidance for Assessment and Response at National and Project Levels., Environment and Natural Resources Working Paper, Rome, p. 64.

[3] Tilman, D., Socolow, R., Foley, J.A., Hill, J., Larson, E., Lynd, L., Pacala, S., Reilly, J., Searchinger, T., Somerville, C., Williams, R., 2009. Beneficial Biofuels—The Food, Energy, and Environment Trilemma. Science 325, 270-271.

[4] Nassi o Di Nasso, N., Guidi, W., Ragaglini, G., Tozzini, C., Bonari, E., 2010. Biomass production and energy balance of a twelve-year-old short-rotation coppice poplar stand under different cutting cycles. Global Change Biology Bioenergy 2, 89–97.

[5] Sartori, F., Lal, R., Ebinger, M.H., Parrish, D.J., 2006. Potential Soil Carbon Sequestration and CO2 Offset by Dedicated Energy Crops in the USA. Critical Reviews in Plant Sciences 25, 441-472.

[6] Londo, M., Dekker, J., Ter Kreus, W., 2005. Willow short-rotation coppice for energy and breeding birds: An exploration of potentials in relation to management. Biomass and Bioenergy 28, 281-293.

[7] Bellamy, P.E., Croxton, P.J., Heard, M.S., Hinsley, S.A., Hulmes, L., Hulmes, S., Nuttall, P., Pywell, R.F., Rothery, P., 2009. The impact of growing miscanthus for biomass on farmland bird populations. Biomass and Bioenergy 33, 191-199.

[8] Dickmann, D., 2006. Silviculture and biology of short-rotation woody crops in temperate regions: Then and now. Biomass and Bioenergy 30, 696-705.

[9] Drew, A.P., Zsuffa, L., Mitchell, C.P., 1987. Terminology relating to woody plant biomass and its production. Biomass 12, 79-82.

[10] McAlpine, R., Brown, C., Herrick, A., Ruark, H., 1966. ''Silage''sycamore. Forestry Farmer 26 7-16.

[11] Herrick, A.M., Brown, C.L., 1967. A new concept in cellulose production: silage sycamore. Agricultural Science Review 5, 8-13.

[12] Auclair, D., Bouvarel, L., 1992. Intensive or extensive cultivation of short rotation hybrid poplar coppice on forest land. Bioresource Technology 42, 53-59.

[13] Kenney, W., Sennerby-Forsse, L., Layton, P., 1990a. A review of biomass quality research relevant to the use of poplar and willow for energy conversion. Biomass 21, 163-188.

[14] Sims, R.E.H., Senelwa, K., Maiava, T., Bullock, T., 1999. Eucalyptus for energy in New Zealand-Part II: coppice performance. Biomass and Bioenergy 17, 333-343.

[15] Grünewald, H., Böhm, C., Quinkenstein, A., Grundmann, P., Jörg Eberts, J., von Wühlisch, G., 2009. Robinia pseudoacacia L.: A Lesser Known Tree Species for Biomass Production. Bioenergy Research 2, 123-133.

[16] Rytter, L., Šlapokas, T., Granhall, U., 1989. Woody biomass and litter production of fertilized grey alder plantations on a low-humified peat bog. Forest Ecology and Management 28, 161-176.

[17] Keoleian, G.A., Volk, T.A., 2005. Renewable energy from willow biomass crops: Life cycle energy, environmental and economic performance. Critical Reviews in Plant Sciences 24, 385-406.

[18] Grislis K, Labrecque L., 2009. Proliferating Willow for Biomass. Silviculture Magazine summer 2009, 12-15.

[19] Mosseler, A., 1990. Hybrid performance and species crossability relationships in willows (*Salix*). Canadian Journal of Botany 68, 2329-2338.

[20] Kenney, W.A., Sennerby-Forsse, L., Layton, P., 1990b. A review of biomass quality research relevant to the use of poplar and willow for energy conversion. Biomass 21, 163-188.

[21] Labrecque, M., Teodorescu, T., Cogliastro, A., Daigle, S., 1993. Growth patterns and biomass productivity of two Salix species grown under short-rotation intensive culture in southern Quebec. Biomass and Bioenergy 4, 419-425.

[22] Lavoie, F., Savoie, P., D'Amours, L., Joannis, H., 2008. Development and field performance of a willow cuttershredder-baler. Applied Engineering in Agriculture 24, 165-172.

[23] Dickmann, D.I., Kuzovkina, J., 2008. Poplars and willows of the world, with emphasis on silviculturally important species. FAO Rome, Italy

[24] Persson, G., 1995. Willow stand evapotranspiration simulated for Swedish soils. Agricultural Water Management 28, 271-293.

[25] Jackson, M., Attwood, P., 1996. Roots of willow (Salix viminalis L) show marked tolerance to oxygen shortage in flooded soils and in solution culture. Plant and Soil 187, 37-45.

[26] Ledin, S., 1996. Willow wood properties, production and economy. Biomass and Bioenergy 11, 75-83.

[27] Argus, G.W., 1999. Classification of Salix in the New World. Botanical Electronic News 227, http://www.ou.edu/cas/botany-micro/ben227.html.

[28] Labrecque, M., Teodorescu, T., Daigle, S., 1997. Biomass productivity and wood energy of salix species after 2 years growth in SRIC fertilized with wastewater sludge. Biomass and Bioenergy 12, 409-417.

[29] Labrecque, M., Teodorescu, T.I., 2005b. Field performance and biomass production of 12 willow and poplar clones in short-rotation coppice in southern Quebec (Canada). Biomass and Bioenergy 29, 1-9.

[30] Labrecque, M., Teodorescu, T.I., Babeux, P., Cogliastro, A., Daigle, S., 1994. Impact of herbaceous competition and drainage conditions on the early productivity of willows under short-rotation intensive culture. Canadian Journal of Forest Research 24, 493-501.

[31] Defra, 2004. Growing Short Rotation Coppice - Best Practice Guidelines For Applicants to Defra's Energy Crops Scheme. London (UK).

[32] Bergkvist, P., Ledin, S., 1998. Stem biomass yields at different planting designs and spacings in willow coppice systems Biomass and Bioenergy 14, 149-156

[33] Mitchell, C., 1995. New cultural treatments and yield optimization. Biomass and Bioenergy 9, 11-33.

[34] Proe, M., Craig, J., Griffiths, J., 2002. Effects of spacing, species and coppicing on leaf area, light interception and photosynthesis in short rotation forestry Biomass and Bioenergy 23, 315-326

[35] Sennerby-Forsse, L., Ferm, A., Kauppi, A., 1992. Coppicing ability and sustainability. In: Mitchell, C., Ford-Robertson, J., Hinckley, T., Sennerby-Forsse, L. (Eds.), Ecophysiology of short rotation forest crops. Elsevier, London, pp. 146–184.

[36] Willoughby, I., Clay, D.V., 1996. Herbicides for Farm Woodlands and Short Rotation Coppice. London.

[37] Sage, 1999. Weed competition in willow coppice crops: the cause and extent of yield losses. Weed Research 39, 399-411.

[38] Mitchell, C.P., 1992. Ecophysiology of short rotation forest crops. Biomass and Bioenergy 2, 25-37.

[39] Kopp, R.F., White, E.H., Abrahamson, L.P., Nowak, C.A., Zsuffa, L., Burns, K.F., 1993. Willow biomass trials in Central New York State. Biomass and Bioenergy 5, 179-187.

[40] Potter, C.J., 1990. Coppiced trees as energy crops. Final report to ETSU for the DTI on contract ETSU B 1078.

[41] Caslin, B., Finnan, J., McCracken, A.R., 2010. Short Rotation Coppice Willow Best Practice Guidelines.

[42] Hytönen, J., 1995. Ten-year biomass production and stand structure of Salix [`]aquatica' energy forest plantation in Southern Finland. Biomass and Bioenergy 8, 63-71.

[43] Labrecque, M., Teodorescu, T., 2003. High biomass yields achieved by Salix clones in SRIC following two 3-years coppice rotations on abandoned farmland in southern Quebec, Canada. Biomass and Bioenergy 25, 135-146.

[44] Adegbidi, H., Volk, T., White, E., Abrahamson, L., Briggs, R., Bickelaupt, D., 2001. Biomass and nutrient removal by willow clones in experimental bioenergy plantation in New York State. Biomass and Bioenergy 20, 399-411.

[45] Ericsson, T., 1994. Nutrient cycling in energy forest plantations. Biomass and Bioenergy 6, 115-121.

[46] Alriksson, B., Ledin, S., Seeger, P., 1997. Effect of nitrogen fertilization on growth in a Salix viminalis stand using a response surface experimental design. Scandinavian Journal of Forest Research 12, 321-327.

[47] Ledin, S., Willebrand, E., 1996. Handbook on How to Grow Short Rotation Forests. Swedish University of Agricultural Sciences, Department of Short Rotation Forestry, Uppsala.

[48] Buchholz, T., Volk, T., 2011. Improving the Profitability of Willow Crops—Identifying Opportunities with a Crop Budget Model. BioEnergy Research 4, 85-95.

[49] Perttu, K., 1999. Environmental and hygienic aspects of willow coppice in Sweden. Biomass and Bioenergy 16, 291-297.

[50] Martin, P., Stephens, W., 2006. Willow growth in response to nutrients and moisture on a clay landfill cap soil. I: Growth and biomass production. Bioresource Technology 97, 437-448.

[51] Larsson, S., 2003. Short rotation Willow biomass plantation irrigated and fertilized with wastewaters- Results form a 4-year multidisciplinary field project in Sweden, France, Northern Ireland and Greece supported by the EU-FAIR Programme (FAIR5-CT97-3947) Final Report. Svalov, Sweden.

[52] Jonsson, M., Dimitriou, I., Aronsson, P., Elowson, T., 2006. Treatment of log yard run-off by irrigation of grass and willows. Environmental Pollution 139, 157-166.

[53] Labrecque, M., Teodorescu, T., 2001. Influence of plantation site and wastewater sludge fertilization on the performance and foliar nutrient status of two willow species grown under SRIC in southern Quebec (Canada). Forest Ecology and Management 150, 223-239.

[54] Cavanagh, A., Gasser, M.O., Labrecque, M., 2011. Pig slurry as fertilizer on willow plantation. Biomass and Bioenergy 35, 4165-4173.

[55] Hall, R., Allen, S., Rosier, P., Hopkins, R., 1998. Transpiration from coppiced poplar and willow measured using sap-flow methods. Agricultural and Forest Meteorology 90, 275-290.

[56] Guidi, W., Piccioni, E., Bonari, E., 2008. Evapotranspiration and crop coefficient of poplar and willow short-rotation coppice used as vegetation filter Bioresource Technology 99, 4832-4840.

[57] Aronsson, P., Bergstrom, L., 2001. Nitrate leaching from lysimeter- grown short rotation willow coppice in relation to N- application, irrigation and soil type. Biomass and Bioenergy 21, 155-164.

[58] Carlander, A., Schönning, C., Stenström, T.A., 2009. Energy forest irrigated with wastewater: a comparative microbial risk assessment. Journal of Water and Health 7 413-433

[59] Labrecque, M., Teodorescu, T., Daigle, S., 1995. Effect of wastewater sludge on growth and heavy metal bioaccumulation of two *Salix* species. Plant and Soil 171, 303-316.

[60] Bauer, L.S., 1992. Response of the Imported Willow Leaf Beetle to Bacillus thuringiensis var. san diego on Poplar Willow. Journal of Ivertebrate Pathology 59, 330-331.

[61] Sage, R.B., Tucker, K., 1997. Invertebrates in the canopy of willow and poplar short rotation coppices. Aspects of Applied Biology 49, 105-112.

[62] Blackman, R.L., Eastop, V.F., 1994. Aphids on the world's trees:an identification and information guide.

[63] Collins, C.M., Rosado, R.G., Leather, S.R., 2001. The impact of the aphids Tuberolachnus salignus and Pterocomma salicis on willow trees. Ann. Appl. Biol. 138, 133-140.

[64] Aradottir, G.I., Karp, A., Hanley, S., Shield, I., Woodcock, C.M., Dewhirst, S., Collins, C.M., Leather, S., Harrington, R., 2009. Host selection of the giant willow aphid (Tuberolachnus salignus). Proceedings of the 8th International Symposium on Aphids. REDIA, XCII, , 223-225.

[65] Seago, A., Lingafelter, S.W., 2003. Discovery of Crepidodera Bella Parra (Coleoptera: Chrysomelidae: Alticini) in Maryland and redescription. Journal of the New York Entomological Society 111, 51-56.

[66] Pei, M.H., McCracken, A.R. (Eds.), 2005. Rust Diseases of Willow and Poplar CABI Publishing, CAB International Wallingford, Oxfordshire OX10 8DE UK.

[67] Vujanovic, V., Labrecque, M., 2002. Biodiversity of pathogenic mycobiota in Salix bioenergy plantations, Québec. Canadian Plant Disease Survey 82 138 -139.

[68] Pei, M., Lindegaard, K., Ruiz, C., Bayon, C., 2008. Rust resistance of some varieties and recently bred genotypes of biomass willows. Biomass and Bioenergy 32, 453-459

[69] Parker, S.R., Pei, M.H., Royle, D.J., Hunter, T., Whelan, M.J., 1995. Epidemiology, population dynamics and management of rust diseases in willow energy plantations. Final Report of Project ETSU B/W6/00214/REP. Energy Technology Support Group, Department of Trade and Industry, UK.

[70] McCracken, A.R., Dawson, W.M., 1997. Growing clonal mixtures of willow to reduce effect of Melampsora epitea var. epitea. European Journal of Forest Pathology 27, 319-329.

[71] McCracken, A.R., Dawson, W.M., 2003. Rust disease (Melampsora epited) of willow (Salix spp.) grown as short rotation coppice (SRC) in inter- and intra-species mixtures. Ann. Appl. Biol. 143, 381-393.

[72] Labrecque, M., Teodorescu, T., 2005a. Preliminary evaluation of a living willow (Salix spp.) sound barrier along a highway in Québec, Canada. Journal of Arboriculture 31, 95-98.

[73] Teodorescu, T.I., Guidi, W., Labrecque, M., 2011. The use of non-dormant rods as planting material: A new approach to establishing willow for environmental applications. Ecological Engineering 37, 1430-1433.

[74] Jirjis, R., 1995. Storage and drying of wood fuel. Biomass and Bioenergy 9, 181-190.

[75] Gigler, J.K., van Loon, W.K.P., van den Berg, J.V., Sonneveld, C., Meerdink, G., 2000. Natural wind drying of willow stems. Biomass and Bioenergy 19, 153-163.

[76] Ceulemans, R., McDonald, A., Pereira, J., 1996. A comparison among eucalyptus, poplar and willow characteristics with particular reference to a coppice, growth-modelling approach. Biomass and Bioenergy 11, 215-231.

[77] Mola-Yudego, B., Aronsson, P., 2008. Yield models for commercial willow biomass plantations in Sweden. Biomass and Bioenergy 32, 829-837.

[78] Volk, T., Kiernan, B., Kopp, R., Abrahamson, L., 2001. First and second-rotation yield of willow clones at two sites in New York State. Proceeding of the 5th Biomass Conference of the Americas., Orlando, FL.

[79] Liu, T.T., McConkey, B.G., Ma, Z.Y., Liu, Z.G., Li, X., Cheng, L.L. 2011. Strengths, Weaknessness, Opportunities and Threats Analysis of Bioenergy Production on Marginal Land. Energy Procedia 5: 2378–2386

[80] Licht, L.A., and Isebrands, J.G. 2005. Linking phytoremediated pollutant removal to biomass economic opportunities. Biomass Bioenergy 28: 203-218.

[81] O'Neill, G.J., and Gordon, A.M. 1994. The nitrogen filtering capability of Carolina poplar in artificial riparian zone. J. Env. Quality 23: 1218-1223.

[82] P. Giasson, A. Jaouich, Giasson P. et A. Jaouich. 1998. La phytorestauration des sols contaminés au Québec. Vecteur environnement 31(4):40-53.

[83] Ministère de l'environnement du Québec. 1994. Dix ans de restauration des terrains contaminés (1983-1993). Gouvernement du Québec, Québec, 30 pp.

[84] McEldowney, S., D. J. Hardman, and S. Waite. 1993. Pollution: Ecology and biotreatment. Longman Scientific & Technical, Singapore Publishers, Singapore.

[85] Vandecasteele, B., E. Meers, P. Vervaeke, B. D. Vos, P. Quataert, and F. M. G. Tack. 2005. Growth and trace metal accumulation of two Salix clones on sediment-derived soils with increasing contamination levels. Chemosphere 58: 995-1002.

[86] Kuzovkina, Y.A., Volk, T.A. 2009. The characterization of willow (*Salix* L.) varieties for use in ecological engineering applications: coordination of structure, function and autecology. Ecological Engineering 35, 1178–1189.

[87] Dowling, D. N., and S. L. Doty. 2009. Improving phytoremediation through biotechnology. Current Opinion in Biotechnology 20: 204-206.

[88] Jiménez, E. M., J. M. Peñalosa, R. Manzano, R. O. C. Ruiz, R. Gamarra, and E. Esteban. 2009. Heavy metals distribution in soils surrounding an abandoned mine in NW Madrid (Spain) and their transference to wild flora. Journal of Hazardous Materials 162: 854-859

[89] Harada, E., A. Hokura, S. Takada, K. Baba, Y. Terada, I. Nakai, and K. Yazaki. 2010. Characterization of Cadmium Accumulation in Willow as a Woody Metal Accumulator

Using Synchrotron Radiation-Based X-Ray Microanalyses. Plant and Cell Physiology 51: 848-853.

[90] Pitre, F. E., Teodorescu T. I., and Labrecque M. 2010. Brownfield Phytoremediation of Heavy Metals using Brassica and Salix supplemented with EDTA: Results of the First Growing Season. Journal of Environmental Science and Engineering 4: 51-59.

[92] Karp, A., Hanley, S. J., Trybush, S. O., Macalpine, W., Pei, M. and Shield, I. (2011), Genetic Improvement of Willow for Bioenergy and Biofuels. Journal of Integrative Plant Biology, 53: 151–165.

[91] Guidi W., Kadri H., Labrecque M. (2012) Establishment techniques to using willow for phytoremediation on a former oil refinery in southern Quebec: achievements and constraints. Chemistry and Ecology 28 (1): 49-64

Chapter 18

Microbial Biomass in Batch and Continuous System

Onofre Monge Amaya, María Teresa Certucha Barragán
and Francisco Javier Almendariz Tapia

Additional information is available at the end of the chapter

http://dx.doi.org/10.5772/55303

1. Introduction

Microorganism is a microscopic organism, commonly used term to describe a cell or more, including viruses [1]. This definition includes all prokaryotes, as the eukaryotic unicellular: protozoa, algae and fungi. Bacteria are an extremely diverse group of organisms with extensive variation of morphological, ecological and physiological, which due to their diversity, are regularly found in heterogeneous communities [2].

The microorganisms are widely used in the biological treatment of waste solids and liquids [3]. Importantly, the microorganisms used in biological treatments as a partnership a community or consortium [2].

Nowadays there is a current increasingly important in the sense of using microorganisms, especially bacteria, algae and fungi, for decontamination and to help recovery from natural environments and for treating municipal or industrial effluents. It is estimated that the best microorganisms for the removal of toxins present in a place, are initially isolated in their own area where they have been naturally selected, although in a second genetic manipulation of these can significantly strengthen the capacity of microorganisms culture collections and isolates from the different environments of interest. This is supported by the observation that microorganisms capable of living in a polluted environment, and thus perform vital functions, in cellular metabolism have highly effective devices for decontamination. To know the details of these mechanisms could be exploited to purify the water, such as using bacteria capable of capturing heavy metals on their cell wall [4].

Thus, this rapid advance of science and technological development has allowed decontamination using microorganisms in the water. Using the metabolism of microorganisms has enabled the construction of biological reactors at much lower cost than physicochemical, and also the construction of treatment plants mixed system for greater efficiency [5]. Developed countries currently use these biomining processes through the involvement of bacteria such as *Acidithiobacillus ferrooxidans.*

Some species of *Klebsiella* and *Pseudomonas* are capable of degradation of reactive pollutants. It also recognizes the ability of some microorganisms or their enzymes to degrade under certain conditions to cyanide reagent, employed in the leaching of gold and silver recovery [6]. Several studies have shown that the biomass of different species of bacteria, fungi and algae are capable of concentrating metal ions in their structures that are found in aquatic environments [7].

Bioremediation is defined as a natural process, during in which different microorganisms are capable of removing organic and inorganic contaminants in a given environment [8]. In the bioremediation process, there are different objectives to assess; mainly seeks to avoid a long term noxious effects to other organisms as well as natural resources; it seeks to recover the ecological balance that exists in the environment; and finally, it seeks to achieve that the contaminated area, with a subsequent treatment by biological processes, can be reused for recreation or productive purposes [9].

Different types of biological treatment systems are used in the field of environmental engineering. The biochemical reactions leading to the oxidation of organic matter are conducted in reactors that can be classified as aerobic or anaerobic, suspended growth or biofilm, with mechanical or without mechanical mixing, etc. In order to design an appropriate reactor for a given wastewater treatment system, both the microbial kinetics of substrate removal and the fundamental properties of different reactors have to be understood [10].

The use of microorganisms as tools of decontamination is fairly recent. The biggest advances in the field were made after the oil spill of the Exxon Valdez on the coast of Prince William, in Alaska (1990). Since this ecological disaster lot of oil left in the water, sought alternative ways of dealing with pollution.

The scientists who developed the first successful experiences of bioremediation of oil a large scale in Alaska, were based on the premise that all natural ecosystems have organisms capable of metabolizing toxic compounds and xenobiotics, although these are often found in proportions less than 1% microbial community. This premise was fulfilled in Alaska and in almost all cases studied later [5].

Several studies have shown that the biomass of different species of bacteria, fungi and algae are capable of concentrating metal ions in their structures that are found in aquatic environments [7]. Bacteria as the genus *Pseudomonas* of mining environments have been identified that are resistant to heavy metals such as cadmium (Cd), copper (Cu) and lead (Pb) [11]. Some species of marine microalgae, *Staphylococcus saprophyticus* and fungi have been reported the biosorption of cadmium, chromium, lead and copper from wastewater [7, 12, 13].

In other studies it is known that microbial strains are able to bioremediate contaminated soils with different metals and organic compounds. It is known that *Escherichia coli* is able to bioaccumulate cadmium concentrations of 5 mg/L as well as copper and zinc which are taken from the culture medium by a process in which occurs a binding peptides secreted by the bacteria [14]. Other studies reported bacterial consortia isolated from mining effluents that adsorbed copper [15], as well as anaerobic consortia that in continuous system showed high percentages of copper and iron removal [16, 17].

It is well-known that the use of the water drainage basins are the primary sources for anthropogenic unloads; it represents a risk for the human health, particularly the pollution caused by the high concentrations of some heavy metals, such as, zinc (Zn), nickel (Ni), chromium (Cr), lead (Pb) and copper (Cu) [18, 19, 20]. Those metals go through the aquatic environment principally by direct loads of industrial sources, soils and sediments and are distributed in the water, biota, being the mining industry one of the most important source [21, 22, 23]. An example of contamination by heavy metals in water and sediment, it is the San Pedro River basin, one of the most important rivers of the north of Sonora, Mexico. This river has been severely contaminated by the wastewater discharges with high concentration of copper generated during the mining activity of the region. In addition, wastewaters discharges untreated raw sewage coming from the city of Cananea, Sonora [24]; also contribute with the pollution of this source of water. Considering the importance and impact of this problem on the community of Cananea city, different kinds of researches had been developing. In reports the sampling stations in the San Pedro River and the propagation and isolation of the bacteria and the copper biosorption in an aerobic bioreactor. However, it is necessary to make more studies that allow suggesting other alternatives for the treatment of acid mine drainages (AMD), and consequently, reducing the concentration of the heavy metals in the San Pedro River, until acceptable levels according to Mexican regulations (NOM-001])[15).

2. Anaerobic and aerobic process

2.1. Anaerobic processes

Opposite to the aerobic are anaerobic processes, which are performed in the absence of oxygen by groups of heterotrophic bacteria, which in a process of liquefaction/gasification in two stages, becomes a 90% organic matter present at first in intermediate (partially finished products stabilized that include organic acids and alcohols) and then to methane and gaseous carbon dioxide:

$$organic\ mater \xrightarrow{acid\ forming\ bacteria} intermediaries + CO_2 + H_2S + H_2O \tag{1}$$

$$organic\ acids \xrightarrow{methane\ forming\ bacteria} CH_4 + CO_2 \tag{2}$$

The process is applied universally in hot anaerobic digesters, where in the primary and biological sludge is maintained for about 30 days at 35 ºC to reduce its volume (about 30%)

and their ability to putrefaction, there by simplifies the removal of sludge. The advantage of this type of digestion is that generates energy in the form of methane and the production of sludge is only 10% [5, 25].

2.2. Anaerobic treatment

Anaerobic treatment processes require the presence of a diverse closely dependent group of bacteria to bring about the complete conversion of complex mixtures of substrates to methane gas. It is puzzling that single species of bacteria have not evolved to convert at least simple substrates such as carbohydrates, amino acids, or fatty acids all the way to methane [26].

Conventional phase and high-rate two-phase anaerobic digestion processes have frequently been employed in order to treat both soluble and solid types of domestic and industrial wastes. The most significant outcome of anaerobic digestion processes is that they generate energy in the form of biogas namely, methane and hydrogen. Therefore, due to current imperative environmental issues such as global warming, ozone depletion, and formation of acid rain, substitution of renewable energy sources produced from biomass, such as methane and hydrogen, produced through anaerobic digestion processes will definitely affect the demand and consumption of fossil-fuel derived energy [27].

The anaerobic treatment is a biological process widely used in treating wastewater. When these have a high organic load, presents itself as the only alternative would be an expensive aerobic treatment, due to the oxygen supply. The anaerobic treatment is characterized by the production of "biogas" consisting mainly of methane (60-80%) and carbon dioxide (40-20%) and capable of being used as fuel for generating thermal energy and / or electric. Furthermore, only a small fraction of COD treated (5-10%) is used to form new bacteria, compared to 50-70% of an aerobic process. However, the slow anaerobic process requires working with high residence times, so it is necessary to design reactors or digesters with a high concentration of microorganisms [28, 29]. Actually is a complex process involving several groups of bacteria, both strictly anaerobic and facultative, which, through a series of stages and in the absence of oxygen, flows mainly in the formation of methane and carbon dioxide. Each stage of the process, described below, is carried out by different groups of bacteria, which must be in perfect balanced. Figure 1 shows a schematic representation of the main conversion processes in anaerobic digestion, suggested by Gujer [30].

(1) Hydrolysis. In this process complex particulate matter is converted into dissolved compounds with a lower molecular weight. The process requires the mediation of exo-enzymes that are excreted by fermentative bacteria. Proteins are degraded via (poly) peptides to amino acids, carbohydrates are transformed into soluble sugars (mono and disaccharides) and lipids are converted to long chain fatty acids and glycerine. In practice, the hydrolysis rate can be limiting for the overall rate of anaerobic digestion. In particular the conversion rate of lipids becomes very low below 18°C.

(2) Acidogenesis. Dissolved compounds, generated in the hydrolysing step, are taken up in the cells of fermentative bacteria and after acidogenesis excreted as simple organic compounds like volatile fatty acids (VFA), alcohols and mineral compounds like CO_2, H_2, NH_3, H_2S, etc. Acidogenic fermentation is carried out by a diverse group of bacteria, most of which are obligate anaerobe. However, some are facultative and can also metabolize organic matter via the oxidative pathway. This is important in anaerobic wastewater treatment, since dissolved oxygen (DO) otherwise might become toxic for obligate anaerobic organisms, such as methanogens [31].

(3) Acetogenesis. The products of acidogenesis are converted into the final precursors for methane generation: acetate, hydrogen and carbon dioxide. As indicated in Figure 1, a fraction of approximately 70% of the COD originally present in the influent is converted into acetic acid and the remainder of the electron donor capacity is concentrated in the formed hydrogen. Naturally the generation of highly reduced material like hydrogen must be accompanied by production of oxidized material like CO_2.

(4) Methanogenesis. Methanogenesis may be the rate liminting step in the overall digestion process, especially at high temperatures (> 18°C) and when the organic material in the influent is mainly soluble and little hydrolysis is required. Methane is produced from acetate or from the reduction of carbon dioxide by hydrogen using acetotrophic and hydrogenotrophic bacteria, respectively:

$$\textit{Acetotrophic methanogenesis}: CH_3COOH \rightarrow CH_4 + CO_2 \tag{3}$$

$$\textit{Hydrogenotrophic methanogenesis}: 4H_2 + CO_2 \rightarrow CH_4 + 2H_2O \tag{4}$$

Different from aerobic treatment where the bacterial mass was modeled as a single bacterial suspension, anaerobic treatment of complex wastewaters, with particulate matter in the influent, is only feasible by the action of a consortium of the four mentioned groups of bacteria that each have their own kinetics and yield coefficients. The bacteria that produce methane from hydrogen and carbon dioxide grow faster than those utilizing acetate that the acetotrophic methanogens usually are rate limiting for the transformation of acidified wastewaters to biogas [32].

The different groups of bacteria involved in the conversion of influent organic matter all exert anabolic and catabolic activity. Hence, parallel to the release of the different fermentation products, new biomass is formed associated with the four conversion processes described above. For convenience, the first three processes often are lumped together and denominated acid fermentation, while the fourth step is referred to as methanogenic fermentation.

The removal of organic matter-COD during the acid fermentation is limited to the release of hydrogen only 30% of the organic matter is converted into methane via the hydrogenotrophic pathway. Hence, a necessary condition for efficient organic matter removal in an anaerobic treatment system is that a sufficient mass of acetotrophic methanogens develops.

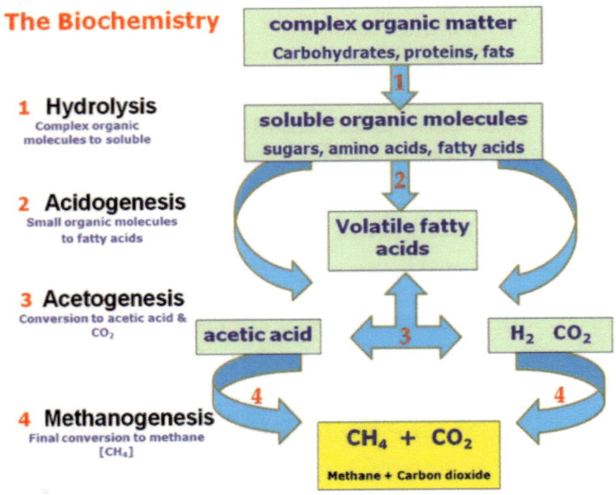

Figure 1. Schematic representation of the main conversion processes in anaerobic digestion

Acid fermentation tends to cause a decrease in the pH because of the production of VFA and other intermediates that dissociate and produce protons. As methanogenesis will only develop well at a neutral pH values, instability may arise, if for some reason the rate of acid removal by methane production falls behind the acid production rate: the net production of acid will tend to cause a decrease in pH, and thus may reduce the methanogenic activity further. In practice, this so called "souring" of the anaerobic reactor contents is the most common cause for operational failure of anaerobic treatment systems. The danger of souring can be avoided, by maintaining the proper balance between acid and methanogenic fermentation which in fact means that both the methanogenic digestion capacity and buffer capacity of the system should be sufficiently high [29, 33].

2.3. Chemical oxygen demand (COD)

Natural organic detritus and organic waste from wastewater treatment plants, failing septic systems, and agricultural and urban runoff, acts as a food source for water-borne bacteria. Bacteria decompose these organic materials using dissolved oxygen. The determination of Chemical Oxygen Demand (COD) is widely used in municipal and industrial laboratories to measure the overall level of organic contamination in wastewater. The contamination level is determined by measuring the equivalent amount of oxygen required to oxidize organic matter in the sample.

In the COD method, the water sample is oxidized by digesting in a sealed reaction tube with sulphuric acid and potassium dichromate in the presence of a silver sulphate catalyst. The amount of dichromate reduced is proportional to the COD. A reagent blank is prepared for each batch of tubes in order to compensate for the oxygen demand of the reagent itself.

Over the range of the test a series of colors from yellow through green to blue are produced. The color is indicative of the chemical oxygen demand and is measured using a photometer. The results are expressed as milligrams of oxygen consumed per liter of sample [34].

Test Procedure

i. Mixed reagent. Potassium dichromate solution. To a 1000 mL volumetric flask, add 42.256 ± 0.001 g of potassium dichromate (previously dried for one hour at 140 - 150 °C). To the flask, add approximately 500 mL of water and mix the contents to dissolve. Then add 33.3 g de $HgSO_4$ to the potassium solution. Add in an ice bath slowly 167 mL pure H_2SO_4. When the mixture has cooled, stir the mixture until the solid dissolves and dilute to one liter.
ii. Silver sulphate in sulphuric acid (10 g). To a glass bottle, add 10.0 ± 0.1 g of silver sulphate and 1000 ± 10 mL of sulphuric acid and stopper. To obtain a satisfactory solution, swirl the initial mixture and allow it to stand overnight. Swirl the contents again until all the silver sulphate dissolves. This solution may be stored in the dark at room temperature for up to an indefinite period.
iii. Pipette 2.0 mL sample into cuvette with 2.0 mL of the mixed reagent solution and 1.0 mL of the silver sulphate in sulphuric acid solution. Invert cuvettes carefully.
iv. Heating reactor for determination of COD for about 30 minutes.
v. Heat cuvettes for 2 h at 150 °C.
vi. read cuvettes in the spectrophotometer at 620 nm.

2.4. Technique to determine alkalinity

Typical control strategy in methanogenic anaerobic reactors is to maintain a relatively low concentration of volatile fatty acids (VFA) and a pH range of $6.6 < pH < 7.4$. Normally in such reactors the carbonate system forms the main weak-acid system responsible for maintaining the pH around neutrality, while the VFA systems (acetic, propionic, and butyric acids) are the major cause for pH decline. Under stable operating conditions, the H_2 and acetic acid formed by acidogenic and acetogenic bacterial activity are utilized immediately by the methanogens and converted to methane. Consequently, the VFA concentration is typically very low, carbonate alkalinity is not consumed and the pH is stable. Conversely, under overload conditions or in the presence of toxins or inhibitory substances, the activity of the methanogenic and acetogenic populations is reduced causing an accumulation of VFA which in turn increases the total acidity in the water, reducing pH. The extent of the pH drop depends on the H_2CO_3 alkalinity concentration. In medium and well-buffered waters (typically the case in anaerobic digestion), high concentrations of VFA would have to form in order to cause a detectable pH drop, by which time reactor failure would have occurred. Therefore, pH measurement cannot form the sole control means, and direct measurement of either (or both) VFA or H_2CO_3 alkalinity concentration is necessary.

The most used technique for the determination of alkalinity for the control of the system anaerobic is described below:

25 mL of sample are taken and placed on a plate with stirring to a solution titrated with 0.02 N sulfuric acid, it initial pH is measured and the acid is added until the pH changes to 5.75 volume of spent acid, followed by titrating until the pH changes to 4.3 and the volume of spent acid is taken and is determined the alpha value.

Alpha = acid vol. (5.75)/ acid vol. (4.3) if this value is greater than 0.55 the bioreactor is acidified and must add a buffer, on the contrary, if it is less, acid must be added [35].

2.5. Methanogenic activity determination

The specific methanogenic activity (SMA = $gDQO\text{-}CH_4 \cdot gVSS^{-1} \cdot d^{-1}$) is defined as the rate of methane production, expressed as COD, regarding biomass expressed as the content of volatile suspended solids (VSS). In anaerobic degradability test measures the rate of degradation of a compound relative to a standard compound that is acetic acid determining [36].

$$SMA = \frac{m}{\gamma_{\frac{CH_4}{COD}} \cdot X} \tag{5}$$

Where: Slope = m= $\frac{LCH_4}{d}$; Biomass = X [=] gVSS/L; Methane conversion $\gamma \frac{CH_4}{COD} = 0.35 \frac{LCH_4}{gCOD}$

Methanogenic activity and toxicity

Methanogenic activity were performed using the pressure transducer technique, which involves the monitoring of the pressure increase developed insealed vials fed with non-gaseous substrates or pressure decrease in vials pressurised with gaseous substrates. Strict anaerobic conditions must be maintained. The same technique can be used to perform the methanogenic toxicity tests. The fifty percent inhibition concentration (IC_{50}) was defined as the methanogenic concentration that caused a 50% relative activity loss [35, 37].

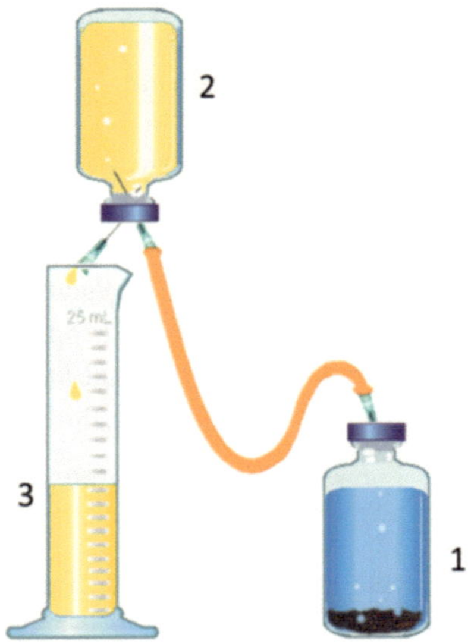

Figure 2. Schematic representation of methane measuring displacement of a solution of 3% NaOH. 1) serological bottle, 2) serological bottle with 3%NaOH 3) test tube to measure the displaced NaOH

The technique is as follows:

1. The sludge is left in mineral medium for 24 hours at 30-35 ºC in order to consume the entire carbon source which may have been brought into water of the plant.
2. Methanogenic activity tests were conducted in 160 mL in serology bottles with an operating volume of 150 mL. The volume of volatile suspended solids was set at 2 g/L and COD concentration used was varied from 0.25, 0.5, 1, 2, 3 y 5 g/L using acetate as a carbon source, and staying a relationship of 0.125, 0.25, 0.5, 1, 1.5 y 2.5 gCOD/gVSS respectively.
3. The bottles were sealed with rubber stoppers, and incubated 24 h at 35°C.
4. Methane was determined by the displacement volume of a solution of 3% NaOH [38].

Figure 2 show a schematic representation of methane measuring.

2.6. Characterization of the sludge (total suspended solids (TSS), volatile suspend solids (VSS), fixed solids (FS))

The total solids (TS) contents of sludge are used in the design and process control of wastewater treatment facilities. Total dissolved solids (TDS) are used to evaluate the suitability of water for both domestic supplies and industrial purposes. The total suspended solids (TSS), including the volatile fraction (VSS), are commonly monitored to evaluate the degree of pollution in natural waters and serves as a key process control parameter for wastewater treatment operation.

The measurement of solids is by means of the gravimetric procedure. The various forms of solids are determined by weighing after the appropriate handling procedures.

1. Placing a clean porcelain crucible for 60 min at 550 °C in the muffle then is passed to a desiccator to cool and then weighed (noted as weight of the crucible).
2. Add with a blunt pipette 10 mL of anaerobics sludge. In a warming rack remove all the water possible. Go to the muffle furnace at 100 °C within 2 h; after this time is cooled and weighed (weight recorded as 100 °C).
3. Place the crucible in the muffle at 550 °C for 1 h and pass it to the desiccator and record the weight of the crucible (as weight at 550 °C).
4. Finally returned to the desiccator. After obtaining the three weights, and proceeded to determine the total solids, volatile and fixed as follows: [35]

$$\textit{Total suspended solids} \qquad TSS: \frac{\textit{Weight to } 100^{o}C - \textit{Weight of the crucible}}{\textit{Sample volume}} [=] \frac{g}{L} \tag{6}$$

$$\textit{Fixed solids} \qquad FS: \frac{\textit{Weight to } 550^{o}C - \textit{Weight of the crucible}}{\textit{Sample volume}} [=] \frac{g}{L} \tag{7}$$

$$\textit{Volatile suspended solids } VSS: TSS - FSS[=] \frac{g}{L} \tag{8}$$

2.7. Setting sludge volume index (SVI)

The sludge volume index is defined as 'the volume in mL occupied by 1 g of sludge after it has settled for a specified period of time' generally ranging from 20 min to 1 or 2 hr in a 1 – or 2 L cylinder. One-half hour is most common setting time allow the mixed liquor to settle for 30 min. (larger cylinder is desirable to minimize bridging of sludge floe and war effects). SVI is 50-150 mL/mg, the sludge settle ability if good.

SVI typically is used to monitor settling characteristics of activated sludge and other biological suspension. Although SVI is not supported theoretically, experience has shown it to be useful in routine process control. The SIV determination consisted of:

1. Place in the imhoff cone of 1000 mL, 100 mL of sludge and diluted to 1000 mL with phosphate buffer.
2. Was allowed to stand for 45 minutes and then stir the contents with a glass rod.
3. The volume occupied by the mud was measured by sedimentation after 30 minutes.
4. The SIV was calculated by dividing this volume by the present VSS g in 100 mL of sludge (Sludge/gVSS) [34].

2.8. Granule density

Among all the different types of anaerobic digesters applied at full scale, UASB (Upflow Anaerobic Sludge Blanket) reactors present the best commercial acceptance. The success of these reactors is related to their capacity for biomass accumulation by settling without the need of a carrier. Good settling properties are obtained through the flocculation of the biomass in the form of dense granules with diameters up to several millimetres. Actually, as individual cells and granules have similar densities, the greater settling velocity of the latter is only related to its larger particle size. The study of this phenomenon has lead to the development of several techniques for characterizing the resistance of the granules, their porosity, settling properties, bacterial composition and organization, activity, nature and composition of exopolymers, as well as their size distribution. This last parameter is particularly useful for studying the physico-chemical factors promoting sludge granulation [38].

1. Search about 6 stainless steel screens with an aperture of about 2 to 0.149 mm for the test.
2. In a vertical stack such that always at the top this larger diameter with respect to the bottom.
3. Take and pass a sample of 25 mL of sludge by the sieves.
4. Washing the sludge with a buffer and phosphate to make them pass through the screens; separated by size.
5. Retrieve the granules, separately, to be retained in the meshes with a backwash of phosphate buffer solution (Table 1) and then determine VSS, FS and TSS [38].

Compound	g/L
K_2HPO_4	4
$Na_2HPO_4{\cdot}7H_2O$	5.09
KH_2PO_4	1.08
pH solution 7.5	

Table 1. Phosphate buffer solution

3. Aerobic processes

Among the biological processes used for the construction of bioreactors, there are two fundamental types of processes: the aerobic and anaerobic.

Aerobic processes are those that need oxygen. There are strict aerobic processes, which are those that can only work if there is oxygen, and facultative aerobic processes, which are those that can switch to anaerobic, according to the concentration of oxygen available.

In general, the aerobic processes have the following reaction:

$$organic\ materia + O_2 \rightarrow CO_2 + H_2O + new\ cells \quad (9)$$

As can be seen in the above reaction, essentially, aerobic metabolism is responsible for catalyzing larger molecules into carbon dioxide, water and new cells. It is noteworthy that the different groups of microorganisms have different metabolisms, and therefore are able to catalyze a wide range of substances, although sometimes other secondary products are obtained in addition to those mentioned.

Aerobic processes are very efficient, operate at a wide range of possible substances to degrade, and in relatively simple cycles are stable; there is rapid conversion of organic pollutants in microbial cells and their operation relatively free of odors [38].

3.1. Growth of microorganisms

3.1.1. Isolation

Most of classical and clinical microbiology depends on the isolation of a pure culture that consists of only one species. This isolate is then later used for characterization such as species determination or an antibiotic resistance profile. For many applications it is essential to isolate and maintain a pure culture of the organism of interest. The goal is to obtain isolated colonies of the organism of interest. These colonies arise from one single cell and are therefore a clone of that original cell. The original cell is called a colony-forming unit (CFU). To obtain a pure culture, it is crucial to maintain a sterile environmental.

To accomplish the microbiological analysis and isolation of strains, water samples are collected, this should be in sterile plastic containers of 500 mL. Is performed enrichment in nutrient broth, to ensure a favorable conditioning bacteria that may be stressed by environmental conditions and ensure better isolation, for which they are placed 10 mL of sterile nutrient broth sample, duplicate it.

Incubate at 37 ° C for 24 hours and reseed poured plate technique or grooves in the selected specific culture media: nutrient agar for *Bacillus sp* and *Pseudomonas sp*; agar EMB (eosin methylene blue) for *Enterobacteriaceae*; agar M17 for *Enterococcus.* Also plantings from water samples directly on agar PDA (Potato Dextrose Agar) acidifying the medium with tartaric acid for the isolation of fungi and yeast, and incubated at 22 °C [38].

3.2. Macroscopic and microscopic characterization

All cells contain the same mayor macromolecules in approximately the same proportions which perform essentially to same functions. Cell shape and arrangement are the initial steps for identifying bacteria. Cell grouping is related to the number of division and whether the cells remain together or separate after division. The Gram stain is an important toll for the classification and eventual identification of bacterial species.

The macroscopic characterization of the colonies was determined according to the observation of general appearance: shape, color, size, texture, elevation and range. The microscopic characterization to observe shape, color and size was performed by Gram stain for bacteria, with lactophenol blue stain for fungi and solution saline for yeast [37, 39].

3.3. Characterization of strains

3.3.1. Biochemical characterization and selection of strains

To grow, microorganisms from the environment should take all the substances required for the synthesis of their cellular materials and power generation. These substances are known nutrients. A culture medium should contain, therefore, all the necessary nutrients in appropriate amounts in the specific requirements of the microorganisms to what has been devised. At selected strains were performed the following biochemical tests, in order to know their identification: catalase production, nitrate reduction, mobility, indole production, the use of citrate as a carbon source, production of urease, methyl red, Voges-Proskauer, carbohydrate fermentation, starch hydrolysis, gelatin hydrolysis and hydrolysis of esculin [40,41]. To determine resistance to low pH, this is changed in the culture media from 3 to 6. The pH was measured with a potentiometer, is adjusted with 10 M NaOH (sodium hydroxide) and HCl (hydrochloric acid).

3.4. Growth kinetics

The relatively large surface area of microorganisms exposed to the environment where they live enables them to take up and assimilate nutrients readily and a to multiply at impressive rates. An important parameter is the specific growth rate, wich is often expressed as the mean doubling time, defined as the time required by the microbial population to douyble its cellular protein content. The doubling time varies widely depending on the microbial species, the nature of the substrate and the degree of adaption to the substrate.

The growth of a bacterial culture can be determined by measuring the increase of turbidity in the medium as optical density or OD. The most common way to do this is to compare the absorbance of the culture to inoculated medium by shining light with a wavelength of 600 nm through the culture. The more growth occurs, the more turbid the culture will become and more light will be absorbed. Using optical density gives indirect measurement of bacterial growth. It doesn't tell anything about how many living cells are in the culture. This becomes especially important in stationary phase. Dead cells still absorb light. To determine the actual number of live bacteria the broth is diluted and plated on appropriate media and incubated. In theory, colonies arising on a plate originate from one single bacterium and give therefore an accurate number of the live cells in the culture at that time point.

Procedure:

The kinetics of selected bacterial growth, it is used 10 mL of 24 h culture of the strains and inoculated into 60 mL of nutrient broth, the following conditions: 35 °C and 100 rpm agitation. Samples were read every 30 minutes in a Spectronic 20D$^+$ visible spectrophotometer at 600 nm, substituting readings transmittance (% T) in the following equation:

$$A = 2 - \log_{10}(\%T) \qquad (10)$$

Where A, is the absorbance.

The different phases of growth kinetics can be observed by plotting the log (% T) versus time [1].

3.5. Biosorption process with bacteria in batch system

Microbial growth and substrate utilization expressions can be incorporated into mass balances to yield equations that can be used to predict effluent microorganism and substrate concentrations, and thus process efficiency. Continuous flow systems are grouped into two broad categories, suspended-growth and attached-growth processes, depending on whether the process microorganisms are maintained in suspension, or are attached to an inert medium (e.g., rocks, sand, granular activated carbon, or plastic materials). Attached-growth processes are also called fixed-film processes or biofilm processes.

Biosorption has provided an alternative process to the traditional physico-chemical methods, utilizing inexpensive biomass to sequester toxic heavy metals. In the 80 last decades, many researchers have focus on the treatment of wastewater containing heavy metals by the use of living organisms and/or their biomass. Many types of organisms such as bacteria, fungi, yeast and algae or their biomasses, have been used for metal uptake [42].

Biosorption tests batch system are carried out with each strain selected in 500 mL erlenmeyer flask, add 90 mL of a solution containing the metal to study, at an initial concentration established and adding 10 mL culture of 24 h of each strain, with a biomass concentration of 1 g/L. Target used 100 mL of metal solution without bacteria. Samples were analyzed in duplicate for each strain. Is used nephelometer of McFarland to estimate the number of cells/mL [40, 42]. The conditions established are: pH between 4 and 5, if the metal precipitates at neutral pH, 37 °C and 100 rpm agitation [44, 45]. To read the metal concentration is done by atomic absorption spectrophotometry, taking 5 mL sample every 15 minutes and prepared as described by [23, 46] and the concentration is calculated from the calibration curve prepared with standard solution of each metal studied. The detection limits can be 0.02 mg/L, analyzing in duplicate.

3.6. Affecting parameters

Within the anaerobic and aerobic environment, various important parameters affect the rates of the different steps of the process, i.e. pH and alkalinity, temperature, and hydraulic retention times.

Each group of microorganisms has a different optimum pH range. Methanogenic bacteria are extremely sensitive to pH with an optimum between 6.5 and 7.2 pH, alkalinity and volatile acids/alkalinity ratio. The fermentative microorganisms are somewhat less sensitive and can function in a wider range of pH between 4.0 and 8.5. at a low pH the main products are acetic and butyric acid, while at a pH of 8.0 mainly acetic and propionic acid are produced.

The temperature has an important effect on the physicochemical properties of the components found in the digestion substrate. It also influences the growth rate and metabolism of microorganisms and hence the population dynamics in the anaerobic reactor. Acetotrophic methanogens are one of the most sensitive groups to increasing temperatures. The degradation of propionate and butyrate is also sensitive to temperatures above 70 °C. The temperature has moreover a significant effect on the partial pressure of H_2 in digesters, hence influencing the kinetics of the syntrophic metabolism. Thermodynamics show that endergonic reactions (under standard conditions), for instance the breakdown of propionate into acetate, CO_2, H_2, would become energetically more favourable at higher temperature, while reactions which are exergonic (e.g. hydrogenotrophic methanogenesis) are less favoured at higher temperatures.

The solids retention time (SRT) is the average time the solids spend in the digester, whereas the hydraulic retention time (HRT) is the average time the liquid sludge is held in the digester. The subsequent steps of the digestion process are directly related to the SRT. A decrease in the SRT decreases the extent of the reactions and viceversa. Each time sludge is withdrawn, a fraction of the bacterial population is removed thus implying that the cell growth must at least compensate the cell removal to ensure steady state and avoid process failure [28].

3.7. McFarland nephelometer

The McFarland nephelometer was described in 1907 by J. McFarland as an instrument for estimating the number of bacteria in suspensions used for calculating the bacterial opsonic index and for vaccine preparation.

Another important factor is known that cells per milliliter are taken at a given time, and a known way is through the % of transmittance and that can be determined by the technique of McFarland nephelometer, which is described below.

McFarland Nephelometer Standards

a. Set up 10 test tubes or vials of equal size and quality: Use new hoses washed and completely dry.
b. Prepare H_2SO_4 1% chemically pure.
c. Prepare an aqueous solution of barium chloride, 1% chemically pure.
d. Add to the tubes the designated amounts of the two solutions as shown in Table 2 for a total of 10 mL/tube.
e. Close the tubes or vials. The suspension of barium sulphate corresponding to an approximately homogeneous precipitate of density of the cells per milliliter in the standard variable, as shown in the Table 2.
f. In the Figure 3(a) shows the % of transmittance against the number. From the tube which can be removed if there is not a spectrophotometer to read the transmittance, and in Figure 3(b) shows the number of cells that are depending on number.

Tube's Number	**1**	**2**	**3**	**4**	**5**	**6**	**7**	**8**	**9**	**10**
Barium chloride (mL)	0.1	0.2	0.3	0.4	0.5	0.6	0.7	0.8	0.9	1
Sulfuric Acid (mL)	9.9	9.8	9.7	9.6	9.5	9.4	9.3	9.2	9.1	9
Aprox. Cell density ($X10^8$/mL)	3	6	9	12	15	18	21	24	27	30

Table 2. McFarland Nephelometer Standards

Population density is monitored by taking readings of % of transmittance, which compared to the McFarland nephelometer; transmittance readings should be less than 10% in order to maintain the population density. This technique is used to know the time and turbidity of most practical way to achieve the desired amount of biomass for biosorption experiments batch system and continuous [39].

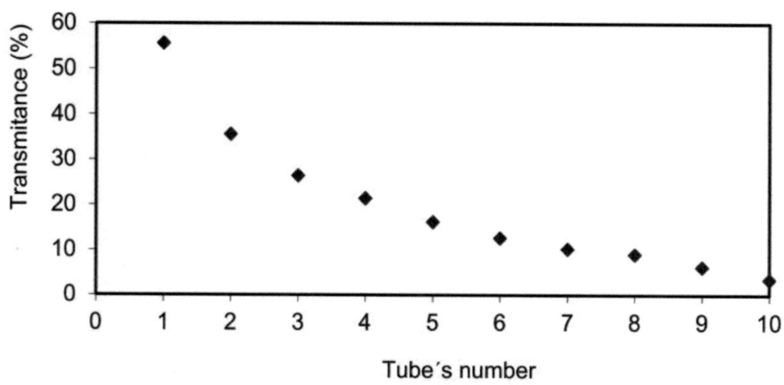

(a)

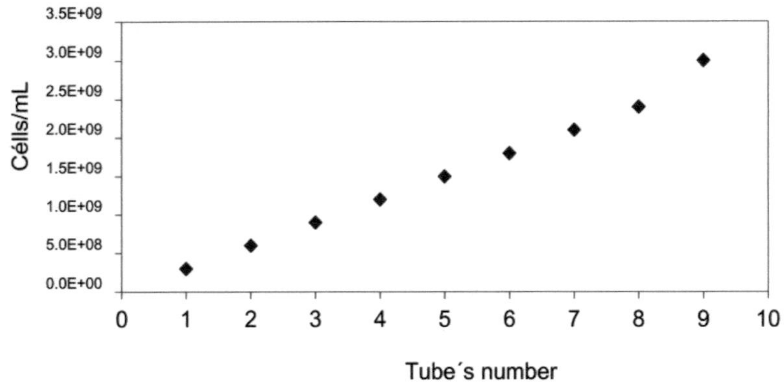

(b)

Figure 3. Standard curves of nephelometer with transmittance (%) and cell density

4. System in batch and continuous culture

The construction of bioreactors is based on a simple principle: make the pollutants are converted into the substrate (food) of microorganisms, and that these, while feeding and increases its population, decontaminated water. For the construction of a bioreactor is necessary to know the type of microorganisms with which they are going to work and as well as the growth curve characteristic of them [5].

The key factor of a bioreactor is to maintain microorganisms in the growth stage most of the time as possible, i.e. keep the microbial population to its maximum level, to optimize the efficiency of the degradation processes. This is achieved by controlling the environmental conditions (temperature, pH, aeration and nutrient availability) and the flows in and out, so never lack food and do not reach the death phase or endogenous [2, 5].

The teams that are made homogeneous reactions can be of three general types: discontinuous (batch), continuous flow steady and unsteady flow semicontinuous.

Batch reactors are simple to operate and industrially used when small amounts are to treat substance. Continuous reactors are ideal for industrial purposes be treated when large quantities of substance and can achieve good control of product quality. Semicontinuous reactors are more flexible systems, but more difficult to analyze and operate than previous; in them the reaction rate can be controlled with a good strategy at the dosage of the reactants [48].

In a perfect batch reactor there is no entry or exit of reactant. It is further assumed that the reactor is well stirred, i.e. that the composition is the same at all points of the reactor for a given time instant. Since the input and output are zero the material balance is:

$$\begin{pmatrix} \textit{Disappearance of} \\ \textit{reactant by} \\ \textit{chemical reaction} \end{pmatrix} = - \begin{pmatrix} \textit{Reactant} \\ \textit{accumulation in the} \\ \textit{control volume} \end{pmatrix}$$

All points have the same composition; the volume control to perform the balance is the entire reactor. Evaluating the terms:

$$r_A V = -\frac{dN_A}{dt} \tag{11}$$

And given that: $N_A = N_{Ao}(1 - X_A)$ results:

$$r_A V = N_{A0}\frac{dX_A}{dt} \tag{12}$$

Integrating gives the equation for the design for the batch reactor:

$$t = N_{Ao}\int_0^{X_A} \frac{dX_A}{r_A V} \tag{13}$$

If the reaction volume remains constant may be expressed in function of the concentration of reagent $C_A = N_A / V$

$$t = C_{Ao}\int_0^{X_A} \frac{dX_A}{r_A} = -\int_{C_{Ao}}^{C_A} \frac{dC_A}{r_A} \tag{14}$$

This intermittent or batch reactor is characterized by the variation in the reaction's degree and the properties of the reaction mixture with the lapse of time [49]

A batch reactor has no inflow or outflow reagents of the reaction products while being performed. In almost every batch reactors, the longer that a reactant in the reactor, most of it becomes product to reach equilibrium is exhausted or the reagent [50].

The reactor of the continuous flow type, in which the degree of reaction can vary with respect to the position in the reactor, but not a function of time. Therefore, one of the classifications of the reactors is based on the operation method [49].

Normally, the conversion increases with time that the reagents remain in the reactor. In the case of continuous flow systems, this time usually increases with increasing reactor volume; therefore, the conversion X is a function of reactor volume V [50].

The tubular reactor plug flow (RTFP) is characterized in that the flow is directed, without any element of the exceeding or being mixed with any other element located before or after that, i.e. no mixing in the flow direction (axial direction). As a result, all fluid elements have the same residence time within the reactor [48].

As fluid composition varies along the reactor, material balance must be performed in a differential volume element transverse to the direction of flow.

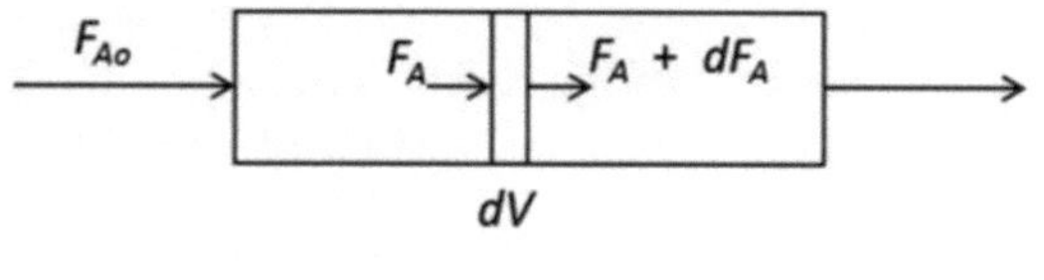

$$Inlet = Outlet + Disappearance\ of\ reaction$$

$$F_A = F_A + dF_A + r_A dV$$

Given that $dF_A = d[F_{Ao}(1\text{-}X_A)] = -F_{Ao}dX_A$ by substitution is

$$F_{Ao}dX_A = r_A dV \tag{15}$$

That integrated is

$$\int_o^V \frac{dV}{F_{Ao}} = \int_0^{X_{Af}} \frac{dX_A}{r_A} \tag{16}$$

Or:

$$\tau = \frac{V}{V_o} = C_{Ao}\int_o^{X_{Af}} \frac{dX_A}{r_A} \tag{17}$$

In some cases it is convenient to divide the output current of a plug flow reactor by returning part of it to the reactor inlet.

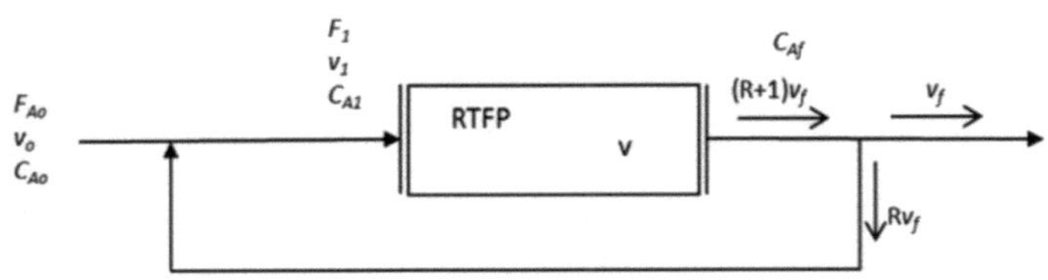

The recirculation ratio is defined:

$$R = \frac{flow\ which\ is\ recycled}{flow\ out} \tag{18}$$

Raising the design equation for the reactor (within the recycle loop) without expansion

$$\frac{V}{v1} = -\int_{C_{A1}}^{C_{Af}} \frac{dC_A}{r_A} \tag{19}$$

If its considers that there is no expansion or contraction in the reactor, raising the junction of the inlet and the recirculation $V_1 = (R + 1)\ V_o$ and furthermore $C_{A1} = (C_{Ao} + C_{Af})\ /\ (R + 1)$, therefore the equation of reactor design is:

$$\tau = \frac{V}{V_o} = -(R+1)\int_{\left(\frac{C_{Ao}+RC_{Af}}{R+1}\right)}^{C_{Af}} \frac{dC_A}{r_A} \tag{20}$$

Another classification relates to the shape. If a laboratory vessel is equipped with a stirrer efficient, composition and temperature of the reaction mass will tend to be equal in all areas of the reactor. A container in which there is uniformity of properties is called a stirred tank reactor (or well mixed) or STR [48].

4.1. Anaerobic continuous studies

Generally, the design rules of biological reactors are all based on high removal efficiency of degradable organic matter. Consequently, if the substrate composition and strength of a wastewater are known, a basic design of a high-rate anaerobic system can be established. According to [52], the main design criteria of UASB reactors are, among others: applicable organic load, upflow velocity, three-phase separator, and influent distribution system.

The UASB reactors are generally designed based on the organic volumetric load (OVL) (kgCOD/m^3-day) that is defined as follows:

$$OVL = QSo / V \tag{21}$$

where Q: influent flow rate (m^3/day), So: influent COO $(kgCOD/m^3)$, and V: volume of reactor (m^3), From Equation (21) the volume of the reactor, V, can be obtained:

$$V = QSo / OVL \tag{22}$$

For most industrial wastewaters, the OVL (based on degradable COD) is the critical factor for the reactor volume. Its value depends on the quantity and quality of the granular sludge; the nature, type, and concentration of the pollutants; the temperature; the required treatment efficiency and the desired safety regarding peak loads [16, 53].

4.2. Studies of biosorption of heavy metals with aerobic bacteria using biomass support in batch system

For studies of biosorption of heavy metals using aerobic bacteria and a support for the biomass, using 500 mL Erlenmeyer flasks, which are placed 5 g of the support for the immobilization of selected biomass, 90 mL of a solution containing the metal study established at an initial concentration, 10 ml of biomass with a density of 1 g/L and as target, 100 mL of metal with 5 g of support material. The flasks were plugged with a cotton swab having aeration, then is placed in an incubator with shaking at 100 rpm and temperature established for mesophilic bacteria to 45 °C. Samples are taken at set times to analyze the concentration of metals by atomic absorption. The conditions are the same for the studies using only bacteria without carrier material. All experiments were performed in duplicate and the efficiency of biosorption (E) is calculated using the equation:

$$E = \left(\frac{C_o - C_f}{C_o} \right) * 100 \tag{23}$$

Where: C_o y C_f initial and final amounts of the metal (mg/L).

4.3. Studies of heavy metal biosorption in a continuous system with aerobic biomass using biomass support

Biosorption is a rapid phenomenon of passive metal sequestration by the non-growing biomass. To carry out the studies of metal biosorption in a continuous system is conditioned first a reactor column, which has side ports for sampling. The reactor is packed with carrier material of biomass with a particle size between 1 and 6 mm, to avoid clogging. Both mineral medium such as air are fed through the bottom of the reactor to promote the growth of bacteria and the pH is controlled if the metal to be studied could precipitate at neutral pH. The mineral medium is inoculated with 10% biomass that develop in this medium for the time given the growth kinetics and the reactor is kept in recirculation until the development of biomass (1 g/L) and that adheres to support material. Biomass concentration was estimated by measuring the percent of transmittance to an optical density at 600 nm (Spectronic 20D$^+$) and for the amount of biomass produced in cells/mL was determined using the Table 2 of McFarland nephelometer, described above.

When produced in the reactor is 1 g/L of biomass and this is immobilized in the zeolite, are set constant conditions of operation of the reactor as: air flow 10 times the feed flow of the contaminated medium, hydraulic retention time (HRT) of a day and ambient temperature of 30 °C. The tests are performed to set conditions of initial concentrations of metals and pH set, and takes days to the input samples at different heights of the reactor and output to meet the metal concentration, biomass is recycled to make more time contact between the bacteria and the metal being studied. It can be perform a second and third experimental run at different initial concentrations of feeding and at the same pH, maintaining the feed stream and recirculation same. Other studies may be changing the pH, keeping other conditions constant.

In the experimental runs carried out is analyzed for pH, metal concentration by atomic absorption and to determine the concentration of cells/mL of biomass, measures the percentage of transmittance in the spectronic 20D$^+$ and compared by the technique of Nephelometer of Mc Farland. The support used is analyzed by the technique of sludge digestion, are performed analyzes of biomass produced per day and chemical oxygen demand (COD). At the end of the experiments are performed technical analyses of the medium used X-ray Diffraction (XRD), scanning electron microscopy (SEM) and Energy Dispersive Spectroscopy X-ray (EDS) at different column heights to see if deposited on the support certain amount of heavy metals or all was absorbed by the biomass.

4.4. Support the immobilization of biomass

Immobilization of cells as biocatalysts is almost as common as enzyme immobilization. Immobilization is the restrict ion of cell mobility within a defined space. Immobilized cell cultures have the following potential advantages over suspension cultures.

- Immobilization provides high cell concentrations.
- Immobilization provides cell reuse and eliminates the costly processes of cell recovery and cell recycle.
- Immobilization eliminates cell washout problems at high dilution rates.
- The combination of high cell concentrations and high flow rates (no washout restrictions) allows high volumetric productivities.

- Immobilization may also provide favorable microenvironmental conditions (i.e., cell-cell contact, nutrient-product gradients, pH gradients) for cells, resulting in better performance of the biocatalyst» (e.g., higher product yields and rates).
- In some cases, immobilization improves genetic stability.
- For some cells, protection against shear damage is important.

The major limitation on immobilization is that the product of interest should be excreted by the cells. A further complication is that immobilization often leads to systems for which diffusional limitations are important. In such cases the control of microenvironmental conditions is difficult, owing to the resulting heterogeneity in the system. With living cells, growth and gas evolution present significant problems in some systems and can lead to significant mechanical disruption of the immobilizing matrix.

The primary advantage of immobilized cells over immobilized enzymes is that immobilized cells can perform multistep, cofactor-requiring, biosynthetic reactions that are not practical using purified enzyme preparations.

Adsorption of cells on inert support surfaces has been widely used for cell immobilization. The major advantage of immobilization by adsorption is direct contact between nutrient and support materials. High cell loadings can be obtained using microporous support materials. However, porous support materials may cause intraparticle pore diffusion limitations at high cell densities, as is also the case with polymer-entrapped cell systems. Also, the control of microenvironmental conditions is a problem with porous support materials. A ratio of pore to cell diameter of 4 to 5 is recommended for the immobilization of cells onto the inner surface of porous support particles. At small pore sizes. Accessibility of the nutrient into inner surfaces of pores may be the limiting factor, whereas at large pore sizes the specific surface area may be the limiting factor. Therefore, there may be an optimal pore size, resulting in the maximum rate of bioconversion.

Adsorption capacity and strength of binding are the two major factors that affect the selection of a suitable support material. Adsorption capacity varies between 2 mg/g (porous silica) and 250 mg/g (wood chips). Porous glass carriers provide adsorption capacities (10^8 to 10^9 cells/g) that are less than or comparable to those of gel-entrapped cell concentrations (10^9 to 10^{11} cells/mL). The binding forces between the cell and support surfaces may vary, depending on the surface properties of the support material and the type of cells. Electrostatic forces are dominant when positively charged support surfaces (ion exchange resins, gelatin) are used. Cells also adhere on negatively charged surfaces by covalent binding or H bonding. The adsorption of cells on neutral polymer support surfaces may be mediated by chemical bonding, such as covalent bonding, H bonds, or van der Waals forces. Some specific chelating agents may be used to develop stronger cell-surface interact ions. Among the support materials used for cell adsorption are porous glass, porous silica, alumina, ceramics, gelatin, chitosan, activated carbon, wood chips, polypropylene ion-exchange resins (DEAE-Sephadex, CMC-), and Sepharose [54].

Various reactor configurations can be used for immobilized cell systems. Since the support matrices used for cell immobilization are often mechanically fragile, bioreactors with low hydrodynamic shear, such as packed-column, fluidized-bed, or airlift reactors, are preferred. Mechanically agitated fermenters can be used for some immobilized-cell systems if the support matrix is strong and durable. Any of these reactors can usually be operated in a perfusion mode by passing nutrient solution through a column of immobilized cells [54].

Since the design of reactors for the removal of heavy metals from liquid effluent must consider optimum contact between these and the biomass, it has been considered the use of different types of support for the immobilization of the biomass with the aim of achieving greater efficiency in removing heavy metals. This achieves prevent biosorbent is removed from the reactor in the output current and at the same time is obtained a greater mechanical stability thereby reducing the shear stresses that could damage the structure of the microorganism which affects removal efficiency heavy metals [53].

Living biomass immobilized, must first take the form of biofilm on supports prepared from a variety of inert materials. One of the materials that have been studied as biomass support is activated carbon by porosity and high surface area, besides being an abundant product is obtained as a byproduct of the production of oil from coconut, olive and processing sugarcane [53]. Other materials have been used as biomass support such as silica, polyacrylamide gel and polyurethane include agar, cellulose, alginates, polyacrylamides, the silica gel, sand, textile fibers, calcium alginate, polysulfone, glutaraldehyde and other organic compounds, and have been used for removing heavy metals [55,56].

There are other materials that could be used for biomass carriers; such as the natural zeolites are known important industrial applications due to its high affinity for water and that the cavities only allow passage of molecules of a certain size. Have been used as additives in animal feed, such as soil improvers in agriculture due to increased nitrogen retention and soil moisture, and as catalysts in industrial processes of refining, petrochemicals and fine chemicals [57].

4.4.1. Activated charcoal

The name of activated charcoal is applied to a series of artificially prepared porous carbons to exhibit a high degree of porosity and a high inner surface. These characteristics are responsible for their adsorptive properties, which are used widely in many applications in gas phase and liquid phase. Chemically it is composed of carbon, oxygen, hydrogen and ash. The activated carbon adsorbent is a very versatile, because the size and distribution of pores in the carbonaceous structure can be controlled to meet the current and future technology. The pore sizes ranging from smaller called micropores (2.0 nm) until the mesopores (2 - 50 nm) and macropore (<50 nm). It should be borne in mind that most adsorption occurs in the micropores (greater than 90% of the surface area) the mesopores and macropores are extremely important because in the activated charcoal are those which facilitate access of the species will adsorb to the interior of the particle and of the micropores [58].

One of the materials that have been studied as biomass support is activated charcoal. Its high porosity and high surface area activated charcoal make it an ideal material to be carried out the process of adsorption of heavy metals. Another reason why activated charcoal is used for the adsorption is its low cost, since it is an abundant product is obtained as a byproduct of the production of oil from coconut, olive and processing of sugar cane [55].

4.4.2. Glutaraldehyde

Microbial cells can be immobilized by cross-linking between cells, using bi or multifunctional reagents as glutaraldehyde or toluene di isocyanate.

Glutaraldehyde is a colorless liquid with a pungent odor used to sterilize medical and dental equipment is also used in water treatment industry and as a chemical preservative. However, it is toxic and can cause severe eye irritation, nose, throat and lungs, along with headaches, drowsiness and vomiting. Glutaraldehyde monomer can polymerize by aldol condensation, giving poliglutaraldehído alpha, beta unsaturated reaction typically occurs at alkaline pH [59].

4.4.3. Silica

The mechanism involved is based on the formation of covalent bonds between the inorganic support (silica) and cells in the presence of crosslinking agents. A said joint is needed for the modification of the support surface. The reaction requires the introduction of reactive organic groups on the silica surface for the attachment of cells to the support. As coupling agent generally used aminopropyl triethoxy silane; this organic functional group condenses with hydroxyl groups of the silica and the group as a result becomes available for covalent bond formation on the surface. Covalent bonds can also be established by treating the silica surface with glutaraldehyde or isocyanate.

The advantage of this method is that the support can be generated without the limitations on physical and chemical conditions imposed by the biocatalyst, which can be optimized by the characteristics of mechanical stability, porosity, strength of the support, etc.

When you want to form covalent bonds between the substrate and cells, the problem is how to promote adhesion of cells to relatively large surface without damaging its stability and resistance to washing. The support may have pores of greater diameter than the cell to allow the latter to penetrate the internal surfaces. Porous supports are used which are embedded by immersion in cell suspensions [60].

Another important matrix being used for immobilization for metal removal is silica. Silica-immobilized preparations offer advantage in terms of reusability and stability. The silica immobilized product is mechanically strong and exhibits excellent flow characteristics [68]. A silica immobilized algal preparation AlgaSORBR (Bio-Recovery Systems, Inc., Las Cruces, NM 80003, USA) which is being used commercially retains approximately 90% of the original metal uptake efficiency even after prolonged use (> 18 months) [57].

4.4.4. *Polyacrylamide gel*

Polyacrylamide gels are formed by polymerization of acrylamide by the action of a crosslinked agent, is chemically inert, uniform properties, able to be prepared quickly and reproducibly. Thus, in addition, transparent gels with mechanical stability, water insoluble, relatively non-ionic and allow good visualization of the bands for a long time. Also has the advantage that by varying the concentration of polymers can be modified in a controlled manner the pore size, there sometimes is used in diagnosis least because of their neurotoxocidad [59].

Whole cell immobilization within a polyacrylamide gel also provides a useful laboratory scale system and has been used to biosorb and recover a number of heavy metal(s). Good results have been obtained in the case of polyacrylamide immobilized cells of Citrobacter where a very high removal of uranium, cadmium and lead was observed from solutions supplemented with glycerol $-2PO_4$. Rhizopus arrhizus biomass immobilized on polyacrylamide gel was effective in almost completely removing Cu^{2+}, Co^{2+} and Cd^{2+} from synthetic metal solution [26, 55].

4.4.5. *Polyurethane*

Inert materials such as polyurethane, impregnated with a suitable culture medium provided a homogeneous aerobic condition in the fermenter and impurities do not contribute to the final product. An additional advantage of using inert supports is the easy recovery of the product of interest, ease of performing balances because all nutrient concentrations in the middle of production are known, so one can study the effect of a given component of medium [60].

Recent studies have shown the superiority of polyurethane and polysulfone as immobilization support in comparison to polyacrylamide and alginate matrices. It has been reported a novel polyurathane gel bead fabrication technique for immobilizing Pseudomonas aeruginosa CSU. Preliminary studies conducted by them revealed that the P. aeruginosa CSU biomass immobilized within the polyurethane gel beads were effective in the removal of hexavalent uranium from low concentration acidic waters. Other authors have been immobilized phormidium laminosum on polysulfone and epoxy resins. They were successful in reusing the polysulfone immobilized biomass for ten consecutive biosorption/desorption cycles without apparent loss of efficiency after reconditioning it with 0.1 M NaOH. Immobilization of Citrobacter biomass in polysulfone matrix increased its metal loading capacity for lead, cadmium and zinc metals [57].

4.4.6. *Alginate*

There are many studies on the composition of alginate and its advantages to cell immobilization. Because the chemical composition of this polymer and as a consequence that the same reactions can be obtained in reaction, alginate gels are recommended for cells sensitive to environmental conditions. Recent studies of the diffusional characteristics of the immobilized system, have improved our understanding of the environment surrounding the immobilized cells, optimize protocols and improve the stability of alginate gels [60].

One of the matrices that have been used in metal recovery by both viable and non-viable cells is the entrapment in the matrix of insoluble Ca-alginate. Fluidized beds of Ca-entrapped cells of Chlorella vulgaris and Spirulina platensis were successfully used to recover gold from a simulated gold-bearing process solution containing $AuCl_4$, $CuCl_2$, $FeCl_2$ and $ZnCl_2$. The Ca-alginate immobilized cells of Chlorella salina also showed greater binding of cobalt, zinc and manganese than the free cells. *Rhizopus arrhizus* entrapped on alginate beads was successfully used for the removal of uranium over multiple biosorption and desorption cycles. Accumulation was also dependent on cell density in alginate beads with greater uptake of cobalt at the highest cell densities [57].

4.4.7. Natural zeolite

Zeolites are crystalline aluminosilicates, three-dimensional, microporous, based on framework structure with a rigid anion, with well-defined channels and cavities. These cavities contain exchangeable metal cations (Na^+, K^+, etc.) And can also retain removable and replaceable guest molecules (water in natural zeolites). To date about 40 have been characterized structures of natural zeolites and have developed more than 130 synthetic structures. The most important natural zeolites are analcime, chabazite, clinoptilolite, erionite, ferrierite, heulandite, laumontite, and phillipsite mordonita [58].

Zeolites are composed of aluminum, silicon, sodium, hydrogen and oxygen. The crystal structure is based on the three network addresses with SiO_4 tetrahedral shaped with four oxygens shared with adjacent tetrahedra. The physical properties unique aspects provide for a wide variety of practical applications. Figure 4 shows the basic structure of the zeolite tetrahedral [63].

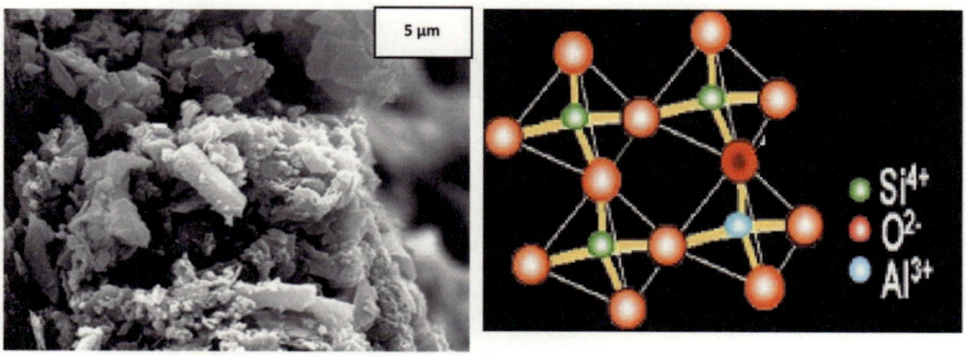

Figure 4. Basic tetrahedral structure of the zeolites.

The physical properties of the zeolite are that they possess features bright, hardness and wear resistance. Applications of natural zeolites make use of one or more of its chemical properties, usually including adsorption, ion exchange and catalysis. These properties are a function of the crystal structure of each species, cationic structure and composition [63].

Clinoptilolite is from the zeolite minerals are best known for its uses and applications. It is a natural zeolite formed from volcanic ash in lakes and marine waters millions of years ago. Clinoptilolite, is the most studied and is considered more useful, since it is known as an adsorbent of certain toxic gases such as hydrogen sulfide and sulfur dioxide. In fact few countries that have had deposits in operation, including: Japan, Italy, USA, Russia, Hungary, Bulgaria, Cuba, Yugoslavia and Mexico [39]. Recognizes the capacity of the zeolites natural adsorb heavy metals and other contaminants from water. In certain cases, it requires a pretreatment of the zeolite to modify or improve its adsorption properties [62].

5. Conclusion

This chapter has provided the results of research of aerobic and anaerobic biomass in the batch and continuous system using a support for the biomass: silica, polyacrylamide gel, polyurethane, calcium alginate, glutaraldehyde, charcoal and zeolites. The use of support for the biomass increases the development of the microorganisms. Also describes the affecting parameters: time, pH, temperature, HRT, toxicity and stirring speed. Also in this chapter describe the techniques for determination of parameters for anaerobic such as, chemical oxygen demand (COD), alkalinity, methane production, total solids (TSS), volatile solids (VSS), volatile fatty acids (VFA) concentrations, and for the aerobics, the biomass concentration using % of transmittance, McFarland nephelometer, isolation, macroscopic and microscopic characterization, growth kinetics, in batch and continuous system.

Author details

Onofre Monge Amaya, María Teresa Certucha Barragán
and Francisco Javier Almendariz Tapia
University of Sonora, Department of Chemistry and Metallurgy. Hermosillo, Sonora. México

Acknowledgement

This work was made possible through support provided by the University of Sonora, through the Department of Chemical Engineering and Metallurgy, and Engineering Division. The authors would like to thank: The National Council for Science and Technology (CONACyT), well as students Gisel Figueroa, Gonzalo Figueroa, Guadalupe López, Karla Hernandez, Hiram Bañuelos, Carlos Jaramillo, Luis Carlos Platt, Axel Valenzuela, Glenda Duarte.

6. References

[1] Madigan M.T., Martinko J. M., Parker J. (2004) Brock Biología de los microorganismos, ED. Prentice Hall. 647 p.

[2] Eweis J. B., Ergas S. J., Chang D. P., Schroeder E. D., Tejero M. I., Amieva del Val J.J. (1999) Principios de biorrecuperación (Bioremediation). Primera Edición. Mac Graw Hall Interamericana. 4, 12, 21, 94.

[3] Atlas. M.R. y Bartha R. (2002) Ecología microbiana y microbiología ambiental. 4a Ed. Pearson educación. Madrid, España. pp. 3-24.

[4] Steciow M. (2005) Microbiología Ambiental http:www.cricyt.edu.ar/enciclopedia/términos/Biorremed. Htm

[5] Fontúrbel R. F. e Ibáñez N.C. (2004). Empleo del Metabolismo Microbiano para la Descontaminación de Aguas. Biología y Ciencias de la Salud. No. 17.

[6] Saigí F. y López A. (2004) Las ciencias de la vida y la biotecnología en la nueva sociedad del conocimiento. La base de la nueva economía. http://www.uoc.edu/dt/esp/saigi1104.pdf. Agosto-2008.

[7] Basso, M.C., Cerella, E.G.,Cukierman, A.L (2002) Empleo de Algas Marinas para la Biosorción de Metales Pesados de Aguas Contaminadas. Avances en Energías Renovables y Medio Ambiente. Argentina. 6:1

[8] Cotoras T. D. (2003) Bacterias para descontaminar aguas residuales de la minería y la industria. Facultad de Ciencias Químicas y Farmacéuticas. Universidad de Chile. Induambiente 13(14): 27-29. http://www.estrucplan.com.ar/Secciones/Noticias/Noticia.asp. Abril 2012

[9] Mijares M. A. P. (2003) Aislamiento y Caracterización de Bacterias con Capacidad para Inmovilizar Cadmio. Tesis Licenciatura. Biología con área en Biotecnología. Universidad de las Américas, Puebla. Capítulo dos.

[10] Cervantes Carlos (2006) Las Relaciones de las Bacterias y los Metales. Carisma de la Ciencia 3: 1-4.

[11] Rivas B.G.A., Gutierréz, S., Merino F. (2004) Bioremoción de Metales Pesados en Solución por *Pseudomonas fluorescens* M1A-45 Aisladas de Ambientes Minero. Segunda Semana de Ciencia y Tecnología.

[12] Acosta I., Moctezuma-Zárate M. G., Cárdenas J. F. y Gutiérrez C. (2007) Bioadsorción de Cadmio (II) en Solución Acuosa por Biomasa Fúngicas. Información Tecnológica. 18:1: 9-14.

[13] Ilhan S., Nourbakhsh M.N., Kilicarslan S., Ozdag H. (2004) Removal of Chromium, Lead and Copper Ions From Industrial Waste Waters by *Sthaphlococcus saprophyticus* Turkish Electronic Journal of Biotechnology. 2, 50-57.

[14] Lasat M. M. (2002) Phytoextraction of toxic metals: A review of biological mechanisms. J. Environ. Qual. 31,109-120.

[15] Monge, O., Valenzuela, J., Acedo, E., Certucha, M., & Almendáriz, J. (2008) Biosorción de cobre en sistema por lote y continuo con bacterias aerobias inmovilizadas en zeolita natural (clinoptilolita). Rev. Int. Contam. Ambient. 24 (3) 107-115.

[16] Certucha-Barragán M.T, Acedo-Félix E, Monge-Amaya O, Valenzuela-García J.L, Almendariz-TapiaF.J, Leal-Cruz A.L (2009) Copper Bioaccumulation in an Upflow Anaerobic Sludge Blanket (UASB) Reactor Chemical Speciation and Bioavailability. 21:3: 161-164.

[17] Certucha-Barragán M. T, Acedo-Félix E, Almendariz-Tapia F. J, Pérez-Moreno R, Monge-Amaya, Valenzuela Garcia J. L, and O. Monge-Amaya (2011) Iron Influence on Copper Biosorption Using Anaerobic Sludge and its Microstructural Characterization, Mineral Processing & Extractive Metall. Rev., 32: 1 - 8.

[18] Topalián M.L., Castané P.M, Rovedatti, M.G. and Salibián A. (1999) Principal Component Analysis of Dissolved Heavy Metals in Water of the Reconquista River (Buenos Aires, Argentina). Bull. Environ. Contam. Toxicol 63: 484 – 490.
[19] Karri S, Sierra-Alvarez, R and Field J.A. (2006) Toxicity of Copper to Acetoclastic and Hydrogenotrophic Activities of Metanogens and Sulfate Reducers in Anaerobic Sluge. Chemosphere, 62: 121 – 127.
[20] Upadhyay, A.K., Gupta, K.K., Sircar, J.K., Deb, M.K. and Mundhara, G.L. (2006) Heavy Metals in Freshly Deposited Sediments of the Rivers Subernarenkha, India: an Example of Lithogenic and Anthtopogenic Effects Environ. Geol., 50:397–403.
[21] Elder, F.J. (1988) Metal Biogeochemistry in Surface-Water Systems A Review of Principles and Concepts, U. S. Geologycal Survey Circular 1013, U.S.A
[22] Alloway, B.J. and Ayres, D.C. (1993) Chemical Principles of Environmental Pollution. Blackie Academic & Professional. New York, USA. pp. 291.
[23] Gomez-Alvarez et al., 2007 Gómez–Álvarez, A., Valenzuela-García, J. L., Aguayo-Salinas, S., Meza-Figueroa, D., Ramirez-Hernández, J., Ochoa-Ortega, G. (2007) Chemical partitioning of sediment contamination by heavy metals in the San Pedro River, Sonora, Mexico. Chemical Speciation and Bioavailability. 19, 25-36.
[24] Gómez-Álvarez, A. (2001). Evaluación de la calidad física y química del agua y sedimento del Río San Pedro, Sonora, México, Durante el Periodo 1997-1999. Tesis de Maestría en Ciencias de la Ingeniería. Universidad de Sonora
[25] Henry J. G. y Heinke W.G. (1999) Ingeniería Ambiental. Ed Prentice Hall. pp. 195, 196, 431.
[26] Speece R.E. (1996) Amaerobic Biotchnology.Archae Press.pp.24-63
[27] Burak Demirel, Paul Scher ,Orhan Yenigun, and Turgut Onay (2010) Production of Methane and Hydrogen from Biomass through Conventional and High-Rate Anaerobic Digestion Processes Critical Reviews in Environmental Science and Technology. 40:116–146.
[28] Lise Appels , Jan Baeyens , Jan Degre`ve , Raf Dewil (2008) Principles and Potential of the Anaerobic Digestion of Waste-Activated Sludge Progress in Energy and Combustion Science. 34: 755–781.
[29] A.c. van Haanadel 2006 Advanced Biological Treatment Processes for Industrial Wastewaters Principles and Applications.In: Francisco 1.Cervantes, Spyros G. Pavlostathis and Adrianus C. van Haandel: in IWA.pp. 67-70.
[30] Gujer, W. and Zehnder, AJ. (1983) Conversion Processes in Anaerobic Digestion Wat.Sci.Techn. 15,127.
[31] Chiu-Yue LIinm and Chin-Chao Chen (1999) Effect of Heavy Metals on the Methanogenic UASB Granule Wat. Res. 33, 2, pp. 409-416.
[32] Henzen, M. and Harremoes, P. (1983) Anaerobic Treatment of Waste Water in Fixed Film Reactors – a Literature Review Wat.Sci. Techn. 15, l.
[33] Sandoval C, Carreño M. De, Castillo E F, Vergara M. (2007) Sludge Anaerobic Microbiologic Characterization Used in the Organic Fraction of Urban Solid Waste Treatment Scientia et Technica Año XIII, 35:509-514.

[34] APHA. *Standard Methods for the Examination of Water and Wastewater* (1995) Washington DC, USA: American Public Health Association/American Water Works Association/Water Environment Federation.

[35] Bouvier J.C., Steyer J.P. and Delgenes J.P. (2002) On-line titrimetric sensor for the control of VFA and/or alkalinity in anaerobic digestion processes treating industrial vinasses, VII Latin American Workshop and Symposium on Anaerobic Digestion, Merida, Mexico, October 23-25/2001, 65-68.

[36] Almendariz F. J. (2005) Tratamiento de Sosas Gastadas en un Reactor de Lecho Expandido de Lodo Granular Anaerobio Tesis Doctoral. Universidad Autónoma Metropolitana. México, D.F.

[37] Alves M., Cavaleiro A.J., Ferreira E.C., Amaral A.L, Mota M., da Motta M., Vivier H. and Pons M-N. (2000) Characterisation by image analysis of anaerobic sludge under shock conditions. Water Science and Technology Vol 41 No IWA Publishing 12. 207–214.

[38] Laguna A., Ouattara A., González R. O., Barón O., Fama G., El Mamouni R., Guiot S., Monroy O. y Macarie H. (1999) A Simple and Low Cost Technique for Determining the Granulometry of Upflow Anaerobic Sludge Blanket Reactor Wat. Sci. Technol 40 (8): 1-8.

[39] Onofre Monge, Leobardo Valenzuela y Evelia Acedo (2010) *Biosorción de Cobre con Bacterias Aerobias Inmovilizadas en Zeolita. Aislamiento, Caracterización, Selección de Microorganismos y Cinética de Biosorción.* Lambert Academic Publishing.

[40] Aquiahuatl R. M.A. y Pérez Ch. M.L. (2004) *Manual de prácticas del laboratorio de microbiología general.* UAM. México. Pp. 37-49.

[41] Koneman E. W., Allen S., Janda W., Schrenberger P., Winn W. (1999) Diagnóstico microbiológico. Médica Panamericana. México. pp. 1-349; 1258-1357.

[42] Mac Faddin J. F. (2003) Pruebas bioquímicas para la identificación de bacterias de importancia clínica. Médica Panamericana. Pp. 54 -421.

[43] Abuzer Çelekli a, Mehmet Yavuzatmacab, Hüseyin Bozkurtc (2010). An ecofriendly process: predictive modelling of copper adsorption from aqueous solution on *Spirulina platensis*. Journal of Hazardous Materials. Vol.173, 123–129.

[44] Monge A. O. (2003) Biorrestauración de suelos contaminados por cianuro, utilizando Bacillus sp. Tesis de Maestría en Ciencias de la Ingeniería. Universidad de Sonora.

[45] Duarte S. D. (1997). Evaluación de la capacidad de biosorción de Ni(II) por Thiobacillus Ferrooxidans. Tesis de Licenciatura en Ingeniería Química. Universidad de Sonora.

[46] Oliveira M. J. (2003). Estudio de la biosorción de cobre por perlas de alginate de calcio. Tesis de Licenciatura. Universidad Nacional Mayor de San Marcos. Perú.

[47] Agemian H. y Chau A. S. Y. (1975) An atomic absorption method for the determination of 20 elements in lake sediments after acid digestion. Analytica Chemica Acta 80:61-66.

[48] Borzacconi L. y López I. (2003) Cinética de ingeniería de reacciones. Notas de curso. Facultad de Ingeniería. http://www.fing.edu.uy/iq/reactores/cursos/reactores1.pdf consulta: Abril-2012.

[49] Smith J. M. (1991) Ingeniería de la Cinética Química. ED. Continental. México D. F. pp. 369-377.

[50] Fogler, H. S., Escalona G. R. L. y Ramírez S. J. F. (2001) Elementos de ingeniería de las reacciones químicas. Pearson Educación. México. pp. 8-11; 34-39.

[51] Jawed, M. and Tare, V (1999) Microbial Composition Assessment of Anaerobic Biomass Through Methanogenic Activity Tests Water S.A 25. 3: 345 – 350.

[52] Lettinga, G., van Velsen, A.F.M., Hobma, S.W., De Zeeuw, W. y Klapwijk, A. (1980) Use of the Upflow Sludge Blanket (UASB) Reactor Concept for Biological Wastewater Treatment, Especially for Anaerobic Treatment. Biotechnol. Bioeng .22: 699-734.

[53] Certucha-Barragán M. T, Duarte-Rodríguez G. R, Acedo-Félix E, Almendariz-Tapia F. J, Monge-Amaya, Valenzuela Garcia J. L, Leal-Cruz A. (2010) Estudio de la Biosorción de Cobre Utilizando Lodo Anaerobio Acidogénico. Rev. Int. Contam. Ambient. 2: 26-36.

[54] Shuler, M. L. and Kargi, F. (2002) *Bioprocess Engineering: Basic Concepts, 2nd edition,*Prentice-Hall. Inc .

[55] Reyes E. D., Cerino C. F. y Suárez H. M. A. (2006) Remoción de metales pesados con carbón activado como soporte de biomasa. Ingenierías. IX, 31.59-64.

[56] Cañizares- Villanueva, R. O. (2000) Biosorción de metales pesados mediante el uso de biomasa microbiana. Revista Latinoamericana de Microbiología. 42:131-143.

[57] Gupta R., Prerna Ahuja, Seema Khan, R. K. Saxena and Mohapatra H. (2000) Microbial biosorbents: Meeting challenges of heavy metal pollution in aqueous solutions. CURRENT SCIENCE, 78, 8, 967-973.

[58] Leyva Ramos R., Medellín C. N. A., Guerrero C. R.M., Berber M. M.S., Aragón P. A. y Jacobo A. A. (2005) Intercambio iónico de plata (I) en solución acuosa sobre clinoptilolita. Rev. Int. Contam. Ambient. 21 (4) 193-200.

[59] Rodríguez, H. (2001). Estudio de la contaminación por metales pesados en la Cuenca de Llobregat. Tesis Doctoral. Barcelona.

[60] Vega B. (2008) Inmovilización de células. 1º Curso Iberoamericano de Biocatálisis Aplicada a Química Verde. Facultad de Química. Universidad Nacional de Quilmes. Argentina.

[61] Krause, U., Thomson-carter, F.M. and Pennington, T.H. (1996) Molecular Epidemiology of *Escherichia coli* O157:H7 by Pulsed-Field Gel Electrophoresis and Comparison with That by Bacteriophage Typing. J.Clin. Microbiol. 34:959-961.

[62] Núñez Gaona O. (2004) Producción de invertasa por *Aspergillus niger* en fermentación en medio sólido. Tesis maestria en biotecnología. UAM-Iztapalapa, México.

[63] Servin R. L. (2006) Metalurgia de minerales no metálicos, zeolitas. http://www.monografias.com/trabajos/zeolitas/zeolitas.htm. Consulta: Abril 2012.

[64] Aguayo, S.S. y Mejía, Z. F. A. (2006) Adsorción de arsénico en zeolitas naturales pre-tratadas. Memorias del XVl Congreso Internacional de Metalurgia Extractiva. Saltillo, Coahuila. 303-310.

Chapter 19

Development of Sustainable Willow Short Rotation Forestry in Northern Europe

Theo Verwijst, Anneli Lundkvist, Stina Edelfeldt and Johannes Albertsson

Additional information is available at the end of the chapter

http://dx.doi.org/10.5772/55072

1. Introduction

Modern willow short rotation forestry is based on traditional woodland management which uses the ability of certain tree species to grow new shoots from the stump after being cut down. Depending on site fertility, growing season length, initial planting density and species, willows may be coppiced from once a year to every fifth year, and the stands may remain productive over several decades. Traditionally, small-scale willow plantations have been used for fuel, fodder, convenience wood, basked making, bee keeping, and for horticultural purposes. Willows also may be used for erosion control, including wind and water erosion, and to avoid snow drift along roads. While the traditional use of willow is declining rapidly in Europe, the use of willow as an alternative crop for farmers has led to an increasing interest in willow breeding and cultivation [1]. A renewed research effort on short rotation willow coppice plantations in Sweden commenced in the late 1960's due to a predicted shortage of raw materials for the pulp and paper industries, which turned out to be a false alarm. However, the 1970's energy crisis constituted a new driver to continue research on willows as a source of biomass for energy purposes. Additional drivers, such as employment issues in the Swedish country side, and environmental concerns also influenced research funding rates and directions towards willow short rotation coppice. In the late 1980's willow growing for energy was implemented at a larger scale and commercialized in Sweden. A tax on carbon dioxide emissions for the combustion of fossil fuel in heat production was introduced by the Swedish government during the 1990's and created more favorable market conditions for investment in and implementation of biofuel systems [2]. In 1996, Sweden joined the European Union, which employed an agricultural policy in which subsidy levels to farmers constantly were altered and adapted to short term market situations. As willow growing is a long term commitment which requires longer term investments, this EU-policy promoted the use of annual crops, and the exponential increase of areas under willow cultivation leveled out after 1996 and even started to decline.

In the meantime, the Swedish concept of large-scale willow cultivation for bioenergy purposes was exported to several EU-countries, notably to the UK and Poland, and a development of similar growing systems also was pursued overseas, in New Zealand and in the USA [3].

It was recognized early that willow growth concurs with potentially high evapotranspiration rates [4] and high nitrogen retention rates [5]. Willow species also may exhibit selective uptake of heavy metals [6], which underlies the potential to use willow as a phytoextractor for e.g. Cd from polluted soils [7]. These special traits of willow have allowed a further development of short-rotation willow coppice systems for environmental purposes [8]. Willow growing systems may be used as vegetation filters for purification of waste water [9], for cleaning of polluted drainage water from agricultural land [10] and as a recipient of nutrients from municipal sludge [11]. As willow stands are harvested at regular intervals, the pollutants are removed from the soil-plant system, while added nutrients and water enhance the systems' biomass production. These systems then function as multi-purpose systems, simultaneously aiming at biomass production for energy purposes and provision of environmental services, while producing clean water and neutralizing potentially hazardous compounds. Several efforts have been made to assess the economic gains of such multi-purpose systems [e.g. 12, 13], and Volk et al. [3] concluded that the economic valuation of the environmental benefits is necessary for a further deployment of woody crops.

In the following sections, a brief overview will be given of the plant material and growing system used in willow short- rotation forestry (SRF) and of the history of willow research, with a focus on the developments in Sweden. We then continue with a description of the development and implementation of willow SRF in commercial practice, and with the current guidelines for commercial willow growing. We also present an update of recent research, performed to improve the productivity and sustainability of willow short rotation forestry as an agricultural crop for bioenergy purposes, and include some results of ongoing research projects.

2. Species characteristics and natural distribution of willows

The genus *Salix* comprises about 350 to 500 different species worldwide [14] and is taxonomically complex and difficult to arrange in distinct sub-groups, probably due to intersectional and intersubgeneric polyploidy [15]. About 10% of the willow species consist of deciduous tree species, some of which may attain a height of > 20 meter. However, the vast majority consists of multiple stemmed trees and shrubs, and also a number of very short procumbent species can be found, not exceeding the height of the herb-layer in which they reside. Willow mainly is a boreal-arctic genus, with its natural distribution primarily in the northern hemisphere. Most willow species are found in China and in the former Soviet Union, and some indigenous species are present in India and Japan. The genus also occurs naturally in the southern hemisphere in Africa and in Central- and South America [14], and has been introduced in Australasia and New Zealand. Many species have been transferred beyond their natural range. The short rotation coppice systems currently in use in Sweden are mainly based on *Salix viminalis*, which was introduced in the 1700's from continental Europe for the purpose of basket making, and on their hybrids with *S. burjatica* and *S. schwerinii*, recently introduced from Siberia.

Early records of willow cultivation date from 2000 years ago in the Roman Empire and in modern times willow breeding and selection programs have been recorded from Sweden, the UK, Belgium, France, Croatia, Poland, Hungary, former Yugoslavia, Romania, Bulgaria and China, but also outside Eurasia in New Zealand, Argentina, Chile, Canada and in the USA. The development of molecular methods in plant breeding is likely to speed up the selection of new and viable material [16] and is envisaged to lead to a willow crop which is less prone to pests and diseases and which can be managed with lower inputs than the current systems [17].

The widespread interest in the willow genus is due to the fact that many of its species, which are light demanding pioneer trees, exhibit a very high growth rate in their juvenile stage. Many willow species can easily be propagated by means of cuttings, and most species and their hybrids will generate new shoots abundantly after cutting down older shoots and stems [18]. Under Swedish conditions, willow has a very high and well documented growth potential [19] which, though, is not completely realized in commercial short rotation forestry [20]. To fully exploit the growth potential of willows, a soil fertility level is required which is comparable with those found on conventional agricultural soils in Sweden. To maintain growth in the long term, dry sites have to be avoided and nutrients have to be added at a rate which balances nutrient removal by harvest. Compared to conventional forestry, willows require a relative intensive management, but compared to conventional agricultural practice, management input is much lower.

3. Growing systems & population dynamics

Given the huge range in size, growth form and coppice ability in the willow genus, production systems for willow may vary from single-stemmed systems with less than 500 trees ha^{-1} and a rotation period of over 20 years, to systems which contain over $4{\times}10^4$ plants plant which generate over half a million shoots ha^{-1} in a one-year coppice cycle. In the remainder of this chapter, we focus on growing systems which are generated from cuttings, at a planting density of $1{\times}10^4$ to $1.5{\times}10^4$ cuttings ha^{-1}, and treated as a coppice system, undergoing multiple cutting cycles. In Scandinavian conditions, one season may be too short to replenish carbohydrate reserves in willow stubs after harvest, and a one-year harvest cycle may deplete a plantation and compromise its viability [21]. Cutting cycle lengths in Swedish practice have been 3 to 5 years, and with the introduction of faster growing clones, cutting cycle lengths now are being decreased to 2 to 4 years. In commercial practice, a double row system is employed (Figure 1). However, Bergkvist and Ledin [22] showed that planting design could be adjusted, within certain limits without losing yield potential, to the requirements of tractors and machines used in managing the *Salix* stand.

Figure 1. Machine planting of willow by means of a Woodpecker 601, using long rods and planting three double rows at a time (Photo: Nils-Erik Nordh).

The development of a population of willow stems is constrained by competitive interactions which lead to self-thinning, yield-density effects and to skewed size-frequency distributions of stems [23, 24]. Those effects of competitive interactions need to be accounted for when determining optimal plant spacing and harvest frequency. Especially in dense willow coppice, not only shoot mortality but also extensive stool mortality may occur [25], thereby leading to lasting gaps and production losses [26]. Studies on the long term dynamics of willow coppice have shown that an initial variability in plant size becomes enlarged over time, that self-thinning leads to mortality of the initially smallest stools [27], and that the competitive hierarchy between stools is preserved over harvest [26]. As soil factors are known to be important determinants of willow growth [28, 29, 30, 31], differences in soil at field scale likely underlie the initial size variability between plants. Differences in cutting quality also may cause an initial variability in growth performance between plants (see section 4.2). To be able to detect possible effects of cutting quality and to separate those from soil factors, it is advantageous to perform controlled experiments which allow the relative variation to be attributed to only a few factors. Verwijst et al. [32] compared the relative variation in shoot height of willow populations grown in the field with the relative variation of populations grown in boxes which had a standard soil and were treated as similar as possible with regard to fertilization and irrigation (Figure 2). The controlled experiments showed a decreased relative variation and enhanced the detection of cutting quality traits with relevance for early establishment success.

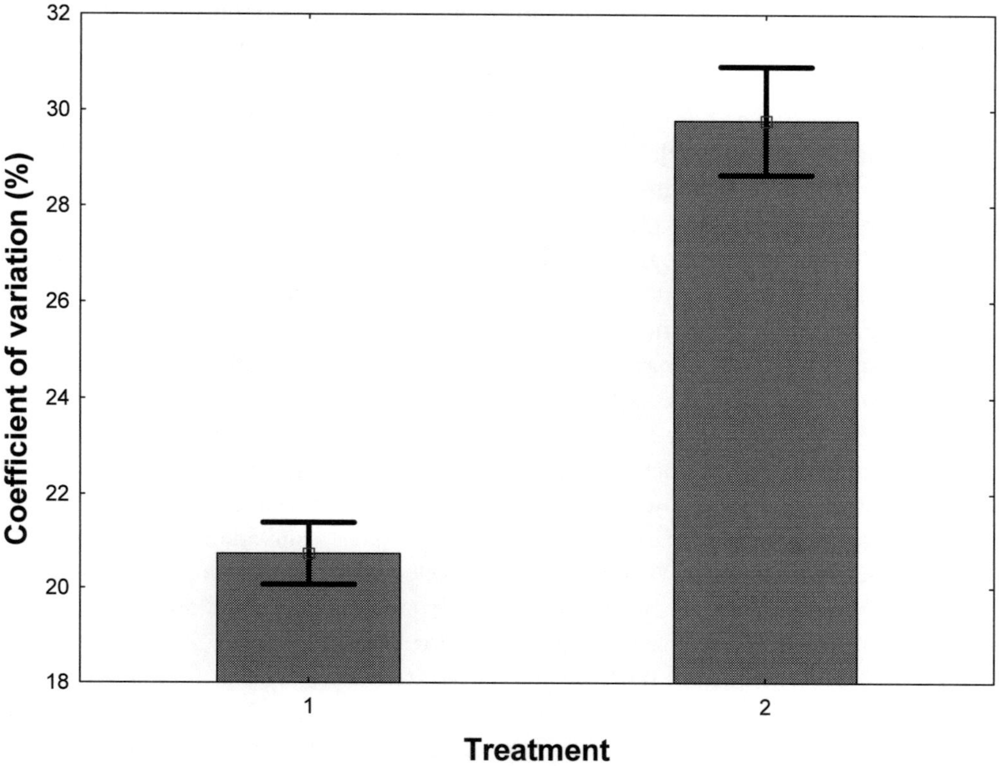

Figure 2. Relative variation (%) and its standard error in plant height for shoots from cuttings planted in a controlled environment (Treatment 1) versus shoots planted in the field (Treatment 2).

As willow is a relatively new crop, advances in willow breeding generate a steady increase in potential and attainable yield [33, 34]. This increase in biomass yield is estimated to be 50 to 100% since the 1970s'. This means that spacing, harvest frequency and fertilization have to be adapted to the rapidly evolving new plant material, in order to avoid mortality and ensure a high productivity also during the later cutting cycles. Most of the planted willow stands in Sweden consist of monoclonal stands or blocks of monoclonal units. However, such monoclonal stands are vulnerable to pathogen adaptations [35] and it has been shown that clone mixtures may be effective against the spread of diseases [36]. However, the relative competitive power of willow clones does differ, which means that certain clones may be outcompeted by other ones in mixtures of clones. If a mixture consists of only a few clones and one of the components is attacked by a pathogen, the susceptible clone is likely to be outcompeted by the others, thereby causing gaps, a delayed stand closure and lower productivity in later cutting cycles [37]. Furthermore, as clone-site interactions have been reported for willow, and the performance of clones in mixtures can not be predicted from their performance in pure stands [24], successful clone combinations are expected to be highly site specific.

3.1. Site choice and preparation

Many willow species do have abroad ecological amplitude. However, to obtain a high productivity, willow has specific site requirements. Being a pioneer species, willow is light demanding, and a rapid establishment can only be achieved without competition by weeds for light. Once established, the leaf area index of a willow canopy will exceed 6 $m^2 \times m^{-2}$ [4, 38] and will suppress weed growth. Willow thrives on most agricultural soils, as long as the pH is in the range of 5 to 7 [39]. Water use efficiency of a willow crop is about 4 to 6 $g \times kg^{-1}$ [4]. This is a high value compared to values of other tree species, but given the potentially high biomass production of willow, water availability is conceived as a critical factor in willow SRF [40]. Consequently, lighter soils, especially in drier areas, should be avoided for willow growing. A low precipitation during the growing season can be compensated for if winter precipitation is abundant and soils have a good water holding capacity or do have access to groundwater. While many willow species have a boreal-arctic origin and are native to northern temperate regions, fast-growing hybrids may be susceptible to frost damage from bud-burst and onwards. If planted at frost-exposed sites, a single night frost may decrease a single year's productivity by 50% [41] and will also impact negatively on the biomass production in the following years. Therefore, sites prone to late spring frost should be avoided and it is important to choose clones which have a site-adapted phenology with regard to timing of bud burst. Willow can be harvested with a reasonable cost-efficiency on sites which are 5 ha or larger, and even on slightly smaller sites if willow is harvested on adjacent sites. Planting and harvest equipment for willow requires a relatively widely spaced headland (10 to 12 m in width), which means that single willow fields should not be smaller than 2 ha, and easily could be reached by the harvest machines [42]. Larger stones also should be removed from the soil surface, as they may damage harvest equipment. As planting (see section 4.2) requires a well prepared seed bed, autumn plowing and early spring seedbed preparation are common measures prior to planting. Such preparation has to go along with adequate weed control (see section 4.3). Another selection criterion for willow growing sites is the proximity to a consumer, usually a combined heat and power plant. As moist willow chips do have low energy content per volume, transportation distances by road should be minimized [43]. Finally, willow growing is a form of land-use, and as such, it may interfere with a range of other interests than sheer biomass production. Short rotation forests may affect landscape views, the environment and biodiversity in a positive or negative way, depending on the functions that we require from a semi-natural landscape element, and on how we choose to integrate such functions in a single growing system [1, 44].

3.2. Planting & cutting quality

One of the large advantages of most willows is that they can be propagated vegetatively by means of cuttings. Traditionally, cuttings of about 20 cm in length were produced manually from 1-year old long rods. These cuttings were taken during the winter period, when willow is dormant, and could be stored in a fridge until planting in spring. During commercialization of the growing system in Sweden in the late 1980s, manual planting was replaced by machine planting. Establishment costs for short rotation willow coppice decreased substantially during the initial phase of commercialization in Sweden [45]. This was mainly achieved by mechanisation of planting, employing equipment which, in one process, cuts willow rods (1.8 – 2.4 m. long) into cuttings and then plants them (Figure 1). These cuttings are around 18 to 20 cm long, and the cutting is pressed down into the prepared soil so that only 1-2 cm protrudes above soil surface. This is believed to provide the cutting with good soil contact, thereby minimizing the risk of drying out [46]. Field storage of cuttings can result in water loss and reduce shoot survival and biomass production. This problem has partly been overcome by the use of entire shoots, which are considered to be more resistant to desiccation than cuttings [47]. Volk et al. [48] also pointed to risks of desiccation and showed that a prolonged time of field storage after cold storage may lead to a decrease in survival and growth rate.

Stage	Description
1	No sign of bud swelling, the tip of the bud is tightly pressed to the shoot.
2	The tip of the bud starts to bend from the stem, bud scales are starting to open and the length of the shoot tip is 1–4 mm.
3	The shoot tip is 5 mm or longer, protruding leaves are put together.
4	New leaves start to bend from each other.
5	One or more new leaves are perpendicular to the shoot axis.

Table 1. Assessment criteria for bud burst stages.

Cutting size (length and diameter) has positive effects on subsequent willow growth. The positive effects of cutting size on growth and survival decline with increasing sizes ([49, 50, 51], and Rossi [52] found that the differences in cutting length with relevance for establishment in practice are to be found between lengths of 10 and 20 cm. Positive effects of cutting size generally are attributed to the size of the carbohydrate pool available for allocation to roots and shoots [53]. The effect of cutting length may also be associated to the ability of longer cuttings to withstand soil desiccation [54]. The phenological development of buds and shoots is affected by cutting size and also by the height above ground from where the cuttings were taken [51]. Using the simple assessment criteria for bud development as described in Table 1, bud development, a few weeks after planting, is a function of the diameter size of the planted cutting (Figure 3). However, cuttings derived from apical positions along shoots display for a given diameter a higher shoot biomass production than cuttings derived from the more basal parts (Figure 4). As willow rods display a taper, the question arises which of the two factors (cutting size or position) is the strongest determinant of shoot biomass production during early establishment.

A further evaluation of produced shoot biomass on the cuttings showed that cutting size by far is the single most important determinant of early biomass production, which led to the recommendation to employ thicker cuttings and to discard the thinner apical parts from long rods. While the introduction of planting machines has increased the speed of planting and reduced planting costs, ongoing research indicates that planting machines may cause damage to cuttings, especially when planted in compacted soils. Preliminary results by Verwijst et al. [32] and by Edelfeldt et al. [55], suggest that that undamaged cuttings had a better growth performance than visibly damaged cuttings. Planting by machine on hard soil resulted in a relatively large number of cuttings landing on the soils surface. Soil compaction and machine planting interacted with cutting dimensions, the poorer performance of thinner cuttings being more pronounced in compacted soil (Figure 5).

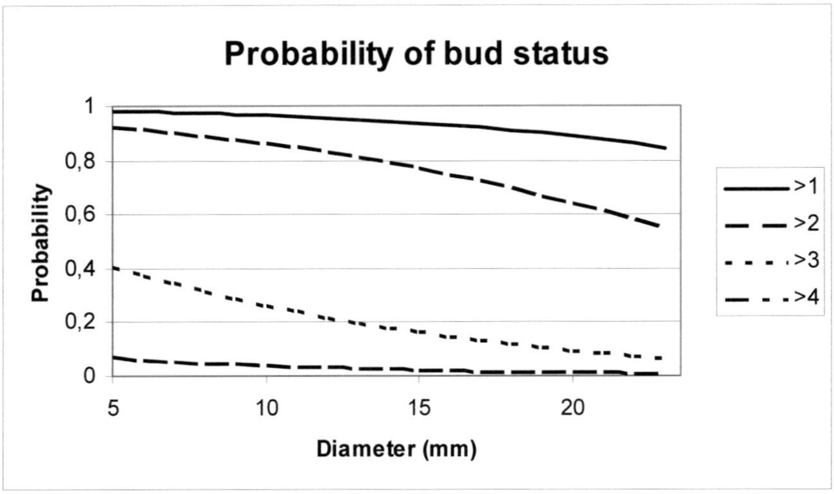

Figure 3. Probability of bud status (see Table 1) at average values for five clones and cuttings derived from a position of 95 cm above soil surface, a few weeks after planting. Probability of high bud status decreases with diameter.

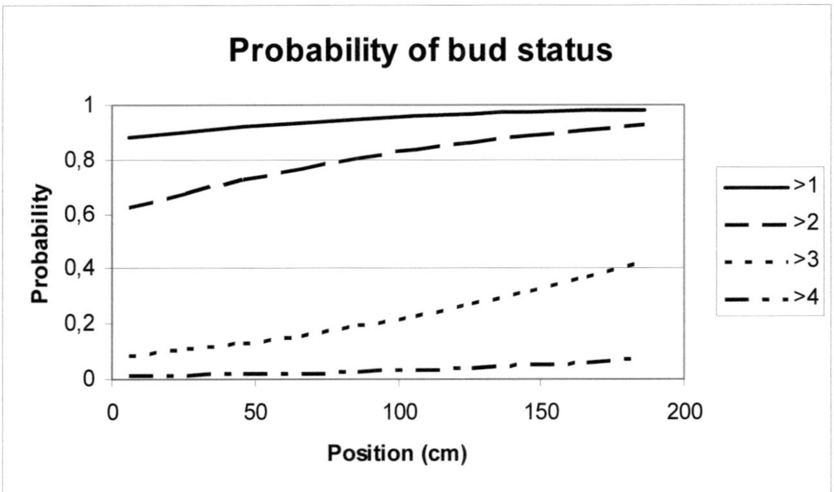

Figure 4. Probability of bud status (see Table 1) at average values for five clones and diameter 12.5 mm, a few weeks after planting. Probability of high bud status increases with the original height position of the cutting along the rods from which is was derived.

Furthermore, machine planting also increased the relative variation of shoot height (Figure 6) compared to hand prepared and planted cuttings. Consequently, to obtain a faster and more even establishment of willows, Edelfeldt et al. [55] recommend thorough soil cultivation prior to planting, further development of planting machines to minimise damage to cuttings at planting, and the use of cuttings with a diameter of at least 10-11 mm.

Figure 5. Cuttings planted by machine in a hard soil were transformed to a soft soil to isolate the effect of machine planting from other factors. The thinner cuttings were visually damaged and displayed a lower sprouting performance than the thicker ones (Photo: Nils-Erik Nordh).

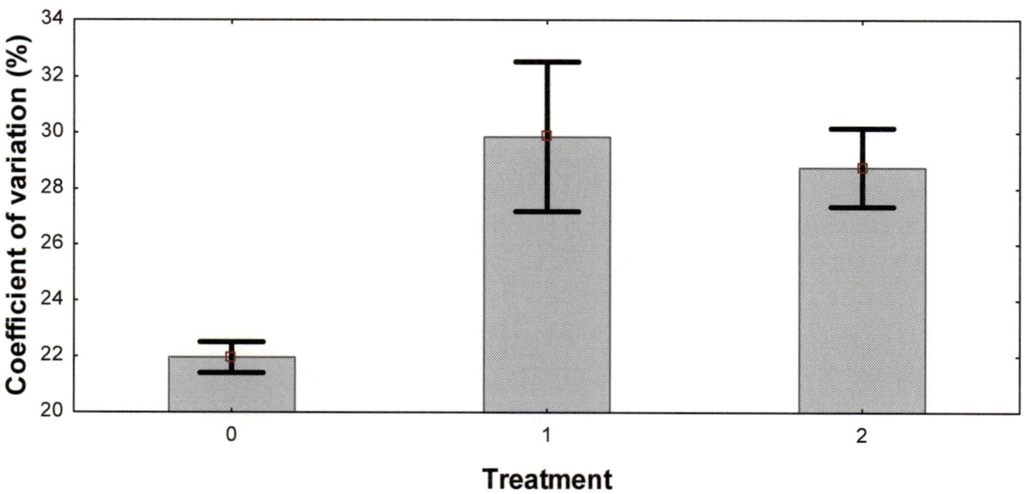

Figure 6. Relative variation (%) and its standard error in plant height for manually planted cuttings (Treatment 0) versus machine planted cuttings in Soft and Hard soil (Treatments 1 and 2, respectively).

3.3. Weed control

Weed control is necessary when establishing willows from cuttings, because its takes a relatively long time for willow cuttings to develop attributes which make them competitive against weeds. Competition is an interaction between plants which require the same limited resources like nutrients, water and light. Harper [56] defines competition as 'An interaction between individuals brought about by a shared requirement for a resource in limited supply and leading to a reduction in the survivorship, growth and/or reproduction of the individuals concerned', and thereby points to the effects of competition. The aim of weed control is to ensure that as much resources as possible are accessible for the crop and not for the weeds, and to reduce or delay growth and development of the weed flora [57].

Willows under establishment from cuttings have a relatively low competitive power against weeds because it takes a while for willows to develop roots needed for the uptake of nutrients and water. Consequently, perennial weeds, which have a developed root system prior to the onset of leaves, have to be removed completely before planting willows. This commonly is done by means of one or two applications of a glyphosate-based herbicide, applied at the appropriate rate, during the summer/autumn prior to spring planting. If the area has not been used for agricultural purposes for a number of years before planting, it is recommended to grow cereals there for at least one season to ensure an adequate weed control [42]. The relative competitive ability is also affected by seed rate (plant density), which is low for willow (between 1 and 2 cuttings m^{-2}), in comparison to the amount of germinating annual weeds triggered by seed bed preparation. Such weeds may germinate only a few days after seed bed preparation, while it may take a week or more for willow cuttings to exhibit a first bud burst after planting. This implies that the time between the last seed bed preparation and willow planting should be minimized. To counteract the effects of the inherent differences in relative emergence time between willow and weeds, soil cultivation by different types of cultivators, rototillers or harrows are recommended as a weed control measure during willow establishment [46, 58]. There are also different soil-applied herbicides that are permitted to be used at planting or shortly thereafter. Given the low planting density of willow cuttings, a full canopy closure, which for willow implies a leaf area index > 6 $m^2 \times m^{-2}$ [4, 38] is hardly ever reached during the establishment year, which means that if weeds are not kept back during the establishment year, they may establish and compete with willows for light. The use of mechanical weeding may therefore proceed even after bud burst and early shoot formation in willows. As the cuttings are well fixed in the soil and young willow shoots are flexible, they will not become damaged by this treatment. The current recommendation is to perform these control measures at least three times during the first year [46].

Weed control might also be necessary to perform the year after planting depending on weed management success the first year, clones and site conditions. As the willow plants will be better established by then, it is usually enough to perform mechanical weeding two times early in the season [46]. Another possibility is to spray a soil-applied herbicide well before bud burst [42] or to use a selective herbicide during spring or early summer [46]. Weed control the second year usually requires that the first year shoots are cut back. This practice has been questioned [59] and is no longer recommended in Sweden [42]. If weed control has been efficient during the establishment phase, no additional measures are required to control the weeds the following years. If early plant mortality has led to gaps in the stand, weeds may establish and maintain themselves below canopy gaps (Figure 7). In case weeds survived below such gaps, weeds may be controlled directly after each harvest.

If the weeds are not controlled during the establishment phase, willow growth might be dramatically reduced. Field experiments conducted in Southern Sweden by Albertsson in 2010-2012, with 10 modern willow varieties, grown both with- and without weeds, have shown that weeds can increase plant mortality, and reduce growth the first year by more than 95% [42], see Figure 8. Several other studies have also shown that willow, in the establishment phase, is very sensitive to competition from other plants [60, 61, 62]. Preliminary data from the Swedish study also suggest that there is an interaction between voles and weediness, since plots with weeds show more damage by voles than plots without weeds, thereby making weed control even more important.

Figure 7. Poor establishment of willow leads to gaps in which weeds may establish, thereby causing the need for prolonged weed control after a first harvest (Photo: Nils-Erik Nordh).

Weeds in willow short rotation coppice might, in the future, be controlled with other measures than the above mentioned. Studies are ongoing to investigate if willow clones differ in their ability to compete with weeds. Fast initial growth, early bud burst, fast canopy closure and the ability to tolerate or release allelophatic substances might be favorable weed competing traits. If differences exist, it might be possible to breed for these traits or to use competitive willow varieties that combine well with a specific weed control measure.

Different cover crops such as rye (*Secale cereale* L.), dutch white clover (*Trifolium repens* L.), buckwheat (*Fagopyrum esculentum* Moench) and caragana (*Caragana arborescens* Lam.) have been studied as a way of controlling weeds and improve nutritional management in willow [63, 64, 65]. However, there is still more research to be conducted in this area before a suitable willow cover crop system is ready for commercial use. Mechanical weeding techniques are under constant development and recent results indicate that automatic intra-row weeding is possible [66]. Hence, these techniques may be further developed to be used in willow since weeds within the rows are hard to control mechanically with conventional equipment.

Figure 8. Weeds were removed mechanically and by hand in the willow stand to the left while no weed control measures were performed in the willow stand to the right. The photo was taken five months after planting (Photo: Johannes Albertsson).

3.4. Fertilization

Most field-based cropping systems do have an actual production which is well below their potential production. The potential production of a crop is determined genetically by its nutrient-, water- and light use efficiency. But given those efficiencies, a field environment hardly ever constantly provides optimal supply of water, nutrients and light to the crops. The production which is attained after restriction by abiotic factors such as light, water and nutrients is called attainable production, and can be regulated by site choice and fertilization. Actual production is usually lower than the level of attainable production, being utterly restricted by the effects of biotic agents, such as herbivores and pathogens.

Consequently, plant breeding and selection partly strive to generate plant material with a high resistance against pests and diseases, but also to generate material with a positive response for treatments such as fertilization. From a farmers' perspective, fertilization may be applied if it enhances profitability of the cropping system. Profitability then is a function of costs for fertilizers, the net value of the crop, and of the fertilization effect, i.e. the additional biomass increment per unit added fertilizer. The willow clones that have been released during the last decades in Sweden display a higher actual productivity that the earlier ones [45, 67], and this seems amongst others to be the result of a clonal selection towards a higher shoot/root allocation patterns, resulting in a higher harvestable biomass increment per unit fertilizer. While selection thus promotes a positive response to fertilization and irrigation, it may also increase the susceptibility of clones for incidental drought periods [68].

Recommendations for farmers with regard to fertilization of willow coppice on agricultural land during the last decades in Sweden have been subject to a great deal of confusion, due to the fact that fertilizer costs, net crop revenue and fertilization response of the crop all rapidly have been changing through time. Early recommendations by Ledin [69] were based on fertilization trials with older willow clones and on economic calculations which accounted for projected crop values which were not met by the market. Net values for different fertilization strategies under different scenarios with regard to fertilization costs and actual net crop values recently have been calculated [70] after field based parameterizations of the fertilization response of more recent willow clones. It was found that fertilization responses differed widely between clones and sites and that fertilization should be adapted to the local conditions. Under current market conditions and using recently released willow clones, fertilization can greatly enhance profitability. The need for fertilization of modern clones in a first cutting cycle could not be assessed due to lack of data. However, fertilization during the first year may positively affect weed growth, and is therefore not recommended. Plantations with modern willow clones should be fertilized with at least 220 kg N ha^{-1} during the second and consecutive cutting cycles. Annual fertilization in willow stand would require a further machine development, as conventional machinery cannot enter tall willow stands.

Fertilization may also be performed with nutrient-rich residues such as municipal wastewater and sludge to willow short rotation coppice [71] and may render a more cost-effective and sustainable cultivation. Rosenqvist and Ness [72] provide an economic analysis of leached purification through willow coppice vegetation filters and showed that economic gains were made compared to conventional purification, while an increased biomass production led to additional economic gains. It also is concluded that willow vegetation filters are more cost-effective than conventional treatment methods and may facilitate recycling of valuable products in society [5]. This conclusion is sustained by other assessments of the economic gains of such multi-purpose systems [e.g. 12, 13]. Volk et al [3] even concluded that the economic valuation of the environmental benefits is necessary for a further deployment of woody crops.

3.5. Control of pests and diseases

Attainable biomass production of a crop, as determined by its genetics and actual resource levels provided in a particular field situation, is usually reduced by the action of pathogens and herbivores. Especially in genera with species that hybridize easily, such as willow, the relationship between plant breeding and pest and disease control is strong, because such genera in general attract many kinds of insects and pathogens. Plants may be well adapted to a specified range of abiotic conditions, which display a site specific variation. However, pests and diseases are biotic factors which not only vary in space and time, but may also co-evolve with plants. Consequently, potentially pathogenic organisms may be present and may do little harm for longer periods in a willow stand, until virulent strains develop which may be very clone specific. For instance, susceptibility to defined pathotypes of leaf rust (*Melampsora epitea*) is rather clone specific [73]. Consequently, it is important that new clones are released constantly by breeding programs and that a broad genetic base is used, targeting a broad tolerance to a range of pathogens. Poplar breeding programs in Western Europe previously have underrated this issue, resulting in the destruction of many poplar stands by leaf rust varieties that managed to adapt to the poplar clones [74]. In willow breeding, this issue was acknowledged early. Development of new high producing willow clones was initiated in Sweden in 1987 by Svalöf-Weibull AB [33]. The main purpose of the breeding program was to develop high yielding clones resistant towards pests, frost, and diseases, and with morphology suitable for mechanical harvesting. From 1996 to 2002 several new clones were developed in cooperation between Svalöf-Weibull and Long-Ashton research in UK, also with a strong focus on pest and disease resistance [34]. Strong advances were made early with regard to leaf rust in willow [75] and resistance of willow to several insect species has also been exploited [76, 77]. Production losses between 20 and 40% have been recorded in willow after defoliation by insects [78]. Willow, however, usually recovers well after defoliation, and as the population dynamics of many insects is erratic, and under control of very many factors, damage prevention by means of breeding towards resistance has been chosen, instead of the use of pesticides. Salix has probably the best environmental profile among the arable bioenergy crops available today, partly because neither fungicides nor insecticides are used in the production. This environmental profile is largely an outcome of plant breeding because resistance to pests and diseases, such as leaf rust and certain insects, has been highly prioritized since commercial breeding started in Sweden 25 years ago [79, 80].

3.6. Harvest and logistics

During early commercialization of the willow coppice system as an agricultural crop in Sweden, funding agencies made the decision to put the far majority of the development costs for harvest machines on the account of commercial machine developers. This resulted in a situation in the early 1990s where many willow stands needed to be harvested before self-thinning would lead to an irreversible mortality among willow stools and long-term production losses, while harvest machines still had to be developed and assembled. This is one of the reasons for the early commercial yields to be disappointingly low (see section 4.7).

Fortunately, a variety of willow harvest machines are on the market now, and recent technical improvements greatly enhance harvesting speed while lowering the costs for willow harvesting. In Sweden, willow is usually harvested during the winter, when the soil is able to carry heavy machinery and when willow chips can be transported to district heating plants for direct use, without long-term storage (Figure 9).

Figure 9. Willow harvest by means of a self-propelled chipper which blows the willow chips in an adjacent container (Photo: Nils-Erik Nordh).

However, mild and wet winters may prohibit the use of heavy harvesters, which means that either lighter equipment has to be developed or that the harvest season has to be extended. Expanding the harvesting season for willow biomass crops would expand the time period over which it can be a part of the fuel supply and increase the number of acres that a single harvesting machine could cover in a single year. This would likely increase the demand for willow and certainly reduce harvesting costs, because capital expenditures for a harvester would be spread across more tons of biomass. Nordh [81] investigated the possibility to extend the harvest season, focusing on the re-growth capacity of willow coppice after harvesting , and found that willow (clone Tora) could be harvested from autumn, prior to the onset of dormancy, until late spring, when bud burst already had commenced. Early and late harvest did not increase plant mortality, but it could result in a slight production decrease in the consecutive season.

Apart from direct chipping (Figure 9), willow biomass can be baled (Figure 10) and fragmented in a later stage, possibly after storage, which will decrease moisture content of the willow biomass.

Figure 10. Willow harvest may be performed by means of a machine which produces bales that can be transported by conventional machines. Bales may be stored to obtain biomass with lower moisture content (Photo: Nils-Erik Nordh).

To harvest willow rods for conventional planting by means of a machine, equipment has been developed which can harvest entire one-year old shoots. Mature stands can also be harvested by means of a whole-shoot harvester (Figure 11) which may carry its load to the headland for further transportation. Special equipment has been developed to make bundles from a pile of whole shoots, thereby improving further transportation logistics. As willow is a low-density fuel, willow should preferably be cultivated in the proximity of the consumer, to decrease transportation distances and costs.

3.7. Yield levels

Biomass productivity of short rotation coppice has been studied for several fast growing species in many places of the world, showing an average annual production of 10 to 20 oven dry tonnes (odt) ha^{-1} in most places [82]. In intensively irrigated and fertilized willow plots

Figure 11. A tractor-pulled whole shoot harvester, unloading willow shoots at the headland (Photo: Nils-Erik Nordh).

in southern Sweden, growth rates of > 30 odt ha^{-1} yr^{-1} have been recorded [83]. The potential production of a certain genotype can only be reached if resources (light, water and nutrients) are permanent available and without limitations, and in the absence of pests and diseases. An analysis of short rotation coppice yields in Sweden over the period 1989-2005 showed disappointingly low mean annual production figures of 2.6, 4.2 and 4.5 odt ha^{-1} during the first, second and third cutting cycles, respectively [20]. These low figures can partly be explained by the use of old clones, which have a much lower potential production than those which were released later [34] and which have a relatively high susceptibility to pathogens. Other reasons for this low productivity are site choice, as farmers have been reluctant to use the better soils for willow plantations, and a very poor management. Many of the early plantations never received fertilizer and suffered from a poor establishment due to inadequate weed control. However, annual average yields over 10 odt ha^{-1} have been reached in commercial plantations if fertilization was applied and adequate weed control performed [84], and did not require more than an average availability of water. Taking account of the water use efficiency of willow and precipitation during the growing season, Lindroth & Båth [85] calculated the annual maximum yield to be 8–9 odt ha^{-1} for north-eastern, 9–10 odt ha^{-1} for eastern and 11-17 odt ha^{-1} for southern and south-western Sweden. Studies confined to the new willow clones which have been developed in cooperation between Svalöf-Weibull and Long-Ashton research in UK between 1996 and 2002 confirm that willow breeding has been leading to higher yields in commercial practice. For the new clones, reported yields vary between 5 and 12 odt ha^{-1}, with extremes between 2 and 18 odt ha^{-1} yr^{-1} [34, 86, 87, 88]. This large variation seems to be related to interactions between clones and sites [33, 89].

4. Conclusion

Willow short rotation coppice systems are relatively new as a farm crop and both farmers and extension workers in Sweden have gone through a learning process which is now leading to higher yields in commercial plantations. Traditional willow breeding and selection are already greatly contributing to increasing yields, and it is expected that future improvements of the willow varieties will result in a significant increase of the yields in the near future. Many of the early field research results are currently extended with more controlled experiments, and help to improve short rotation coppice management. Although the early commercial implementation of willow coppice did not meet the expectations with regard to yield, profitability and areal expansion of willow coppice, analyses of the early commercial fields contribute to the improvement of stand management, and of the planting, harvest and transport logistics. Further developments of willow coppice as multi-purpose systems, including environmental functions, are promising. Current research suggests that there is room for further improvements with regard to cutting quality, planting, weed control and fertilization, all of which will contribute to higher future yields.

Author details

Theo Verwijst*, Anneli Lundkvist and Stina Edelfeldt
Department of Crop Production Ecology, Swedish University of Agricultural Sciences, Uppsala, Sweden

Johannes Albertsson
Department of Plant Breeding and Biotechnology, Swedish University of Agricultural Sciences, Alnarp, Sweden

Acknowledgement

We kindly acknowledge the financial support from The Swedish Research Council for Environment, Agricultural Sciences and Spatial Planning, Stockholm, Sweden; The Swedish University of Agricultural Sciences (SLU), and The Thermal Engineering Research Association (Värmeforsk), Sweden. We thank Nils-Erik Nordh for many of the photographs which illustrate this chapter. Inger Åhman, Nils-Ove Bertholdsson, David Hansson, Sten Segerslätt, Gunnar Henriksson, Stig Larsson, Gabriele Engqvist, Bertil Christensson and Sven Erik Svensson all are acknowledged for their advice and constructive co-operation in

* Corresponding Author

the different phases of our willow work. Erik Rasmusson, Eskil Kemphe, Fatih Mohammad, Vehbo Hot, Ingegerd Nilsson, Nils-Erik Nordh and Richard Childs are kindly acknowledged for practical help with the experiments. Finally we thank all the agriculturally skilled and hard working students that have helped us coping with all the experiments through the years.

5. References

[1] Verwijst T (2001) Willows: An Underestimated Resource for Environment and Society. The forestry chronicle 77: 281-285.

[2] McCormick K, Kåberger T (2005) "Exploring a Pioneering Bioenergy System: The Case of Enköping in Sweden". Journal of cleaner production 13: 1004-1005.

[3] Volk TA, Verwijst T, Tharakan PJ, Abrahamson LP, White EH (2004a) Growing Fuel: A Sustainability Assessment of Willow Biomass Crops. Frontiers in Ecology and the Environment 2: 411–418.

[4] Lindroth A, Verwijst T, Halldin, S (1994) Water-use efficiency of willow: variation with season, humidity and biomass allocation. Journal of hydrology 156: 1-19.

[5] Aronsson P, Perttu K (2001) Willow Vegetation Filters for Wastewater Treatment and Soil Remediation combined with Biomass Production. The forestry chronicle, 77: 293-299.

[6] Landberg T, Greger M (2002) Interclonal Variation of Heavy Metal Interactions in *Salix viminalis*. Environmental toxicology and chemistry 21: 2669-2674.

[7] Klang-Westin E, Eriksson J (2003) Potential of *Salix* as a Phytoextractor for Cd and Moderately Contaminated Soils. Plant and soil 249: 127-137.

[8] Mirck J, Isebrands JG, Verwijst T, Ledin S (2005) Development of Short Rotation Willow Coppice Systems for Environmental Purposes in Sweden. Biomass & bioenergy 28: 219-220.

[9] Perttu KL, Kowalik PJ (1997) '*Salix* Vegetation Filters for Purification of Water and Soils.' Biomass & bioenergy 12: 9-19.

[10] Elowson S (1999) Willow as a Vegetation Filter for Cleaning of Polluted Drainage Water from Agricultural Land. Biomass & bioenergy 16: 281-290.

[11] Perttu KL (1999) Environmental and Hygienic Aspects of Willow Coppice in Sweden. Biomass & bioenergy 16: 291-297.

[12] Rosenqvist H, Aronsson P, Hasselgren K, Perttu K (1997) Economics of Using Municipal Wastewater Irrigation of Willow Coppice Crops. Biomass & bioenergy 12: 1-8.

[13] Lewandowski I, Schmidt U, Londo M, Faaij A (2006) The Economic Value of the Phytoremediation Function – Assessed by the Example of Cadmium Remediation by Willow (*Salix* ssp). Agricultural systems 89: 68-89.

[14] Argus GW (1999) Classification of *Salix* in the New World. Botanical Electronic News 227. 6p.

[15] Argus GW (1997) Infrageneric Classification of *Salix* (*Salicaceae*) in the New World. Systematic Botany Monographs 52. 121 p.

[16] Kopp RF, Smart LB, Maynard CA, Isebrands JG, Tuskan GA, Abrahamson LP (2001) The Development of Improved Willow Clones for Eastern North America. The forestry chronicle 77: 287-292.

[17] Weih M, Bonosi L, Ghelardini L, Rönnberg-Wästljung AC (2011) Optimizing Nitrogen Economy under Drought: Increased Leaf Nitrogen is an Acclimation to Water Stress in Willow (*Salix* spp.). Annals of botany 108: 1347-1353.

[18] Mitchell CP, Ford-Robertson JB, Hinckley T, Sennerby-Forsse L (Editors) (1992) Ecophysiology of Short Rotation Forest Crops. Elsevier, London.

[19] Christersson L (1986) High Technology Biomass Production by *Salix* Clones on a Sandy Soil in Southern Sweden. Tree Physiology 2: 261-277.

[20] Mola Yudego B, Aronsson P (2008) Yield Models for Commercial Willow Biomass Plantations in Sweden. Biomass & bioenergy 32: 829-837.

[21] Willebrand E, Ledin S, Verwijst T (1993) Willow Coppice Systems in Short Rotation Forestry: Effects of Plant Spacing, Rotation Length and Clonal Composition on Biomass Production. Biomass & bioenergy 4: 323-331.

[22] Bergkvist P, Ledin S (1998) Stem Biomass Yields at Different Planting Designs and Spacings in Willow Coppice Systems. Biomass & bioenergy 14: 149–156.

[23] Verwijst T (1991) Shoot Mortality and Dynamics of Live and Dead Biomass in a Stand of *Salix viminalis*. Biomass & bioenergy 1: 35-39.

[24] Willebrand E, Verwijst T (1993) Population Dynamics of Willow Coppice Systems and their Implications for Management of Short rotation Forests. The forestry chronicle 69: 699-704.

[25] Verwijst T (1996a) Stool Mortality and Development of a Competitive Hierarchy in a *Salix viminalis* Coppice System. Biomass & bioenergy 10: 245-250.

[26] Verwijst T (1996b) Cyclic and Progressive Changes in Short Rotation Willow Coppice Systems. Biomass & bioenergy 11:161-165.

[27] Nordh N-E (2005) Long Term Changes in Stand Structure and Biomass Production in Short Rotation Willow Coppice. Ph.D. Thesis. Acta Universitatis Agriculturae Sueciae 120. Swedish University of Agricultural Sciences, Uppsala, Sweden. 26 p.

[28] Tahvanainen L, Rytkonen VM (1999) Biomass Production of *Salix viminalis* in Southern Finland and the Effect of Soil Properties and Climate Conditions on its Production and Survival. Biomass & bioenergy 16: 103-117.

[29] Stolarski M, Szczukowski S, Tworkowski J, Bieniek A (2009) Productivity of Willow Coppice *Salix* spp. under Contrasting Soil Conditions. Electronic journal of polish agricultural universities 12: 1-10.

[30] Alriksson B (1997) Influence of site factors on *Salix* growth with emphasis on nitrogen response under different soil conditions. Ph.D. Thesis. Acta Universitatis Agriculturae Sueciae Silvestria 46. Swedish University of Agricultural Sciences, Uppsala, Sweden. 28p.

[31] Schaff SD, Pezeshki SR, Shields FD (2003) Effects of Soil Conditions on Survival and Growth of Black Willow Cuttings. Environmental management 31: 748-763.

[32] Verwijst T, Nordh N-E, Lundkvist A (2010) Effekter av sticklingsparametrar på grobarhet och tillväxt i salix. Grödor från åker till energiproduktion. Rapport 1144. Värmeforsk Service AB (Thermal Engineering Research Association), Stockholm,

Sweden. 47 p. Available: http://www.varmeforsk.se/rapporter?action=show&id=2522. Accessed 2012 April 18.

[33] Larsson S (1998) Genetic Improvement of Willow for Short Rotation Coppice. Biomass & bioenergy 15: 23-26.

[34] Karp A, Hanley SJ, Trybush SO, Macalpine W, Pei M, Shield I (2011) Genetic Improvement of Willow for Bioenergy and Biofuels. Journal of integrated plant biology 53: 151-165.

[35] McCracken AR, Dawson WM (1996) Interaction of Willow (*Salix*) Clones grown in Polyclonal Stands in Short Rotation Coppice. Biomass & bioenergy 10: 307-311.

[36] McCracken AR, Dawson WM (2003) Rust Disease (*Melampsora epitea*) of Willow (*Salix* spp.) Grown as Short Rotation Coppice (SRC) in Inter- and Intra-species Mixtures. Annals of applied biology 143: 381-393.

[37] Verwijst T (1993) Influence of the Pathogen *Melampsora epitea* on Interspecific Competition in a Mixture of *Salix viminalis* Clones. Journal of vegetation science 4: 717-722.

[38] Merilo E, Heinsoo K, Kull O, Koppel A (2006) Above-ground Production of Two Willow Species in Relation to Radiation Iterception and Light Use Efficiency. Proceedings of Estonian Academy of Sciences. Biology, Ecology 55: 341–54.

[39] Caslin B, Finnan J, McCracken A (2010) Short Rotation Coppice Willow - Best Practice Guidelines. Teagasc, Crops Research Centre, Oak Park, Carlow, Ireland. 72 p. Available: http://www.teagasc.ie/forestry/docs/grants/WillowBestPracticeManual_2012.pdf. Accessed: 2012 April 18.

[40] Lindroth A, Cienciala E (1996) Water Use Efficiency of Short rotation *Salix viminalis* at Leaf, Tree and Stand Scales. Tree Physiology 16: 257-262.

[41] Verwijst T, Elowson S, Li X, Leng G (1996) Production Losses due to a Summer Frost in *Salix viminalis* Short Rotation Forest in Southern Sweden. Scandinavian journal of forest research 11: 104-110.

[42] Hollsten R, Arkelöv O, Ingelman G (2012) Handbok för Salixodlare. Jordbruksverket (Swedish Board of Agriculture), Jönköping, Sweden. 24 p. Available: http://www2.jordbruksverket.se/webdav/files/SJV/trycksaker/Pdf_ovrigt/ovr250.pdf. Accessed 2012 April 18.

[43] Börjesson P (1996). Energy Analysis of Biomass Production and Transportation. Biomass & bioenergy 11: 305-318.

[44] Weih M, Karacic A, Munkert H, Verwijst T, Diekmann M (2003) Influence of Young Poplar Stands on Floristic Diversity in Agricultural Landscapes. Basic and applied ecology 4: 149-156.

[45] Rosenqvist H (1997) Salixodling - Kalkylmetoder och Lönsamhet. Dissertation. Acta Universitatis Agriculturae Sueciae Silvestria 24. Swedish University of Agricultural Sciences, Uppsala, Sweden. 241p.

[46] Gustafsson J, Larsson S, Nordh NE (2009) Manual for SRC Willow Growers. Lantmännen Agroenergi, Sweden. 18 p. Available: http://www.agroenergi.se/PageFiles/135/ENG_Odlarmanual%20Salix%20(Revidering).pdf?epslanguage=sv. Accessed 2012 April 18.

[47] Danfors B, Ledin S, Rosenqvist H (1997) Short Rotation Willow Coppice - Growers Manual. Swedish Institute of Agricultural Engineering. JTI-informerar No. 1. 40 p.

[48] Volk T, Ballard B, Robinson DJ, Abrahamson LP (2004b) Effect of Cutting Storage Conditions during Planting Operations on the Survival and Biomass Production of Four Willow (*Salix* L.) Clones. New forest 28: 63-78.

[49] Burgess D, Hendrickson Q, Roy L (1990) The Importance of Initial Cutting Size for Improving the Growth Performance of *Salix alba* L. Scandinavian journal of forest research 5: 215-24.

[50] Shield IF, Macalpine W, Karp A (2008) The Effect of the Size of the Cuttings Planted on the Subsequent Performance of Three Contrasting Willow Cultivars for Short Rotation Coppice. Aspects of applied biology 90: 225-231.

[51] Verwijst T, Lundkvist A, Edelfeldt S, Forkman J, Nordh NE (2012) Effects of Clone and Cutting Traits on Shoot Emergence and Early Growth of Willow. Biomass & bioenergy 37: 257-264.

[52] Rossi P (1999) Length of Cuttings in Establishment and Production of Short Rotation Plantations of *Salix* 'Aquatica'. New forests 18: 161–77.

[53] Carpenter LT, Pezeshki SR, Shields FD (2008) Responses of Nonstructural Carbohydrates to Shoot Removal and Soil Moisture Treatments in *Salix nigra*. Trees 22: 737-748.

[54] Gage EA, Cooper DJ (2004) Controls on Willow Cutting Survival in a Montane Riparian Area. Journal of range management 57: 597-600.

[55] Edelfeldt S, Verwijst T, Lundkvist A, Forkman J (2012) Effects of Mechanical Planting on Establishment and Early Growth of Willow. Submitted to Biomass & bioenergy.

[56] Harper JL (1977) Population Biology of Plants. Academic Press: London and New York. 892 pp.

[57] Lundkvist A, Verwijst T (2011) Weed Biology and Weed Management in Organic Farming. In: Nokkoul R, editor. Research in Organic Farming. Rijeka: InTech. pp. 157-186. Available: http://www.intechopen.com/books/research-in-organic-farming/weed-biology-and-weed-management-in-organic-farming. Accessed: 2012 April 18.

[58] Abrahamson LP, Volk TA, Kopp RF, White EH, Ballard JL (2002) Willow Biomass Producer's Handbook. State University of New York College of Environmental Science and Forestry, Syracuse, NY, US. 31 p. Available: http://www.esf.edu/willow/pdf/2001%20finalhandbook.pdf. Accessed 2012 April 18.

[59] Verwijst T, Nordh N-E (2010) Effekter av skottnedklippning efter etableringsåret på produktionener första och andra omdrevet i salixodlingar. Grödor från åker till energiproduktion. Rapport 1136. Värmeforsk Service AB (Thermal Engineering Research Association), Stockholm, Sweden. 33 p. Available: http://www.varmeforsk.se/rapporter?action=show&id=2543. Accessed 2012 April 18.

[60] Labrecque M, Teodorescu TI, Babeux P, Cogliastro A, Daigle S (1994) Impact of Herbaceous Competition and Drainage Conditions on the Early Productivity of Willows under Short rotation Intensive Culture. Canadian journal of forest research 24: 493-501.

[61] Clay D, Dixon F (1997) Effect of Ground-cover Vegetation on the Growth of Poplar and Willow Short rotation Coppice. Aspects of applied biology 49: 53-60.

[62] Sage R (1999) Weed Competition in Willow Coppice Crops: the Cause and Extent of Yield Losses. Weed research 39: 399-411.

[63] Volk T (2002) Alternative Methods of Site Preparations and Coppice Management during the Establishment of Short rotation Woody Crops. Ph.D. Thesis. State University of New York, College of Environmental Science and Forestry, Syracuse, NY. 284 p. Available: http://en.scientificcommons.org/31526556. Accessed 2012 April 18.

[64] Smart L (2012) Cropping Systems for Shrub Willow Bioenergy Crops. Willowpedia, Cornell University, College of Agricultural and Life Science, Geneva, NY, US. Available: http://willow.cals.cornell.edu/Research/Cropping%20Systems.html. Accessed 2012 April 18.

[65] Moukoumi J, Farrell RE, Van Rees KJC, Hynes RK, Bélanger N (2012). Intercropping *Caragana arborescens* with *Salix miyabeana* to Satisfy Nitrogen Demand and Maximize Growth. BioEnergy Research: 1-14.

[66] Van Der Weide RY, Bleeker PO, Achten VTJM, Lotz LAP, Fogelberg F, Melander B (2008) Innovation in Mechanical Weed Control in Crop Rows. Weed research 48: 215-224.

[67] Helby P, Börjesson P, Hansen AC, Roos A, Rosenqvist H, Takeuchi L (2004) Market Development Problems for Sustainable Bioenergy in Sweden. (The BIOMARK Project). Report no. 38. Environmental and Energy System Studies, Lund University, Sweden. 193 p. Available: http://www.energimyndigheten.se/Global/Forskning/Energisystemstudier/Biomark%20project%20Projekt%2012010-1.pdf . Accessed 2012 April 18.

[68] Weih M (2001) Evidence for Increased Sensitivity to Nutrient and Water Stress in a Fast-growing Hybrid Willow Compared with a Natural Willow Clone. Tree Physiology 21: 1141-1148.

[69] Ledin S (ed.) (1994) Gödsling av Salixodlingar: Ramprogram Energiskog. Rapport 1994: 25. Närings- och teknikutvecklingsverket (NUTEK), Stockholm, Sweden. 51p.

[70] Aronsson P & Rosenqvist H (2011) Gödslingsrekommendationer för Salix 2011. Slutrapport. Sveriges lantbruksuniversitet, Uppsala, Sweden, Stiftelsen lantbruksforskning, Stockholm, Sweden. 30 p. Available: http://pub.epsilon.slu.se/8466/1/aronsson_p_111125.pdf . Accessed 2012 April 18.

[71] Dimitriou I, Aronsson P (2011) Wastewater and Sewage Sludge Application to Willows and Poplars grown in Lysimeters – Plant Response and Treatment Efficiency. Biomass & bioenergy 35: 161-170.

[72] Rosenqvist H, Ness B (2004) An Economic Analysis of Leached Purification through Willow Coppice Vegetation Filters. Bioresource technology 94: 321-329.

[73] Pei MH, Royle DJ, Hunter T (1996) Pathogenic Specialization in *Melampsora epitea* var. *epitea* on *Salix*. Plant pathology 45: 679-690.

[74] Newcombe G (1996) The Specificity of Fungal Pathogens of *Populus*. In: Stettler RF, Bradshaw HD, Heilman PE, Hinckly TM, editors. Biology of *Populus*. Ottawa: NRC Research Press. pp. 223-246.

[75] Ramstedt M (1999) Rust Disease on Willows – Virulence Variation and Resistance Breeding Strategies. Forest ecology and management 121: 101-111.
[76] Åhman I (2000) Breeding for Resistane to Leaf Beetles attacking Biomass Willow in Europe. In: Isebrands JG, Richardson J, editors. 21st Session of the International Poplar Commission (IPC-2000): Poplar and Willow Culture: Meeting the Needs of Society and the Environment. USDA Forest Service. General Technical Report NC-215. St. Paul, MN: U.S. Dept. of Agriculture, Forest Service, North Central Forest Experiment Station. 226 p. Available: http://ncrs.fs.fed.us/pubs/gtr/gtr_nc215.pdf. Accessed 2012 April 18.
[77] Glynn C, Larsson S (2000) Rapid Gall Midge Adaptation to a Resistant Willow Genotype. Agricultural and forest entomology 2: 115-121.
[78] Björkman C, Höglund S, Eklund K, Larsson S (2000) Effects of Leaf Beetle Damage on Stem Wood Production in Coppicing Willow. Agricultural and forest entomology 2:131-139.
[79] Åhman I, Larsson S (1994) Genetic Improvement of Willow (*Salix*) as a Source of Bioenergy. Norwegian journal of agriculture sciences 18: Supplement 47-56.
[80] Åhman I, Larsson S (1999) Resistensförädling i *Salix* för Energiproduktion. Växtskyddsnotiser 63: 17-19.
[81] Nordh N-E (2010) Effekter på överlevnad och tillväxt vid förlängd skördesäsong av salix. Grödor från åker till energiproduktion. Rapport 1147. Värmeforsk Service AB (Thermal Engineering Research Association), Stockholm, Sweden. 31 p. Available: http://www.varmeforsk.se/rapporter?action=show&id=2490. Accessed 2012 April 18.
[82] Ceulemans R, McDonald A, Pereira JS (1996) A Comparison Among Eucalypt, Poplar and Willow Characteristics with Particular Reference to a Coppice, Growth Modeling Approach. Biomass & bioenergy 11: 215-231.
[83] Christersson L (1987) Biomass Production by Irrigated and Fertilized *Salix* Clones. Biomass 12: 83-95
[84] Larsson S, Lindegaard K (2003) Full Scale Implementation of Short Rotation Willow Coppice, SRC, in Sweden. Örebro, Sweden: Agrobränsle AB.
[85] Lindroth A, Båth A (1999) Assessment of Regional Willow Coppice Yield in Sweden on Basis of Water Availability. Forest ecology and management 121: 57-65.
[86] Larsson S (2001) Commercial Varieties from the Swedish Willow Breeding Programme. Aspects of applied biology 65: 193-198.
[87] Lindegaard K, Parfitt RI, Donaldson G, Hunter T, Dawson WM, Forbes EGA, Carter MM, Whinney CC, Whinney JE, Larsson S (2001) Comparative Trials of Elite Swedish and UK Biomass Willow Varieties. Aspects of applied biology 65: 183-198.
[88] Walle IV, Van Camp N, Van de Casteele L, Verheyen K, Lemeur R (2007) Short Rotation Forestry of Birch, Maple, Poplar and Willow in Flanders (Belgium) I - Biomass Production after 4 Years of Tree Growth. Biomass & bioenergy 31: 267-275.
[89] Aylott MJ, Casella E, Tubby I, Street NR, Smith P, Taylor G (2008) Yield and Spatial Supply of Bioenergy Poplar and Willow Short Rotation Coppice in the UK. New Phytologist 178: 358-370.

Chapter 20

Artemia, a New Source of Animal Protein Ingredient in Poultry Nutrition

A. Zarei

Additional information is available at the end of the chapter

http://dx.doi.org/10.5772/53610

1. Introduction

In the nutritional behavior of single stomach animals, the origin of protein is important and its quality varies between different sources and animal origin is better than plant origin [7]. From the standpoint of salmonella contamination and due to high microbial potential, however, some of these proteins such as meat and bone meal ought to be used with caution. If heat treatment is not enough during fish meal processing, thiaminase will remain in fish meal and will cause harmful effects .Some researches indicated that thiamin will be reduced by thiaminase under special storage condition of ingredient before feeding animal [6,7] . Also, severe heating during meat and bone meal and fish meal processing to be confident that poultry by-product is safe, however some amino acids will probably degenerate and their bioavailability will decrease [5,7, 8,9] .

Although, meat & bone and fish meal have been used in poultry feeding, exclusively, but artemia biomass is also one of the animal proteins with high nutritional value which can be used in aquaculture and animal nutrition [1,4] .

The aim of this study is a survey of biology and characteristics of artemia and possibility of its usage in poultry nutrition as a protein source.

2. What is artemia?

Artemia or brine shrimp belongs to the animal kingdom, phylum of *arthropoda* , subphylum of *crustacean*, class of *branchiopoda*, order of *anostraca*, family of *artemidae* and genus of *artemia*. Linnaeus (1758) and Leach (1819) called it "*Cancer salinus*" and "*Artemia salina*", respectively. The latter name is because of the effect of salinity on morphological growth and development of artemia. Two species of artemia in Iran are: *Artemia urmiana* and

Artemia parthenogenetica. The first is native of Urmia lake and the second was observed in 12 regions of artemia habitats in Iran.

Artemia spreads in tropical and sub tropical regions in saline environments of the world, and over 500 artemia regions are discovered around the globe. Nine species of artemia were recognized in these regions.

More than two million kilograms of dried cysts of artemia with 0.4mm diameter are transacted in world markets every year. It is used as an aquaculture feed for hatched naplius. Uniformity of cysts and embryos with diapose has made artemia a unique source of aquaculture feeds. Artemia cysts can spread by wind and migratory birds.

Artemia contains 40-60% crude protein (dry matter basis) [11] .

3. Morphology and ecology

Morphologically, artemia has fragmented body with leave and wide shaped appearance. It's body consists of three compartments; head, thorax and abdomen, with total length of 8-10 mm and 10-12 mm in male and female respectively (Figure 1). Width of body is 4mm in both sexes. The exoskeleton of artemia is extremely thin (0.3-1μ) and flexible that called" chitin" and it is connected to muscle from inner surface.

Figure 1. Artemia morphology a) female and b) male

The blood circulatory system of artemia is open.

This animal is euryhalin and can tolerate high concentration of salty habitat. There is a glandular organ in back of artemia neck that named; "salt gland" or "neck organ" .This organ exudates extra salt from the body to environment. Salty organ extinct at maturity and then this function is performed by exopodits of legs.

Although there are some limitations in the living environment of artemia, like high temperature and high salinity and drought, this animal can tolerate these conditions by producing cysts and going to diapose until the condition become suitable, and then it will continue its living.

Salinity and temperature are two important factors for growth and survival of artemia.

Artemia can tolerate salinity even more than 250 g/l and suitable range of temperature is 6-35°C.This crustacean has adapted itself with hard environmental conditions. In hypoxia, artemia increases the oxygen carrying capacity through the increase of the amount of hemoglobin. In this situation, body color turns red from original pale brown.

Physiological adaptation of artemia in high salinity is an effective defense method against predators using following mechanisms:

1. A powerful and effective osmoregulation system
2. Overcome high hypoxia in high salinity condition by higher pigmentation (inhalation pigments)
3. Production of embryos in diapose stage in cysts that can tolerate environmentally unfavorable condition.

4. Nutrition

From nutritional standpoint, artemia is non- selective filter feeder, and eats algae, bacteria, protozoa and yeasts, as long as feed particles diameter is not over 50-70 μm. Artemia feed can be alive or dead in artificial culture system. Artemia can uses bacteria and protozoa as feed sources, which grow in artemia culture medium. This protozoa (such as; *Candida, Rhodotorula*) can also be directly swallowed by artemia. The best algae's for artemia nutrition includes: *Dunaliella salina*, *Spirulina* and *Scenedesmus*. For artemia culture, agricultural products can be utilized such as; rice, corn, wheat, barley flours and their bran.

5. Reproduction

All bisexual species holds 42 chromosomes (2n = 42). *A.persimilis* holds 44 chromosomes (2n = 44) and *Artemia partenogenic* is diploid, triploid, tetraploid and even pantaploid. As a general rule, artemia populations are defined on the basis of the number of their chromosomes. However, contrary to mammalian, female artemia is heterogametic [3]. Artemia is produced by two ways: sexual and parthenogenesis (development of a new individual from an unfertilized egg). The mature female ovulates each 140 hours.

According to strain of artemia or method of living, it selects one of the following conditions; oviparous or ovoviviparous. In suitable situation of rising, reproduction trend is as larvae production (ovoviviparous) and in unsuitable situation of growth (salinity >50gm/lit and oxygen <5mg/lit) oviparous will occur. In the latter condition, growth of embryo will stop and enters diapauses. In suitable saline and nutritional conditions, females can produce 75 naplius each day and over its lift cycle (50 days), it reproduces 10-11 times.

In extreme hypoxia, due to increasing hemoglobin production, the color of artemia will change from light brown to yellow and then red.

Artemia cysts are spread by wind and birds. Earth pond or region of high salty water is suitable for culture and reproduction of artemia.

6. Different kinds of artemia from the nutritional point of view

From the nutritional point of view ,different kinds of artemia are:

- Decapsulated cysts
- Newly hatched nauplii
- Metanauplii and juvenile and adult
- Frozen and freeze – dried artemia

These forms of artemia are commonly used for newly hatched shrimps, sturgeons, trout, aquarium fishes and some crustaceans. Artemia biomass (consist of cysts and different living stages of artemia) is a suitable protein resource for other animals like poultry that consists of different stages of artemia growth.

7. Methods of artemia harvesting

According to artemia habitat, different biomass harvesting is utilized. In breeding pools and lake beaches, artemia is collected using a lace net that is fastened to two large floaters from each side (figure 2).

Because of phototropism characteristic, artemia can be collected easily by light source at night.

After harvesting, artemia biomass can be dried and cured under sunlight (Figure 3). Then the dried artemia will be milled before using in poultry diet.

Figure 2. Artemia biomass harvest

Figure 3. Flaked artemia

8. Chemical composition of artemia meal

The chemical composition of different kinds of artemia meal (dried at 50-60°C as sun cured or oven dried) is shown in Table 1. As shown in table 1, the chemical composition of 3 kinds of artemia meal (collected from different regions of Iran) is not identical. The quality of those ,depends on region, species, time of harvest and percentage of artemia mixture (artemia in different stages of living shows different compositions). So prior to using this ingredient, it must be analyzed for main nutrients.

Chemical composition		Kind of Artemia meal		
		ULAM	EPAM	GSLAM
Dry matter	g/kg	928	924	938
Crude Protein	g/kg	401.9	390.8	423.5
Gross Energy	MJ/kg	16.86	16.32	14.98
Crude Fat	g/kg	136	85.5	206.5
Crude Fiber	g/kg	36	18	28
Crude Ash	g/kg	240	287	284
Calcium (Ca)	g/kg	23.4	20.2	26.1
Phosphorus (P)	g/kg	11.1	8.6	14.2
Sodium (Na)	g/kg	12.1	9.6	16.4
Magnesium (Mg)	g/kg	3.3	4.1	3.1
Potassium (K)	g/kg	16.5	20.9	13.9
Iron (Fe)	mg/kg	1147.25	1642.75	437.75
Manganese (Mn)	mg/kg	53.78	132.45	84.08
Copper (Cu)	mg/kg	3.5	3.55	5.05
Zinc (Zn)	mg/kg	52.75	46.75	59

1- Zarei,A (2006) ,2- Urmia Lake Artemia Meal , 3- Earth Pond Artemia Meal , 4- Ghom Salt Lake Artemia Meal

Table 1. Chemical composition of three kinds of artemia meal (ULAM[2], EPAM[3], GSLAM[4]) (as g/kg , MJ /kg or mg/kg – DM basis)1

9. Metabolizable energy of artemia meal

An experiment was designed to determine different classes of metabolizable energy (AME, AMEn, TME, TMEn) in artemia meals [13] .For determination of metabolizable energy of artemia meal and comparison with fish meal, samples gathered from 3 regions include : Urmia Lake Artemia Meal(ULAM), Earth Ponds Artemia Meal (beside urmia lake)(EPAM) and Ghom Salt Lake Artemia Meal(GSLAM). Then samples dried, milled and used in a biological experiment with fish meal. 20 Rhode Island Red cockerels with approximately same live weight used in Sibbald assay with completely randomized design with 5 treatments and 4 repetitions for determination of AME, AMEn, TME and TMEn.

Results showed there were significant differences between treatments from standpoints of metabolizable energy ($P<0.05$). ULAM and FM had highest ME and EPAM and GLAM had lowest ME. The highest TME belong to FM and the lowest TME pertained to EPAM. Except to EPAM that had the lowest TMEn, other treatments didn't have any differences between them.

10. Protein and amino acids digestibility of artemia meal

Result from *in vitro* and *in vivo* experiments showed that this ingredient has high quality of protein and the amount of digestibility was more than 90% [12] .

In order to determination of artemia meal's amino acid digestibility, five-week old male broiler chicks were given a semi-purified diet in which artemia meal was the sole source of protein. Apparent amino acid digestibility values of the assay diet, using ileum and excreta contents, were calculated using chromic oxide as indigestible marker. True digestibility values were calculated using endogenous output determined by feeding a nitrogen-free diet. The results showed (Table 2) that in determination of apparent amino acid digestibility of excreta, serine had the lowest (0.80) and methionine had the highest (0.92) digestibility, while glycine had the lowest (0.88) and arginine and leucine had the highest (0.95) apparent ileal digestibility. In measuring true excreta and ileal amino acid digestibility, alanine and glycine had the lowest (0.90 and 0.93) and methionine had the highest (0.96 and 0.99) digestibility, respectively. In general, the site of measurement had no effect on apparent or true amino acid digestibility of artemia meal [2] .

11. Artemia meal in broiler diets

In another experiment, different levels of protein from two kinds of artemia meal include artemia meal from Urmia lake and artemia meal from earth ponds beside Urmia lake with levels of 0, 25, 50, 75, 100 percent replaced to prue fish meal protein [12]. The experimental design was completely randomized with factorial method; include 10 treatments and 3 repetitions that in each repetition there were 10 one day-old male broilers from Ross 308 strain. This experiment was performed in 7 weeks and during and end of it, traits that related to broiler performance and carcass, was measured and analyzed. Results showed that effect of kind of artemia meal and effect of level of protein replacement weren't significant for feed intake. But interaction between these two was significant for this trait ($P<0.05$). The highest feed intake belong to Urmia lake artemia meal treatment with 50% level of replacement and the lowest feed intake related to treatment of without artemia meal (contain 5% fish meal). For body weight gain and feed conversion ratio, effect of kind of artemia meal and effect of level of protein replacement and effect of interaction between these two weren't significant. These effects weren't significant for all carcass traits and gastro intestinal parts exception for femur percent that treatment of without artemia meal (contain 5% fish meal) had a lowest percent to comparison with other treatments for this trait.

Amino acids	Apparent digestibility				True digestibility			
	Excreta	Ileal	SEM[1]	P[2]	Excreta	Ileal	SEM	P
Methionine	0.92	0.94	0.004	NS	0.96	0.99	0.004	0.09
Lysine	0.88	0.92	0.007	NS	0.92	0.96	0.007	NS
Threonine	0.85	0.90	0.013	NS	0.93	0.98	0.011	NS
Tryptophan	0.88	0.94	0.014	NS	0.90	0.97	0.017	NS
Arginine	0.89	0.95	0.008	0.09	0.93	0.98	0.008	NS
Isoleucine	0.88	0.94	0.011	NS	0.92	0.98	0.011	NS
Leucine	0.89	0.95	0.009	0.06	0.94	0.98	0.009	NS
Valine	0.87	0.93	0.011	NS	0.93	0.98	0.010	NS
Histidine	0.89	0.93	0.007	NS	0.95	0.97	0.007	NS
Phenylalanine	0.87	0.94	0.009	0.09	0.92	0.97	0.009	NS
Glycine	0.81	0.88	0.015	NS	-	0.93	-	-
Serine	0.80	0.89	0.018	NS	0.91	0.97	0.017	NS
Alanine	0.85	0.91	0.014	NS	0.90	0.94	0.014	NS
Aspartic acid	0.86	0.91	0.010	NS	0.91	0.94	0.005	0.09
Glutamic acid	0.87	0.93	0.014	NS	0.93	0.95	0.013	NS
Total	0.85	0.92	0.010	0.09	0.94	0.96	0.011	NS
CP(N×6.25)[3]	0.81	0.89	0.013	NS	0.89	0.94	0.012	NS

NS – Non Significant ; [1]- Standard Error of Mean ; [2] – Probability ; CP- Crude Protein ;N – Nitrogen [3] – The values (protein digestibility) were not corrected for uric acid.

Table 2. Apparent and true digestibility (coefficients) of artemia meal determined by sampling either excreta or ileum contents

12. Conclusion

Results of this studies revealed that artemia meal can be used as a feedstuff in poultry and other farm animal's diets because it has high level of protein and high protein digestibility. Compared with other animal proteins, artemia does not contain any feather, bone, hair or gastrointestinal tract components. In addition, in artemia production there is no requirement for high pressure and high temperature treatments which can influence protein quality. Artificial culture of artemia is easy and is possible everywhere.

Author details

A. Zarei*
Department of Animal Science, College of Agriculture and Natural Resources, Islamic Azad University- Karaj Branch, Karaj, Iran

13. References

[1] Abatzopoulos, TH.J, J.A. Beardmore., J.s.Clegg. and P.Sorgeloos. 2002. Artemia: Basic and Applied Biology. Kluwer Academic Publishers.
[2] Aghakhanian, P., A. Zarei, H. Lotfollahian and N. Eila. 2009. Apparent and true amino acid digestibility of artemia meal in broiler chicks.South African Journal of Animal Science.39 (1).
[3] Bowen, S.T.1964: The genetics of *Artemia salina*. IV. Hybridization of wild population with mutant stocks. Biol. Bull., 126, 333.
[4] Gilbert, V.S.1995.Introduction, biology and ecology of artemia. laboratory of aquaculture and artemia reference center university of Gent, Belgium.
[5] Johnson, M.L. and C.M. Parsons., G.C. Fahey, Jr., N.R. Merchen., C.G. Aldrich. 1998. Effects of species raw material source, ash content, and processing temperature on amino acid digestibility of animal by product meals by cecectomized roosters and ileally cannulated dogs. Journal of Animal Science.76:1112-1122.
[6] Klasing, K.C. 1998.Comparative Avian Nutrition.CAB International.
[7] Leeson. S. J.D.Summers. 2001. Nutrition of the Chicken.4 th edition. university Books.
[8] Parsons, C.M.1999. Protein quality and amino acid digestibility. Multi-state. Poultry meeting, May 25-27.
[9] Shirley, R.B. and C.M. Parsons. 2000. Effect of pressure processing on amino acid digestibility of meat and bone meal for poultry. Poultry Science.79:1775-1581.
[10] Sibbald, I.R., 1976.The true metabolizable energy values of several feedingstuffs measured with roosters, laying hens, turkeys and broiler hens. Poultry Science. 55:1459 -1463.
[11] Sorgeloos, P.1989. Two strains of Artemia in Urmia lake (Iran), Artemia Newsl., 13, 5.
[12] Zarei, A. (1).2006. Use of artemia meal as a protein supplement in broiler diet. International Journal of Poultry Science .5(2).142-148.
[13] Zarei, A. (2). 2006. Apparent and true metabolizable energy of artemia meal. International Journal of Poultry Science .5(7):627-628.
[14] Zarei, A. 2007. Determination of protein digestibility of animal protein feeds under condition of *in vitro* and *in vivo*, BSAS Annual Conference, 2-4 April 2007, Southport, England (oral presentation).

---

* Corresponding Author

Chapter 21

Biomass from the Sea

Ernesto A. Chávez and Alejandra Chávez-Hidalgo

Additional information is available at the end of the chapter

http://dx.doi.org/10.5772/54520

1. Introduction

In the world oceans there is large amount of biomass suspended in the photic zone of water column. Part of the living part is of plant origin, the phytoplankton and other is the animal component or zooplankton. There is also large proportion of particulate organic matter composed by remains of dead animals and feces. They represent the basis of the food webs with three or four trophic levels where all the consumers are animals in whose top the carnivores or top predators, are found. In all aquatic trophic webs, many species are exploited.

2. The primary producers

Nearly 0.3% of solar energy incident on the sea surface is fixed by phytoplankton the tiny plant organisms suspended in natural waters, over the 40-60 meters of the upper water column, accounts to 75% of primary productivity of an area of the word oceans near to $3.5 \cdot 10^{14}$ square meters; the remaining 25% is produced by macro algae. The amount of biomass of all the consumers is based upon primary production by phytoplankton, which range between 0.05 – 0.5 $gCm^{-2}d^{-1}$, but in some very productive upwelling zones or in some grass beds, it can be as high as 5 $gCm^{-2}d^{-1}$ (Russell-Hunter 1970; Margalef 1974; Cushing and Walsh 1976). As a result of this photosynthetic process, the carbon gross production of the sea amounts to 15.5 x 10^{10} mt of Carbon per year, equivalent to a net production of 1.5 x 10^{10} mt, most of it in shore waters. By having in mind the energetic efficiency, these figures amount to 8 per cent of global aquatic primary production (Pauly and Christensen 1995; Friedland et al. 2012), meaning that there is a maximum limit to fisheries production.

Biological production through the fixation of light is a process interacting with the degradation or dissipation of energy by all organisms; in other words, the persistence of life as we know it, depends on a permanent input of energy, which after being fixed and transformed in chemical energy by the plants, is dissipated constantly by all organisms on Earth. Human beings have been able to simplify the food webs channelizing the production of a few species which are exploited by man; agriculture systems are a typical example of this process. However, this implies a limit to the maximum potential production of biomass by organisms (Pauly and Christensen 1995; Friedland et al. 2012; Botsford 2012).

3. The secondary producers and food webs

The thermodynamics of the biomass flow and secondary production indicates that the transfer efficiency of carbon in the sea webs may approach to 15 per cent; however, many other authors (in Christensen and Pauly 1993; Pauly and Christensen 1995) adopted the value of 10 per cent. All of the consumers depend from the chemical energy to subsist; this energy is synthesized by the primary producers and transferred to other trophic levels trough consumption by herbivores and then passed to several levels of animals through predation.

Zooplankton, the free-living animals suspended at the water column, are the kind or organisms which make use of the primary production. The main component of this food webs is the group of copepods. Apart o being composed mostly by herbivores, zooplankton also contains many predators of first order, like jelly fish and other crustaceans as larval stages of benthic organisms spending in most cases, from a few days to several months suspended in the water column as predators of micro zooplankton, then being recruited to the benthic communities as they grow.

Caloric value of organisms indicates very uniform qualities through the food web, being higher in those animals storing lipids in their bodies. In sugars and proteins, the caloric value is 4,100 cal g^{-1}, whilst in lipids this value amounts to 9,300 cal g^{-1}, but when these substances are not totally oxidized, the calories available are nearly 90% of their total caloric values. A high production of biomass from the primary producers would be uptaken by the herbivores and transferred to upper levels of the food web. This means that a high primary production will imply high biomass of consumers in proportion following the rule of 10 per cent; this is, for each ton of top predators, there will be 10 mt of predators of first order, and 100 mt of herbivores. The biomass of the carnivores ranges between 0.5 and 2 g C m^{-2} and follows the 10% rule respecting to the lower level. The biomass of primary producers, mainly phytoplankton, may be lower than the herbivores because of their high turnover rate. It is pertinent to mention that upwelling zones of the sea, like in Peru on the west coast of America and West Africa, significant amounts of nutrients are flowing up from the deep sea enriching the surface waters in the photic zone and stimulating the primary productivity. In these zones, the process of evolution has allowed the organization of short food chains, where the sardine and anchovies take advantage exploiting much of this production, allowing the growth of large schools which are exploited by human beings, with levels of exploitation of more than 12 Million mt, as occurred in Peru in the early seventies.

4. The fisheries

The exploitation of aquatic populations by human beings through fisheries, leads to a change in the trophic structure of ecosystems, allowing that opportunistic species, formerly infrequent, to become abundant and reducing the biodiversity; this seems to be the case of squids and jellyfishes. This process determines an increase of the primary production/biomass ratio in the ecosystem. The most productive ecosystems are those associated to upwelling, where the fast growing predators with short life spans, plankton feeders determining the existence of short food chains, allow the existence of very productive fisheries as in the case of anchovy and sardine fisheries. In other natural communities, where the ecosystem usually imposes high environmental stability, top predators usually are animals with long life span in relatively long food chains; in this case, the potential biomass production is low, because the evolutionary forces are oriented towards the density dependent processes, leading to the organization of ecosystems with high biodiversity as occurs in coral reefs. In this kind of communities, the surplus production is almost nule, because the production/consumption ratio approaches zero, severely reducing the capacity of commercial exploitation.

4.1. The logistic curve approach

According to Graham (1939), the maximum yield that can be extracted from a wild stock is found at the half of the virgin size of that population, as seen in Fig. 1A, B. A similar view is commented by Zabel et al. (2003). After this premise, a simplistic approach can be adopted by assuming that when the catch trend shows a maximum, followed by a decline, then that

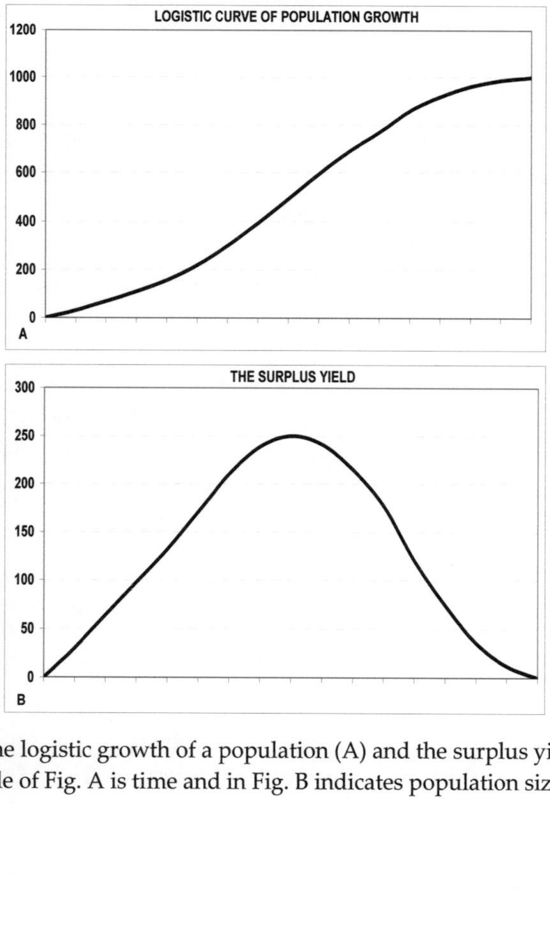

Figure 1. Principles of the logistic growth of a population (A) and the surplus yield of an exploited stock (B). Horizontal scale of Fig. A is time and in Fig. B indicates population size.

maximum yield corresponds to the half of the population size at the virgin stock. Stock assessment based upon this approach is very limited and despite that its ecological principles as background are valid, there are many factors constraining the validity of this procedure and therefore other approaches more accurate and based upon age structure have been adopted over time.

By following the former statements, a simple approach to roughly estimate the stock biomass is by just fitting a parabola to the catch records of some fisheries or regions, even deliberately ignoring a relationship of the stock density of populations, just by usually using the catch per unit of effort as an indicator of stock density. In this case, time was used as an indirect indicator of fishing effort, because the information on this variable is not easily available and because it is beyond the scope of this paper. Therefore, second degree regressions were used to several fisheries and regions just to have an idea on when the maximum yields, presumably equivalent to the Maximum Sustainable Yields (MSY), were attained. It is assumed that the stock biomass is at least twice bigger than the maximum yield attained in a certain time, and in that point is supposed that the exploitation rate E, is 50 per cent. This approach is conservative, because the intrinsic growth rate is not provided, given that many populations are involved. In the stock assessment process, the E value is usually lower than 0.5; however, by considering that many species are involved in the procedure is analysis, is likely to expect that in this collection there may be species which are overexploited, as well as others which may be underexploited. For this reason, it is reasonable to adopt a conservative criterion instead of being too optimistic assuming that the biomass could reach higher values. It is pertinent to mention that most of the regressions applied and described in the following paragraphs excepting three, provided high and significant R^2 coefficients.

On being consistent with this idea, estimations of the MSY by applying a parabola were fitted to catch data of the world fisheries exploited and recorded for different regions as shown in Fig. 2 (A - F) and in Table 1. The time scale of catch extracted from FAO (2010), data goes from 1950 to 2010. It is evident that in most cases the catch has attained a maximum yield, which for practical purposes; it can be considered as equivalent to the MSY level.

4.2. Biomass and fish production

The FAO (1995, 2005) is involved in the task of recording the world statistics of food production and often publishes assessments accounting for the status of world fisheries (FAO 1995, 2005; Froese and Pauly 2012). The catch records are grouped by statistical regions subdivided in 17 sub regions and in the following paragraphs, some highlights on the current status of the fisheries of these regions and sub regions is given, as well as some rough estimations of the biomass on which the exploitation of fish resources is based.

4.2.1. The Atlantic

In the Atlantic North-eastern, the MSY level was attained in the middle eighties (Fig. 2A), with 11.6 million (M) mt; this catch implies a biomass of 23.2 M mt with a significant

REGION	MSY	BIOMASS	Mean 2008 - 10 YIELD	Mean 2008 - 10 BIOMASS
ATLANTIC NORTHEASTERN	11,600,000	23,200,000	8,600,000	17,200,000
ATLANTIC EASTERN CENTRAL	3,700,000	7,400,000	3,750,000	7,500,000
ATLANTIC SOUTHEASTERN	2,700,000	5,400,000	1,300,000	2,600,000
ATLANTIC NORTHWESTERN	3,500,000	7,000,000	2,400,000	4,800,000
ATLANTIC SOUTHWESTERN	2,650,000	5,300,000	1,840,000	3,680,000
GULF OF MEXICO*	800,000	1,600,000	550,000	1,100,000
TOTAL ATLANTIC	**24,150,000**	**48,300,000**	**17,890,000**	**35,780,000**
PACIFIC NORTHEASTERN	2,950,000	5,900,000	2,440,000	4,880,000
PACIFIC NORTHWESTERN	22,550,000	45,100,000	20,900,000	41,800,000
PACIFIC WESTERN CENTRAL	12,000,000	24,000,000	12,000,000	24,000,000
PACIFIC EASTERN CENTRAL	2,000,000	4,000,000	2,000,000	4,000,000
PACIFIC SOUTHEASTERN	14,500,000	29,000,000	10,900,000	21,800,000
PACIFIC SOUTHWESTERN	800,000	1,600,000	600,000	1,200,000
TOTAL PACIFIC	**54,800,000**	**109,600,000**	**48,840,000**	**97,680,000**
ANTARCTIC INDIAN OCEAN	90,000	180,000	10,000	20,000
INDIAN OCEAN EASTERN	7,000,000	14,000,000	6,800,000	13,600,000
INDIAN OCEAN WESTERN	4,500,000	9,000,000	4,500,000	9,000,000
TOTAL INDIAN OCEAN	**11,590,000**	**23,180,000**	**11,310,000**	**22,620,000**
ANTARCTIC TOTAL	40,000	80,000	5,000	10,000
MEDITERRANEAN & BLACK SEA	1,700,000	3,400,000	1,500,000	3,000,000
OUTSIDE THE ANTARCTIC	80,000	160,000	20,000	40,000
TOTAL MARINE REGIONS	**99,710,000**	**199,420,000**	**79,565,000**	**159,130,000**

*Included in the Atlantic Southwestern region

Table 1. Maximum yields, equivalent to the MSY, of catch data recorded in FAO statistics for the seventeen statistical areas. Biomass estimates of total yields per area within a region and the total for the whole region are indicated. Current average yields, for the years 2008-2010 and their corresponding biomass are also shown on the two right side columns. Values are rounded, in mt.

decrease in biomass of 6 million mt in the last three years (Table 1). In Fig. 2B, the maximum catch of the Atlantic Eastern Central is displayed, and corresponds to 3.7 M mt, attained in the year 2000; this figure corresponds to a biomass of 7.4 M mt, but at the end of the period displays an increase of 100,000 mt. In the Atlantic South eastern, the maximum yield was obtained in the early eighties, with 2.7 M mt (Fig. 2C); the corresponding biomass is 5.4 M mt, with a significant decrease in biomass during the last three years to only 2.6 M mt. The catch trend of the Atlantic North western (Fig. 2D) is declining, with a maximum of 3.5 M mt attained in the early seventies; to this figure corresponds a biomass of 7 M mt (Table 1). The low biomass estimated for the years 2008-2010, with somewhat more than 4.8 M mt, is something to be concerned. The catch trend of the Atlantic South western (Fig. 2E) is not very clear, because it seems to attain a maximum followed by a decline, but the projection of the regression line suggests that the maximum yield will be reached until the year 2030 with

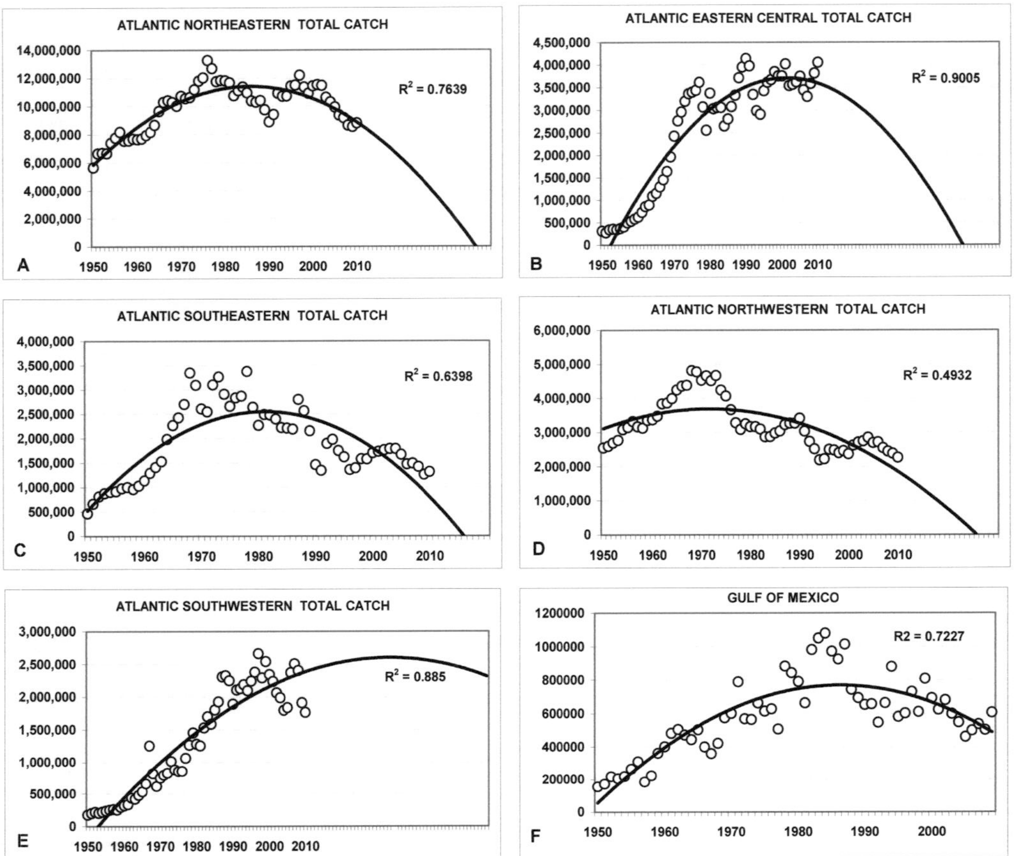

Figure 2. Trend of total catches extracted from several regions of the Atlantic in the period 1950 – 2010. A. Atlantic north eastern; in this region the maximum catches were obtained in the late eighties. B. Atlantic eastern central; the maximum yields were obtained around the year 2000. C. Atlantic south eastern; the maximum yield was obtained in the early eighties. D. Atlantic north western; the maximum yield was obtained by the year 1970, with a declining trend afterwards. E. Atlantic south western. It is not clear whether the maximum yield was attained by the early 2000's, or it still may grow to a maximum near the year 2030. In the Gulf of Mexico, whose data are included in those of Fig. 2.E, more than 60 species caught and recorded in the statistics, are included in this analysis; here, the MSY was attained in the middle 80's.

2.65 M mt. The corresponding biomass will be 5.3 M mt (Table 1); the stock current biomass is 3.68 M mt. It was possible to examine with some detail the catch trend of the Gulf of Mexico (Fig. 2F), whose values are part of those for the Atlantic South western; in this case, the maximum yield was obtained in the late eighties with 800,000 mt, with a corresponding biomass of 1.6 Million mt; the current biomass is only 1.1 M mt. The global MSY for the Atlantic Ocean is 24.15 M mt, corresponding to a biomass of 48.3 M mt but these values do not correspond to the same year; unfortunately in all cases but one, current yields were left behind and the current biomass is considerably lower than the figures provided. The current biomass estimated for the Atlantic Ocean amounts to 35.78 M mt (Table 1).

4.2.2. *The Pacific*

The catch obtained from this region at the maximum yield level, accounts to 62 per cent of world catch, with 54.85M mt (Table 1). The biomass from which this catch was extracted is 109.6 M mt. The current biomass is 89 per cent of the one at the MSY level. In Fig. 3A the catch trend of the Pacific north eastern is displayed; here the maximum yield was recorded by the year 2000, with almost 3 M mt, extracted from a stock biomass of 5.9 M mt; current biomass is unfortunately one Million lower and the trend is declining. In the Pacific north western, a similar trend is displayed (Fig. 3B), the maximum yield was obtained also by the year 2000, with nearly 22.6 M mt corresponding to a biomass of 45.1 M mt. The current

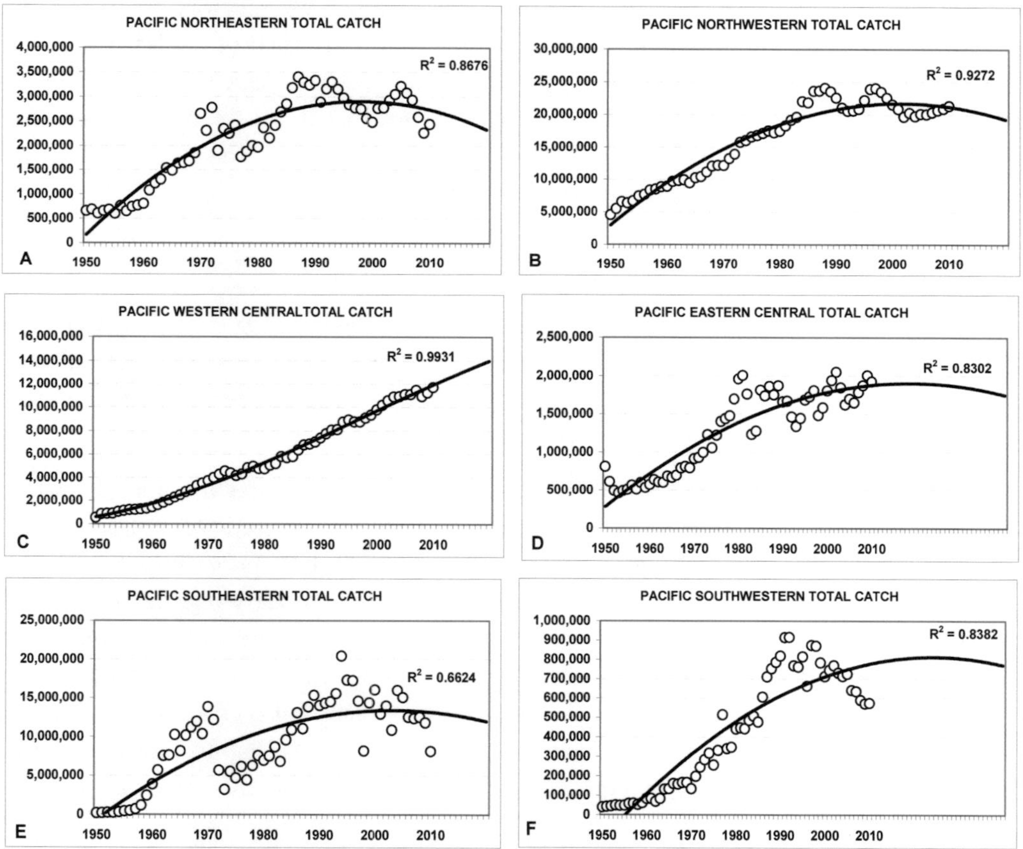

Figure 3. Trend of total catches extracted from several regions of the Pacific Ocean in the period 1950 – 2010. A. Pacific north eastern; in this region the maximum catches were obtained in the late nineties. B. Pacific north western; the maximum yields were obtained around the year 2000. C. Pacific western central; the maximum yield has not been reached and the fisheries seem to be in the eumetric phase. D. Pacific eastern central; the maximum yield seems that will be obtained in the near future. E. Pacific south eastern, the maximum yield was attained in the middle 2000's, but the catch of the last three years suggests a decline. F. Pacific south western, the trend suggests that the maximum yield has not been reached yet, but the catch has been declining since the last fifteen years.

biomass is 41.8 M mt. The catch of the Pacific western central displays a growing trend with 12 M mt in the last three years (Fig. 3C). To this catch corresponds a stock biomass of 24 M mt; no signs of stabilization of the catch are perceived, which is encouraging. In the Eastern central region, the yield seems to be attaining a maximum with around 2 M mt and a biomass of 4 M mt (Fig. 3D). These values are considered the current ones. The catch of the Pacific western central displays a growing trend with 12 M mt in the last three years (Fig. 3C). To this catch corresponds a stock biomass of 24 M mt; no signs of stabilization of the catch are perceived, which is encouraging. In the Eastern central region, the yield seems to be attaining a maximum with around 2 M mt and a biomass of 4 M mt (Fig. 3D). These values are considered the current ones. The south eastern region displays large variability, and the trend suggests that the maximum was attained a few years before, with a catch of 14.5 M mt corresponding to a biomass of 29 M mt. The mean catch of the last three years indicates a biomass decline to 21.8 M mt. The south western region suggests that the maximum was already attained, but the trend indicates that it will be reached within 15 years or so, with a catch of 800,000 mt and a biomass of 1.6 M mt; the current biomass is only 1.2 M mt.

4.2.3. The Indian Ocean, the Antarctic and the Mediterranean-Black Sea

These regions hardly attain a yield of 14 M mt at the MSY level, being the Indian Ocean the most productive of this group with 12 M mt caught in the whole area. The stock biomass approaches to 24 M mt at the MSY level but at the current exploitation level, this variable implies a reduction of almost 2 M mt, with almost 22 M mt.

The three regions in which it is divided show remarkable differences implying important characteristics in the fishing intensity applied; for instance, the Antarctic zone seems to have been completely overexploited and probably collapsed since 1992 (more recent catch data are not available) and the maximum catch was attained by the early eighties with nearly 100,000 mt as mean trend (Fig. 4A). The same as the Pacific western central, in the Indian ocean eastern the yield describes an increasing trend, with nearly 7 M mt in the last three years, with no signs of stabilization in the near future, which is also encouraging (Fig. 4B). The catch in the western region is also growing, but it seems to be stabilizing currently; the catch is 4.5 M mt corresponding to a maximum stock biomass of 9 M mt (Fig. 4C).

The catch at the Antarctic shows a declining trend, with a maximum yield of nearly 40,000 mt recorded in the middle fifties (Fig. 4D) and a stock biomass of 80,000 mt. Same as the Antarctic Indian Ocean, these fisheries seem to be completely collapsed since the late eighties.

The Mediterranean and Black sea display a quite stable catch trend through the last 30 years and the MSY was attained by the year 1990 with 1.6 M mt, from a biomass of 3.2 M mt. After that year there has been a slow declining trend, such that the current stock biomass is no higher than 3 M mt (Fig. 4E).

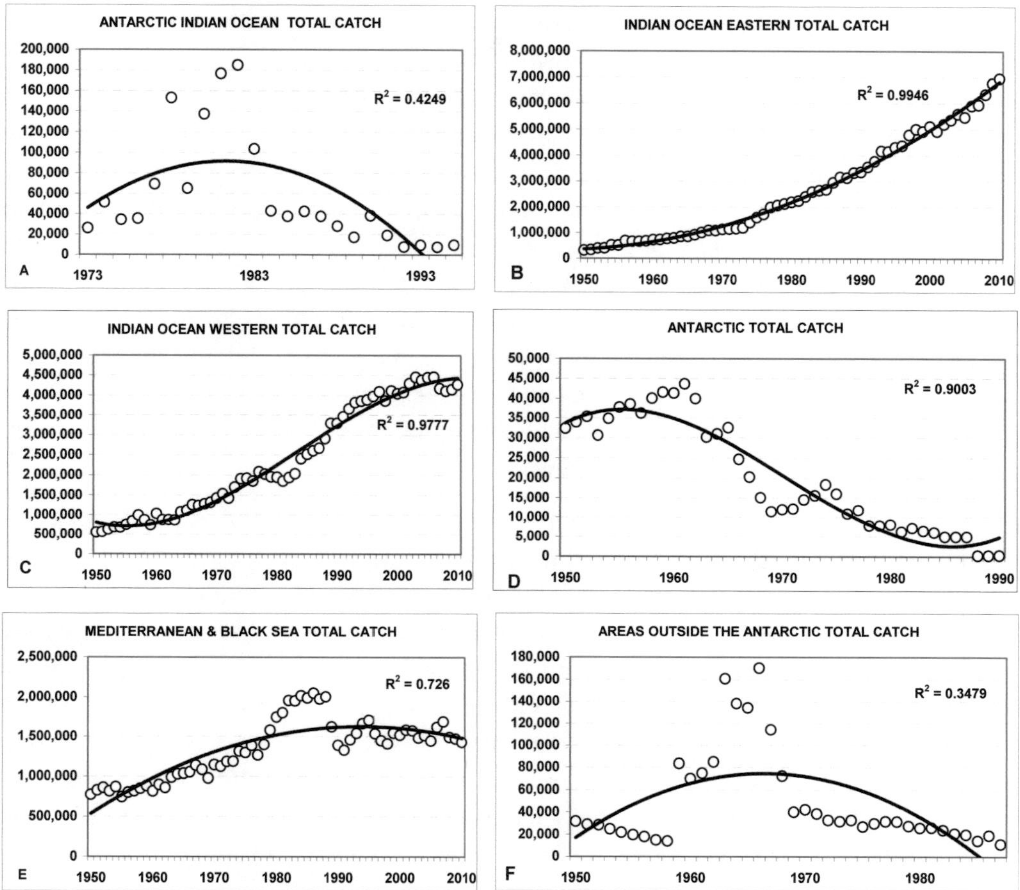

Figure 4. Trend of total catches extracted from several regions of the Indian Ocean, Antarctic, and Mediterranean-Black Sea in the period 1950 - 2010. A. Antarctic Indian ocean; the maximum yield was obtained in the year 1980 and the fisheries collapsed by the early nineties. B. Indian Ocean eastern; the fisheries are in a eumetric stage of growth nowadays. C. Indian Ocean western; after a sustained growth since the fifties, these fisheries appear to be approaching their MSY level nowadays. D. Antarctic ocean; all information related to this ocean confirms a collapse of its fisheries, as occurred in this case. E. Mediterranean-Black sea; after a period of slow but consistent growth, the MSY appears to have attained the MSY level in the late nineties, followed but a slow decline nowadays. F. The areas outside the Antarctic show the same trend as the Antarctic itself, with an apparent collapse nowadays.

Finally, the areas outside the Antarctic, apart from a peak of the catch in the middle sixties, display low yields that currently are above 11,000 mt from a stock biomass of 40,000 mt. This fishing region appears to be collapsed too.

5. Concluding remarks

The use of surplus yield models for assessment of exploited fish stocks, has becoming an tool hardly used nowadays, because the use of age structured methods with the aid of computing techniques, allow more powerful and more accuracy in the assessments. There were times when fisheries researchers devoted their efforts into that approach and more sophisticated variations of the original statements were made (Walter 1975, 1978; Csirke and Caddy 1983; Arreguín-Sánchez and Chávez 1986; Polacheck et al. 1993; Fréon and Yáñez 1995); however, this approach has became obsolete over time, despite its background ecological principles are still valid. However, the large variance implicit in the estimations caused by several factors, contributed in a great deal to its current lack of use. Despite this consideration, it was decided to adopt that approach in this paper, for several reasons, the first one is it accessibility and easy way to just fitting a second degree curve in the spreadsheet where a bunch of catch data involving as many species as they are exploited in the world oceans, just to have a guideline on the maximum yield level and the year when it was reached. It also provided a minimum basic requirement for the estimation the stock biomass on which fisheries of each region were based.

It is remarkable to realize that the maximum yield of the world oceans approaches very close to 100 M mt and the biomass of all the exploited stocks is near to 200 M mt. Another important point to call the attention is that in most cases, the MSY was attained more than a decade ago and that the current yield and stock biomass are nearly 40 per cent below those maxima. This is something to concern and is a possible indicator of excessive pressure on the fish stocks and in this respect those on the Antarctic seem to be the most heavily impacted by fishing activities. Evidently the over exploited fisheries have passed by several stages (Pauly et al. 1998) already pointed by other authors (Harding 1968; Feeny et al. 1990; Myers and Worm 2003) and unfortunately the perspective suggests that other world oceans apart from the Antarctic, will follow the same steps if no action is taken by the nations to ensure exploiting the sea in a sustainable way (Jorgensen et al. 2007).

Author details

Ernesto A. Chávez* and Alejandra Chávez-Hidalgo
Centro Interdisciplinario de Ciencias Marinas, Instituto Politécnico Nacional, La Paz, México

6. References

Arreguín-Sánchez F. and Chávez E.A. 1986. Influencia del reclutamiento sobre el rendimiento pesquero. IOC Workshop Report No. 44:95-104

Botsford L.W., Castilla J.C., and Peterson C.H. 2012. The management of fisheries and marine ecosystems. Science 277:509-515

* Corresponding Author

Csirke J. and Caddy J.F. 1983. Production modeling using mortality estimates. Can. J. Fish. Aquat. Sci. 40: 43-51.

Christensen, V. and Pauly D. 1993. Trophic models of aquatic ecosystems. International Center for Living Aquatic Resources Management. Manila, Philippinnes.

Cushing D.H. and Walsh J.J. 1976. The ecology of the seas. Blackwell.

FAO. 1995. http://www.fao.org/docrep/006/ac442s/AC442s31.htm# (Declaration Kyoto Conference/Fisheries FAO Fisheries Department).

FAO. 2005. Review of the state of world marine fishery resources. Marine resource service. Fishery resource division. FAO Fisheries Department. Technical paper 457.

Feeny D., Berkes F., MacCay B.J., and Acheson J.M. 1990. The tragedy of the commons: twenty-two years later. Human Ecology, 1(18):1-19.

Fréon P. and Yáñez E. 1995. Influencia del medio ambiente en evaluación de stock: una aproximación con modelos globales de producción. Invest. Mar. Valparaíso, 23:25-47.

Friedland K.D., Stock C., Drinkwater K.F., Link, J.S., Leaf R.T., Shank B.V., Rose J.M., Pilskain C.H., and Fogarty M.J. 2012. Pathways between primary production and fisheries yields of large marine ecosystems. PloS One 7(1):e28945.

Froese, R. and D. Pauly Editors 2012. FishBase. World Wide Web electronic publication. http://www.fishbase.org, version (08/2011).

Graham M. 1939. The sigmoid curve and the overfishing problem. Rapp. Conseil. Explor. Mer. 110:15-20.

Harding G. 1968. The tragedy of the Commons: The population problem has no technical solution; it requires a fundamental extension in mortality. Science 162: 1243-1248.

Jorgensen C., Enberg K., Dunlop E.S., Arlinghaus R., Boukal D.S., Brander K., Ernande B., Gårdmark A., Johnston F., Matsumura S., Pardoe H., Raab K., Silvia A., Vainikka A., Dieckmann U., Heino M., and Rijnsdrop A.D. 2007. Managing evolving fish stocks. Science 318:1247.

Margalef R. 1974. Ecología. ed. Omega, 951 pp.

Myers R.A. and Worm B. 2003. Rapid worldwide depletion of predatory fish communities. Nature 423:280-283.

Pauly D. and Christensen, V. 1995. Primary production required to sustain global fisheries. Nature 376:255-279.

Pauly D, Christensen V., Dalsgaard J., Froese R., and Torres F.Jr. 1998. Fishing down marine food webs. Science 279:860-863.

Polacheck T., Hilborn R., and Punt A.E. 1993. Fitting surplus production models: comparing methods and measuring uncertainty. Can. J. Fish. Aquat. Sci. 50: 2597-2607.

Russell-Hunter W.D. 1970. Aquatic productivity. Macmillan Pub. Co., NY., 306 pp.

Walter G.G. 1975. Graphical method for estimating parameters in simple models of fisheries. J. Fish. Res. Board Can. 32: 2163-2168.

Walter G.G. 1978. A surplus yield model incorporating recruitment and applied to a stock of Atlantic mackerel (*Scomber scombrus*). J. Fish. Res. Board Can. 35:229-234.

Zabel R.W, Harvey C.J, Katz S.L, Good T.P., and Levin P.S. 2003. Ecologically sustainable yield. American Scientist. 91(2):150-157.

Chapter 22

Characeae Biomass: Is the Subject Exhausted?

Carlos E. de M. Bicudo and Norma C. Bueno

Additional information is available at the end of the chapter

http://dx.doi.org/10.5772/54685

1. Introduction

Popularly known as stoneworts, brittleworts, muskgrass, muskworts or bass-weeds, Characeae are among the largest and most complex green algae. All common names come from some characteristics these plants may exhibit, such as the brittle, limestone (calcium carbonate) exoskeleton that can form on the external surfaces of the plant (e.g. *Chara vulgaris* and *Chara globularis*) and, particularly, from the distinctive smell of stale garlic emitted by the plant when crushed. It is important to note, however, that most charophyte species do not accumulate lime in observable amounts. The widespread misinterpretation that *Chara* plants generally form lime is understandable, since the two very common species above thrive in shallow waters where they are readily collected, forming spectacular extensive growths, and may solidify directly into a marl layer or onto a curious tufa rock, a porous limestone formed by deposits from springs. Plants accumulating lime become gray or whitish and quite opaque, whereas the many species without evident lime are generally soft, nearly transparent, with a glassy brilliance and rich green color.

Charophytes forms a significant part of the submerged vegetation of both natural and artificial systems represented by lakes, ponds, ditches, streams, canals, bog-pools, concrete tanks, reservoirs and excavations such as gravel pits, and are found on all continents except Antarctica. They are common in the littoral region of oligotrophic to moderately eutrophic water bodies (Kufel & Kufel 2002), and some authors (e.g. Krause 1985) consider these macrophytes indicators of water quality. *Nitella* specimens predominate in mildly acid water as in igneous rock areas, whereas *Chara*'s predominate in hard waters, but this is not a rule. They are characteristic of a disturbed habitat where periodic drastic changes create less favorable conditions for the growth of other algal species. They are often the first plants to colonize newly dug or cleared ponds and ditches, and some species are characteristic of ephemeral water bodies which dry up completely in the summer. The fast maturing charophytes have an advantage over the slow-growing macrophytes in such habitats. Charophytes are usually at a competitive disadvantage in shallow, moderately productive habitats, but tend to dominate in deeper water at low light intensities, particularly where the water has a high pH value. They are more often found in mesotrophic and eutrophic, hard water, calcium rich and low in phosphate waters. Charophytes may grow in silt, mud, peat or sand and they often form a dense carpet, known as a charophyte meadow, which restricts colonization by other macrophytes. The more common charophyte species do not die down during the winter. They have been recorded growing down to 60 m deep in clear water, but usually prefer depths between 1 and 10 m. In tropical countries such as Brazil, charophytes grow best in shallow water bodies, mostly in small reservoirs built for cattle disedentation, where they form dense carpets at the littoral zone of the reservoirs, usually at 20-40 cm depths. They may often grow intermingled with other macrophytes, mainly with water lilies (*Nymphaea* spp.), whose floating leaves they use to cut down the light intensity.

In size, they are generally moderately large, average shoots varying from 15 to 30 cm in height, but they may range from 5 mm to 2 m at extremes. Specimens of *Chara hornemannii* collected from the Rodrigo de Freitas Pond, in the city of Rio de Janeiro, ranged between 1.9-2 m tall.

The charophyte 'plant' or thallus is erect, central axis or 'stem' is branched and differentiated into a regular succession of nodes and internodes (Figure 1). Each node bears a whorl of branches of limited growth (the 'leaves' or branchlets), but branches capable of unlimited growth may arise axillary to the leaves. The axis consists of a chain of alternating

Figure 1. *Chara braunii* specimen showing the central axes branched and differentiated into a regular succession of nodes and internode, and oospores (black little rice-like structures at the verticelate branchlets) (source: Gutza Wikipedia)

long and short cells, the single long cells forming the internodes and the short, discoid cells forming the nodes. The single axial intermodal cell is commonly 1-4 cm long, but they may reach 50-60 cm in *Nitella cernua* and *Nitella translucens*. The intermodal cell is commonly 0.1-0.3 mm broad, but in the last two species, it may reach 2-3 mm broad. The plant is anchored by non-pigmented, single-celled processes, with or without a differentiation into nodes and internodes, the rhizoids, which penetrate the soil or substrate.

The importance of charophytes is indirect, as food for migratory waterfowl, protection of fish fry, and as a nuisance in shallow waters of reservoirs and recreational areas. They may also be used for sulfur baths, cattle food, fertilizers, scouring and filtering agents, and even for supposed control of mosquito larvae.

2. Methods for biomass estimation

Papers dealing with charophytes biomass are not numerous worldwide, and methods to measure that attribute are more or less standard.

Two boat-based and one in-water sampling method were used by Rodusky et al. (2005) to collect submersed aquatic macrophytes (SAV) as part of a long term monitoring program in Lake Okeechobee, Florida, U.S.A. The boat-based methods consisted of a ponar dredge used only to collect *Chara*, and an oyster tongs-like apparatus to collect all other SAV. The in-water method involved use of a 0.5 m^2 PVC quadrat frame deployed by a diver. Comparison of the three methods above showed no consistent pattern to the significant differences found in sampling precision between the three sampling methods, regardless of the geographical location, sediment type, SAV species or density.

To estimate charophytes biomass, the quadrat method is the most used one. According to the method, first a quadrate shall be delimited in the field, e.g. a 25 cm^2 (Westlake 1965, 1971; Krebs 1989). Within this quadrat, a 5 cm diameter (area 19.7 cm^2) and 50 cm tall PVC tube is inserted. Tube wall must be perforated throughout the first basal 25 cm to allow water circulation and the gathering of the plants.

Once collected, material must be stored in glass vials (e.g. 50 ml volume) and taken to the laboratory. In the laboratory, charophytes must be gently washed and if necessary scrapped with a very soft brush to remove other algal material and sediments adhered to the plants. After washed and/or scrapped, the excess water must be dried with some paper towel and finally placed in a porcelain melter.

For the analytic procedure, the charophyte material must be calcinated at 550ºC during 1 hour, then cooled in a desiccator and weighted using an analytical scale to have P_0. Immediately after, plants must be taken to an aerated oven at 65-70ºC until no further weight change is observed for quantification of its dry weight (P_1), and after 1 hour calcination at 550ºC for determination of its ash dry weight (P_2) (Hunter 1976). Determination of the ash free dry mass (AFDM) (P_3) is done using the mathematics $P_3 = (P_1 - P_0) - (P_2 - P_0)$. If total phosphorus (TP) is required, Strickland & Parsons (1965) method is to be used, i.e. the calcinated material is washed with 25 ml of HCl 1N, crushed and heated in a water-bath for 1 hour. After cooling, samples are diluted with 50-250 ml deionized water depending on the amount of calcinated material.

Palmer & Reid (2010) proposed a method they called 'invention' for the production of macroalgae to provide a sustained, economical source of biomass that may be used in various end-uses processes, including energy production. Their method provides specific combinations of macroalgae types, saltwater growth media compositions, and open pond water containers that resulted in biomass production beyond what may occur naturally without the required manipulation. Specifically, macroalgae that produce an exosqueleton in the presence of brackish water (e.g. stoneworths) have been found to provide excellent biomass production of at least 10 metric tons and up to 200 metric tons per acre per year under their method conditions.

Total phosphorus concentration is determined using a spectrophotometer. Another possibility for TP determination is by the molybdenum blue colorimetric method (Murphy & Riley 1962) after digestion with $K_2S_2O_8$ in an autoclave at 120ºC for 30 minutes (APHA 1995). Total nitrogen (TN) can be determined using spectrophotometry, based on the Koeofell colorimetric method. Calcium and magnesium can also be determined using a spectrophotometer, however, based on the Calmagite colorimetric method.

3. Charophytes biomass

The vast majority of papers published on charophytes worldwide, deals with their taxonomy and systematics. Comparatively, very few papers were published dealing with their biology, including citology, genetics, ecology and physiology.

Measure of biomass is one possibility to estimate the macrophyte's capacity to photosynthetize (Wetzel 1964) and the most used. Other possibilities include population density and biovolume. According to Wetzel (2001), the submersed macrophytes biomass is low if compared to that of other plants. The importance of the charophytes living at the littoral zone of lakes is directly related to the amount of submersed biomass, spatial structure and these plants association with other submersed and emerged macrophytes.

Literature regarding charophytes biomass is not rare neither profuse worldwide and dates mostly from 1980 on, when eutrophication was recognized to be one of the most important events of the century. While not profuse, literature available consents a pretty good overview on the subject.

3.1. Seasonal variation

In the temperate region of the World, aquatic macrophytes show very sharp annual variation, with a growth season of their aerial biomass during the spring and summer, and another season of the underground biomass and detritus accumulation during the fall and winter (Esteves & Camargo 1986). In the tropical region, however, deterministic of the aquatic macrophytes biomass seasonality are the rainy and dry periods (Esteves 2011). Very little was done, however, up to now regarding the charophytes biomass seasonality in the tropics. Perhaps the only contribution in this regard is the work by Carneiro et al. (1994), who studied the extensive *Chara hornemanii* beds prospering in the Piratininga Lagoon, State of Rio de Janeiro, southeast Brazil at the depth from 0.30 cm to 1 m, and realized that N and P inputs, low water turnover and low water column depth favored growth of phytoplankton, macroalgae and aquatic macrophytes, including charophytes. The same authors also observed a very clear seasonal behavior of the charophyte population that stared during the winter and lasted until the beginning of summer, when the alga covered about 60% of the lagoon sediments. During the summer, the alga biomass reached 500 mg m^{-2} (Carneiro et al. 1994).

Using aerial photographs and field work in brackish water lagoons of Åland Island, Finland, Berglund et al. (2003) observed seasonal and interannual growth, distribution and biomass variation of some charophyte species. According to the last authors, filamentous green algae contributed with 45-70% of the total biomass studied, charophytes with 25-40% and vascular plants with 3-18%. The biomass peak was reached in July and August, and the average biomass was negatively correlated with the charophytes exposition to direct sun light, i.e. the charophyte coverage was greater when their exposition to solar radiation was low, being highly affected by the presence of filamentous algae.

Seasonal changes in the biomass of a monospecific community (*Chara globularis*) and of several communities with high charophyte coverage (*Chara globularis–Myriophyllum alterniflorum*, *Chara globularis–Potamogeton gramineus* and *Nitella translucens–Potamogeton natans*) were studied monthly, from May 1996 to June 1997, by Fernández-Aláez et al. (2002) in three shallow lakes in northwest Spain. Weather and hydrological regime strongly influenced the seasonal biomass patterns and the between-the-year differences in the biomass of the macrophytes. The *Chara globularis* community biomass showed a bimodal pattern, with maximum in mid-July (128 g DW m^{-2}) and late autumn (165 g DW m^{-2}). *Chara globularis* overwintered as a green plant and during the subsequent growth period characterized by high temperature and low rainfall reached a maximum of 305 g DW m^{-2} in June 1997. The highest biomass of *Chara globularis* in the *Chara–Myriophyllum* community was reached in July (Lake Sentiz 160 g m^{-2}, Lake Redos 204 g m^{-2}), while the minimum (Lake Sentiz 10 g m^{-2}; Lake Redos 3 g m^{-2}) was recorded in February or March. *Myriophyllum alterniflorum* (average biomass 95 g m^{-2}) was a better competitor than *Chara globularis* in Redos lake and appeared to be favored by the early beginning of the growing season in 1997 and by the later increase in the water level. *Nitella translucens* biomass (average 64 g m^{-2}) showed a high stability during the entire study period, but lacked a well-defined seasonal pattern. *Potamogeton natans* had a marked maximum biomass in August (426 g m^{-2}). Although the stability of the *Potamogeton natans* population was low, shading did not have a significant influence on the development of *Nitella translucens* biomass.

Torn et al. (2006) measured the seasonal dynamics of the biomass, elongation growth and primary production rate of *Chara tomentosa* in Rame Bay, NE Baltic Sea, a shallow and semi-enclosed sea inlet on the western coast of the Estonian mainland, during the vegetation period of 2002. Their measurements showed extremely high plant heights (up to 1.42 m) and biomass values (5.2 kg (w.w.) m^{-2}) indicating the importance of the charophyte for the aquatic ecosystem. Torn et al. (2006) observed that the apical part of the plants grew more intensively from early spring to midsummer, whereas that of the subapical one was very low during the entire study period. The plant's net primary production rate peaked in July (43.4 mgO $g(d.w.)^{-1}$ 24 h^{-1}), remarkably lower rates being measured in May and September. The elongation growth and primary production were not correlated with the water nutrient concentrations and temperature. As the active growth of *Chara tomentosa* takes place during a relative short period at the beginning of summer, the amount of available solar radiation and the temperature levels during this sensitive time may have had a significant effect on the community in the same year (Figure 2).

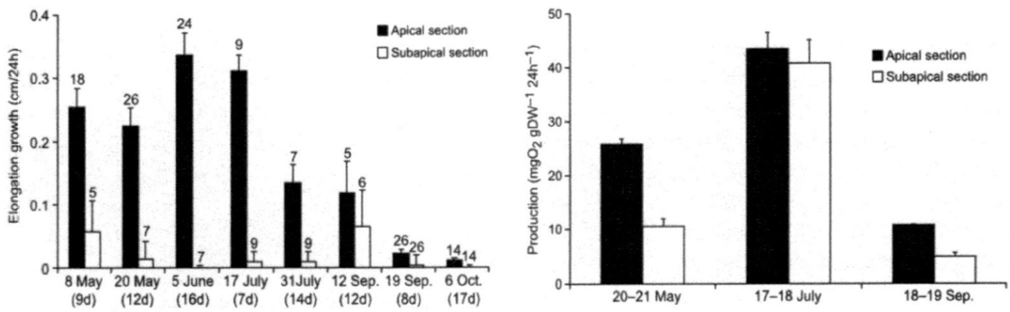

Figure 2. Left: seasonal variation in average elongation growth + S.E. of *Chara tomentosa*. Date of measurement and duration of the experimental period are indicated in between parenthesis. Number of replicates is indicated at the top of each bar. **Right**: Seasonal variation in average diurnal net primary production rate + S.E. of *Chara tomentosa*. In each period photosynthesis was measured in the period of 24 h replicated 3-fold (Torn et al. 2006).

Seasonal growth of *Chara globularis* var. *virgata* caused a regular summer depletion of Ca^{2+} and HCO_3^- by associated $CaCO_3$ deposition, and a more extreme and unusual depletion of K^+ was followed over three years (1985-1987) by Talling & Parker (2002) in a shallow upland lake (Malham Tarn) in northern England. Chemical analysis of the *Chara globularis* var. *virgata* biomass and of the underlying sediments indicated a large benthic nutrient stock that far surpassed that represented by the phytoplankton. Growth in the *Chara globularis* var. *virgata* biomass and the magnitude of the water-borne inputs influenced removals of Ca^{2+}, K^+ and inorganic N. According to Talling & Parker (2002), several features of Malham Tarn are suggestive in relation to the general case of phytoplankton-phytobenthos interaction and possible long-term change. So, the low P concentrations in the open water are probably linked to the fairly low phytoplankton abundance and influenced by the dense benthic *Chara globularis* var. *virgata* with a major capacity for P uptake. Also, the additional *Chara globularis* var. *virgata* capacity for K^+ uptake led to a major seasonal reduction of concentration in the lake water and outflow, of a magnitude rarely if ever recorded elsewhere. The annual growth of *Chara globularis* var. *virgata* seemed to involve further translocation of N, P and K from stocks in the sediments.

3.2. Impact of climatic fluctuations on the biomass

Sender (2008) studied the long term changes of the macrophytes structure in the Lake Moszne located in the Poleski National Park in Poland. Lake Moszne is a relatively small (17.5 ha), distrophic and shallow (1 m) water body. The lake is not connected with the size, nor with the depth of the reservoir, thus depending on the climatic conditions as well as on the economic and recreational activities, and on the hydro-technical changes imposed to the lake (Sender 2008). As a result, a distinct decrease of the plant association variety was observed, as well as changes in their qualitative composition. In fact, changes in qualitative and quantitative structure of lake Moszne macrophytes were probably caused by both abiotic and biotic factors. The macrophytes structure was subject to fluctuation, the changes indicating notable growth of water trophy. The biomass of macrophytes also showed an increase tendency. Nowadays, the structure of vegetation of the lake does not show the typical features for distrophic lakes.

It is well known that algal populations are often present in considerable and varying densities within shallow lakes, as both planktonic and benthic components (Talling & Parker 2002), and that shallow lakes have become the archetypical example of ecosystems with alternative stable states (Scheffer & van Nes 2007). Moreover, that shallow lakes may switch from a state dominated by submersed macrophytes to a phytoplankton-dominated state when a critical nutrient is exceeded (Kosten et al. 2011). Last authors explored how climate change affected that critical nutrient concentration by linking a graphical model to data from 83 lakes along a large climate gradient in South America. Their data indicated that in warmer climates, submersed macrophytes may tolerate more underwater shade than in cooler lakes, although the relationship between phytoplankton biomass and nutrient concentrations did not change consistently along the climate gradient. According to Kosten et al. (2011), in several lakes in the warm and intermediate regions, submersed macrophytes were found until relatively greater depths than in the cool regions, taking the available light at the sediments surface into account.

Rip et al. (2007) is an excellent case-study of how temporal pattern of precipitation and flow from land to water, may give a coherent, quantitative explanation of the observed dynamics in P, phytoplankton, turbidity and charophytes. Studying the external P load to a wetland with two shallow lakes in the Botshol Nature Reserve, The Netherlands the above authors observed that P load reduction resulted in a rapid decrease of phytoplankton biomass and turbidity, and after four years in an explosive charophyte growth. Such a clear water state, however, was unstable and the ecosystem alternated between clear, high-vegetation and turbid, low-vegetation states. Rip et al. (2007) used a water quality processes' model in conjunction with a 14-year nutrient budget for Botshol to determine if fluctuations in precipitation and nutrient load effectively caused the ecosystem instability. Their results indicated that during wet winters when groundwater level rose above surface water level, P from runoff was stored in the lake sediments and banks (Figure 3). Stored P was released the following spring and summer under anaerobic sediments conditions, thus resulting in an increase of phytoplankton density and light attenuation in the water column. Also, in years with high net precipitation, flow from land to surface water also transported humic acids, further increasing light attenuation. Conversely, in years with dry winters, P and humic acid loads to surface water were reduced, and growth of submersed macrophytes enhanced by clear water. Rip et al. (2007) concluded by stating that global warming caused winters in the Netherlands to become warmer and wetter during the last 50 years, consequently increasing flow from land to water of humic acids and P and, ultimately, enhancing instability of charophyte populations. Finally, in the first half of the 20th Century interannual variation in precipitation was not sufficient to cause large changes in the internal P flux in Botshol, and submersed macrophytes population were stable.

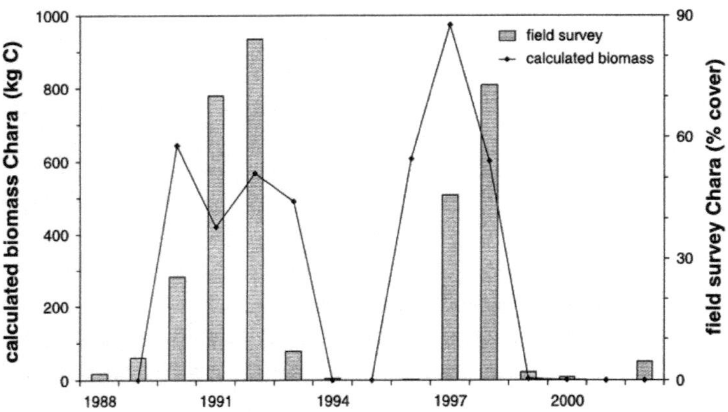

Figure 3. Calculated *Chara* biomass as model results and field surveys at subarea I of the Botshol Natural Reserve for 1989-2002 (Rip et al. 2007).

Recently, Salmaso et al. (2012) studied the combined effects of nutrient availability and temperature on phytoplankton in large and deep lakes of the Alps, lakes Garda, Iseo, Como, Lugano and Maggiore. A significant effect of temperature fluctuations and trophic status on the development of the main groups of cyanobacteria and eukaryotic phytoplankton was observed. However, high positive relationships of nutrient availability with temperature were found only in a few algal groups including charophytes, chlorophytes, dinophytes and, partly, cyanobacteria. Their results have implications in the evaluation of the impact of different climatic scenarios in lakes of different trophic status, suggesting a net increase of only selected eutrophic- or eurytrophic sensitive groups with increasing water temperature in more enriched systems.

3.3. Influence of depth and transparency

Once established, aquatic macrophytes have a positive effect on the transparency of water through several buffer mechanisms (Stephen et al. 1998). Furthermore, the presence of charophytes has been associated with the maintenance of clear water, and changes from a state of clear to turbid water have been associated with the eutrophication of the environment (e.g. Blindow et al. 1993, Kufel & Kufel 1997).

Steinman et al. (1997) studied the influence of water depth and transparency on the charophyte biomass distribution in the southern end of the subtropical Lake Okeechobee, U.S.A. Their first survey (August 1994) was conducted on 47 stations within the 3-Pole Bay. Subsequent surveys (November 1994-December 1996) were conducted on a monthly or bimonthly basis on 7 stations. According to the authors, the distribution and abundance of *Chara* population in the lake showed a marked seasonal phenology, although there were notable differences in biomass among the years and stations. *Chara* plants were observed only in August, September and October, and in 1996 also in November. Also, biomass never exceeded 20 g AFDM m^{-2} and declined significantly from 1994 to 1996. The charophyte biomass was inversely related to the water depth and positively related to the Secchi disc depth, suggesting that irradiance strongly influenced the charophyte distribution in the lake, a hypothesis that was confirmed by data they collected from photosynthetic measurements and phytosynthesis-irradiance curves (Steinman et al. 1997).

The role of charophytes in increasing the water transparency was also studied by Nõges et al. (2003). Under the frame of the EC project ECOFRAME, last authors worked out the water quality criteria for two shallow lakes of the Vooremaa landscape protection area, Central Estonia. Lake Prossa is a macrophyte-dominated system with an area of 33 ha and a mean depth of 2.2 m. Most of its bottom is covered by a thick mat of charophytes all year round. Lake Kaiavere is located 10 km far from Lake Prossa, is much larger (250 ha, mean depth 2.8 m) and is phytoplankton-dominated. Nevertheless, the nutrient dynamics was very similar in the two lakes (Figure 4). The first vernal phytoplankton peak was expressed in reduced Secchi depth in both lakes. After that peak, the water became clear in Lake Prossa, but remained turbid in Lake Kaiavere. Towards fall, the individual mean weight of zooplankton decreased in Lake Prossa, the *Chara*-lake, but remained smaller than in the plankton-dominated one (Lake Kaiavere) (Figure 5). Therefore, zooplankton grazing would initiate the clear-water phase in the *Chara*-lake, but other factors were needed for its maintenance. Another factor that showed a clear difference between the two lakes was the carbonate alkalinity that was rather stable or even increased during the spring in the phytoplankton-dominated lake, while it decreased by nearly 50% between April and July in the *Chara*-lake. The reduced sediment resuspension and the possible allellopathic influence of charophytes on phytoplankton remain the main explanations for the maintenance of the extensive clear-water period in the *Chara*-lake.

Blindow & Schütte (2007) worked with material from fresh and brackish water in Sweden and found out that both turbidity and salinity acted as stress factors on *Chara aspera*. According to the last authors, in clearwater lakes the species can occur in high densities and reach deep water, where the ability to hibernate as a green plant together with shoot elongation may further extend the lower depth limit. In turbid lakes, the plants can still form dense mats, but are restricted to shallow water due to the poor light availability, although shoot elongation may allow a certain extension of the depth range (Figure 6).

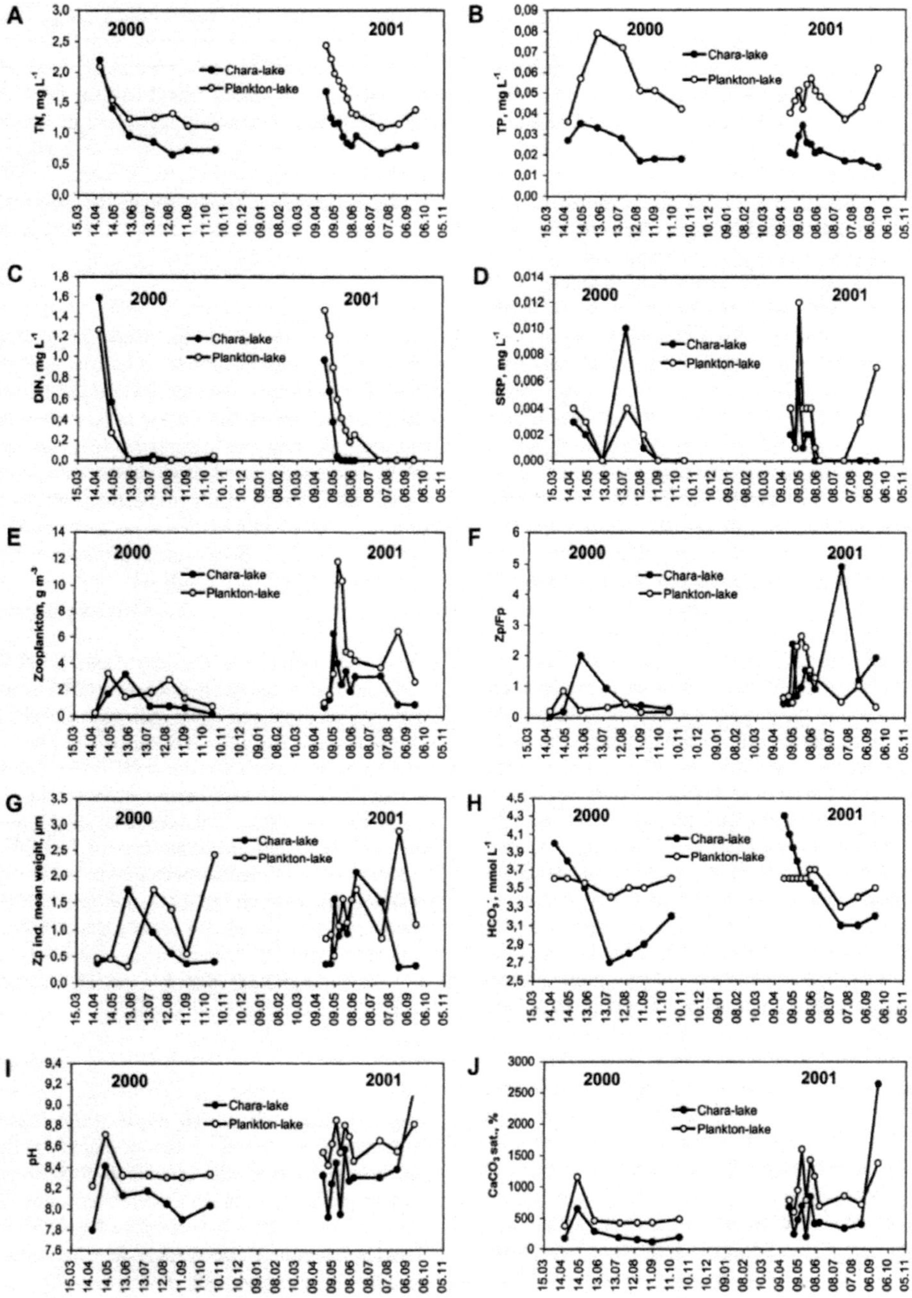

Figure 4. Seasonal dynamics of some chemical and biological features in lakes Prossa (*Chara*-lake) and Kaiavere (plankton-lake). A – total nitrogen, B – total phosphorus, C – dissolved organic nitrogen, D – soluble reactive phosphorus, E – zooplankton/phytoplankton biomass ratio, G – zooplankton mean individual weight, H – hydrocarbonate concentration, I – pH, and J – calcite saturation level (Nõges et al. 2003).

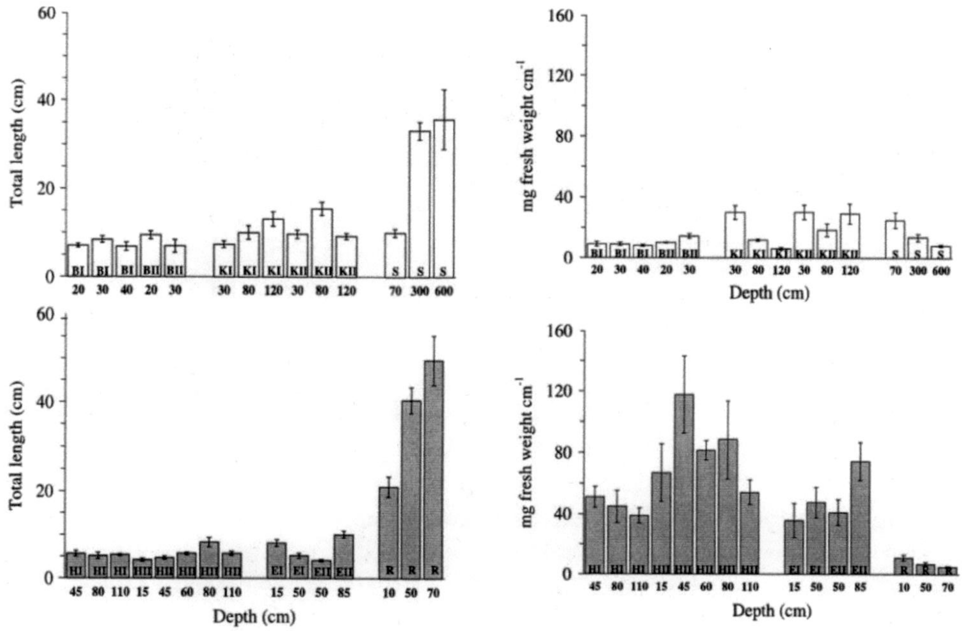

Figure 5. Left: total length of *Chara aspera* (mean values + S.E., n = 5), determined at all sites, depth ranges and sampling occasions. Above freshwater sites, bellow brackish water sites. K – Lake Krankesjön, B – Lake Börringesjön, S – Lake Storacksen, E – Edenryd bay, H – Höllviken bay, R – Redensee bay. I and II – first and second sampling occasions, respectively. Right: fresh weight:total length ratio of *Chara aspera* determined at all sites, depth ranges and sampling occasions (Blindow & Schütte 2007).

A field study conducted from July 2003 to May 2005 in the Myall Lake, a brackish shallow lake in New South Wales, Australia, revealed that *Chara fibrosa* var. *fibrosa* and *Nitella hyalina* occurred in areas of the entire lake that were deeper than 50 cm. Also, more fresh shoots were obtained during the winter (water temperature 13-16ºC), thus suggesting that winter may be their preferred growing season. Their biomass varied from 0 to 321 g DW m^{-2}, their maximum biomass being displayed between 1 and 2.5 m depth (Asaeda et al. 2007). These authors also observed that charophyte's shoots were longer in deeper waters, varying from c. 30 cm at 1 m depth to 60-90 cm between 2 and 4 m depth. Plants growing in shallow depths had shorter internodes implying a shorter life cycle of shoots. Also, nodal spacing was relatively regular in contrast to its deeper water counterparts although spacing tended to increase at locations farther from the apex (Figure 7). Finally, numbers of oospore and antheridia were higher in shallower water reaching their maximum at around 80 cm.

Chambers & Kalff (1985) used original data from eight lakes in southern Quebec, Canada and literature data from other lakes throughout the World to predict the maximum depth of charophytes colonization and the irradiance over the growing season at the maximum depth of colonization, concluding that the depth distribution of the aquatic macrophyte communities is quantitatively related to Secchi depth. According to regression models proposed in Chambers & Kalff (1985), natural distribution of aquatic macrophytes is restricted to depths of less than 12 m, whereas charophytes can colonize to great depths and up to a predicted 42 m in the very clearest lakes (Secchi depth 28 m).

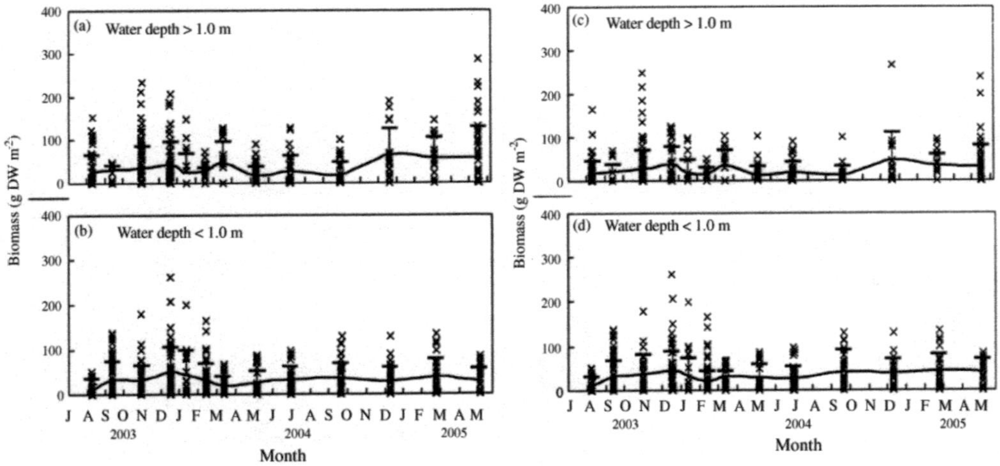

Figure 6. Seasonal variation of (a and b) *Chara fibrosa* and (c and d) *Nitella hyalina* biomass at location deeper than and shallower than 1 m. "X" markers denote individual measurements, the thick solid line represents monthly means, and short flat bars indicate standard deviations (mean + 1 S.D.) (Asaeda et al. 2007).

3.4. Nutrients

The concentrations of N, P and C in the above-ground biomass of 14 dominant macrophyte species (including *Chara globularis* and *Nitella translucens*) in seven shallow lakes of NW Spain were measured by Fernández-Aláez et al. (1999) that found significant differences for the three nutrients among the species and among the groups of macrophytes. The charophytes showed the lowest P (0.053% dry weight) and C (35.24% dry weight) content. Also, only the charophytes exhibited a strong association between N and P ($r = 0.734$, $p < 0.0001$), reflecting an important biochemical connection in these species.

Phosphorus was established as a limiting factor of all the macrophytes (N:P = 35:1), especially charophytes, in which it was below the critical minimum. Siong et al. (2006) used sequential P fractionation to study the nutrient speciation in three submersed macrophytes species, *Chara fibrosa, Najas marina* and *Vallisneria gigantea,* and the implications for P nutrient cycling in the Myall Lake, New South Wales, Australia. The mean TP of both *Najas marina* and *Vallisneria gigantea* was significantly higher than that of *Chara fibrosa,* even when the comparison made was based on the ash-free dry weight (AFDW). However, P co-precipitation with calcite ($CaCO_3$) induced during intense periods of photosynthesis occurs in hard water lakes, and this indirect mechanism of reducing P bioavailability in the water column may have been underestimated in assessing *Chara* beds acting as nutrient sink in shallow lakes. According to their results, besides the indirect mechanism above, P in the water column was also directly co-precipitated with encrusted calcite along the charophyte intermodal cell, and such a calcification should be regarded as a positive feedback in stabilizing *Chara* dominance in lakes. Siong & Asaeda (2009) studied the effect of Mg on the charophyte calcite encrustation, and assessed whether charophytes growing on the non-calcareous sediments of the Myall Lake could function as an effective nutrient sink for P in a similar manner to charophytes growing on the calcareous sediments of freshwater calcium-rich hard water systems. According to the last authors, calcification of *Chara fibrosa* was significantly inhibited by Mg in the water column and, consequently, reduced the formation of Ca-bound P that has a potential sink for P. However, a large percentage of non-bioavailable forms of P in the lake sediments suggested that P sink was through burial of dead organic matter and subsequent mineralization process.

The inorganic phosphorus concentration was not yet significantly related to the charophyte biomass. Palma-Silva et al. (2002) observed that the charophyte community (*Chara angolensis* and *Chara fibrosa*) sometimes occupied the entire benthic region in the Imboacica coastal lagoon in Brazil, and presented a large variation in C:N:P ratio. Results of their investigation (samples taken in March, April, May, July and October 1997) indicated that the charophytes fast growth may have absorbed a great amount of the nutrients entering the lagoon. Values of nutrient concentrations in the charophytes biomass were, according to those authors, within the expected range for the group, with the most eutrophic sampling station in the lake showing the highest N and P values. C:N:P ratios presented high values, and the biomass values were higher in the less eutrophic areas. The biomass reached maximum values of between 400 and 600 g DW m^{-2}, and the C:N:P ratio varied from 51:7:1 to 1603:87:1, indicating that the two *Chara* species may grow in a wide range of nutrient concentration. The same authors concluded that the charophyte community would be responsible by the nutrient decrease in the water column and keeping the water clear after drawdowns (Palma-Silva et al. 2002).

Several authors concluded that the nutrient kinetics favor the phytoplankton growth over *Chara*, thus assuming a P-limited condition. Therefore, although nutrient concentration may influence the charophyte phenology and abundance, light appeared to be a stronger regulator in the Okeechobee Lake. Schwarz & Hawes (1997) also observed the influence of the water transparency on the variation of the charophyte biomass in the Coleridge Lake, New Zealand. In the latter lake, total algal biomass did not surpass 180 g DW m^{-2} between 5 and 10 m depth. Pereyra-Ramos (1981) worked with seven charophyte species collected from Polish lakes and observed an increase of their fresh dry weight during the summer (July): *Chara rudis* 2.07 kg m^{-2}, *Chara vulgaris* 1.61 kg m^{-2}, *Chara contraria* 0.54 kg m^{-2}, *Chara fragilis* 0.39 kg m^{-2}, *Chara jubata* 0.37 kg m^{-2}, *Chara tomentosa* 0.28 kg m^{-2} and *Nitellopsis obtusa* 0.24 kg m^{-2}. Together, the charophytes represented 53% of the total submersed macrophytes biomass, 28% of *Elodea* sp. and 8% of *Ceratophyllum demersum*, two submersed macrophytes. According to Howard-Williams et al. (1995), *Chara corallina* biomass in deep (average 90 m depth) New Zealand lakes ranged around 300 g DW m^{-2}. Bakker et al. (2010) registered a strong decline of the *Chara* sp. biomass under the nutrient enriched condition of Lake Loenderveen, Norway. Similar situation was already detected by Blindow et al. (1993) and van de Bund & van Donk (2004) for other water bodies.

3.5. Trace contaminants

The Anthropocene period is characteristic by rapid urbanization, industrialization, mining activities, metal ore refining, agricultural chemicals, liquid and solid wastes, resulting in heavy metal pollution of water and land resources. There has been an increasing load of heavy metals (Cu, Zn, Cd, Cr, Hg and Ni) in the aquatic ecosystems, which in turn are being assimilated and transferred within food chains by the process of biomagnification. The problem with the heavy metals is their non-biodegradable nature. The conventional methods used to remove metal ions include chemical precipitation, lime coagulation, ion exchange, reverse osmosis solvent extraction, aeration, chemical oxidation, electrodialysis, ultra filtration, and chlorination (Rich & Cherry 1987).

Research was carried out recently to evaluate the metal accumulation in charophytes. Hence, Bibi et al. (2010) investigated the effects of Cd, Cr and Zn on the growth of *Nitella graciliformis* and their bioaccumulation in the plant under laboratory conditions. Charophyte specimens were exposed to different Cd, Cr and Zn concentrations, and it was observed that the heavy metals concentrations in the plant increased with the increasing metals concentrations in the mediums. As a result, negative growth occurred and the internode elongation was reduced when exposed to these metals at any concentration, however, intracellular *Nitella gracilliformis* has a potential for accumulating Cd, Cr and Zn. Bibi et al. (2010) concluded their investigation by stating that their study should be an integral part of the sustainable development of ecosystems and pollution assessment programs.

Absorption processes are being widely used for the removal of heavy metals from aqueous solutions. According to Shaikh Parveen & Bhosle Arjun (2011), use of various products has been widely investigated in the recent years as an alternative for the currently expensive methods of water treatment, and some natural products can be effectively used as a low cost absorbent. The above mentioned authors conducted batch studies of *Hydrilla* sp. and *Chara* sp. to evaluate the uptake of Cr from aqueous solutions. They found out that about 91.7% removal was obtained with 2 mg L^{-1} of *Chara* sp. at 2 mg L^{-1} Cr concentration after a period of seven days at pH 4. Their results also indicated that the metal removal increased as the days were extended, however, with the increasing contact time *Hydrilla* sp. proved to be better than *Chara* sp. in the Cr removal.

4. Final remarks

As it was mentioned before, literature on charophytes biomass is not rare neither profuse worldwide and dates mostly from 1980 on, when eutrophication was recognized to be one of the most important events of the century. Despite of not being profuse, literature available consents a pretty good overview on the subject.

In the temperate region of the World, aquatic macrophytes show very sharp annual variation, with a growth season of their aerial biomass during the spring and summer, and another season of the underground biomass and detritus accumulation during the fall and winter. Very little, however, was done up to now regarding the charophytes biomass seasonality in the tropics. The single paper published based on charophytes from the tropical region defined, however, deterministic of the aquatic macrophytes biomass seasonality the rainy and dry periods. Water temperature and rain precipitation are, nevertheless, somewhat connected to each other, since the rainy season in the tropics somewhat coincides with the high temperature season.

A climate gradient in South America was studied, indicating that in warmer climates, submersed macrophytes may tolerate more underwater shade than in cooler lakes. Moreover, in several lakes in the warm and intermediate regions, submersed macrophytes were met until relatively greater depths than in the cool regions, taking the available light at the sediments surface into account. According to a very detailed long term study, global warming has been causing winters in the Netherlands to become warmer and wetter during the last 50 years, consequently increasing flow of humic acids and P from land to water that, ultimately, has been enhancing instability of charophyte populations. Such studies conclusion is that in the first half of the 20th Century interannual variation in precipitation was not sufficient to cause large changes in the internal P flux, and submersed macrophytes population was stable.

The presence of charophytes has been associated with the maintenance of clear water, and changes from a state of clear to turbid water have been associated with the eutrophication of the environment. Original data from eight lakes in southern Quebec, Canada and literature data from other lakes throughout the World were used to predict the maximum depth of charophytes colonization and the irradiance over the growing season at the maximum colonization depth, concluding that the depth distribution of the aquatic macrophyte communities is quantitatively related to the Secchi depth. Regression models using the same information above, defined that natural distribution of aquatic macrophytes is restricted to depths of less than 12 m, whereas charophytes can colonize to great depths and up to a predicted 42 m in the very clearest lakes.

The inorganic phosphorus concentration was not yet significantly related to the charophyte biomass. Concentrations of N, P and C in the above-ground biomass of 14 dominant macrophyte species (*Chara globularis* and *Nitella translucens* included) in seven shallow lakes of NW Spain pointed to significant differences for the three nutrients among the species and among the macrophytes groups, the charophytes showing the lowest P and C content. Also, only the charophytes showed a strong association between N and P.

Only recently some research has been carried out to evaluate the metal accumulation in charophytes. Therefore, charophyte specimens were exposed in laboratory experiments to different Cd, Cr and Zn concentrations, showing that the heavy metals concentrations in the plant increased with the increasing metals concentrations in the cultivation mediums used. As a result, negative growth occurred and the internode elongation was reduced when exposed to these metals at any concentration, however, intracellular *Nitella gracilliformis* revealed a potential for accumulating Cd, Cr and Zn.

Summarizing, all research done up to now on the charophytes biomass is still very punctual, i.e. they most often focused one special environment under very specific conditions. There are very few studies focusing a larger time scale and comparing several localities. In the last cases, results are much more consistent. The scientific community needs much more studies, to be able to formulate generalizations. In other words, despite of producing some important information, study of charophytes biomass is far from being exhausted, on the contrary they have just started.

Author details

Carlos E. de M. Bicudo
Instituto de Botânica, São Paulo, SP, Brasil

Norma C. Bueno
Universidade Estadual do Oeste do Paraná, Cascavel, PR, Brasil

Acknowledgement

CEMB is very much indebted to CNPq, Conselho Nacional de Desenvolvimento Científico e Tecnológico for partial financial support (Grand nº 309474/2010-8).

5. References

[1] APHA, American Public Health Association (1998) Standard methods for the examination of water and waste water. American Public Health Association, Washington, D.C. 20th edition.

[2] Asaeda T, Rajapakse L & Sanderson B (2007) Morphological and reproductive acclimatation to growth of two charophyte species in shallow and deeper water. Aquatic Botany 86: 393-401.

[3] Bakker ES, van Donk E, Declerck SAJ, Helmsing NR, Hidding B & Nolet BA (2010) Effect of macrophyte community composition and nutrient enrichment on plant biomass and algal blooms. Basic and Applied Ecology 11: 432-439.

[4] Berglund J, Mattila J, Rönnberg O, Keikkila J & Bondsdorff E (2003) Seasonal and inter-annual variation in occurrence and biomass of rooted macrophytes and drift algae in shallow bays. Estuarine Coastal Shelf Science 56: 1167-1175.

[5] Bibi MH, Asaeda T & Azim E (2010) Effects of Cd, Cr, and Zn on growth and metal accumulation in an aquatic macrophyte, *Nitella graciliformis*. Chemistry and Ecology 26(1): 49-56.

[6] Blindow I, Andersson G, Hargeby A & Johansson S (1993) Long-term pattern of alternative stable states in two shallow eutrophic lakes. Freshwater Biology 30: 159-167.

[7] Blindow I & Schütte M (2007) Elongation and mat formation of *Chara aspera* under different light and salinity conditions. Hydrobiologia 584: 69-76.

[8] Carneiro MER, Azevedo C, Ramalho NM & Knoppers B. (1994) A biomassa de *Chara hornemannii* em relação ao comportamento físico-químico da lagoa de Piratininga (RJ). Anais da Academia Brasileira de Ciências 66(2): 213-222.

[9] Chambers PA & Kalff J (1985) Depth distribution and biomass of submersed aquatic macrophyte communities in relation to Secchi depth. Canadian Journal of Fisheries and Aquatic Sciences 42: 701-709.

[10] Esteves FA (Coord.) (2011) Fundamentos de Limnologia. Editora Interciência, Rio de Janeiro. (3rd edition).

[11] Esteves FA & Camargo AFM (1986) Caracterização de sedimentos de 17 reservatórios do Estado de São Paulo com base no teor de feopigmentos, carbono orgânico e nitrogênio orgânico. Ciência e Cultura 34(5): 669-674.

[12] Fernández-Aláez M, Fernández-Aláez C & Bécares E (1999) Nutrient contents in macrophytes in Spanish shallow lakes. Hydrobiologia 408-409: 317-326.

[13] Fernández-Aláez M, Fernández-Aláez C & Rodríguez S (2002) Seasonal changes in biomass of charophytes in shallow lakes northwest of Spain. Aquatic Botany 72(3-4): 335-348.

[14] Howard-Williams C, Schwarz AM & Vincent WF (1995) Deep-water aquatic plant communities in an oligotrophic lake: physiological responses do variable light. Freshwater Biology 33: 91-102.

[15] Hunter RD (1976) Changes in carbon and nitrogen content during decomposition of three macrophytes in freshwater and marine environments. Hydrobiologia 51(2): 119-128.

[16] Kosten S, Jeppesen E, Huszar VLM, Mazzeo N, van Nes EH, Peeters ETHM & Scheffer M (2011) Ambiguous climate impacts on competition between submerged macrophytes and phytoplankton in shallow lakes. Freshwater Biology 56: 1540-1553.

[17] Krause W (1985) Über die Standortsansprüche und das Ausbreitungs-verhalten der Stern-Armleucheralge *Nitellopsis obtusa* (Desvaux) J. Groves. Carolinea 42(4): 31-42.

[18] Krebs CJ (1989) Ecological methodology. Harper & Row Publishers, New York.

[19] Kufel I & Kufel L (1997) Eutrophication processes in a shallow macrophyte-dominated lake: nutrient loading to and flow through lake Luknajno (Poland). Hydrobiologia 342-343: 387-394.

[20] Kufel L & Kufel I (2002) *Chara* beds acting as nutrient sinks in shallow lakes: a review. Aquatic Botany 72: 249-260.

[21] Murphy J & Riley J (1962) A modified single solution method for determination of phosphate in natural waters. Analytica Chemica Acta 27: 31-36.

[22] Nõges P, Tuvikene L, Feldmann T, Tõnno I, Künnap H, Luup H, Salujõe J & Nõges T (2003) The role of charophytes in increasing water transparency: a case study of two shallow lakes in Estonia. Hydrobiologia 506-509: 567-573.

[23] Palma-Silva C, Albertoni EF & Esteves FA (2002) Clear water associated with biomass and nutrient variation during the growth of a charophyte stand after a drawdown, in a tropical coastal lagoon. Hydrobiologia 482: 79-87.

[24] Palmer M & Reid BR (2010) Biomass production and processing and methods of use thereof. Patentdocs: patent application number 20100240114. Available: http://www.faqs.org/patents/app/20100240114.

[25] Pereyra-Ramos E (1981) The ecological role of Characeae in the lake littoral. Ekologia Polska 29(2): 167-209.

[26] Rich G & Cherry K (1987) Hazardous waste treatment technologies. Pudvan Publishers, New York.

[27] Rip WJ, Ouboter MRL & Los HJ (2007) Impact of climatic fluctuations on Characeae biomass in a shallow, restored lake in The Netherlands. Hydrobiologia 584: 415-424.

[28] Rodusky AJ, Sharfstein B, East TL & Maki RP (2005) A comparison of three methods to collect submerged aquatic vegetation in a shallow lake. Environmental Monitoring and Assessment 110: 87-97.

[29] Salmaso, N, Buzzi F, Garibaldi L, Morabito G & Simona M (2012) Effects of nutrient availability and temperature on phytoplankton development: a case study of large lakes south of the Alps. Aquatic Science (on line first).

[30] Scheffer M & van Nes EH (2007) Shallow lakes theory revisited: various alternative regimes driven by climate, nutrients and lake size. Hydrobiologia 584: 455-466.

[31] Schwarz AM & Hawes I (1997) Effects of changing water clarity on characean biomass and species composition in a large oligotrophic lake. Aquatic Botany 56: 69-181.

[32] Sender J (2008) Long term changes of macrophytes structure in the lake Moszne (Poleski National Park). Teka Komisji Ochromy i Ksztaltowania Środowiska Pryzyrodniczego 5: 154-163.

[33] Shaikh Parveen R & Bhosle Arjun B (2011) Bioaccumulation of chromium by aquatic macrophytes *Hydrilla* sp. & *Chara* sp. Advances in Applied Sciences Research 21(1): 214-220.

[34] Siong K & Asaeda T (2009) Effect of magnesium on charophytes calcification: implication for phosphorus speciation stored in biomass and sediment in Myall Lake (Australia). Hydrobiologia 632: 247-259.

[35] Siong K, Asaeda T, Fujino T & Redden A (2006) Difference characteristics of phosphorus in *Chara* and two submerged angiosperm species: implications for phosphorus nutrient cycling in an aquatic ecosystem. Wetlands Ecology and Management 14: 505-510.

[36] Stephen D, Moss B & Phillips G (1998) The relative importance of top-down and bottom-up control of phytoplankton in a shallow macrophyte-dominated lake. Freshwater Biology 39: 699-713.

[37] Steinman AD, Meeker RH, Rodusky AJ, Davis WP & Hwang S-J (1997) Ecological properties of charophytes in a large subtropical lake. Journal of the North American Benthological Society 16(4): 781-793.

[38] Strickland JDH & Parsons TR (1965) A manual of sea water analysis. Bulletin of the Fisheries Research Board of Canada 125: 1-185.

[39] Talling JF & Parker JE (2002) Seasonal dynamics of phytoplankton and phytobenthos, and associated chemical interactions, in a shallow upland lake (Malham Tarn, northern England). Hydrobiologia 487: 167-181.

[40] Torn K, Martin G & Paalme T (2006) Seasonal changes in biomass, elongation growth and primary production rate of *Chara tomentosa* in the NE Baltic Sea. Annales Botanici Fennici 43: 276-283.

[41] van de Bund WJ & van Donk E (2004) Effects of fish and nutrient additions on food-web stability of a charophyte-dominated lake. Freshwater Biology 49(12): 1565-1573.

[42] Westlake DF (1965) Some basic data for investigations of the productivity of aquatic macrophytes. Memorie dell'Istituto Italiano di Idrobiologia Dottore Marco de Marchi 18: 229-248.

[43] Westlake DF (1971) Macrophytes, In: Vollenweider RA. A manual of methods for measuring primary production in aquatic environments. Blackwell Scientific Publications, Oxford (IBP Handbook 12). p. 25-32 (9th edition).

[44] Wetzel RG (1964) A comparative study of the primary productivity of higher aquatic plants, periphyton and phytoplankton in a large, shallow lake. Internationale Revue der Gesamten Hydrobiologie 49(1): 1-61.

[45] Wetzel RG (2001) Limnology: lake and river ecosystems. Academic Press, Orlando, Florida (3rd edition).